Finite Population Sampling and Inference

WILEY SERIES IN PROBABILITY AND STATISTICS
SURVEY METHODOLOGY SECTION

Established by WALTER A. SHEWHART and SAMUEL S. WILKS

Editors: *Robert M. Groves, Graham Kalton, J. N. K. Rao, Norbert Schwarz, Christopher Skinner*

A complete list of the titles in this series appears at the end of this volume.

Finite Population Sampling and Inference

A Prediction Approach

RICHARD VALLIANT
ALAN H. DORFMAN
RICHARD M. ROYALL

A Wiley-Interscience Publication
JOHN WILEY & SONS, INC.
New York • Chichester • Weinheim • Brisbane • Singapore • Toronto

This book is printed on acid-free paper. ♾

Published simultaneously in Canada.

For ordering and customer service, call 1-800-CALL-WILEY.

Library of Congress Cataloging-in-Publication Data:

Valliant, Richard, 1950–
Finite population sampling and inference : a prediction approach / Richard Valliant, Alan H. Dorfman, Richard M. Royall.
p. cm. -- (Wiley series in probability and statistics. Survey methodology section)
Includes bibliographical references and index.
ISBN 0-471-29341-5 (cloth : alk. paper)
1. Sampling (Statistics) 2. Prediction theory. I. Dorfman, Alan H. II. Royall, Richard M. III. Title. IV. Series.
QA276.6 . V35 2000
519.5′2--dc21 00-026383

Printed in the United States of America.
10 9 8 7 6 5 4 3 2 1

To Carla and Joanna,
Jane, Isaac, and Rachel,
Martha, Patty, and Betsy

Contents

Preface

Among the branches of statistics, survey sampling is notable for its public importance and its theoretical isolation. It should perhaps be an important component of every statistician's education, but by and large is neglected, and when not neglected, found to be an alien subject having its own rules and orientation at odds with standard methods of statistical inference. Students of statistics catch a glimpse, shudder, and pass on.

The nub of its separateness is this: mainstream statistics focuses on data as realizations of random processes inherent in their generation; survey sampling has come to regard itself as reliant on the randomness the sampler injects into the process of selecting what data to look at, out of a larger body of data. The finite population is 'out there' complete and fixed, a mammoth parameter, and the sampler does, as it were, a dance around it, picking this, omitting that. Survey design and inference has become the study of that dance.

The premise of *Finite Population Sampling and Inference: A Prediction Approach* is that this dichotomy is *not* well founded, but because of the above developments, it is written for those who assume it is: the vast majority of survey samplers/theoreticians, or practitioners on the front lines carrying out surveys for governments or private institutions for whom probability sampling and inference resting on the probabilities attached to that sampling process, are a *sine qua non*; statisticians as a whole, who would prefer perhaps to ignore this step-child altogether, out of the notion that it has little to contribute to statistics as a whole, and mainstream statistics, to it.

This text presents the view that survey sampling properly proceeds from the same underpinning as does statistical thought generally, namely, that the data can be understood as realizations of random variables, and that models constructed to mirror the random process can be usefully employed for inference. Outside of survey sampling, statisticians typically are interested in estimating the parameters or other aspects of the model. Model-based survey sampling is no different, except that estimation of model parameters now (usually) plays a role subordinate to the dominant task of inference about the realized finite population itself.

The component of statistics that has been most intensively applied to survey sampling is regression theory and general linear models, and much of this text dwells on that application. However, we see no limitations in principle to the applications to sampling of a great variety of statistical theory and methodology, and hope, with the publication of this text, to entice those outside the sampling field into bringing their knowledge to bear on survey sampling. The alert reader will note many open questions, explicit and implicit, regarding sampling inference and design, throughout this text.

Finite populations are good laboratories. Ideas on model selection, parameter estimation, diagnostics, etc. can be well tested by sampling from known actual finite populations. Sound methods should yield good estimates of known quantities. This is a distinct advantage over the simple estimation of model parameters, which for real-world data are rarely known.

For the survey sampler, even if, out of habit, loyalty, or prudence he or she adheres to the design-based approach, the results in this text can serve as a flashlight, illuminating standard practice and sometimes suggesting non-standard improvements. Indeed, the model-based approach has had, over the last several years, a good deal of subterranean impact. Witness, for example, the spread of "model-assisted" sampling. (The current acceptance of the "sandwich variance estimator" may well also have been influenced by important empirical work in model-based sampling—see Chapter 5.) Naturally, however, it is our hope that the ideas embodied in this text will begin to assume their proper role, directly, in the sampling world.

For those who would use this as a textbook, we note that some sections are starred, and can be skipped without breaking the continuity. Each chapter includes numerical examples and, where appropriate, simulation studies to illustrate theoretical ideas. Exercises vary in difficulty and kind. Software and actual populations are listed in the Appendices, and can readily be accessed to give students experience with actual sampling situations, and an opportunity to do simulation work of their own. Some exercises require the proofs of theorems, to deepen understanding. Software and data that accompany this book can be accessed online through a browser at: ftp://ftp.wiley.com/public/sci_tech_med/finite populations. You can also access the files from a link on the Wiley Mathematics and Statistics web site's software supplements at: http://www.wiley.com/products/subjects/mathematics/feature/software_supplem_math.html.

Prerequisites for a course based on this text are standard courses in mathematical statistics such as *Mathematical Statistics: Basic Ideas and Selected Topics* (1977) by P. Bickel and K. Doksum and in regression theory such as in S. Weisberg's *Applied Linear Regression.* An introductory course in standard population sampling theory is useful, but not essential.

We wish to thank Ray Chambers for detailed comments, which proved extremely helpful on earlier versions of portions of the text, Hyunshik Lee for important insights in the difficult area of outlier robust sampling, and Stephen Roey for providing the data on the third grade population in Appendix B.

We would especially like to thank our wives and children for their long enduring patience with a work that for the most part had to proceed outside of the conventional working hours.

Richard Valliant
Alan H. Dorfman
Richard M. Royall

January 2000

CHAPTER 1

Introduction to Prediction Theory

1.1. SAMPLING THEORY AND THE REST OF STATISTICS

A finite population is a collection of distinct units such as people, businesses, states, farms, accounts, file cards, or schools. Finite population sampling theory, sometimes called *survey sampling*, is concerned with selecting samples or subsets of the units, observing features of the sample units, then using those observations to make inferences about the entire population. For example, the population units might be all of the short-stay hospitals in the United States. For each hospital in a sample we might determine the number of patients who had heart bypass surgery during a particular calendar quarter, with the goal being to estimate the total number of bypasses performed in the entire United States. Or, a sample of businesses might be selected from the population of all businesses having 50 or more employees in order to estimate the number of employees who receive paid vacation as a benefit.

Finite population sampling is distinguished from the rest of statistics by its focus on the actual population of which the sample is a part. In other parts of statistics, observations are typically represented as realizations of random variables, and the inferences refer not to any actual population of units, but to the probability law that governs the random variables. For example, a sample of parts coming off an assembly line could be tested to determine how many meet engineering specifications. A statistician might then represent the results probabilistically, with each part having the same unknown probability θ of being defective, and seek to estimate θ or test whether it is less than some acceptable level. The problem becomes one of finite population sampling when attention shifts from the probability θ to the actual proportion of defectives in a day's production, (number of defectives)/(number produced).

There are five general steps in a sampling investigation of a finite population:

1. Define the scope and objectives of the study, including
    - Population to be studied;
    - General information to collect (e.g., the state of nation's workforce).

2. Choose tools and techniques for making observations, for example:
   - A questionnaire containing attitude scales or questions asking for factual data like a person's income or the number of employees in a business;
   - Physical measurements such as height, weight, and blood pressure;
   - Expert inspection by Customs officials of import shipments to determine whether contraband is being smuggled.
3. Choose a sample.
4. Gather data on the sample.
5. Analyze the data and make inferences.

Sampling theory (and this book) concentrates on step 3 (choosing samples), and step 5 (analyzing data and making inferences). Steps 1, 2, and 4 are critically important to a successful investigation but are relatively more dependent on informal judgment, experience, and administrative skill.

1.2. PREDICTION THEORY

Suppose that the number of units N in the finite population is known and that with each unit is associated a number y_i. The general problem is to choose some of the units as a sample, observe the y's for the sample units, and then use those observations to estimate the value of some function $h(y_1, y_2, \ldots, y_N)$ of all the y's in the population. The function $h(y_1, y_2, \ldots, y_N)$ can be a simple combination of the y's like their total or mean or may be something more complex like a quantile. In most of this book we will concentrate on functions that are linear combinations of the y's, with special emphasis on the population total, $\Sigma_{i=1}^{N} y_i$.

The prediction approach treats the numbers $y_1, \ldots, y_N$ as realized values of random variables $Y_1, \ldots, Y_N$. After the sample has been observed, estimating $h(y_1, y_2, \ldots, y_N)$ entails predicting a function of the unobserved Y's. Relationships among the random variables are expressed in a model for their joint probability distribution, and predictions are made with reference to this model. Notice that we use the term "prediction" in the sense of making a statistical guess about the Y's we have not seen—not in the literal sense of forecasting future values. We are merely guessing the values we *would* see if we were to gather data beyond the sample, which we usually have no intention of doing. In most applications, the values of Y will have been realized for all of the finite population units before we ever select a sample. After we select and observe a sample, we will know the y's for the sample units, but the y values for the nonsample units remain unknown. Our ignorance of the nonsample y values means that we must mathematically predict some function of those values in order to have an estimator or predictor for the full population.

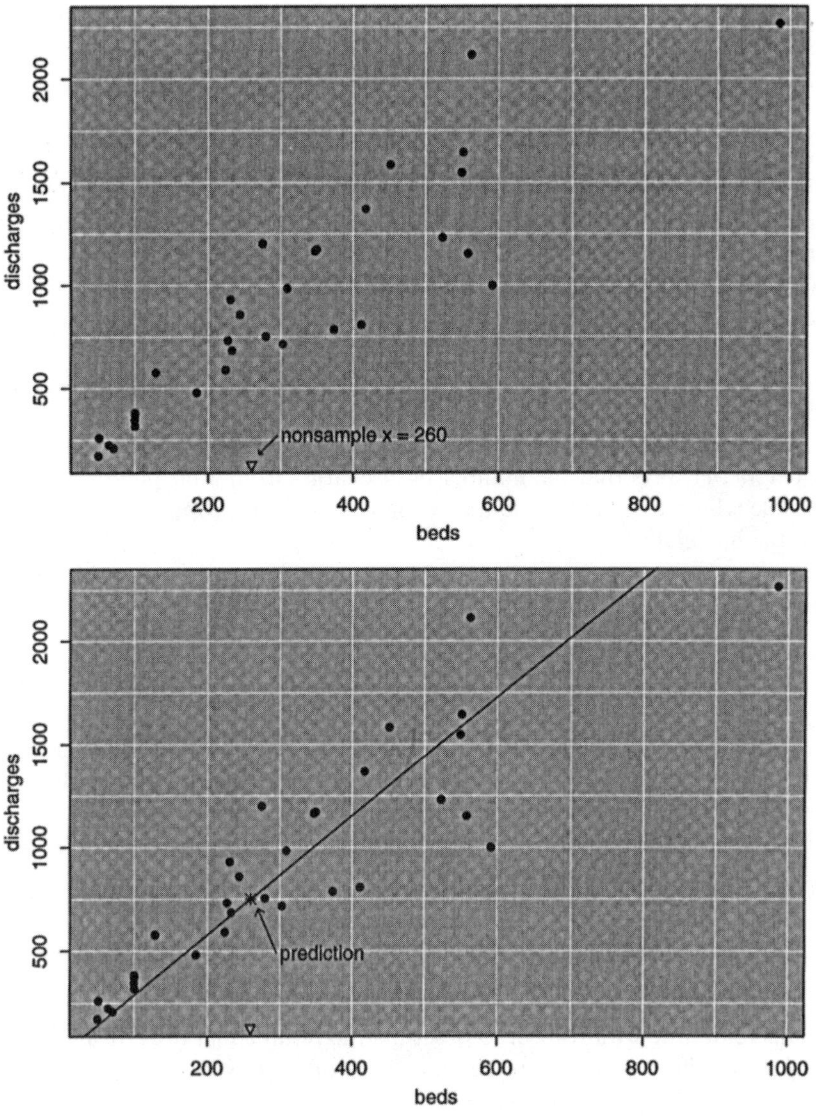

Figure 1.1 Number of patients discharged and number of beds in 33 short-stay U.S. hospitals.

The upper panel of Figure 1.1 shows the number of patients discharged (y) plotted versus the number of beds (x) in a sample of 32 short-stay hospitals. Suppose that this is actually a sample from a population of $N = 33$ hospitals and that we want to estimate $T = \Sigma_{i=1}^{N} y_i$, the population total of discharges. Denote the set of sample units by s. The population total can be written as the sum of the discharges in the 32 sample hospitals, $\Sigma_{i \in s} y_i$, plus the number of discharges y from the single nonsample hospital:

$$T = \sum_{i \in s} y_i + y.$$

The first component is known, and estimating T is equivalent to predicting the number of discharges y from the one nonsample hospital. If the number of beds x in the nonsample hospital is known, then a natural way to do the predicting is to use a regression model that treats the 33 numbers, $y_1, y_2, \ldots, y_{33}$, as realized values of random variables. We can fit the regression model to the observed sample data and then use the fitted model and the known number of beds in our nonsample hospital to predict the value of the unobserved random variable Y.

In view of Figure 1.1, we might consider the simple proportional regression model M that says that the number of discharges from a hospital is a random variable whose expected value and variance are both proportional to the size x of the hospital:

$$E_M(Y_i) = \beta x_i \qquad i = 1, 2, \ldots, N;$$
$$\operatorname{cov}_M(Y_i, Y_j) = \begin{cases} \sigma^2 x_i & i = j \\ 0 & \text{otherwise.} \end{cases} \tag{1.2.1}$$

The subscript M is used to denote statistical calculations made with respect to a prediction model M to distinguish them from calculations made with respect to the probability sampling plans used in conventional sampling theory (see Section 1.3). The best linear unbiased estimator of β under this model is found by minimizing the weighted sum of squares

$$\sum_{i \in s} \frac{1}{\sigma^2 x_i} (y_i - \beta x_i)^2.$$

Differentiating with respect to β, setting the result equal to zero, and solving for β gives the best linear unbiased estimator as $\hat{\beta} = \Sigma_{i \in s} y_i / \Sigma_{i \in s} x_i$. For the sample of 32 hospitals in Figure 1.1, $\hat{\beta} = 28{,}641/9{,}938 = 2.882$ (the data are completely listed in Table 1.1). In a usual regression analysis, we might now estimate the standard error of $\hat{\beta}$ and possibly give a confidence interval or test a hypothesis about the value of the parameter β. But, for the finite population estimation problem, $\hat{\beta}$ is just an intermediate step on the way to estimating T.

Now it happens the number of beds in the nonsample hospital is $x = 260$, and so its predicted number of discharges is $\hat{\beta} x = (2.882)(260) \cong 749$. Then the estimate for the population total is $\hat{T} = \Sigma_{i \in s} y_i + \hat{\beta} x = 28{,}641 + 749 = 29{,}390$.

How close is this realization of $\hat{T}$ to T? One of the nice aspects of the field of finite population sampling is that often one can test ideas on actual

Table 1.1. Illustrative Population of 33 Hospitals Taken From the Hospital Population in the Appendix

Unit	Unit No. in Appendix Population	No. of Beds x	No. of Discharges y
1	44	49	173
2	45	50	260
3	62	64	225
4	71	70	209
5	91	99	346
6	94	100	383
7	95	100	318
8	96	100	373
9	127	128	577
10	168	184	481
11	187	224	590
12	192	227	732
13	196	231	931
14	197	233	684
15	203	244	858
16	214	260	*1076*
17	226	275	1201
18	230	279	754
19	245	303	715
20	251	309	985
21	269	347	1166
22	272	350	1173
23	286	373	787
24	303	411	808
25	304	417	1369
26	314	451	1584
27	337	523	1232
28	348	549	1547
29	351	551	1645
30	353	558	1152
31	354	562	2116
32	361	591	999
33	393	986	2268
Totals for $N=33$		10,198	*29,717*
Totals for sample $n=32$		9,938	28,641

Unit 16 is the single nonsample unit.

populations, and see precisely how closely a method leads to the truth. As it happens, the unsampled unit (unit 16 in Table 1.1) has $y = 1{,}076$. Thus the actual total for the $N = 33$ hospitals is $T = 29{,}717$. We are naturally not exactly on the mark, because the actual value of Y does not fall exactly on the regression line. The absolute value of the relative error $(\hat{T} - T)/T$ is about 1%, which would usually be considered good. An estimate of the variance of $\hat{T} - T$ or, equivalently, of $\text{var}_M(x_{16}\hat{\beta} - Y_{16})$, yields a value for the standard error of about 208 (see Exercise 1.1), so that a 95% confidence interval of the form $\hat{T} \pm 1.96\sqrt{\text{var}(\hat{T} - T)}$ does, in fact, include the true T.

This same line of reasoning extends to the case of more than one nonsample unit. If r denotes the set of nonsample units (r for remainder), the population total is $T = \Sigma_{i \in s} y_i + \Sigma_{i \in r} y_i$ and the task of estimating T becomes one of predicting the value $\Sigma_{i \in r} y_i$ for the unobserved random variable $\Sigma_{i \in r} Y_i$. If we let $\hat{Y}_i$ denote the predictor for nonsample unit i, then an estimator of T is $\hat{T} = \Sigma_{i \in s} y_i + \Sigma_{i \in r} \hat{Y}_i$. The estimation error is

$$\hat{T} - T = \sum_{i \in r} \hat{Y}_i - \sum_{i \in r} Y_i$$

the difference between the prediction of the nonsample total and its actual value.

Suppose we consider predictors of the nonsample total that are linear combinations of the sample y's, $\hat{T}_r = \Sigma_{i \in s} a_i Y_i$, and find the coefficients $a_1, a_2, \ldots, a_n$ that are in some sense "best." One approach is to find the linear unbiased predictor $\hat{T}_r$ of the nonsample total that has the smallest error variance, defined as $\text{var}_M(\hat{T}_r - \Sigma_{i \in r} Y_i)$. Under model (1.2.1), this predictor is $\hat{T}_r = (\Sigma_s Y_i / \Sigma_s x_i)\Sigma_r x_i$, as we shall see in Section 2.1. The corresponding estimator of T is $\hat{T}_R = \Sigma_s Y_i + (\Sigma_s Y_i / \Sigma_s x_i)\Sigma_r x_i = (\Sigma_s Y_i / \Sigma_s x_i)\Sigma_{i=1}^{N} x_i$ and is known as the *ratio estimator*. By converting summations to means—$\bar{Y}_s = \Sigma_{i \in s} Y_i / n$, $\bar{x}_s = \Sigma_{i \in s} x_i / n$, and $\bar{x} = \Sigma_{i=1}^{N} x_i / N$—the ratio estimator can also be expressed as

$$\hat{T}_R = N \bar{Y}_s \frac{\bar{x}}{\bar{x}_s}.$$

We can use the asymptotic normality of the standardized error, $(\hat{T}_R - T)/\sqrt{\text{var}_M(\hat{T}_R - T)}$, under model (1.2.1), and a consistent variance estimator to set approximate confidence intervals for T. The theory behind this and other results will be covered in detail beginning in Chapter 2. For now we simply emphasize the equivalence, after sampling, between estimating T and predicting $\Sigma_{i \in r} Y_i$, and the central role of probability models in providing a framework for this prediction.

Model (1.2.1) seems to be compatible with the observations in Figure 1.1, but it is certainly not unique in this respect. There may be other models with, say, a quadratic term in x or a nonzero intercept, that would fit the data as well or better than (1.2.1). Thus it might be reasonable to adopt (1.2.1) as a *working model*, using it to generate and evaluate estimators, confidence intervals, and so forth, while being fully aware that other models might be more appropriate. Robustness (insensitivity to failure of the working model) is a key consideration in the prediction approach to finite population sampling; it will be studied in detail in Chapters 3 and 4.

Within the model-based framework, methods of prediction exist other than the one just described. With some additions to (1.2.1) we can use Bayesian prediction techniques (Bolfarine and Zacks, 1992; Datta and Ghosh, 1991; Ericson, 1969; Ghosh and Meeden, 1997; Royall and Pfeffermann, 1982), calculating the posterior distribution of T given the sample y's. Or, we could try fiducial inference (Kalbfleisch and Sprott, 1969) or likelihood prediction (Bjornstadt, 1990; Royall, 1976a). These methods tend to be more complicated and in the main will be beyond the scope of this book. What all these approaches have in common is the recognition that (i) after the sample units have been chosen and their y-values observed, estimating T is equivalent to predicting the value of the sum of the nonsample y's; and (ii) the model for the joint probability distribution of $Y_1, \ldots, Y_N$ provides the basis for the prediction.

By contrast, the approach that has dominated finite population sampling for the last half-century is not based on prediction models at all but on the probability distributions created when the choice of which units will comprise the sample is left to chance. This approach, like the prediction approach based on linear regression models, produces a theory in which inferences are made in terms of bias, variance, and approximate normality. We examine it briefly in the next section.

1.3. PROBABILITY SAMPLING THEORY

A *probability sampling plan* (or *probability sampling design*) is a scheme for choosing the sample so that every subset s of units has a known probability $p(s)$ of selection. For a given population vector $\mathbf{y} = (y_1, \ldots, y_N)'$ and a given estimator, such as the ratio estimator $\hat{T}_R$ in Section 1.2, every possible sample gives a value to the estimator. The probability that the estimator will assume a particular value is the probability that one of the samples giving that value will be selected. Definitions of bias and variance are stated in terms of the selection probabilities $p(s)$. For instance, the bias of the ratio estimator is

$$E_\pi(\hat{T}_R - T) = \sum_s p(s)[\hat{T}_R(s) - T], \tag{1.3.1}$$

where the summation is over all possible samples s that can be drawn using a particular probability sampling plan. The subscript π indicates that the expectation is taken with respect to a probability sampling plan and is not based on a model as in Section 1.2. The variance is defined as

$$\text{var}_\pi(\hat{T}_R) = \sum_s p(s)[\hat{T}_R(s) - E_\pi(\hat{T}_R)]^2. \tag{1.3.2}$$

Because the bias, variance, and other statistical quantities are computed by averaging over all the samples that might be drawn under a particular sampling design, these are sometimes referred to as *repeated sampling* or *design-based* properties. An estimator $\hat{T}$ for which $E_\pi(\hat{T} - T) = 0$ is called *design unbiased*. The quantity $\text{var}_\pi(\hat{T})$ is the *design variance*.

In this approach the only random element is the set s, showing which units are to comprise the sample. Sometimes, in order to decide on a sampling plan or estimator, the vector $\mathbf{y}$ is represented as a realization of a random vector $\mathbf{Y}$ (just as in the prediction approach), but here it is the selection probabilities, not the probability distribution of $\mathbf{Y}$ that are used in the definitions of bias and variance. Thus the probability sampling approach can be used in the presence of a prediction model, just as the prediction approach can be used with a probability sample. The key distinction between the two approaches—probability sampling and prediction—concerns which probability distribution is used in inference, not what procedure is used for choosing the sample on which that inference is based.

A *simple random sampling* plan of size n assigns to each of the $\binom{N}{n}$ sets of n different units the same probability of selection $1\Big/\binom{N}{n}$. In this "without replacement" sampling plan, which will be denoted by *srswor*, no unit can appear in the sample more than once. If repetitions of units are allowed in the sample, then the total number of possible samples of size n grows to N^n. Under a *simple random sampling with replacement* (*srswr*) plan, each of these N^n samples is given the same selection probability, $1/N^n$.

If y is a zero-one variable, then, when an *srswor* plan is used, the probability that a sample containing k 1's will be selected is

$$\binom{N\bar{y}}{k}\binom{N(1-\bar{y})}{n-k}\Big/\binom{N}{n},$$

where $\bar{y}$ is the proportion of 1's in the population. But, when an *srswr* plan is used the probability is

$$\binom{n}{k}\bar{y}^k(1-\bar{y})^{n-k}.$$

Thus the number of 1's in the sample has a hypergeometric probability distribution for *srswor* but a binomial distribution under *srswr*.

In many analyses, it is unnecessary to actually calculate the probability $p(s)$ of selecting the entire sample. What is needed is simply, for each sample unit i, the probability of selecting that particular unit, π_i. For an *srswor* every one of the n units has the same selection probability, $\pi_i = n/N$, but with more complicated plans each sample unit can have a different value of π_i. Estimators, like the Horvitz-Thompson estimator that will be discussed in Section 1.3.2, often depend on the selection probability π_i for each sample unit.

The estimators used in most practical problems have far more complicated sampling distributions than the hypergeometric and binomial. Probability sampling theory has concentrated on examining the expected values and variances of estimators and showing that in some important instances the distributions are approximately normal when both n and N are large.

For example, with a *srswor* plan, the expansion estimator, defined as the product of the population size and the sample mean, $\hat{T}_0 = N\bar{y}_s$, is an unbiased estimator of the population total T, that is, $E_\pi(\hat{T}_0) = T$. Its variance is $\text{var}_\pi(\hat{T}_0) = N^2(1 - n/N)S^2/n$, where $S^2 = \Sigma_{i=1}^N (y_i - \bar{y})^2/(N - 1)$. An unbiased estimator of the variance is obtained when S^2 is replaced by $v_s = \Sigma_s(y_i - \bar{y}_s)^2/(n - 1)$. These statements refer to the probability distributions created by the simple random sampling plan and are valid whenever such a plan is used. No assumptions about the population vector $\mathbf{y}$ are needed. However, if $\mathbf{y}$ is considered to belong to a hypothetical sequence of vectors obtained by increasing N, then under some mild assumptions about this sequence, the sampling distribution of $\sqrt{n}(\bar{y}_s - \bar{y})/\sqrt{(1 - n/N)v_s}$ converges to a standard normal distribution as n and N grow (see the theorem in Appendix A.9; Hájek, 1960; Madow, 1948).

1.3.1. Techniques Used in Probability Sampling

The use of simple random sampling and the sample mean for estimating a population mean is intuitively appealing when the population is relatively homogeneous. But, in many applications some information is available about differences among population elements. If the elements are villages, for example, each village's location and approximate number of inhabitants may be known. If the elements are businesses, the number of employees at an earlier time period and the type of product or service each business provides may be known. Efforts to use this information to get better estimates at lower cost lead to many different techniques for sampling and estimation. These include stratification, cluster sampling, probability proportional to size sampling, and adjustments to estimators through the use of models. These elementary methods can be combined to yield rather complex sampling plans and estimators.

Stratification refers simply to dividing the population into mutually exclusive groups or *strata*, each of which is sampled. Businesses, for example, might be stratified by type (manufacturing, retail trade, wholesale trade, service, etc.), or they might be stratified according to their employment size with, say, businesses having 50 employees or less in one stratum, 51–100 in another, 101–200 in third, and so on. If independent simple random sampling plans (with or without replacement) are used within each of the strata, the overall scheme is known as a *stratified simple random sampling plan.*

Cluster sampling also divides the population into disjoint groups. Whether the groups are considered strata or clusters depends on how the sample is drawn. When at least one sample element is selected from each, the groups are strata. But, if a sample of the groups is selected, so that not all are represented in the sample, then the groups of units are, by definition, clusters.

Frequently, to reduce costs or for other practical reasons, instead of observing all elements in a sample cluster, we might observe only a sample of its elements. Then, the sampling plan is carried out in two stages: (1) a sample of clusters is selected, and (2) a subsample of units within each of the sample clusters is drawn. When the population is organized in natural geographic clusters, for example, the households in a country, two or more stage sampling procedures can have important logistical and statistical advantages.

When the elements vary in size and a measure x of the size of each element is available, stratification according to size is one of the techniques for using this information. Another way is to make the inclusion probability of an element i—that is, the probability that the element i will be included in the sample—proportional to the measure of size x_i. Horvitz and Thompson (1952) showed that, for any sampling plan that gives every element a nonzero inclusion probability π_i, an unbiased estimator of the population total is

$$\hat{T}_\pi = \sum_{i \in s} \frac{y_i}{\pi_i}.$$

The mathematics of this and related facts are given in Section 1.3.2 below. For a probability-proportionate-to-size (*pps*) sampling plan, $\pi_i = nx_i/(N\bar{x})$, where $\bar{x}$ is the population mean of x, and the Horvitz-Thompson estimator becomes

$$\hat{T}_\pi = N\bar{x} \sum_{i \in s} \frac{y_i}{nx_i}.$$

In repeated sampling, the variance of $\hat{T}_\pi$ is small when the ratios y_i/x_i are nearly constant. Such *pps* sampling plans are sometimes used in selecting the first-stage sample in two-stage cluster sampling, with the number of units in each cluster serving as the measure of cluster size x.

In addition to these methods, an array of estimators arises through the use of models. The idea is to construct estimators that are (approximately) design unbiased, whether the model holds or not, and that have small design variance, if the model does hold. The coherent theory of this approach is now referred to as *model assisted* sampling (Särndal, Swensson, and Wretman, 1992). We here merely illustrate this approach in two simple, well-regarded estimators.

Consider the ratio estimator $\hat{T}_R$ defined in Section 1.2. In the probability sampling context the ratio $\bar{x}/\bar{x}_s$ is thought of as a "ratio adjustment" that is applied to the expansion estimator $\hat{T}_0 = N\bar{y}_s$. This adjustment increases the expansion estimator when the average size of the sample units is less than the population average and decreases it when the sample average is larger. Under simple random sampling, the ratio estimator has a design bias, that is, $E_\pi(\hat{T}_R - T) \neq 0$. But, if the variance of y_i is roughly proportional to x_i, as is the case, for example, when the proportional regression model (1.2.1) applies, the variance of $\hat{T}_R$ over all possible samples can be much less than that of $\hat{T}_0$.

Another estimator obtained by using the known values of an auxiliary variable x to adjust the expansion estimator is the *regression estimator*

$$\hat{T}_{LR} = N\bar{y}_s + Nb(\bar{x} - \bar{x}_s),$$

where $b = \Sigma_s(x_i - \bar{x}_s)(y_i - \bar{y}_s)/\Sigma_s(x_i - \bar{x}_s)^2$. Under simple random sampling, $\hat{T}_{LR}$ also has a bias, but its repeated sampling variance is small if the N points (x_i, y_i) are concentrated along a straight line.

As sampling plans and estimators become more complicated, so do the technical problems. It becomes hard to calculate the repeated sampling bias or discover when a particular approximation to the bias is useful. It becomes even harder to find and evaluate estimates of the sampling standard error and to determine whether the normal approximation for the sampling distribution is adequate. However, under the probability sampling approach the basic strategy remains simple:

- Choose a probability sampling plan and an estimator that, under that plan, is (approximately) unbiased with small variance.
- Express uncertainty by an estimate of the standard error.

1.3.2. Some Mathematical Details

Calculations made with respect to the probability sampling plan (the randomization distribution) are fundamentally different from those made with respect to a model. For convenience, we summarize some of the basic design-based calculations used in probability sampling theory. First, let $s(i)$ stand for the set of all potential samples that contain unit i for $i = 1, 2, \ldots, N$. To find the expected value of the sample sum $\Sigma_{i \in s} y_i$, write

$$E_\pi\left(\sum_{i \in s} y_i\right) = \sum_{s} p(s)\left(\sum_{i \in s} y_i\right) = y_1 \sum_{s(1)} p(s) + y_2 \sum_{s(2)} p(s) + \cdots + y_N \sum_{s(N)} p(s). \quad (1.3.3)$$

The sum of the probabilities $\Sigma_{s(i)} p(s)$ is just the probability that unit i will be in the selected sample. Or, using our previous notation

$$\pi_i = \sum_{s(i)} p(s).$$

Expression (1.3.3) then becomes

$$E_\pi\left(\sum_{i \in s} y_i\right) = \sum_{i=1}^{N} \pi_i y_i.$$

For *srswor* every unit has the same selection probability, $\pi_i = n/N$. We can show this as follows: There are $\binom{N}{n}$ possible samples when selection is done without replacement, and each of these has the same selection probability $1\Big/\binom{N}{n}$. Given that unit i is in a sample, the other $n - 1$ sample units must be selected from the remaining $N - 1$ units in the population so that there are $\binom{N-1}{n-1}$ different, potential samples that contain unit i. Thus the selection probability of unit i is the number of samples that contain unit i divided by the total number of samples of size n:

$$\begin{aligned}\pi_i &= \binom{N-1}{n-1}\Big/\binom{n}{N} \\ &= n/N.\end{aligned}$$

A quantity needed for calculating the sampling variance is the *joint selection probability* of any pair of units i and j:

$$\pi_{ij} = \sum_{s(ij)} p(s),$$

where $s(ij)$ indexes the samples that contain both units i and j. For example, consider the joint selection probability of any pair of units, i and j, under *srswor*. Given that unit i is selected, the remaining sample is, in effect, a simple random sample of size $n - 1$ from the remaining $N - 1$ units, so that the probability of selecting unit j, given unit i, is $(n - 1)/(N - 1)$. Thus the probability of selecting both i and j is simply

$$\pi_{ij} = \frac{n}{N}\frac{n-1}{N-1}.$$

The general structure in probability sampling theory can be viewed in an alternative way that highlights the paramount importance of random selection in that theory. For without replacement sampling plans, define the random variable

$$\delta_i(s) = \begin{cases} 1 & \text{if unit } i \text{ is in sample } s \\ 0 & \text{if not} \end{cases}$$

for $i = 1, \ldots, N$. For fixed sample size plans, $\Sigma_{i=1}^{N} \delta_i(s) = n$ for all s for which $p(s) > 0$. Key facts about these indicators are

$$P(\delta_i(s) = 1) = 1 - P(\delta_i(s) = 0) = \pi_i$$

$$P(\delta_i(s) = 1, \delta_j(s) = 1) = \pi_{ij}, \quad \pi_{ii} = \pi_i$$

from which it follows that

$$E_\pi[\delta_i(s)] = E_\pi[\delta_i^2(s)] = \pi_i, \quad E_\pi(\delta_i(s)\delta_j(s)) = \pi_{ij}$$

$$\text{var}_\pi[\delta_i(s)] = \pi_i(1 - \pi_i)$$

$$\text{cov}_\pi[\delta_i(s), \delta_j(s)] = \pi_{ij} - \pi_i\pi_j. \tag{1.3.4}$$

To illustrate the use of these indicators in design-based calculations, we consider the case of an *srswor* plan, where $\pi_i = n/N$, and the expansion estimator of the total, $\hat{T}_0 = N\bar{y}_s = N\,\Sigma_{i=1}^N \delta_i(s)y_i/n$. What is random in this expression is the set s indicating which units are to be in the sample. The y_i are simply fixed constants whose values become known for all i in the sample s. The design-expectation of $\hat{T}_0$ is

$$\begin{aligned} E_\pi(\hat{T}_0) &= \frac{N}{n} E_\pi\left(\sum_{i=1}^N \delta_i(s)y_i\right) \\ &= \frac{N}{n}\left[\sum_{i=1}^N E_\pi(\delta_i(s))y_i\right] \\ &= \frac{N}{n}\sum_{i=1}^N \frac{n}{N} y_i \\ &= T. \end{aligned}$$

Thus the expansion estimator is a design unbiased estimator of the population total under an *srswor* plan. Similarly, using the variances and covariances of the indicator functions, one can show that the variance of $\hat{T}_0$ under an *srswor* plan is $N^2(1 - n/N)\Sigma_{i=1}^N (y_i - \bar{y})^2/n(N - 1)$ (see Exercise 1.6).

An important general estimator in probability sampling is the Horvitz-Thompson estimator (or "the π-estimator") defined earlier as

$$\hat{T}_\pi = \sum_{i \in s} \frac{y_i}{\pi_i}. \tag{1.3.5}$$

Using the indicator variables, we can write (1.3.5) as $\hat{T}_\pi = \Sigma_{i=1}^N \delta_i(s)y_i/\pi_i$. The fact that this estimator is design unbiased follows easily from the properties of the δ_i's (see Exercise 1.8). The design-variance of the π-estimator is

$$\text{var}_\pi(\hat{T}_\pi) = \sum_{i=1}^N \sum_{j=1}^N (\pi_{ij} - \pi_i\pi_j)\frac{y_i}{\pi_i}\frac{y_j}{\pi_j}. \tag{1.3.6}$$

When $\pi_{ij} > 0$ for all pairs of units, a design unbiased estimator of this variance is

$$v_{HT} = \sum_{i \in s}\sum_{j \in s} \frac{\pi_{ij} - \pi_i\pi_j}{\pi_{ij}} \frac{y_i}{\pi_i}\frac{y_j}{\pi_j}, \tag{1.3.7}$$

which is attributable to Horvitz and Thompson (1952). A second design unbiased variance estimator, formulated by Yates and Grundy (1953) and Sen (1953), is

$$v_{YG} = -\frac{1}{2}\sum_{i \in s}\sum_{j \in s}\frac{\pi_{ij} - \pi_i \pi_j}{\pi_{ij}}\left(\frac{y_i}{\pi_i} - \frac{y_j}{\pi_j}\right)^2. \tag{1.3.8}$$

Proof of these results on variance is left to Exercise 1.9.

These formulas can be specialized for simple random sampling without replacement, stratified random sampling, and for other sample designs. See Särndal, Swensson, and Wretman (1992), Sukhatme and Sukhatme (1970), or Thompson (1997) for more thorough coverage of design-based results.

1.4. WHICH APPROACH TO USE?

If we have choices for inference—prediction theory, probability sampling theory, or, perhaps, a hybrid of the two—which should we use? There is no doubt of the mathematical validity of either of the two theories. The key question is which is more appropriate for inference from an observed sample and a calculated estimate. We referred to the prediction model earlier as a "working model" to emphasize that it is tentative and approximate. By contrast our control and knowledge of the probability sampling distribution is complete, at least in principle. Thus it is easy to see the appeal of basing inferences on the latter, keeping them independent of fallible prediction models.

The inevitable fallibility of our models has been an important theme of the theory of prediction based sampling, which we discuss in this book in Chapters 3, 4, and elsewhere. But there are some fundamental issues lying in the background of the model-based versus design-based controversy. A number of review articles that compare the theories are available (see, e.g., Brewer, 1994; Hansen, Madow, and Tepping, 1983; Royall, 1976c, 1983, 1988; Smith, 1976, 1984, 1994). We will briefly cover some of the issues here.

Consider the following elementary example, making use of the hospitals population. Suppose we draw a simple random sample, and use the expansion estimator, $\hat{T}_0 = N\bar{y}_s$, to estimate the total number T of discharges. The bias of $\hat{T}_0$ under an *srswor* plan is zero: $E_\pi(\hat{T}_0 - T) = 0$. This is true regardless of whether there is a relationship between the number of discharges y and the number of beds x in a hospital. Suppose the realized sample contains mostly small hospitals; that is $\bar{x}_s \ll \bar{x}$. Given what we know about the relation between x and y from a scatterplot of the sample data like Figure 1.1, we can be confident that, in this particular sample, the expansion estimator yields an underestimate of T. For example, as an extreme case, suppose that the sample consists of the $n = 5$ smallest hospitals in Table 1.1. The sample mean of x for those five is $\bar{x}_s = 242.6$ compared to the population mean for the $N = 33$ of $\bar{x} = 814.6$. Since hospitals with few beds (small x values) tend to discharge few patients (small y's), the expansion estimator will almost inevitably yield an estimate that is too small, even though it is, from the probability sampling viewpoint, unbiased. Getting the smallest five hospitals in a simple random sample might be regarded as supremely bad luck, but we would still say that $\hat{T}_0$ is unbiased under *srswor*.

At the same time, in some important sense, $\hat{T}_0$ is clearly *negatively biased* in this sample. If we try to formalize the sense in which it is biased, we are led to formalizing the relation between the x's and y's. This is most naturally done by representing the y's as outcomes of random variables following a regression model.

The model (1.2.1), for example, says that the expected value of Y is proportional to the size x. Under this model the bias in a given sample s is

$$E_M(\hat{T}_0 - T) = N\beta(\bar{x}_s - \bar{x}).$$

Here β is an unknown parameter, but since discharges surely tend to grow with number of beds, we know that β is positive. Thus, when the sample average of the auxiliary, $\bar{x}_s$, is below the population average, $\bar{x}$, the expansion estimator has—under the *model*—a negative bias. Similarly, when the average of the auxiliaries is above the population average, we have a positive bias. When applied to a particular sample s, these statements seem decidedly more relevant and informative than does the observation that, from the probability sampling viewpoint, $\hat{T}_0$ is unbiased. The design-based calculation of bias averages over all possible samples (those with $\bar{x}_s$ greater than $\bar{x}$, those with $\bar{x}_s$ smaller, and those with $\bar{x}_s$ close to $\bar{x}$).

In fact, in our sample (the five smallest hospitals), the estimate given by the expansion estimator is $\hat{T}_0 = N\bar{y}_s = 33(242.6) = 8{,}006$, far less than the actual total of $T = 29{,}717$. The same point can readily be illustrated with the largest hospitals, where $\hat{T}_0$ will surely overestimate T, as well as with samples less extreme.

A basic idea governing a great deal of statistical inference (the *Conditionality Principle*) is to *condition* on observed random variables whose probability distribution is known and, thus, not dependent on parameters about which we must make inferences. An example showing the difference between conditional and unconditional calculations was given by Royall (1994) and is similar to a classic illustration of Cox and Hinkley (1974, p. 38). Consider a population of 1000 units from which we have chosen a simple random sample of 100 units. We estimate the population mean by the sample mean $\bar{y}_s$ and estimate its variance as

$$v(\bar{y}_s) = \left(1 - \frac{100}{1000}\right)\frac{s^2}{100}, \qquad (1.4.1)$$

where $s^2 = \Sigma_s(y_i - \bar{y}_s)^2/99$.

Now we reveal that, before drawing this sample, we considered doing a complete census of this population to determine its mean precisely. But we had another study that we also wanted to do. Unable to decide whether to invest the additional time and resources needed to do a census of this population or to invest them in the other study, we flipped a coin. Thus the sample size was actually a random variable, n, which by chance turned out to be 100, but with equal probability could have been 1000. The variance of the sample mean is actually

$$\text{var}(\bar{Y}_s) = \tfrac{1}{2}\,\text{var}(\bar{Y}_s|n = 100) + \tfrac{1}{2}\,\text{var}(\bar{Y}_s|n = 1000). \tag{1.4.2}$$

Now, we have two variances, the conditional one, given the sample size actually used, and the unconditional one. Since $\text{var}(\bar{Y}_s|n = 1000) = 0$, the unconditional variance is just $\frac{1}{2}$ the conditional one. Both are correct mathematically.

Which variance should be reported with the point estimate $\bar{y}_s$? The first in (1.4.1), which estimates the conditional variance, or the unconditional one in (1.4.2), that is smaller by 50 percent? We should use the conditional variance. There is nothing wrong with the other one, probabilistically, but as a measure of the uncertainty in the observed $\bar{y}_s$ based on $n = 100$ observations, the unconditional variance is misleading, understating the uncertainty.

The unconditional variance estimator is probabilistically correct but inferentially wrong. It is wrong as a tool for helping to interpret and communicate the uncertainty in our estimate of the population mean. It is wrong because it fails to distinguish between the uncertainty in $\bar{y}_s$ when $n = 100$ and when $n = 1000$, using a common value in both cases.

More generally, if S is the indicator of the result of the coin toss, it is widely agreed that inferences should be made conditionally on the value of S. That is, statements of bias, standard error, p-values, and so forth, should be made with respect to the size of the sample that was actually observed and not averaged between this sample size and the different one that we would have observed if the coin toss had fallen the other way.

When a random sampling plan is used in a finite population, the set s that identifies the selected sample plays the role of the coin toss in the above example—it is an observed random variable whose probability distribution is known. The Conditionality Principle says that inference should be made conditionally on the observed sample s, not averaged over all samples that might have been selected, as the probability sampling approach does. If, for example, the sample consists of low x values, what does it matter that we might have gotten one with high x values?

A simple example due to Lahiri (1968) illustrates another anomaly that can result from adopting the probability-based perspective. Suppose that, from a population of 100 units, a simple random sample of $n = 25$ is selected *with replacement* by statistician A and $\hat{T}_0$ is used to estimate the total T of some variable y. By chance a set of 25 distinct units (no duplicates) is drawn. The appropriate repeated sampling variance formula used by A is

$$\text{var}_\pi(\hat{T}_0) = \frac{N^2}{n} S^2,$$

where $S^2 = \Sigma_{i=1}^N (y_i - \bar{y})^2/(N - 1)$. Now, suppose that independently, statistician B draws a simple random sample *without replacement* and obtains exactly the same set of 25 units. In that case, the probability sampling variance formula is

$$\text{var}_\pi(\hat{T}_0) = (1 - f)\frac{N^2}{n} S^2,$$

where $f = n/N = 0.25$. Thus, in the with-replacement case the variance is 1.33 times the size of the variance in the without-replacement case—despite the fact that the samples consist of exactly the same units and the value of $\hat{T}_0$ is the same in both samples. Is B's estimate of T really more precise than A's? This example emphasizes a distinction that is important in choosing a probability framework for inference from a given set of observations—the repeated sampling variance is a property of the *process* of selecting the sample and not of the *outcome* of that process.

When the randomization approach to finite population inference theory is examined in terms of the tools and concepts that are fundamental elsewhere in statistics, we find further evidence that something is wrong. The Conditionality Principle has already been cited above. Two other areas that should be mentioned are (1) best linear unbiased estimation and (2) likelihoods.

Elsewhere in statistics, best linear unbiased estimators, under regression models such as (1.2.1) have played a prominent role in theory and methods. But under the randomization approach there is no best linear unbiased (*BLU*) estimator in any nontrivial finite population estimation problem (Godambe, 1955). Basu (1971) gave a simple illustration of this result. Take any probability sampling plan and any function $\tau(\mathbf{y})$ of the population vector $\mathbf{y} = (y_1, \ldots, y_N)'$. Suppose that we select a sample s and construct a design-unbiased estimator, $t(s; \mathbf{y})$, of $\tau(\mathbf{y})$. The estimator $t(s; \mathbf{y})$ can depend on $\mathbf{y}$ only through the y's that are in the sample. For any arbitrary point $\mathbf{y}_0 = (y_{01}, \ldots, y_{0N})'$, we can construct a design-unbiased estimator that has a variance of zero as follows: We think of the N-vector $\mathbf{y}_0$ as a "guess" as to what the full population vector is. An estimator with zero variance when the actual population value $\mathbf{y}$ equals $\mathbf{y}_0$ is

$$t_0(s; \mathbf{y}) = t(s; \mathbf{y}) - t(s; \mathbf{y}_0) + \tau(\mathbf{y}_0).$$

For any value of $\mathbf{y}$, $t_0(s; \mathbf{y})$ is design-unbiased since $E_\pi t_0(s; \mathbf{y}) = \tau(\mathbf{y}) - \tau(\mathbf{y}_0) + \tau(\mathbf{y}_0) = \tau(\mathbf{y})$. But, when $\mathbf{y} = \mathbf{y}_0$, $t_0(s; \mathbf{y}_0) = t(s; \mathbf{y}_0) - t(s; \mathbf{y}_0) + \tau(\mathbf{y}_0) = \tau(\mathbf{y}_0)$, that is, exactly equal to the population quantity to be estimated. Thus, if $\mathbf{y} = \mathbf{y}_0$, the estimator $t_0(s; \mathbf{y})$ has zero estimation error in every sample and, consequently, zero variance. If $t(s;\mathbf{y})$ is linear in the sense that $t(s;\mathbf{y}) = \Sigma_{i\in s}\beta_i(s)y_i + \gamma(s)$ for some functions, $\beta(s)$ and $\gamma(s)$, of the sample, then $t_0(s;\mathbf{y})$ is also linear. Consequently, there can be no best linear unbiased estimator.

Godambe (1966) and Basu (1969) exhibited the likelihood function for the randomization model. Again, this function plays a key role in statistics outside of finite populations. Within a specified model and for a given set of observations, the likelihood provides a measure of support for hypotheses about unknown parameters. Of two parameter values the one with the greater likelihood is the better supported by the observations. Again, suppose that the vector of population values is $\mathbf{y} = (y_1, \ldots, y_N)'$, and we observe a randomly selected sample s with probability $p(s)$. Denote the population unit numbers of the units in the sample as $s = (u_1, \ldots, u_n)$ and the values observed on the n sample units as $\mathbf{y}^* = (y^*_{u_1}, \ldots, y^*_{u_n})'$. Then the likelihood generated by the random sampling distribution for that particular set of units can be written as

$$P\{(u_i, y_{u_i}),\ i = 1, \ldots, n|\mathbf{y}^*\} = \begin{cases} p(s) & \text{if } y^*_{u_i} = y_{u_i} \text{ for all } i \in s \text{ and } p(s) > 0 \\ 0 & \text{otherwise} \end{cases}$$

$$= p(s) \prod_{i\in s} \Delta(y^*_{u_i}, y_{u_i}). \tag{1.4.3}$$

where $\Delta(y^*_{u_i}, y_{u_i})$ is equal to 1 if $y^*_{u_i} = y_{u_i}$ and 0 otherwise. The second factor in (1.4.3) is 1 if all the values we observe in the sample are equal to the actual values in the population and is 0 if any of the population and sample values differ. The first factor in (1.4.3), $p(s)$, is a constant for a given sample that does not depend on the parameter, which is $\mathbf{y}$ in the design-based approach. Since the likelihood function is defined up to an arbitrary constant, the likelihood is proportional to $\Pi_{i\in s}\Delta(y^*_{u_i}, y_{u_i})$.

This finite population likelihood, generated by the randomization distribution, is disturbing. It has only two values: 1 if the sample values agree with those in the population and 0 if not. Thus, it says that any two hypotheses about the population that are both logically compatible with the observed sample are equally well-supported by that sample. The likelihood under the randomization model basically allows no useful inference to be made about the y values for the nonsample units.

The design-based approach regards the unsampled and, thus, unknown values of the variable of interest y, to be a *parameter* (typically of very large dimension). From the standpoint of the field of statistics in general, this makes a certain amount of sense, because the nonsampled y's are *unknown* and *fixed*—properties any parameter has. But, as the uninformative likelihood in (1.4.3) shows, the probability sampling plan does not establish any relationship between the y's of sample units and the unobserved y's, except that the nonsample units had a chance to be the ones selected. This is an inadequate basis for inference, as the next example illustrates.

If we weigh either an ax, an ass, or a box of old horseshoes, that weight tells us nothing whatsoever about the weight of the other two objects. This is true regardless of how we might have used random numbers in deciding which object to weigh. If we are to learn about some units (those not in the sample) from the observations we have made on others, then the two groups must have a stronger logical connection than the possibility that the nonsample units might, by chance, have been selected.

The question of whether inferences should be based on distributions created by deliberate randomization or on probability models arises also in interpreting data from designed experiments, clinical trials, and observational studies. The *Randomization Principle*, which asserts that the random sampling distribution provides the only valid inferences, has its proponents in those areas (e.g., Kempthorne, 1955). But there the principle has never achieved the general acceptance that has allowed it to dominate sampling theory (Basu, 1980; Cornfield, 1971; Royall, 1976c). There is no sound reason why finite population inference should rest on principles different from the rest of statistics.

1.5. WHY USE RANDOM SAMPLING?

The Randomization Principle says that random sampling is the *sine qua non* of finite population inference. If we reject that principle, then the question arises: "Why use random sampling?" There are various arguments for randomization that we will review here, following Royall (1976c). Some are sound whereas others are not.

An extreme (and unsound) position is the one stated in the Randomization Principle, that is, that artificial randomization provides the only basis for rigorous probabilistic inference and that, in the absence of randomization, valid probabilistic inferences are impossible. Stuart (1962), for example, emphatically expressed this viewpoint:

> If you feel at times that the statistician, in his insistence upon random sampling methods, is merely talking himself into a job, you should chasten yourself with the reflection that in the absence of random sampling, the whole apparatus of inference from sample to population falls to the ground, leaving the sampler without a scientific basis for the inferences which he wishes to make.

In contrast to this view, we have seen through various examples that the probability distribution determined by artificial randomization is not appropriate even when it is available. To claim that, in general, probabilistic inferences are not valid when the randomization distribution is not available is simply wrong. This is not to deny that randomization is valuable, but only to deny that it represents the basis for *all* valid, rigorous, probabilistic inference.

One sound reason for randomization is as a means of insuring impartiality. Officials toss a coin to determine which team will receive the ball first. The losing team may curse their luck, but the officials are protected from claims of unfairness. Randomization in choosing samples or allocating treatments can have a similar function, especially when sample results may be used in controversial decisions. Randomization can furnish protection against the charge that a sample was deliberately chosen to support one viewpoint. Randomization can also be an effective shield against many sources of unconscious personal bias that can enter a study when a sample is chosen on the basis of personal judgment. These biases are notoriously insidious and are often impossible to correct by statistical adjustments even when their existence is suspected.

One of the most important roles of randomization in sample surveys appears to be that described by Cornfield (1971) in the context of experimental studies:

> The major function of randomization, either with or without prior matching or stratification on known variables, is to achieve approximate comparability with respect to all variables, whether known or not.

In an experiment, if treatment and placebo groups differ with respect to a key response variable, then we want to attribute the difference to the effect of the treatment. But, if the groups differ with respect to some other important variable, then our conclusion is weakened. The other variable, and not the treatment, might explain the observed difference. Such "other variables" are always present in observational studies, where by definition there is self-selection involved in the composition of groups. Groups of smokers and nonsmokers can be comparable with respect to age, sex, and occupation, but they inevitably differ as regards whatever social, physical, and psychological factors lead to smoking.

Experiments have the advantage that they can avoid the group differences that are created by self-selection. But, even when the units are assigned to comparison groups, say treatment and placebo groups, by the experimenter who sees to it that the groups are comparable with respect to variables A and B, a skeptic need not agree that observed differences represent treatment effects. He might hypothesize that the groups differ with respect to another variable C and develop a plausible explanation of the results in terms of C, not the treatment. If available data do not enable us to examine the groups for the hypothesized C-difference, then the matter is one of dispute.

The experimenter is at an advantage if the units have been divided into the treatment and placebo groups at random so that there was a high probability of getting comparability between the groups on C. Now, the burden of proof shifts to the critic: in the absence of contrary evidence, there is good reason to believe these groups are comparable. There is good reason to believe that what usually happens (comparability on C) has happened in this instance, unless there is evidence to the contrary. The critic must argue otherwise.

In finite population sampling, we need comparability between the sample and nonsample in order to be able to make inferences about the entire population. If there is evidence that the sample and nonsample differ on some known characteristics, this must be considered when constructing estimators, even if randomization is used to select the sample. As Cornfield (1971) noted,

> ... if baseline characteristics arise from bad luck, post-stratification with respect to known variables and statistical analysis can often achieve what randomization failed to do.

In finite population estimation, we would use estimators like the ratio and regression estimators, $\hat{T}_R$ and $\hat{T}_{LR}$, or stratified estimators to account for population structure and imbalances between the sample and nonsample parts of the population.

If we depend entirely on the distribution generated by randomization to make inferences, then we are pushed toward the untenable position of ignoring the particular properties of the one sample we did select, thus violating the Conditionality Principle. If there are variables that affect our estimation target Y, but that we have no knowledge of, then we cannot account for those sample peculiarities. But, we cannot ignore the ones we do know about. As Savage (1962) observed:

> The possibility of...accidentally putting just the junior rabbits into the control group and the senior ones into the experimental group illustrates a flaw in the usual reference-set argument that sees randomization as injecting 'objective,' or gambling-device probabilities into the problem of inference. If the randomization and the experiment were so executed by an automaton that no one knew which...animals had been put into the control group, the argument would, I suppose, apply. But, in fact, this information is not, or ought not to be kept from the experimenter. And he ought not, in principle, to withhold it from those to whom he communicates his results.

Randomization is desirable, but it is neither necessary nor sufficient for rigorous statistical inference. Valid inference can proceed without it, and when randomization is present, it does not necessarily create the proper probabilistic framework for inference. Randomization can provide reasonable assurance of balance with respect to uncontrolled factors—it does not guarantee balance and it does not justify ignoring evidence of imbalance. There is nothing in the nature of randomization that frees experimenters or samplers from the responsibility, which Winsten (1965) assigned to the analyst of observational data, to "identify the individual points in scatter diagrams and get to know why they are there."

EXERCISES

1.1 Consider the hospital population example of Section 1.2 ($N = 33$, $n = 32$). (a) Assuming model (1.2.1) find an explicit expression for $\text{var}(\hat{T} - T)$ in terms of the x's and σ^2. (b) Using the standard expression for an estimate of σ^2 from weighted least squares regression theory, namely, $\hat{\sigma}^2 = \Sigma_s x_i^{-1} r_i^2/(n-1)$, with $r_i = y_i - \hat{\beta}x_i$, derive a variance estimator of $\text{var}(\hat{T} - T)$. Using the data of Table 1.1, derive the value of the variance estimate in this example, and construct an approximate 95% confidence interval for T.

1.2 In the population of Table 1.1, assume the 16th y value is known to be 1,076 and take another unit at random to be the single nonsample unit. Ignoring its y-value, get a 95% confidence interval for the "unknown" total, relying on model (1.2.1). Are there places in the range of x where you might feel more uncomfortable with the result than in others?

1.3 In the population of Table 1.1, assume now the nonsampled units are units 30 and 31. (a) Fitting the remaining 31 points using model (1.2.1), predict Y_{30} and Y_{31} individually. What is the relative error, $(\hat{Y}_i - Y_i)/Y_i$, in each case? (b) Using the same fit, predict the total of the 33 points. What is the relative error of your estimate?

1.4 In the population of Table 1.1, take five nonsample points at random. (a) On the basis of the remaining 28 points, using model (1.2.1) get predicted values for the 5 "unknown" y-values, and the relative error in each instance. (b) Calculate $\hat{T}_R$ and determine the relative error of this estimate.

1.5 For the hospital population example of section 1.2 ($N = 33$, $n = 32$), suppose that instead of the model (1.2.1), the working model is

$$E_M(Y_i) = \alpha + \beta x_i, \quad \text{cov}(Y_i, Y_j) = \begin{cases} \sigma^2 & i = j \\ 0 & i \neq j. \end{cases}$$

Find an estimator of total based on this model. Find a formula for its variance in terms of this model. Construct a 95% confidence interval for T, based on this model. (*Note*: Because of the typically uncertain nature of the variance structure in the working model, it turns out to be considerably safer *not* to rely on the standard approach to variance estimation which this and the preceding problems rely on. Chapter 5 discusses variance estimation in detail.)

1.6 (a) Show that $N^2(1 - n/N)\Sigma_{i=1}^N (y_i - \bar{y})^2/n(N-1)$ is the design-variance of the expansion estimator $\hat{T}_0 = N\bar{y}_s$ under a simple random sampling without replacement (*srswor*) plan. (b) Suppose that an auxiliary variable x is available for each unit in the population. Show that under a *srswor* plan $E_\pi(\bar{x}_s^{(j)}) = \bar{x}^{(j)}$, where $\bar{x}_s^{(j)} = \Sigma_s x_i^j/n$ and $\bar{x}^{(j)} = \Sigma_{i=1}^N x_i^j/N$.

1.7 Suppose for each unit in a population there are two associated numbers x_i and w_i (positive) for $i = 1, \ldots, N$. Show that, under a sampling plan with $\pi_i = nw_i^{1/2}/(N\bar{w}^{(1/2)})$, $E_\pi(n^{-1}\Sigma_s x_i/w_i^{1/2}) = \bar{x}/\bar{w}^{(1/2)}$, where $\bar{w}^{(1/2)} = N^{-1}\Sigma_{i=1}^N w_i^{1/2}$. Samples that satisfy the condition $n^{-1}\Sigma_s x_i/w_i^{1/2} = \bar{x}/\bar{w}^{(1/2)}$ exactly are known as *root*(w)-balanced and will be used extensively in Chapter 3.

1.8 Prove that the Horvitz-Thompson estimator $\hat{T}_{HT} = \Sigma_{i\in s} y_i/\pi_i$ is design unbiased under a sampling plan where the probability of selection of unit i is π_i.

1.9 (a) Show that the variance of the Horvitz-Thompson estimator, under a sampling plan where the probability of selection of unit i is π_i and the joint selection probability of units i and j is π_{ij}, is

$$\text{var}_\pi(\hat{T}_{HT}) = \sum_{i=1}^N \sum_{j=1}^N (\pi_{ij} - \pi_i\pi_j)\frac{y_i}{\pi_i}\frac{y_j}{\pi_j}.$$

(b) When $\pi_{ij} > 0$ for all pairs of units and the total sample size n is fixed, show that the following two estimators of this variance are design unbiased:

$$\text{v}_{HT} = \sum_{i\in s}\sum_{j\in s} \frac{\pi_{ij} - \pi_i\pi_j}{\pi_{ij}}\frac{y_i}{\pi_i}\frac{y_j}{\pi_j} \quad \text{and} \quad \text{v}_{YG} = -\frac{1}{2}\sum_{i\in s}\sum_{j\in s}\frac{\pi_{ij} - \pi_i\pi_j}{\pi_{ij}}\left(\frac{y_i}{\pi_i} - \frac{y_j}{\pi_j}\right)^2.$$

Hint: Use $n = \Sigma_{i=1}^N \delta_i$ and (1.3.4) to show that $\Sigma_{i=1}^N \pi_i = n$ and $\Sigma_{i=1}^N \pi_{ij} = n\pi_i$. Then, show that $\text{var}_\pi(\hat{T}_{HT}) = -\frac{1}{2}\Sigma_{i=1}^N \Sigma_{j=1}^N (\pi_{ij} - \pi_i\pi_j)(y_i/\pi_i - y_j/\pi_j)^2$.

1.10 Specialize the general results for the Horvitz-Thompson estimator to show that with an *srswor* plan, a design unbiased estimator of the variance of $\hat{T}_0 = N\bar{y}_s$ is $N^2(1 - n/N)\text{v}_s/n$, where $\text{v}_s = \Sigma_s(y_i - \bar{y}_s)^2/(n-1)$.

1.11 Suppose that this model holds for all units in the population: $Y_i = \mu + x_i^{1/2}\varepsilon_i$, where $E_M(\varepsilon_i) = 0$, $\text{var}_M(\varepsilon_i) = \sigma^2$, and the errors are independent. (a) Show that both the estimators $\hat{T}_1 = N\Sigma_{i\in s} Y_i x_i^{-1}/\Sigma_{i\in s} x_i^{-1}$ and $\hat{T}_2 = \Sigma_{i\in s} Y_i + (N - n)\Sigma_{i\in r} Y_i x_i^{-1}/\Sigma_{i\in r} x_i^{-1}$ are prediction unbiased in the sense that $E_M(\hat{T}_k - T) = 0$, $k = 1, 2$. (b) Show that neither of these $\hat{T}$'s is design unbiased under an *srswor* plan. (c) Suppose that the sampling plan has $\pi_i = nx_i/(N\bar{x})$. Show that, if the sample is large enough that the design-expectation of $\hat{T}_1$ is approximately equal to the expectation of its numerator divided by the expectation of its denominator, then $\hat{T}_1$ is approximately design unbiased under this sampling plan.

1.12 (Computer experiment) Take 10 random samples of size $n = 5$ from the hospital population of Table 1.1. Calculate $\bar{x}_s$ and the value of the expansion estimator $\hat{T}_0$ corresponding to each of the 10 samples. Is there any correspondence between $\bar{x}_s - \bar{x}$ and $\hat{T}_0 - T$?

1.13 (Computer experiment) Generate the following artificial population: $y_i = (1.1)^i$ for $i = 1, \ldots, 100$. Find T. (a) Would you expect the model

$$E_M(Y_i) = \mu, \ \mathrm{cov}(Y_i, Y_j) = \begin{cases} \sigma^2 & i = j \\ 0 & i \neq j \end{cases}$$

to fit this population well? (b) Show that the prediction estimator of total based on the model in part (a) coincides with the expansion estimator $\hat{T}_0$, and that the variance corresponds to the design-based variance of $\hat{T}_0$, except that S^2 is replaced by σ^2. Based on your answer to (a), would you expect these estimators to perform well? (c) Take, say, $K = 500$ *srswor* samples of size $n = 30$, and get $\hat{T}_0$, estimates of the variance of $\hat{T}_0$, and consequent 95% confidence intervals in each sample. Since we are doing simple random sampling without replacement, we should expect these confidence intervals to cover T what percent of the time? (d) Describe your results by computing summary statistics on bias, variance, average of your variance estimates, and coverage of the confidence intervals across the 500 samples. What condition do you think is violated in Theorem A.9, the central limit theorem for *srswor*?

CHAPTER 2

Prediction Theory Under the General Linear Model

The prediction problem that was introduced in Chapter 1 can be formulated for a general linear model and the best linear unbiased predictor derived under that model. In this chapter the general prediction theorem is proved that gives both the best predictor of any linear combination of the Y's and the variance under the model of that predictor. Before proving the general theorem, we give some details for the ratio estimator in Section 2.1 that will help motivate the more general results in Section 2.2. Sections 2.3 and 2.4 describe some applications of the general prediction theorem and the weights used for estimation that are implied by the theorem.

One of the basic inferential tools is the confidence interval derived from the large sample distribution of an estimator. A theorem in Section 2.5 gives conditions under which a linear unbiased predictor is normally distributed in large samples. This provides the basis for confidence intervals of the form $\hat{T} \pm z_{\alpha/2}\sqrt{v(\hat{T})}$ for a population total T, where $z_{\alpha/2}$ is a percentile of the standard normal distribution, and $v(\hat{T})$ is a consistent estimator of the error variance of $\hat{T}$.

Because the model-based theory has no specific requirement that samples be randomly selected, one might be tempted to think that any selection mechanism can be used. That this is not the case is discussed in Section 2.6. It is quite possible to select samples in such a way that inference is difficult or impossible. Understanding which selection methods should be avoided is critical.

Finally, in Section 2.7 we compare the model-based results for regression estimation to ones that are known as "model-assisted." The model-assisted general regression estimator (GREG) requires that a random sampling plan be used. A model is used in construction of the estimator, but selection probabilities are injected in such a way that the GREG is design-consistent.

2.1. DEFINITIONS AND A SIMPLE EXAMPLE

The finite population consists of N units, each of which has a value of a target variable y associated with it. The population vector of y's is $\mathbf{y} = (y_1, \ldots, y_N)'$ and is treated as the realization of a random vector $\mathbf{Y} = (Y_1, \ldots, Y_N)'$. Our goal will be to estimate a linear combination of the y's, $\boldsymbol{\gamma}'\mathbf{y}$, where $\boldsymbol{\gamma} = (\gamma_1, \ldots, \gamma_N)'$ is an N-vector of constants. If, for example, each $\gamma_i = 1$, then the target is the total; if $\gamma_i = 1/N$, the target is the mean. From the population of N units, a sample s of n units is selected, and the y values of the sample units are observed. As in Chapter 1, we will denote the set of nonsample units, that is, the remainder of the population, by r. For any sample s we can reorder the population vector $\mathbf{y}$ so that the first n elements are those in the sample: $\mathbf{y} = (\mathbf{y}_s', \mathbf{y}_r')'$ where $\mathbf{y}_s$ is the n-vector of observed values and $\mathbf{y}_r$ the $N - n$ unobserved. The vector $\boldsymbol{\gamma}$ can also be partitioned into parts corresponding to the sample and nonsample units, $\boldsymbol{\gamma} = (\boldsymbol{\gamma}_s', \boldsymbol{\gamma}_r')'$. Our estimation target can now be expressed as $\boldsymbol{\gamma}'\mathbf{y} = \boldsymbol{\gamma}_s\mathbf{y}_s' + \boldsymbol{\gamma}_r'\mathbf{y}_r$, which is the realization of the random variable $\boldsymbol{\gamma}'\mathbf{Y} = \boldsymbol{\gamma}_s'\mathbf{Y}_s + \boldsymbol{\gamma}_r'\mathbf{Y}_r$. At this point, it is clear that the problem of estimating $\boldsymbol{\gamma}'\mathbf{y}$ is logically equivalent to that of predicting the value, $\boldsymbol{\gamma}_r'\mathbf{y}_r$, of the unobserved random variable $\boldsymbol{\gamma}_r'\mathbf{Y}_r$.

The types of estimators considered in this chapter are linear combinations of the sample Y's as defined below.

Definition 2.1.1. A *linear estimator* of $\theta = \boldsymbol{\gamma}'\mathbf{Y}$ is defined as $\hat{\theta} = \mathbf{g}_s'\mathbf{Y}_s$ where $\mathbf{g}_s = (g_1, \ldots, g_n)'$ is an n-vector of coefficients.

Definition 2.1.2. The *estimation error* of an estimator $\mathbf{g}_s'\mathbf{Y}_s$ is $\hat{\theta} - \theta = \mathbf{g}_s'\mathbf{Y}_s - \boldsymbol{\gamma}'\mathbf{Y}$.

The estimation error can be rewritten in terms of the sample and nonsample units as

$$\begin{aligned}\mathbf{g}_s'\mathbf{Y}_s - \boldsymbol{\gamma}'\mathbf{Y} &= (\mathbf{g}_s' - \mathbf{y}_s')\mathbf{Y}_s - \boldsymbol{\gamma}_r'\mathbf{Y}_r \\ &= \mathbf{a}'\mathbf{Y}_s - \boldsymbol{\gamma}_r'\mathbf{Y}_r,\end{aligned}$$

where $\mathbf{a} = \mathbf{g}_s - \boldsymbol{\gamma}_s$. The first term depends only on the units in the sample s, and is known after the sample is observed. The second term, which depends on the nonsample units, must then be predicted. That is, using $\mathbf{g}_s'\mathbf{Y}_s$ to estimate $\boldsymbol{\gamma}'\mathbf{Y}$ is equivalent to using $\mathbf{a}'\mathbf{Y}_s$ to predict $\boldsymbol{\gamma}_r'\mathbf{Y}_r$. Consequently, finding a good $\mathbf{g}_s$ is equivalent to finding a good $\mathbf{a}$. We will study this prediction problem under the general linear model M:

$$\begin{aligned}E_M(\mathbf{Y}) &= \mathbf{X}\boldsymbol{\beta} \\ \text{var}_M(\mathbf{Y}) &= \mathbf{V}, \end{aligned} \tag{2.1.1}$$

where $\mathbf{X}$ is an $N \times p$ matrix of auxiliaries, $\boldsymbol{\beta}$ is a $p \times 1$ vector of unknown parameters, and $\mathbf{V}$ is a positive definite covariance matrix. In this development we will assume that all auxiliary values are known for each unit in the population. If the population elements are rearranged so that the first n elements of $\mathbf{Y}$ are those in the sample, and the first n rows of $\mathbf{X}$ are for units in the sample, then $\mathbf{X}$ and $\mathbf{V}$ can be expressed as

$$\mathbf{X} = \begin{bmatrix} \mathbf{X}_s \\ \mathbf{X}_r \end{bmatrix}, \qquad \mathbf{V} = \begin{bmatrix} \mathbf{V}_{ss} & \mathbf{V}_{sr} \\ \mathbf{V}_{rs} & \mathbf{V}_{rr} \end{bmatrix},$$

where $\mathbf{X}_s$ is $n \times p$, $\mathbf{X}_r$ is $(N-n) \times p$, $\mathbf{V}_{ss}$ is $n \times n$, $\mathbf{V}_{rr}$ is $(N-n) \times (N-n)$, $\mathbf{V}_{sr}$ is $n \times (N-n)$, and $\mathbf{V}_{rs} = \mathbf{V}'_{sr}$. We assume that $\mathbf{V}_{ss}$ positive definite.

Definition 2.1.3. The estimator $\hat{\theta}$ is unbiased for θ under a model M if $E_M(\hat{\theta} - \theta) = 0$.

We will also refer to an estimator with this property as being *prediction unbiased* or *model unbiased.*

Definition 2.1.4. The *error variance* (or, equivalently, *prediction variance*) of $\hat{\theta}$ under a model M is $E_M(\hat{\theta} - \theta)^2$.

Theorem 2.2.1 in the next section gives the *best linear unbiased* (*BLU*) predictor of $\boldsymbol{\gamma}'\mathbf{Y}$ under the general model (2.1.1). But, before giving this general result, it will be useful to return to the simple case of the ratio estimator introduced in Chapter 1. The model that leads to the ratio estimator is

$$\begin{aligned} E_M(Y_i) &= \beta x_i \qquad i = 1, 2, \ldots, N \\ \operatorname{cov}_M(Y_i, Y_j) &= \begin{cases} \sigma^2 x_i & i = j \\ 0 & \text{otherwise} \end{cases} \end{aligned} \tag{2.1.2}$$

whereas the estimator itself is

$$\hat{T}_R = N\bar{Y}_s \frac{\bar{x}}{\bar{x}_s}.$$

Given a sample s, the population total of the Y's can be written as $T = \Sigma_{i\in s} Y_i + \Sigma_{i\in r} Y_i$. If we knew the parameter β, we could estimate the total as $T = \Sigma_{i\in s} Y_i + \beta\Sigma_{i\in r} x_i$. Note that any estimator of T can be put in the form

$$\hat{T} = \sum_{i\in s} Y_i + \left[\frac{\hat{T} - \Sigma_{i\in s} Y_i}{\Sigma_{i\in r} x_i}\right] \sum_{i\in r} x_i$$

so that the term in brackets is an implicit estimator of β. In this form, $\hat{T} = \Sigma_{i\in s} Y_i + \hat{\beta}\Sigma_{i\in r} x_i$ and the error is $\hat{T} - T = \hat{\beta}\Sigma_{i\in r} x_i - \Sigma_{i\in r} Y_i$, so that $\hat{T}$ is model-unbiased if

$$[E_M(\hat{\beta}) - \beta]\sum_{i\in r} x_i = 0.$$

Thus $\hat{T}$ is an unbiased estimator of T if and only if its implicit $\hat{\beta}$ is an unbiased estimator of β.

The error variance of $\hat{T} = \Sigma_{i\in s} Y_i + \hat{\beta}\Sigma_{i\in r} x_i$, when the Y's are uncorrelated, is

$$\mathrm{var}_M(\hat{T} - T) = \left(\sum_{i\in r} x_i\right)^2 \mathrm{var}_M(\hat{\beta}) + \mathrm{var}_M\left(\sum_{i\in r} Y_i\right).$$

Consequently, to minimize the error variance of $\hat{T}$, we need to minimize the variance of the estimator $\hat{\beta}$. Suppose we restrict ourselves to unbiased estimators of the form $\hat{\beta} = \Sigma_{i\in s} a_i Y_i$, that is, to estimators that are linear combinations of the Y's. The expectation of this $\hat{\beta}$ under (2.1.2) is $E_M(\Sigma_s a_i Y_i) = \beta\Sigma_s a_i x_i$, and unbiasedness implies that $\Sigma_s a_i x_i = 1$. The variance of $\hat{\beta}$ is $\sigma^2\Sigma_s a_i^2 x_i$ and the minimum variance linear unbiased estimator of $\hat{\beta}$ can be found using the method of Lagrange multipliers. The function to be minimized is

$$\Phi = \sigma^2\sum_s a_i^2 x_i + \lambda\left(\sum_s a_i x_i - 1\right).$$

Differentiating Φ with respect to a particular a_i and equating to 0 gives the set of equations

$$2a_i x_i \sigma^2 + \lambda x_i = 0, \quad i \in s. \tag{2.1.3}$$

Adding these equations and using the condition that $\Sigma_s a_i x_i = 1$, implies that $\lambda = -2\sigma^2/(n\bar{x}_s)$. Substituting this into (2.1.3) yields $a_i = 1/(n\bar{x}_s)$ so that the *BLU* estimator of β is $\hat{\beta} = \bar{Y}_s/\bar{x}$, which is the least squares estimator introduced in Section 1.2.

The *BLU* estimator of the total found by using the *BLU* estimator of β in $\hat{T} = \Sigma_{i\in s} Y_i + \hat{\beta}\Sigma_{i\in r} x_i$ is then the ratio estimator, $\hat{T}_R = \Sigma_{i\in s} Y_i + (\bar{Y}_s/\bar{x}_s)\Sigma_{i\in r} x_i = N\bar{Y}_s\bar{x}/\bar{x}_s$. The error variance of $\hat{T}_R$ can be found by first noting that the estimation error is

$$\hat{T}_R - T = \left(\frac{N\bar{x}}{n\bar{x}_s} - 1\right)\sum_s Y_i - \sum_r Y_i.$$

Because $N\bar{x} - n\bar{x}_s = (N - n)\bar{x}_r$, where $\bar{x}_r$ is the mean of the nonsample x's, the error variance under model (2.1.2) is

$$\begin{aligned}\text{var}_M(\hat{T}_R - T) &= \left[\frac{(N-n)\bar{x}_r}{n\bar{x}_s}\right]^2 n\bar{x}_s\sigma^2 + (N-n)\bar{x}_r\sigma^2 \\ &= \frac{N^2}{n}(1-f)\frac{\bar{x}_r\bar{x}}{\bar{x}_s}\sigma^2. \qquad (2.1.4)\end{aligned}$$

The derivation above is a special case of the more general result for linear models that will be proved in the next section. The general calculations, though, are quite similar to the preceding ones.

2.2. GENERAL PREDICTION THEOREM

The general prediction theorem (Royall, 1976b), giving the *BLU* predictor of $\hat{\theta}$ under model (2.1.1) is

Theorem 2.2.1. Among linear, prediction-unbiased estimators $\hat{\theta}$ of θ, the error variance is minimized by

$$\hat{\theta}_{\text{opt}} = \gamma_s'\mathbf{Y}_s + \gamma_r'[\mathbf{X}_r\hat{\boldsymbol{\beta}} + \mathbf{V}_{rs}\mathbf{V}_{ss}^{-1}(\mathbf{Y}_s - \mathbf{X}_s\hat{\boldsymbol{\beta}})], \qquad (2.2.1)$$

where

$$\hat{\boldsymbol{\beta}} = (\mathbf{X}_s'\mathbf{V}_{ss}^{-1}\mathbf{X}_s)^{-1}\mathbf{X}_s'\mathbf{V}_{ss}^{-1}\mathbf{Y}_s.$$

The error variance of $\hat{\theta}$ is

$$\begin{aligned}\text{var}_M(\hat{\theta} - \theta) &= \gamma_r'(\mathbf{V}_{rr} - \mathbf{V}_{rs}\mathbf{V}_{ss}^{-1}\mathbf{V}_{sr})\gamma_r \\ &\quad + \gamma_r'(\mathbf{X}_r - \mathbf{V}_{rs}\mathbf{V}_{ss}^{-1}\mathbf{X}_s)(\mathbf{X}_s'\mathbf{V}_{ss}^{-1}\mathbf{X}_s)^{-1}(\mathbf{X}_r - \mathbf{V}_{rs}\mathbf{V}_{ss}^{-1}\mathbf{X}_s)'\gamma_r. \qquad (2.2.2)\end{aligned}$$

Proof: The error variance is

$$\begin{aligned}E_M(\mathbf{g}_s'\mathbf{Y}_s - \gamma'\mathbf{Y})^2 &= E_M(\mathbf{a}'\mathbf{Y}_s - \gamma_r'\mathbf{Y}_r)^2 \\ &= \text{var}_M(\mathbf{a}'\mathbf{Y}_s - \gamma_r'\mathbf{Y}_r) + [E_M(\mathbf{a}'\mathbf{Y}_s - \gamma_r'\mathbf{Y}_r)]^2 \\ &= \mathbf{a}'\mathbf{V}_{ss}\mathbf{a} - 2\mathbf{a}'\mathbf{V}_{sr}\gamma_r + \gamma_r'\mathbf{V}_{rr}\gamma_r + [(\mathbf{a}'\mathbf{X}_s - \gamma_r'\mathbf{X}_r)\boldsymbol{\beta}]^2. \qquad (2.2.3)\end{aligned}$$

Since we require unbiasedness, the last term in brackets above is 0. Minimization of the error variance can be done using the method of Lagrange multipliers. The Lagrange function, to be minimized with respect to **a**, is

$$\Phi = \mathbf{a}'\mathbf{V}_{ss}\mathbf{a} - 2\mathbf{a}'\mathbf{V}_{sr}\gamma_r + 2(\mathbf{a}'\mathbf{X}_s - \gamma_r'\mathbf{X}_r)\boldsymbol{\lambda},$$

where $\boldsymbol{\lambda}$ is a vector of Lagrange multipliers. Differentiating with respect to $\mathbf{a}$, setting the result equal to 0, and solving for $\mathbf{a}$ yields two equations

$$\mathbf{X}_s\boldsymbol{\lambda} = \mathbf{V}_{sr}\boldsymbol{\gamma}_r - \mathbf{V}_{ss}\mathbf{a} \tag{2.2.4}$$

$$\mathbf{a} = \mathbf{V}_{ss}^{-1}(\mathbf{V}_{sr}\boldsymbol{\gamma}_r - \mathbf{X}_s\boldsymbol{\lambda}). \tag{2.2.5}$$

Multiplying (2.2.4) on the left by $\mathbf{X}_s'\mathbf{V}_{ss}^{-1}$, solving for $\boldsymbol{\lambda}$, and using $\mathbf{a}'\mathbf{X}_s = \boldsymbol{\gamma}_r'\mathbf{X}_r$ produces $\boldsymbol{\lambda} = \mathbf{A}_s^{-1}(\mathbf{X}_s'\mathbf{V}_{ss}^{-1}\mathbf{V}_{sr} - \mathbf{X}_r')\boldsymbol{\gamma}_r$ where $\mathbf{A}_s = \mathbf{X}_s'\mathbf{V}_{ss}^{-1}\mathbf{X}_s$. Substituting $\boldsymbol{\lambda}$ into (2.2.5) gives $\mathbf{a}_{\text{opt}}$, the optimum value of $\mathbf{a}$, as

$$\mathbf{a}_{\text{opt}} = \mathbf{V}_{ss}^{-1}[\mathbf{V}_{sr} + \mathbf{X}_s\mathbf{A}_s^{-1}(\mathbf{X}_r' - \mathbf{X}_s'\mathbf{V}_{ss}^{-1}\mathbf{V}_{sr})]\boldsymbol{\gamma}_r.$$

It then follows that the best linear unbiased predictor of $\boldsymbol{\gamma}_r'\mathbf{Y}_r$ is $\mathbf{a}_{\text{opt}}'\mathbf{Y}_s = \boldsymbol{\gamma}_r'[\mathbf{X}_r\hat{\boldsymbol{\beta}} + \mathbf{V}_{rs}\mathbf{V}_{ss}^{-1}(\mathbf{Y}_s - \mathbf{X}_s\hat{\boldsymbol{\beta}})]$ and that $\hat{\theta}_{\text{opt}}$ is given by (2.2.1). Next, the error variance (2.2.2) follows from substituting $\mathbf{a}_{\text{opt}}$ into (2.2.3). Details are left to Exercise 2.6. □

Note that the parameter estimator $\hat{\boldsymbol{\beta}}$ is the solution to the problem of minimizing the weighted sum of squares $(\mathbf{Y}_s - \mathbf{X}_s\boldsymbol{\beta})'\mathbf{V}_{ss}^{-1}(\mathbf{Y}_s - \mathbf{X}_s\boldsymbol{\beta})$ with respect to $\boldsymbol{\beta}$. Solving this minimization problem leads to the estimating equations

$$\mathbf{X}_s'\mathbf{V}_{ss}^{-1}\mathbf{X}_s\hat{\boldsymbol{\beta}} = \mathbf{X}_s'\mathbf{V}_{ss}^{-1}\mathbf{Y}_s, \tag{2.2.6}$$

and hence the solution $\hat{\boldsymbol{\beta}} = (\mathbf{X}_s'\mathbf{V}_{ss}^{-1}\mathbf{X}_s)^{-1}\mathbf{X}_s'\mathbf{V}_{ss}^{-1}\mathbf{Y}_s$ as long as $\mathbf{X}_s'\mathbf{V}_{ss}^{-1}\mathbf{X}_s$ is invertible. Expression (2.2.6) constitutes a set of p equations in the p unknowns $\beta_1, \ldots, \beta_p$. These are often called the *normal equations* and the vector solution $\hat{\boldsymbol{\beta}}$ is called the *least squares estimator*. When the x's are qualitative variables that only take on the values 0 and 1, it is possible to set up a regression problem in which $\mathbf{X}_s'\mathbf{V}_{ss}^{-1}\mathbf{X}_s$ is singular, and there is no unique solution for $\hat{\boldsymbol{\beta}}$. Although Theorem 2.2.1 does not apply, such formulations are still of practical importance and will be considered further in Chapter 7.

A feature of the *BLU* predictor in the theorem is that it equals the weighted sum for the sample units, $\boldsymbol{\gamma}_s'\mathbf{Y}_s$, plus a predictor of the weighted sum for the nonsample units, $\boldsymbol{\gamma}_r'[\mathbf{X}_r\hat{\boldsymbol{\beta}} + \mathbf{V}_{rs}\mathbf{V}_{ss}^{-1}(\mathbf{Y}_s - \mathbf{X}_s\hat{\boldsymbol{\beta}})]$. When the sample and nonsample units are uncorrelated, that is, $\mathbf{V}_{rs} = \mathbf{0}$, the *BLU* predictor and its error variance become much simpler as noted in the following corollary to Theorem 2.2.1.

Corollary 2.2.1. Under model (2.1.1), if $\mathbf{V}_{rs} = \mathbf{0}$, the *BLU* predictor is $\hat{\theta}_{\text{opt}} = \boldsymbol{\gamma}_s'\mathbf{Y}_s + \boldsymbol{\gamma}_r'\mathbf{X}_r\hat{\boldsymbol{\beta}}$ with error variance $\text{var}_M(\hat{\theta} - \theta) = \boldsymbol{\gamma}_r'(\mathbf{V}_{rr} + \mathbf{X}_r\mathbf{A}_s^{-1}\mathbf{X}_r')\boldsymbol{\gamma}_r$.

The assumption that $\mathbf{V}_{rs} = \mathbf{0}$ will often be reasonable in situations where single-stage sampling is appropriate. Multistage sampling, where it is often the case that $\mathbf{V}_{rs} \neq \mathbf{0}$, will be considered in Chapters 8 and 9.

To appreciate the formulation of the theorem as one of prediction, rather than estimation, it is instructive to look at the results for the optimum $\hat{\theta}$ if we minimize its variance, $\text{var}_M(\hat{\theta}) = \mathbf{g}_s'\mathbf{V}_{ss}\mathbf{g}_s$ instead of the error variance $\text{var}_M(\hat{\theta} - \theta)$. In that case, the Lagrange function is

$$\Phi = \mathbf{g}_s'\mathbf{V}_{ss}\mathbf{g}_s + 2(\mathbf{g}_s'\mathbf{X}_s - \boldsymbol{\gamma}'\mathbf{X})\boldsymbol{\lambda}.$$

Differentiating with respect to $\mathbf{g}_s$, equating to $\mathbf{0}$, and using similar steps to those in the proof of Theorem 2.2.1, leads to $\mathbf{g}_s = -\mathbf{V}_{ss}^{-1}\mathbf{X}_s\boldsymbol{\lambda}$ and $\boldsymbol{\lambda} = -\mathbf{A}_s^{-1}\mathbf{X}'\boldsymbol{\gamma}$. Substituting the expression for $\boldsymbol{\lambda}$ into that for $\mathbf{g}_s$ gives the optimum $\mathbf{g}_s$ as $\mathbf{g}_s^* = \mathbf{V}_{ss}^{-1}\mathbf{X}_s\mathbf{A}_s^{-1}\mathbf{X}'\boldsymbol{\gamma}$. Consequently, the minimum variance estimator is

$$\hat{\theta}^* = \boldsymbol{\gamma}'\mathbf{X}\hat{\boldsymbol{\beta}}.$$

In other words, the value for each unit in the population is estimated as its expected value from the estimated regression model. Contrast this to $\hat{\theta}_{\text{opt}}$ where the sum for the sample units, $\boldsymbol{\gamma}_s'\mathbf{Y}_s$, is used directly, and the sum for the nonsample units is predicted by the estimated regression mean, $\boldsymbol{\gamma}_r'\mathbf{X}_r\hat{\boldsymbol{\beta}}$, plus an adjustment based on sample residuals, $\boldsymbol{\gamma}_r'\mathbf{V}_{rs}\mathbf{V}_{ss}^{-1}(\mathbf{Y}_s - \mathbf{X}_s\hat{\boldsymbol{\beta}})$.

The error variance of a predictor and the variance of an estimator may also be much different for a particular problem, even though the predictor and the estimator themselves are identical. Exercise 2.11 is an example in which the predictor of the finite population mean and the estimator of the model mean under a specified model are the same, but the error variance of the former is much different from the variance of the latter.

2.3. *BLU* PREDICTOR UNDER SOME SIMPLE MODELS

In some special cases the *BLU* predictors reduce to familiar estimators from probability sampling theory. In the examples below the estimation target is the finite population total $T = \Sigma_{i=1}^{N} y_i$, implying that $\boldsymbol{\gamma} = (1, \ldots, 1)'$. We have already studied one simple case—the ratio estimator—in Section 2.1. In the examples below the notation $\varepsilon_i \sim (a, b)$ denotes a random variable with mean a and variance b. It is instructive to derive the results in these examples from the general formulas in Theorem 2.2.1; see Exercises 2.7–2.10.

Example 2.3.1. Expansion Estimator. Suppose that the model is $Y_i = \mu + \varepsilon_i$ with the ε_i's being uncorrelated and $\varepsilon_i \sim (0, \sigma^2)$. Then, in model 2.1.1 $\boldsymbol{\beta} = \mu$, $\mathbf{X} = \mathbf{I}_N$, $\mathbf{V} = \sigma^2\mathbf{I}_N$, and $\hat{\boldsymbol{\beta}} = \bar{Y}_s \equiv \Sigma_s Y_i/n$. The *BLU* predictor is $\hat{T}_0 = \Sigma_s Y_i + \Sigma_r \bar{Y}_s = N\bar{Y}_s$ with the implied prediction for each nonsample unit being $\bar{Y}_s$. The error variance of the expansion estimator is $\text{var}_M(\hat{T}_0 - T) = N^2(1-f)\sigma^2/n$, where $f = n/N$. This is also the usual, design-based variance formula under simple random sampling without replacement.

Example 2.3.2. Linear Regression Estimator. Under the model with $Y_i = \alpha + \beta x_i + \varepsilon_i$ with the ε_i's being uncorrelated and $\varepsilon_i \sim (0, \sigma^2)$, the *BLU* predictor is $\hat{T}_{LR} = N[\bar{Y}_s + b(\bar{x} - \bar{x}_s)]$, where $b = \Sigma_s(Y_i - \bar{Y}_s)(x_i - \bar{x}_s)/\Sigma_s(x_i - \bar{x}_s)^2$. The error variance is $\text{var}_M(\hat{T}_{LR} - T) = N^2(1 - f)\sigma^2[1 + (\bar{x}_s - \bar{x})^2/\{(1 - f)c_s\}]/n$, where $c_s = \Sigma_s(x_i - \bar{x}_s)^2/n$.

Example 2.3.3. Stratified Expansion Estimator. Suppose that h denotes a stratum and that the model is $Y_{hi} = \mu_h + \varepsilon_{hi}$ with the ε_{hi}'s being uncorrelated and $\varepsilon_{hi} \sim (0, \sigma_h^2)$. The *BLU* predictor is $\hat{T}_{\text{sto}} = \Sigma_h N_h \bar{Y}_{hs}$, where N_h is the number of population units in stratum h, $\bar{Y}_{hs} = \Sigma_{s_h}/n_h$, s_h is the set of sample units in stratum h, and n_h is the number of sample units in the stratum. The error variance is $\text{var}_M(\hat{T}_{\text{sto}} - T) = \Sigma_h N_h^2(1 - f_h)\sigma_h^2/n_h$, where $f_h = n_h/N_h$.

Example 2.3.4. Mean-of-Ratios Estimator. For the model with $Y_i = \beta x_i + \varepsilon_i x_i$, $\varepsilon_i \sim (0, \sigma^2)$, the *BLU* predictor is $\hat{T} = \Sigma_s Y_i + \hat{\beta}\Sigma_r x_i$, where $\hat{\beta} = \Sigma_s Y_i/(nx_i)$. The error variance is $\text{var}_M(\hat{T} - T) = \sigma^2[(N - n)^2\bar{x}_r^2/n + \Sigma_r x_i^2]$. When the sampling fraction is small, the *BLU* is approximated by the mean-of-ratios estimator $\hat{T}_\pi = N\bar{x}\Sigma_s Y_i/(nx_i)$, which, as noted in Chapter 1, is the Horvitz-Thompson estimator when the selection probability of sample unit i is $\pi_i = nx_i/(N\bar{x})$.

Example 2.3.5. Estimating Proportions. Suppose that Y_i can assume the values 0 or 1 only, with probabilities p and $q = 1 - p$, respectively, and that the Y_i's are independent. This type of model is used when a unit either has some characteristic ($y_i = 1$) or not ($y_i = 0$). This also corresponds to a model with $E_M(Y_i) = p$ and $\text{var}_M(Y_i) = pq$. The *BLU* predictor of the population mean $P = \Sigma_{i=1}^N Y_i/N$ (or proportion having the characteristic) is $\hat{P} = \Sigma_s Y_i/n$ with error variance $\text{var}_M(\hat{P} - P) = (1 - f)pq/n$.

2.4. UNIT WEIGHTS

Theorem 2.2.1 gave the $n \times 1$ optimal vector of coefficients in a linear estimator as

$$\begin{aligned} \mathbf{g}_s &= \mathbf{a}_{\text{opt}} + \boldsymbol{\gamma}_s \\ &= \mathbf{V}_{ss}^{-1}[\mathbf{V}_{sr} + \mathbf{X}_s\mathbf{A}_s^{-1}(\mathbf{X}_r' - \mathbf{X}_s'\mathbf{V}_{ss}^{-1}\mathbf{V}_{sr})]\boldsymbol{\gamma}_r + \boldsymbol{\gamma}_s. \end{aligned}$$

The ith component of the vector $\mathbf{g}_s$ is often referred to as the "weight" assigned to sample unit i. The optimal weight depends on the regression component of the model, $E_M(\mathbf{Y})$, the variance specification, $\text{var}_M(\mathbf{Y})$, and on the way that the population is split between the sample and nonsample units. Most surveys collect more than one response variable Y and, in general, each may have a different optimal weight.

When constructing a database from a sample survey, it is often an advantage to have a common "weight" for all Y variables associated with a particular sample unit. This simplifies database construction and facilitates the production of linear estimates using computer software so that users of the data do not have to worry about using a different weight for each Y.

Although not true in general, there are special cases where models for different Y variables do lead to the same weights. Take the case of $\theta = T$ and let $\mathbf{1}_s$ and $\mathbf{1}_r$ be vectors of n and $N - n$ ones. A situation where the same set of weights can be used for estimating the totals of different Y's is the following: Suppose that the mean of Y is a linear combination of the same set of x's for every Y in the survey, that is, $E_M(\mathbf{Y}) = \mathbf{X\beta}$. The value of $\boldsymbol{\beta}$ can be different for each target variable Y. Assume that each Y has its own variance matrix, $\mathbf{V}$, but for constructing estimators we use a single variance structure, for example, $\sigma^2\mathbf{I}_N$. If we use $\sigma^2\mathbf{I}_N$, the vector of weights is

$$\dot{\mathbf{g}}_s = \mathbf{X}_s(\mathbf{X}_s'\mathbf{X}_s)^{-1}\mathbf{X}_r'\mathbf{1}_r + \mathbf{1}_s. \tag{2.4.1}$$

This weight vector would, of course, be optimal if $\text{var}_M(\mathbf{Y}) = \sigma^2\mathbf{I}_N$ in (2.1.1). The estimator of the total using (2.4.1) is

$$\begin{aligned}\dot{T} &= \dot{\mathbf{g}}_s'\mathbf{Y}_s \\ &= \sum_s Y_i + \mathbf{T}_{xr}\dot{\beta},\end{aligned}$$

where $\mathbf{T}_{xr} = \Sigma_r \mathbf{x}_i$ and $\dot{\boldsymbol{\beta}} = (\mathbf{X}_s'\mathbf{X}_s)^{-1}\mathbf{X}_s'\mathbf{Y}_s$. $\dot{T}$ is prediction unbiased under (2.1.1), even though it uses the wrong variance specification (see Exercise 2.15). The weight (2.4.1) depends only on the auxiliary variables and the division between sample and nonsample. In fact, $\dot{\mathbf{g}}_s$ requires only that the population totals, $\mathbf{T}_x = \Sigma_{i=1}^N \mathbf{x}_i$, be known and that the individual x's be known for the sample units. The nonsample totals $\mathbf{T}_{xr}$ can be obtained by subtraction. Note that we require only that the general model $E_M(\mathbf{Y}) = \mathbf{X\beta}$ is the same for each Y, not that the value of $\boldsymbol{\beta}$ is the same for each.

There will be some loss in efficiency under model (2.1.1) from using the suboptimal weights in (2.4.1). How serious the loss is will depend on the population. In a case like the Hospitals population in the Appendix, where the variance of Y clearly depends on the single auxiliary x, the consequence of ignoring that dependence could be a serious increase in variance. In other populations, particularly ones of households where the only auxiliaries may be age, race, and sex indicator variables, the variance structure may not depend on the auxiliaries and may be difficult to estimate (see Chapters 7 and 8). The only practical choices for weights may be ones that are somewhat, but probably not seriously, suboptimal.

Another situation where the same set of weights can also be used for estimating totals of different Y's will be addressed in Chapters 3 and 4. There we will study the properties of estimators in certain types of "balanced samples." When the same regression model, $E_M(\mathbf{Y}) = \mathbf{X\beta}$, holds for all Y variables, and the sample is balanced in a way that will be defined in the later chapters, then the same set of weights can also be used for different Y's.

2.5.* ASYMPTOTIC NORMALITY OF THE *BLU* PREDICTOR

Use of the normal distribution for inference is one of the staples of statistics. This is no less true in sampling than in other areas. In this section we state and prove a theorem giving conditions under which the prediction error, $\hat{\theta} - \theta$, of a linear estimator is asymptotically normal. The proof uses a central limit theorem for independent but not identically distributed random variables, given in Appendix A.8. The notation $x \xrightarrow{d} N(a, b)$, used below, means that a random variable x converges in distribution to a normal random variable with mean a and variance b.

In the remainder of this section, we assume that the elements of $\mathbf{Y}$ in model (2.1.1) are mutually independent so that the covariance matrix $\mathbf{V}$ is diagonal. Theorem 2.5.1, a modification of one due to Royall and Cumberland (1978), describes circumstances under which the *BLU* predictor is asymptotically normal. In order for this property to hold, certain quantities must be "well behaved" in large samples, as described in conditions (i)–(iii) below. First, we write the *BLU* predictor of the nonsample sum, $\boldsymbol{\gamma}_r'\mathbf{Y}_r$, in the form $\mathbf{a}'\mathbf{Y}_s$, where $\mathbf{a}' = (a_1, \ldots, a_n) = \boldsymbol{\gamma}_r'\mathbf{X}_r\mathbf{A}_s^{-1}\mathbf{X}_s'\mathbf{V}_{ss}^{-1}$, as in Section 2.2. The model errors are $\boldsymbol{\varepsilon}_s = \mathbf{Y}_s - \mathbf{X}_s\boldsymbol{\beta}$, $\boldsymbol{\varepsilon}_r = \mathbf{Y}_r - \mathbf{X}_r\boldsymbol{\beta}$, and $\boldsymbol{\varepsilon} = (\boldsymbol{\varepsilon}_s', \boldsymbol{\varepsilon}_r')'$. We suppose that $\varepsilon_1, \ldots, \varepsilon_N$ are independent random variables with distribution functions $F_1, \ldots, F_N$ each of which has mean zero and a finite, positive variance. As the sample and the population become large, so that both the number of sample units, n, and the number of nonsample units, $N - n$, grow without limit, that is, $n \to \infty$ and $(N - n) \to \infty$, assume that three conditions apply:

(i) $\mathbf{A}_s/n = \mathbf{X}_s'\mathbf{V}_{ss}^{-1}\mathbf{X}_s/n \to \mathbf{A}_o$ (positive definite), $\boldsymbol{\gamma}_r'\mathbf{V}_{rr}\boldsymbol{\gamma}_r/(N - n) \to v_o$, a constant, and $\boldsymbol{\gamma}_r'\mathbf{X}_r/(N - n) \to \bar{\mathbf{x}}_o$, a vector of constants;

(ii) $\max_{i=1,\ldots,n} a_i^2/\Sigma_{i=1}^n a_i^2 \to 0$;

(iii) $\max_{1 \leqslant i \leqslant N} \int_{|z|>c} z^2\, dF_i(z) \to 0$ as $c \to \infty$.

The first condition describes characteristics of the sample and nonsample units that remain stable as both groups grow. Condition (ii) says that the squares of the coefficients a_i in the estimator of the nonsample sum are each small relative to their sum. As a practical matter, (ii) means that none of the individual weights in the vector $\mathbf{a}$ should be so large as to dominate the others. The third condition implies that the tails of the distributions generating the model errors are not too heavy.

Theorem 2.5.2. Under conditions (i)–(iii), as $N, n \to \infty$, if $f \to 0$, then

$$\frac{(\hat{\theta}_{\text{opt}} - \theta)}{\sqrt{\text{var}_M(\hat{\theta}_{\text{opt}} - \theta)}} \xrightarrow{d} N(0, 1).$$

Proof: We have $\hat{\theta}_{\text{opt}} - \theta = Z_1 - Z_2$ where $Z_1 = \mathbf{a}'\boldsymbol{\varepsilon}_s$ and $Z_2 = \boldsymbol{\gamma}_r'\boldsymbol{\varepsilon}_r$. The standardized estimation error may then be written as

$$\frac{(\hat{\theta}_{\text{opt}} - \theta)}{\sqrt{\text{var}_M(\hat{\theta}_{\text{opt}} - \theta)}} = \frac{Z_1 - Z_2}{\sqrt{\text{var}_M(Z_1)}} \frac{\sqrt{\text{var}_M(Z_1)}}{\sqrt{\text{var}_M(Z_1) + \text{var}_M(Z_2)}}. \tag{2.5.1}$$

Next, calculate the variances of Z_1 and Z_2 and apply condition (i) to determine their orders of magnitude: $\text{var}_M(Z_1) = \boldsymbol{\gamma}_r'\mathbf{X}_r\mathbf{A}_s^{-1}\mathbf{X}_r'\boldsymbol{\gamma}_r = O(N^2/n)$ and $\text{var}_M(Z_2) = \boldsymbol{\gamma}_r'\mathbf{V}_{rr}\boldsymbol{\gamma}_r = O(N-n)$. Thus $\text{var}_M(Z_1)/[\text{var}_M(Z_1) + \text{var}_M(Z_2)] \to 1$. Since $E_M(Z_2) = 0$ and $\text{var}_M(Z_2/\sqrt{\text{var}_M(Z_1)}) = \text{var}_M(Z_2)/\text{var}_M(Z_1) \to 0$, we have $Z_2/\sqrt{\text{var}_M(Z_1)} \xrightarrow{p} 0$ by Chebyshev's inequality (see Appendix A.4). Finally, applying Theorem A.8 and invoking conditions (ii) and (iii) gives $Z_1/\sqrt{\text{var}_M(Z_1)} \xrightarrow{d} N(0,1)$, which completes the proof. □

Notice that the optimality of the vector $\mathbf{a}' = \boldsymbol{\gamma}_r'\mathbf{X}_r\mathbf{A}_s^{-1}\mathbf{X}_s'\mathbf{V}_{ss}^{-1}$ was not used in the proof. The theorem thus holds for other unbiased, but nonoptimal, estimators of the form $\mathbf{a}'\mathbf{Y}_s$ as long as the vector $\mathbf{a}$ satisfies appropriate conditions, for example, $\mathbf{a}'\mathbf{V}_{ss}\mathbf{a} = O(N^2/n)$ and condition (ii).

If the sampling fraction f does not approach 0 in large samples, we may still have approximate normality of a prediction error, $\hat{\theta} - \theta$, but the variance of the nonsample sum $\boldsymbol{\gamma}_r'\mathbf{Y}_r$ will not be negligible compared to the variance of the sample sum $\mathbf{a}'\mathbf{Y}_s$. This nonnegligibility becomes a matter of concern when, in Chapter 5, we seek variance estimators that are robust to departures from the working model.

The main practical use of Theorem 2.5.1 is to construct approximate confidence intervals for θ of the form $\hat{\theta} \pm z_{\alpha/2}\sqrt{v}$, where v is an estimator of $\text{var}_M(\hat{\theta} - \theta)$ and $z_{\alpha/2}$ is a quantile from the normal distribution. The coverage probability will be approximately $1 - \alpha$ when the sample is large but represents only a small fraction of the population. In small or moderate size samples, or when the sampling fraction is large, the normal approximation can be quite poor in some populations.

2.6. IGNORABLE AND NONIGNORABLE SAMPLE SELECTION METHODS

In the preceding sections of this chapter and in Chapter 1, inference methods have been based on prediction models, not random selection plans. The prediction bias, error variance, and asymptotic normality of the *BLU* predictor were calculated with respect to a model. From that reasoning one might be tempted to conclude that the method of sample selection is completely irrelevant as long as some model holds. This is true in some circumstances, but the fallacy of the general proposition can be easily illustrated.

2.6.1. Examples

Figure 2.1 shows an artificial population and the results of three methods of sample selection. The population was generated to follow the model

$$Y_i = 2x_i + \varepsilon_i x_i^{1/2}, \quad \varepsilon_i \sim N(0, 1) \tag{2.6.1}$$

with the x's coming from the Hospital population in Appendix B. Panel (a) in the figure shows the population plot of Y versus x. The population total (see Table 2.1) is 228,152.4. The optimal estimator under model (2.6.1) is the ratio estimator $\hat{T}_R = N\bar{Y}_s\bar{x}/\bar{x}_s$; however, the way in which the sample is selected can have a bearing on whether the Y's in the sample obey this model or not.

The fact that each unit typically has a realized value of Y before we sample opens up the possibility that the Y's themselves could affect which units are selected as the sample. We might try to select a sample having only large Y's, for example. After all, if we want to estimate the total of the Y's, why not pick a sample with the biggest Y's we can find, so that we can observe as much of the total as possible? Even though we will not know the Y-values before we sample, we could screen units and keep only those with Y greater than some cutoff. Such a quota sample of units is shown as circles in panel (b) of Figure 2.1. The sample was selected by sorting the entire population in a random order and then picking the first 50 units in the randomly sorted list with $Y \geqslant 200$. The least squares fitted line for the model $E_M(Y) = \beta x$, $\text{var}_M(y) = \sigma^2 x$ is drawn as a dashed, line whereas the correct population model $E_M(Y) = 2x$ is the solid line in panel (b). The ratio estimator is 279,059.9, which is 22% too large.

The situation gets even worse if the Y-cutoff is raised. Panel (c) shows a sample of 50 units, selected as in panel (b), but with $Y \geqslant 1000$. The fitted straight line regression through the origin with variance proportional to x is shown as a dashed line and is clearly too high compared to the true line. The ratio estimator for this sample is 306,410.3—a 34% overestimate.

Panels (b) and (c) show samples where the selection criterion depends directly on the target variable Y. The outcome of another method of nonrandom selection is shown in panel (d). There the sample consists of the 52 units with $x \geqslant 525$. The ratio estimator is 234,964.4, which is within 3% of the population total. Panel (d) illustrates that purposive selection based on the covariate x—but not Y itself—can provide an estimator of a total that is unbiased. (Choosing the largest x's is, in fact, the optimal strategy under model (2.6.1). But, unfortunately, it is one that relies too heavily on the exact form of the model. We will cover this type of issue in detail in Chapters 3 and 4.)

2.6.2.* Formal Definition of Ignorable Selection

The selection methods in panels (b) and (c) of Figure 2.1 are called *nonignorable* or *informative* because how the Y's were selected should not be ignored

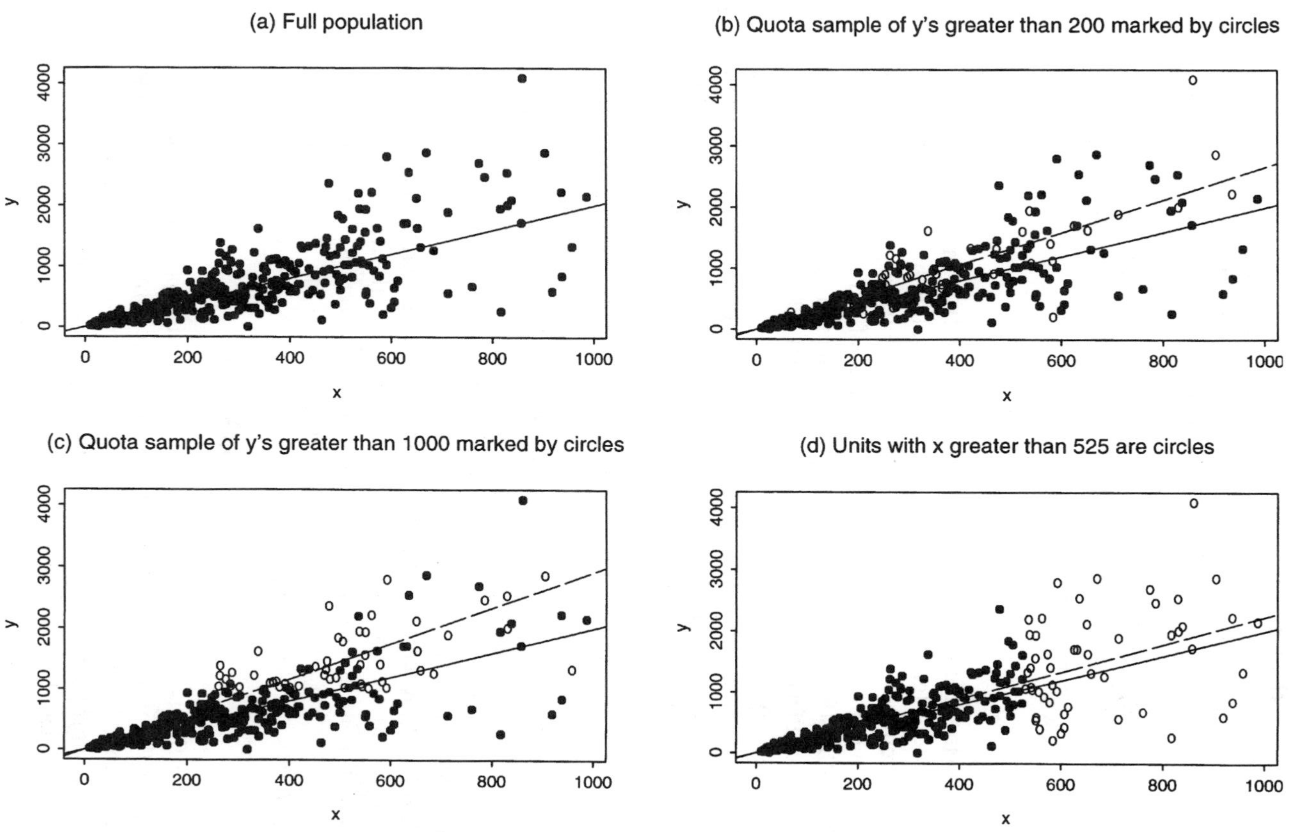

Figure 2.1 Examples of ignorable and nonignorable selection methods.

Table 2.1. Estimated Totals for Samples Selected by Two Nonignorable Selection Methods and One Ignorable Selection Method

Population Total	228,152.4
$\hat{T}_R$ for Y's with $Y \geqslant 200$	279,059.9
$\hat{T}_R$ for Y's with $Y \geqslant 1000$	306,410.3
$\hat{T}_R$ for Y's with $x \geqslant 525$	234,964.4

when making inferences. The method in panel (d) is, in contrast, *ignorable* or *noninformative* Ignorable and nonignorable sampling has received much attention in recent years in the context of estimating parameters of super population models, and more generally, in the analysis of survey data. Skinner, Holt, and Smith (1989) provides a good introduction to analysis. Pfeffermann (1993) and Pfeffermann and Sverchkov (1999) offer methods for testing for nonignorability.

In this section, we formalize this idea, relying on work by Little (1982), Rubin (1976), Smith (1983), and Sugden and Smith (1984). Define the vector of sample indicators for a given sample s as $\boldsymbol{\delta}_s = (\delta_1(s), \ldots, \delta_N(s))'$. An individual indicator $\delta_i(s)$ is the same as defined in Section 1.3.2 and equals 1 if unit i is in the sample and 0 if not. Suppose also that we have a list of prior information (or covariates) on each unit in the population contained in the N-row matrix $\mathbf{X}$. In our earlier example, the covariate data can include the number of beds in each hospital, the number of inpatients treated in a certain year, whether each hospital provided intensive cardiac care or psychiatric care, and many other pieces of information that might be used in sample design.

A sampling scheme is a rule for choosing $\boldsymbol{\delta}_s$. In general, the rule can depend on the covariates $\mathbf{X}$. It might also depend on the target variables $\mathbf{Y} = (Y_1, \ldots, Y_N)'$, as in the examples of the previous section. It could conceivably depend also on an unknown parameter vector $\boldsymbol{\phi}$. Accounting for these possibilities, we write the probability of $\boldsymbol{\delta}_s$ given $\mathbf{X}$ and $\mathbf{Y}$, as

$$f_{\boldsymbol{\delta}_s|\mathbf{X},\mathbf{Y}}(\boldsymbol{\delta}_s|\mathbf{X},\mathbf{Y};\boldsymbol{\phi}).$$

Note that this is just an elaboration of the notation $p(s)$ used in Section 1.3 to represent the probability that sample s is selected. In the standard randomization plans described in Section 1.3.1, like simple random sampling or probability proportional to size sampling, the density of $\boldsymbol{\delta}_s$ does not depend on $\mathbf{Y}$ or $\boldsymbol{\phi}$ and can be written as

$$f_{\boldsymbol{\delta}_s|\mathbf{X}}(\boldsymbol{\delta}_s|\mathbf{X}).$$

In principle, the values of $f_{\boldsymbol{\delta}_s|\mathbf{X}}(\boldsymbol{\delta}_s|\mathbf{X})$ are known to the sampler because $\mathbf{X}$ is known. As we pointed out earlier, it is often adequate for some purposes to keep track of only the individual and joint selection probabilities π_i and π_{ij} rather than the joint probability for the entire sample. The density of the target vector **Y**, given the covariates **X**, can depend on an unknown parameter $\boldsymbol{\beta}$:

$$f_{Y|\mathbf{X}}(\mathbf{Y} \mid \mathbf{X}; \boldsymbol{\beta}).$$

A joint model for both **Y** and the sample indicators is

$$f_{\mathbf{Y},\boldsymbol{\delta}_s|\mathbf{X}}(\mathbf{Y}, \boldsymbol{\delta}_s \mid \mathbf{X}; \boldsymbol{\beta}, \boldsymbol{\phi}) = f_{\mathbf{Y}|\mathbf{X}}(\mathbf{Y} \mid \mathbf{X}; \boldsymbol{\beta}) f_{\boldsymbol{\delta}_s|\mathbf{Y},\mathbf{X}}(\boldsymbol{\delta}_s \mid \mathbf{Y}, \mathbf{X}; \boldsymbol{\phi}). \tag{2.6.2}$$

Since the act of observing a sample partitions the population into sample and nonsample pieces, the joint density of $\mathbf{Y}_s$ and $\boldsymbol{\delta}_s$ can be found by integrating (2.6.2) over the possibilities for the vector of nonsample targets $\mathbf{Y}_r$:

$$f_{\mathbf{Y}_s,\boldsymbol{\delta}_s|\mathbf{X}}(\mathbf{Y}_s, \boldsymbol{\delta}_s \mid \mathbf{X}; \boldsymbol{\beta}, \boldsymbol{\phi}) = \int f_{\mathbf{Y}|\mathbf{X}}(\mathbf{Y}_s, \mathbf{Y}_r \mid \mathbf{X}; \boldsymbol{\beta}) f_{\boldsymbol{\delta}_s|\mathbf{Y},\mathbf{X}}(\boldsymbol{\delta}_s \mid \mathbf{Y}, \mathbf{X}; \boldsymbol{\phi})\, d\mathbf{Y}_r. \tag{2.6.3}$$

Ignoring the sample selection mechanism is the same as working with

$$f_{\mathbf{Y}_s|\mathbf{X}}(\mathbf{Y}_s \mid \mathbf{X}; \boldsymbol{\beta}), \tag{2.6.4}$$

where the sample s is a fixed set of units. For model-based inference, the question is: when is using the density in (2.6.4) justified? If

$$f_{\boldsymbol{\delta}_s|\mathbf{Y},\mathbf{X}}(\boldsymbol{\delta}_s \mid \mathbf{Y}, \mathbf{X}; \boldsymbol{\phi}) = f_{\boldsymbol{\delta}_s|\mathbf{X}}(\boldsymbol{\delta}_s \mid \mathbf{X}; \boldsymbol{\phi}), \tag{2.6.5}$$

then sample selection does not depend on **Y** and (2.6.3) becomes

$$\begin{aligned} f_{\mathbf{Y}_s,\boldsymbol{\delta}_s|\mathbf{X}}(\mathbf{Y}_s, \boldsymbol{\delta}_s \mid \mathbf{X}; \boldsymbol{\beta}, \boldsymbol{\phi}) &= f_{\boldsymbol{\delta}_s|\mathbf{X}}(\boldsymbol{\delta}_s \mid \mathbf{X}; \boldsymbol{\phi}) \int f_{\mathbf{Y}|\mathbf{X}}(\mathbf{Y}_s, \mathbf{Y}_r \mid \mathbf{X}; \boldsymbol{\beta})\, d\mathbf{Y}_r \\ &= f_{\boldsymbol{\delta}_s|\mathbf{X}}(\boldsymbol{\delta}_s \mid \mathbf{X}; \boldsymbol{\phi}) f_{\mathbf{Y}_s|\mathbf{X}}(\mathbf{Y}_s \mid \mathbf{X}; \boldsymbol{\beta}). \end{aligned} \tag{2.6.6}$$

There are then various arguments for why the distribution of $\boldsymbol{\delta}_s$ can be disregarded when making an inference about $\boldsymbol{\beta}$. The factorization theorem for sufficient statistics (Bickel and Doksum, 1977, p. 65), for example, implies that the sufficient statistic for making an inference about $\boldsymbol{\beta}$ is contained in $f_{\mathbf{Y}_s|\mathbf{X}}(\mathbf{Y}_s \mid \mathbf{X}; \boldsymbol{\beta})$, and we can ignore the probability $f_{\boldsymbol{\delta}_s}(\boldsymbol{\delta}_s \mid \mathbf{X}; \boldsymbol{\phi})$ for the selection mechanism. Another quantity used in inference is the likelihood ratio, $f_{\mathbf{Y}_s,\boldsymbol{\delta}_s}(\mathbf{Y}_s, \boldsymbol{\delta}_s \mid \mathbf{X}; \boldsymbol{\beta}_2, \boldsymbol{\phi}) / f_{\mathbf{Y}_s,\boldsymbol{\delta}_s}(\mathbf{Y}_s, \boldsymbol{\delta}_s \mid \mathbf{X}; \boldsymbol{\beta}_1, \boldsymbol{\phi})$, which is a measure of how well the data support one value of the parameter, $\boldsymbol{\beta}_2$, compared to another, $\boldsymbol{\beta}_1$. When sample selection does not depend on **Y** and (2.6.6) holds, the constant, $f_{\boldsymbol{\delta}_s|\mathbf{X}}(\boldsymbol{\delta}_s \mid \mathbf{X}; \boldsymbol{\phi})$, in the ratio of likelihoods cancels out for a given sample, again implying that the selection mechanism can be ignored.

There are many sampling plans that satisfy (2.6.5). The random sampling schemes described in Section 1.3 are examples, but many nonrandom plans also qualify. Selecting the n largest units, as in panel (d) of Figure 2.1, is an example. Another is a plan that purposively selects the sample units from the population to be such that $\bar{x}_s = \bar{x}$. This is an example of a *balanced* sample, and more general versions of balancing will play a prominent role in later chapters. Panels (b) and (c) of Figure 2.1 clearly illustrate plans that violate (2.6.5) since selection depends explicitly on the Y values. One argument that is often presented in favor of randomization is that (2.6.5) is automatically satisfied and awkward situations are avoided where the full model (2.6.3) needs to be considered for inference.

An uncomfortable feature of most surveys is that all units may not respond to data-collection requests. Persons, businesses, or other units may refuse to cooperate, may be impossible to contact, or cannot be surveyed for other reasons. If the nonresponse mechanism depends on the target variable, we may be in a situation where (2.6.5) does not hold, even if the initial sample was randomly selected. An establishment survey may want to estimate the average amount of sales in the last calendar quarter. If the response rate for establishments with large sales is much less than for others, this self-selection mechanism will invalidate (2.6.5) and needs to be accounted for when making inferences.

2.7.* COMPARISONS WITH DESIGN-BASED REGRESSION ESTIMATION

The technique of regression estimation, as a convenient way of using auxiliary data, is also available in the design-based literature. Särndal, Swensson, and Wretman (1992, Chapters 6–8) present that approach, referring to it as *model-assisted*. In the model-assisted approach, models may be used to construct estimators, but randomization must be used to select the sample, and statistical properties are computed with respect to the probability sampling distribution. There are many examples of the use of model-assisted regression estimation in the literature including Estevao, Hidiroglou, and Särndal (1995), Fuller, Loughin, and Baker (1994), and Jayasuriya and Valliant (1996). In this section, we will summarize some of the design-based results and compare them to our earlier model-based ones. Recall that the target variable is treated as a constant in the design-based approach, and we will use y rather than Y where appropriate.

Suppose that each unit in the population has associated with it a vector of auxiliaries $\mathbf{x}_i = (x_{i1}, \ldots, x_{ip})'$. The general regression estimator (GREG) of the finite population total of y is defined as

$$\hat{T}_{yr} = \hat{T}_{y\pi} + \hat{\mathbf{B}}'(\mathbf{T}_x - \hat{\mathbf{T}}_{x\pi}), \tag{2.7.1}$$

where $\hat{T}_{y\pi} = \Sigma_s y_i/\pi_i$ is the Horvitz-Thompson estimator (sometimes called the π-estimator) of the y-total, $\mathbf{T}_x = \Sigma_{i=1}^N \mathbf{x}_i$ is the vector of finite population totals of the auxiliaries, $\hat{\mathbf{T}}_{x\pi}$ is the π-estimator of $\mathbf{T}_x$, and

$$\hat{\mathbf{B}} = \mathbf{A}_{\pi s}^{-1}\mathbf{X}_s'\mathbf{V}_{ss}^{-1}\Pi_s^{-1}\mathbf{y}_s$$

with $\mathbf{A}_{\pi s} = \mathbf{X}_s'\mathbf{V}_{ss}^{-1}\Pi_s^{-1}\mathbf{X}_s$, $\mathbf{V}_{ss} = \text{diag}(v_i)$ and $\mathbf{\Pi}_s = \text{diag}(\pi_i)$, $i \in s$. The GREG can thus be viewed as the π-estimator of T plus an adjustment term. The slope $\hat{\mathbf{B}}$ is considered to be an estimator of the finite population quantity $\mathbf{B} = (\mathbf{X}'\mathbf{V}^{-1}\mathbf{X})^{-1}\mathbf{X}'\mathbf{V}^{-1}\mathbf{y}$, where $\mathbf{y} = (y_1, \ldots, y_N)'$. Since $E_\pi(\mathbf{X}_s'\mathbf{V}_{ss}^{-1}\mathbf{\Pi}_s^{-1}\mathbf{X}_s) = \mathbf{X}'\mathbf{V}^{-1}\mathbf{X}$ and $E_\pi(\mathbf{X}_s'\mathbf{V}_{ss}^{-1}\mathbf{\Pi}_s^{-1}\mathbf{y}_s) = \mathbf{X}'\mathbf{V}^{-1}\mathbf{y}$, $\hat{\mathbf{B}}$ will be approximately π-unbiased for $\mathbf{B}$ when the sample is large. Since $E_\pi(\hat{\mathbf{T}}_{x\pi}) = \mathbf{T}_x$ and $E_\pi(\hat{T}_{y\pi}) = T$, the GREG is approximately design-unbiased for T in large samples. The design-unbiasedness is approximate because $\hat{\mathbf{B}}$ is a product of estimators, and likewise $\hat{\mathbf{B}}'\hat{\mathbf{T}}_{x\pi}$.

The GREG is motivated by a model in which the Y's are independent for $i = 1, 2, \ldots, N$, with

$$\begin{aligned} E_M(Y_i) &= \mathbf{x}_i'\boldsymbol{\beta} \\ \text{var}_M(Y_i) &= v_i \end{aligned} \tag{2.7.2}$$

and the Y's are independent. However, in the model-assisted approach the model is little more than a vehicle for determining the form of $\hat{\mathbf{B}}$. Properties of $\hat{T}_{yr}$ are computed with respect to the randomization distribution associated with the π_i's rather than with respect to the model. The goodness of fit of the model does affect the efficiency of the GREG relative to the π-estimator, but the GREG is still approximately design-unbiased whether the model is correct or not.

The quantity $\hat{\mathbf{B}}$ is a model-unbiased estimator of the parameter $\boldsymbol{\beta}$, as is $\mathbf{B}$ itself, and $E_M(\hat{\mathbf{T}}_{yr} - \mathbf{T}) = 0$ under (2.7.2). However, the GREG is inefficient under this model because, from Theorem 2.2.1, the optimal predictor is

$$\hat{T} = \sum_s Y_i + \hat{\boldsymbol{\beta}}'\mathbf{T}_{xr}, \tag{2.7.3}$$

where $\hat{\boldsymbol{\beta}} = (\mathbf{X}_s'\mathbf{V}_{ss}^{-1}\mathbf{X}_s)^{-1}\mathbf{X}_s'\mathbf{V}_{ss}^{-1}\mathbf{Y}_s$ and $\mathbf{T}_{xr} = \Sigma_r \mathbf{x}_i$. The predicted value for unit i under the model is $\hat{Y}_i = \mathbf{x}_i'\hat{\boldsymbol{\beta}}$. Using this, note that (2.7.3) can also be written as

$$\hat{T} = \sum_{i=1}^N \hat{Y}_i + \sum_s (Y_i - \hat{Y}_i). \tag{2.7.4}$$

We can compare this directly to the GREG which can be written as

$$\hat{T}_{yr} = \sum_{i=1}^N \hat{y}_i + \sum_s (y_i - \hat{y}_i)/\pi_i \tag{2.7.5}$$

with $\hat{y}_i = \mathbf{x}_i'\hat{\mathbf{B}}$ (see Exercise 2.19). Comparing (2.7.4) and (2.7.5), we see that the difference in the two estimators is the use of the selection probabilities in the GREG and their absence in the *BLU* predictor. When the selection probabilities π_i do not differ much among units, the π_i's will essentially cancel out in $\hat{\mathbf{B}}$, and the values of the GREG and the *BLU* predictor will be near each other.

There are a variety of special cases where $\hat{T}$ and $\hat{T}_{yr}$ are exactly equal. If we begin with model (2.7.2) and select an equal probability sample, then $\hat{T} = \hat{T}_{yr}$. For example, if $E_M(Y_i) = \beta x_i$ and $\text{var}_M(Y_i) = \sigma^2 x_i$, both approaches yield the ratio estimator $\hat{T}_R$. When $E_M(Y_i) = \alpha + \beta x_i$ and the model variance is a constant, then both approaches yield the simple linear regression estimator $\hat{T}_{LR}$.

One variance model that fits many populations when a single auxiliary x is present has $\text{var}_M(Y_i) = \sigma^2 x_i^\gamma$. A common probability sampling plan in that situation is to use selection probabilities that are proportional to $x^{\gamma/2}$. The selection probability of unit i is then $\pi_i = n x_i^{\gamma/2} / N\bar{x}^{(\gamma/2)}$, where $\bar{x}^{(\gamma/2)} = \Sigma_{i=1}^N x_i^{\gamma/2}/N$. We will explore this design more in Chapters 3 and 4. If we use the GREG with this combination of v_i and π_i, then elements of the weights $\mathbf{V}_{ss}^{-1}\Pi_s^{-1}$ in the estimator $\hat{\mathbf{B}}$ are proportional to $v_i^{-3/2}$, which is clearly an inefficient choice.

The weights for individual sample units implied by the GREG and the *BLU* predictor can also be compared. The GREG can be written as $\hat{T}_{yr} = \mathbf{g}_{\pi s}'\Pi_s^{-1}\mathbf{y}_s$ with $\mathbf{g}_{\pi s} = \mathbf{V}_{ss}^{-1}\mathbf{X}_s\mathbf{A}_{\pi s}^{-1}(\mathbf{T}_x - \hat{\mathbf{T}}_{x\pi}) + \mathbf{1}_s$ so that the weights are

$$\Pi_s^{-1}\mathbf{g}_{\pi s} = \Pi_s^{-1}[\mathbf{V}_{ss}^{-1}\mathbf{X}_s\mathbf{A}_{\pi s}^{-1}(\mathbf{T}_x - \hat{\mathbf{T}}_{x\pi}) + \mathbf{1}_s],$$

whereas those for the *BLU* predictor under model (2.7.2) are

$$\mathbf{g}_s = \mathbf{V}_{ss}^{-1}\mathbf{X}_s\mathbf{A}_s^{-1}\mathbf{X}_r'\mathbf{1}_r + \mathbf{1}_s.$$

These sets of weights have the property that $\mathbf{g}_{\pi s}'\Pi_s^{-1}\mathbf{X}_s = \mathbf{g}_s'\mathbf{X}_s = \mathbf{T}_x$, which is referred to as being "calibrated" to the population totals of the auxiliary variables (see, e.g., Deville and Särndal, 1992).

We noted in Section 2.4 that the same set of $\mathbf{g}_s$ weights may not be optimal for every Y variable in a survey. In fact, the same set may not even produce model-unbiased estimators for every Y. The GREG, however, will be approximately design-unbiased even if the motivating model (2.7.2) is incorrect. Thus the same set of weights, $\Pi_s^{-1}\mathbf{g}_{\pi s}$, will produce an approximately design-unbiased estimator regardless of the Y variable.

Several authors (Särndal and Wright, 1984; Brewer, 1994, 1995, 1999) have suggested that a reasonable compromise between the *BLU* predictor and the GREG is to use an estimator of the form $\hat{T}_{yr}^* = \hat{T}_{y\pi} + \hat{\mathbf{B}}^{*\prime}(\mathbf{T}_x - \hat{\mathbf{T}}_{x\pi})$ with $\hat{\mathbf{B}}^*$ defined so that $\hat{T}_{yr}^*$ will have the form taken by a predictor, that is,

$$\hat{T}_{yr}^* = \hat{T}_{y\pi} + \hat{\mathbf{B}}^{*\prime}(\mathbf{T}_x - \hat{\mathbf{T}}_{x\pi}) = \sum_s Y_i + \hat{\mathbf{B}}^{*\prime}\mathbf{T}_{xr}. \qquad (2.7.6)$$

An estimator that satisfies (2.7.6) is said to "cosmetically calibrated." As for GREGs in general, the cosmetic estimator $\hat{T}^*_{yr}$ will be design-consistent and model-unbiased for an appropriate choice of slope estimator.

Regression estimators can also be motivated in the design-based approach using asymptotic arguments (Casady and Valliant, 1993; Rao, 1994; Robinson, 1987). An informal argument is the following: In large samples $N^{-1}(\hat{T}_{y\pi}, \hat{\mathbf{T}}_{x\pi})$ will be approximately multivariate normal under appropriate conditions (see, e.g., Krewski and Rao, 1981). The conditional distribution of $N^{-1}\hat{T}_{y\pi}$ given $N^{-1}\hat{\mathbf{T}}_{x\pi}$ can be approximated by a normal distribution with mean $N^{-1}[T_y + \Sigma_{y\mathbf{x}}\boldsymbol{\Sigma}^{-1}_{\mathbf{xx}}(\mathbf{T}_x - \hat{\mathbf{T}}_{x\pi})]$, where T_y is the population total for y, $\boldsymbol{\Sigma}_{y\mathbf{x}}$ is the p-vector of covariances of $N^{-1}\hat{T}_{y\pi}$ with $N^{-1}\hat{\mathbf{T}}_{\mathbf{x}\pi}$, and $\boldsymbol{\Sigma}_{\mathbf{xx}}$ is the $p \times p$ covariance matrix of $N^{-1}\hat{\mathbf{T}}_{\mathbf{x}\pi}$. Replacing T_y by its estimator $\hat{T}_{y\pi}$ and replacing $\boldsymbol{\Sigma}_{\mathbf{xx}}$ and $\boldsymbol{\Sigma}_{y\mathbf{x}}$ by design-unbiased estimators, $\hat{\boldsymbol{\Sigma}}_{\mathbf{xx}}$ and $\hat{\boldsymbol{\Sigma}}_{y\mathbf{x}}$, we obtain another estimator of the total:

$$\hat{T}^*_{yr} = \hat{T}_{y\pi} + \hat{\boldsymbol{\Sigma}}_{y\mathbf{x}}\hat{\boldsymbol{\Sigma}}^{-1}_{\mathbf{xx}}(\mathbf{T}_x - \hat{\mathbf{T}}_{x\pi}).$$

This estimator, like the GREG, is approximately design-unbiased and can have a smaller design-variance than the GREG in large samples. $\hat{T}^*_{yr}$ will also be model-unbiased if $E_M\hat{\Sigma}_{y\mathbf{x}} = \boldsymbol{\beta}\hat{\boldsymbol{\Sigma}}_{\mathbf{xx}}$.

Finally, we note that the GREG is a member of a larger class of estimators defined by Deville and Särndal (1992). If we begin with sample weights defined as inverses of selection probabilities, $w_i = \pi_i^{-1}$, a new set of weights, $\{w_i^*\}_{i\in s}$, can be found that is "close" to the initial weights but are "calibrated" to population totals on some set of auxiliary variables. The estimator of the total is then $\hat{T}^* = \Sigma_s w_i^* y_i$. Formally, we solve the following problem:

$$\text{Find } \{w_i^*\}_{i\in s} \text{ that minimize } G(\mathbf{w}, \mathbf{w}^*) \text{ subject to the constraint } \sum_{i\in s} w_i^*\mathbf{x}_i = \mathbf{T}_x, \tag{2.7.7}$$

where $\mathbf{w}$ is the n-vector of w_i's, $\mathbf{w}^*$ is the n-vector of new weights, and G is a differentiable distance function. If $G(\mathbf{w}, \mathbf{w}^*) = \Sigma_s(w_i^* - w_i)^2/(w_i v_i)$, then the solution vector $\mathbf{w}^*$ leads to the GREG in (2.7.1). Deville and Särndal (1992) list several other distance functions that might be used, including

$$G(\mathbf{w}, \mathbf{w}^*) = \sum_s \left[w_i^* \log\left(\frac{w_i^*}{w_i}\right) - w_i^* + w_i \right] \Big/ v_i,$$

$$G(\mathbf{w}, \mathbf{w}^*) = 2\sum_s (\sqrt{w_i^*} - \sqrt{w_i})^2/v_i,$$

$$G(\mathbf{w}, \mathbf{w}^*) = \sum_s \left[w_i \log\left(\frac{w_i^*}{w_i}\right) + w_i^* - w_i \right] \Big/ v_i,$$

and

$$G(\mathbf{w}, \mathbf{w}^*) = \frac{1}{2}\sum_s (w_i^* - w_i)^2/(w_i^* v_i).$$

Depending on the distance function, the minimization problem in (2.7.7) may have to be solved using iterative methods.

Another interesting feature of their formulation of the calibration approach is the ability to restrict the degree to which the new weights differ from the initial weights. This is achieved by using the constraint $L < w_i^*/w_i < U$ for fixed values of L and U. Recall that Theorem 2.5.2 required that none of the weights in the *BLU* predictor be excessively large relative to the others in order to obtain large sample normality. The (L, U)-bound on the weights in a calibration estimator will help achieve a similar goal as long as none of the initial weights are extreme.

The calibration method can produce estimators that are model-unbiased, although the formulation in (2.7.7) is not based on any model. However, the minimization of a distance between the new weights, w_i^*, and the π-weights, π_i^{-1}, does not seem to be a reasonable starting point for model-based analysis since the π-weights alone often produce estimators with extreme model biases. We will address the topic of bias control more generally in Chapters 3 and 4.

The distance function that we minimized earlier in this chapter to derive the *BLU* predictor was the error variance. There is limited literature on alternative distance measures used with linear models to derive model-based finite population estimators. Chambers (1996) and Bardsley and Chambers (1984), for example, discuss the use of ridge regression to estimate totals. Exploring options other than squared error loss functions with linear models remains a possibility for research.

EXERCISES

2.1 A sample of $n = 10$ units is selected from the Hospitals population in Appendix B. Suppose that the 10 units are numbers 3, 17, 79, 88, 111, 123, 202, 227, 351, and 391. The number of discharges Y and the number of beds x for the sample hospitals are

Y:	41	92	297	377	95	231	601	1063	1645	1894
x:	15	25	80	96	111	125	242	275	551	937.

(a) Plot Y vs. x using the sample data.

(b) For the working model $Y_i = \beta x_i + \varepsilon_i, \varepsilon_i \sim (0, \sigma^2 x_i)$, compute the *BLU* estimate of β and plot the estimated regression on your scatterplot from part (a).

(c) For the working model $Y_i = \alpha + \beta x_i + \varepsilon_i, \varepsilon_i \sim (0, \sigma^2)$, compute the *BLU* estimates of α and β and plot the estimated regression on your scatterplot from part (a).

2.2 Compute the expansion estimator $\hat{T}_0$, ratio estimate $\hat{T}_R$, and the regression estimate $\hat{T}_{LR}$ for the sample in Exercise 2.1. The population total of Y is $T = 320{,}159$. What is the estimation error of each estimate? What is the effect of using the auxiliary x in the ratio and regression estimates compared to ignoring the auxiliary in the expansion estimator?

2.3 Another sample of $n = 10$ units is selected from the Hospitals population in Appendix B. Suppose that this time the 10 units are numbers 34, 73, 112, 151, 191, 230, 269, 309, 348, 387. The number of discharges Y and the number of beds x for the sample hospitals are

Y:	78	315	594	778	410	754	1166	1632	1547	2818
x:	38	70	113	156	227	279	347	437	549	860

(a) Plot Y vs. x using the sample data.

(b) For the working model $Y_i = \beta x_i + \varepsilon_i, \varepsilon_i \sim (0, \sigma^2 x_i)$ compute the *BLU* estimate of β and plot the estimated regression on your scatterplot from part (a).

(c) For the working model $Y_i = \alpha + \beta x_i + \varepsilon_i, \varepsilon_i \sim (0, \sigma^2)$, compute the *BLU* estimates of α and β and plot the estimated regression on your scatterplot from part (a).

2.4 For the sample in Exercise 2.3, compute the expansion estimator $\hat{T}_0$, the ratio estimator $\hat{T}_R$, and the regression estimator $\hat{T}_{LR}$. What is the estimation error of each estimate? What is the effect in this sample of using the auxiliary x in the ratio and regression estimates compared to ignoring the auxiliary in the expansion estimate?

2.5 Suppose that the model $Y_i = \mu + \varepsilon_i, \varepsilon \sim (0, \sigma^2)$ holds in a population of size $N = 1588$. The population *relvariance*, defined as σ^2/μ^2, is equal to 2. If the expansion estimator is used, what sample size is required for $\hat{T}_0$ to have a *coefficient of variation* (*cv*) of 0.10? The *cv* is defined as $\sqrt{\text{var}_M(\hat{T} - T_0)/E_M(T)^2}$.

2.6 Prove that the general formula for the error variance is as given in Theorem 2.2.1.

2.7 Prove that the formula for the error variance for the expansion estimator is as given in Example 2.3.1.

2.8 Prove that the formula for the ratio estimator and its error variance is as given in Section 2.1.

2.9 Prove that the formula for the linear regression estimator and its error variance is as given in Example 2.3.2.

2.10 Prove that the formula for the stratified expansion estimator and its error variance is as given in Example 2.3.3.

2.11 Suppose that the model is $Y_i = \mu + \varepsilon_i$ $(i = 1, \ldots, N)$ with

$$\mathrm{cov}_M(\varepsilon_i, \varepsilon_j) = \begin{cases} \sigma^2 & i = j \\ \sigma^2\rho & i \neq j \end{cases}$$

(a) Find the *BLU* predictor of the finite population mean $\bar{Y} = T/N$ and its error variance.

(b) Find the *BLU* estimator of the model mean μ and its variance. Hint: see Section 2.2.

(c) Discuss the differences in the answers to (a) and (b).

2.12 Suppose the working model is $E_M(Y) = \beta x$, $\mathrm{var}_M(Y) = \sigma^2 v(x)$ for some function $v(x)$. Assume the Y's are mutually independent. Compute the *BLU* predictor of T and its variance under the model when (a) $v(x) = 1$ and (b) $v(x) = x^2$. Evaluate the two estimators for the samples in Exercises 2.1 and 2.3 above.

2.13 Suppose the working model is $E_M(Y) = \beta_0 + \beta_1 x$, $\mathrm{var}_M(Y) = \sigma^2 v(x)$ for some function $v(x)$. Compute the *BLU* predictor of T and its variance under the model when (a) $v(x) = x$ and (b) $v(x) = x^2$. (b) Evaluate the two estimators for the samples in Exercises 2.1 and 2.3 above.

2.14 Compute the prediction bias, defined as $E_M(\hat{T}_0 - T)$, of the expansion estimator under the model $Y_i = \alpha + \beta x_i + \varepsilon_i$ where the errors are uncorrelated and $\varepsilon_i \sim (0, \sigma^2)$. Can you suggest any methods for eliminating the bias?

2.15 Show that (a) the optimal predictor of T under the model $E_M(\mathbf{Y}) = \mathbf{X\beta}$, $\mathrm{var}_M(\mathbf{Y}) = \sigma^2\mathbf{I}$ is also unbiased if $\mathrm{var}_M(\mathbf{Y}) = \mathbf{V}$, and (b) the optimal weights under the model with $\mathrm{var}_M(\mathbf{Y}) = \sigma^2\mathbf{I}$ depend only on the auxiliary x variables.

2.16 Write the following simulation program: (a) Generate a population of size $N = 393$ that follows the model $Y_i = x_i + \varepsilon_i x_i^{1/2}$, where the x's are taken from the Hospitals population in the appendix. Assume that the errors ε_i are independent $N(0, 1)$. (b) Select 100 (or more) *srswor*'s of size $n = 30$ and compute $\hat{T}_R$, the optimal estimator of T, for each sample. (c)

Compute the estimator that would be optimal under the model $Y_i = x_i + \varepsilon_i$. (d) Compare the empirical biases and mean square errors of the estimators of the total. (e) How much efficiency is lost by using the incorrect variance specification?

2.17 Consider these hypothetical survey situations:

(a) A hospital sample is selected by stratifying the population of hospitals in a large state by type of service provided. A stratified simple random sample is selected. Estimates are to be made of the average length of stay per patient classified by type of illness or treatment for a particular calendar quarter. The estimates are to apply to all hospitals in the state, but hospitals in a particular large city will not be surveyed, even if randomly selected, because they are notoriously poor cooperators.

(b) Suppose that in (a) attempts are made to survey any selected hospitals in the large city but 50% of them refuse to participate.

(c) Suppose that in (a) all selected hospitals agree to cooperate in the survey, but the ones in the large city will provide data only during the third week of each month.

Discuss whether the selection mechanisms are ignorable or not. Are there different sets of circumstances where selection can be either ignorable or nonignorable? What data would you need to investigate whether the different sets of circumstances actually hold?

2.18 Prove that the general regression estimator, $\hat{T}_{yr} = \hat{T}_{y\pi} + \hat{\mathbf{B}}'(\mathbf{T}_x - \hat{\mathbf{T}}_{x\pi})$, defined in Section 2.7, is unbiased under the model $E_M(Y_i) = \mathbf{x}_i'\boldsymbol{\beta}$, $\text{var}_M(Y_i) = v_i$.

2.19 Using definitions of terms from Section 2.7, show that the general regression estimator, $\hat{T}_{yr} = \hat{T}_{y\pi} + \hat{\mathbf{B}}'(\mathbf{T}_x - \hat{\mathbf{T}}_{x\pi})$, can also be written in the following forms:

(a) $\hat{T}_{yr} = \Sigma_{i=1}^N \hat{y}_i + \Sigma_s e_i/\pi_i$ with $e_i = y_i - \hat{y}_i$ and $\hat{y}_i = \mathbf{x}_i'\hat{\mathbf{B}}$,

(b) $\hat{T}_{yr} = \Sigma_s g_i y_i/\pi_i$ with $g_i = 1 + (\mathbf{T}_x - \hat{\mathbf{T}}_{x\pi})'\mathbf{A}_{\pi s}^{-1}\mathbf{x}_i/v_i$ and $\mathbf{A}_{\pi s} = \mathbf{X}_s'\mathbf{V}_{ss}^{-1}\boldsymbol{\Pi}_s^{-1}\mathbf{X}_s$, and

(c) $\hat{T}_{yr} = \Sigma_{i=1}^N y_i^0 + \Sigma_s g_i E_i/\pi_i$ with $E_i = y_i - y_i^0$ and $y_i^0 = \mathbf{x}_i'\mathbf{B}$.

2.20 Show that $\mathbf{g}_{\pi s}'\boldsymbol{\Pi}_s^{-1}\mathbf{X}_s = \mathbf{g}_s'\mathbf{X}_s = \mathbf{T}_x$ where $\mathbf{g}_{\pi s}' = (\mathbf{T}_x - \hat{\mathbf{T}}_{x\pi})'\mathbf{A}_{\pi s}^{-1}\mathbf{X}_s'\mathbf{V}_{ss}^{-1} + \mathbf{1}_s'$ is the vector of weights for the general regression estimator and $\mathbf{g}_s = \mathbf{V}_{ss}^{-1}\mathbf{X}_s\mathbf{A}_s^{-1}\mathbf{X}_r'\mathbf{1}_r + \mathbf{1}_s$ is the vector of weights for the *BLU* predictor in the case with $\mathbf{V}_{sr} = \mathbf{0}$. What does $\mathbf{g}_s'\mathbf{X}_s$ equal when the more general set of weights

$$\mathbf{g}_s = \mathbf{V}_{ss}^{-1}[\mathbf{V}_{sr} + \mathbf{X}_s\mathbf{A}_s^{-1}(\mathbf{X}_r' - \mathbf{X}_s'\mathbf{V}_{ss}^{-1}\mathbf{V}_{sr})]\boldsymbol{\gamma}_r + \boldsymbol{\gamma}_s$$

is used?

CHAPTER 3

Bias-Robustness

In Chapters 1 and 2, we have been studying the question of prediction-based estimation from surveys, with little emphasis on an issue generally regarded as fundamental to survey sampling, namely, "how was the sample selected?". With some caveats, noted in Section 2.7, we have proceeded on the idea that if the regression model is correct, then the procedure by which the sample was selected is largely irrelevant. In particular, we have thrown overboard the usual insistence that the sample be selected according to one or another probability sampling plans.

But what if the regression model is *incorrect*? If the model is wrong, it turns out, not surprisingly, that sound inference and efficiency can be undermined.

This chapter and the next consider the question of how to protect inference and maintain efficiency, in light of the all too reasonable supposition that the model one favors is wrong. The major message of the present chapter is that by proper selection of sample and estimator, estimation becomes *robust*. Robust here means "*bias*-robust-against-model-failure."

There are two reasons for placing the main emphasis on the *bias* of the estimator: (1) unchecked, the order of the bias can be larger than that of the root of the variance, and (2) bias can undermine confidence intervals. We generally have no way of accurately estimating the mean square error of a biased estimator, even when we can estimate the variance well.

To see how bias can damage the properties of confidence intervals, recall that the pivot for the usual normal approximation interval is

$$\frac{\hat{T} - T}{\sqrt{\operatorname{var}_M(\hat{T} - T)}} = \frac{\hat{T} - T - E_M(\hat{T} - T)}{\sqrt{\operatorname{var}_M(\hat{T} - T)}} + b,$$

where $b = E_M(\hat{T} - T)/[\operatorname{var}_M(\hat{T} - T)]^{1/2}$ is the standardized bias. The first term on the right-hand side above may be approximately normal in large samples. However, if the bias does not approach 0 faster than the standard error, the confidence intervals based on the pivot will be centered in the wrong place and will not cover the population total at the nominal rate, for example, 95 percent. In practice, a consistent sample estimate of $\operatorname{var}_M(\hat{T} - T)$ is substituted in the denominator of the pivot (see Chapter 5); naturally, this would not rectify the poor behavior of the confidence interval.

As in previous chapters, the focus will be on estimating the total T. Results extend without problem to estimation of the mean. Implications for other population unknowns, like the distribution function and quantiles, will be discussed in Chapter 11.

The outline of Chapter 3 is as follows: Section 3.1 discusses in a general way the ideas of robustness and of representative samples. An example illustrates the "bias" that can arise when the model is incorrect and suggests that a sample that is *representative* of the population, that is, is in some sense a small scale version of the population, can protect against model failure. This leads to the idea of *balanced* samples, discussed in Section 3.2. An empirical study verifies that, from a practical standpoint, balanced samples are an important step toward unbiasedness.

Section 3.3 generalizes the idea of representativeness. This leads to the important idea of *weighted balance* and a corresponding class of estimators that will be bias-robust in weighted balanced samples. Section 3.4 discusses the practical matter of achieving balanced samples, both simple and weighted, relying a good deal on the accumulation of knowledge that has grown through the years in the design-based literature. Section 3.5 reports on results of an empirical study of these estimators. The results, compared to that in section 3.2, suggest that weighted balance may have an advantage over simple balance from the point of view of efficiency. This idea will be more fully investigated in Chapter 4.

3.1 DESIGN AND BIAS

In this section we discuss in general terms two apparently disconnected ideas:

(i) bias-robust estimation;
(ii) representative samples.

The main suggestion is that we achieve (i) through (ii).

To get clarity on (i), it will be necessary to consider *more than one model* for the same data. The reader may be accustomed to the idea, familiar from regression textbooks, of *choosing between* competing models for the same data, where the question is which model best fits the data (see, e.g., Belsley, Kuh, and Welsch, 1980). However, the question regarding *robustness* is what happens if we use one model for inference, when, possibly, another model would have done better. This requires us to entertain two (or more) models *simultaneously* for the same data: a "working model" on which we base the estimator, and a (typically more complicated) "true model" which might fit the data better. Why not just switch to the "true" model if we think it is better? The principal answer is that as a rule we do *not* know it is better, only different, and that, whatever model we rely on, there is always some other model which might have been preferable.

Thus, suppose we had chosen a particular model for a variable of interest Y, and based on that model used the estimator $\hat{T}$ of the target T. What would be the consequence if the model was false? A given model for Y is false if a large collection of data on Y would be seen to fit an *alternative model* better. For example, having selected a through-the-origin model, relating Y to x, a simple linear model with nonzero intercept might fit large samples better. To illustrate points, we return to a somewhat artificial example.

Example 3.1.1. The Expansion Estimator. Suppose that the estimator is the expansion estimator $\hat{T}_0 = N\bar{Y}_s$, which can be justified on the basis of the simple homogeneous model M:

$$Y_i = \mu + \varepsilon_i, \tag{3.1.1}$$

with independent homoscedastic errors ε_i. Suppose, as in Section 1.2, that an auxiliary variable x is available and that after sampling, the result is as in Figure 3.1(a). The sample points are indicated by circles, and the population of x values are marked by an X along the horizontal axis. Casual observation suggests that there is a monotonic, possibly even linear, relation between Y and x. Furthermore, it can be seen that in this sample the x's tend to be bunched toward the lower end of their range. This suggests that the estimate arising from $\hat{T}_0$ will be too low. In some sense, the expansion estimator is, in this sample, "biased low."

To characterize this bias more formally, we might begin by supposing that the actual model M^* is a simple through the origin model of Y on x:

$$Y_i = \beta x_i + \varepsilon_i \tag{3.1.2}$$

This will surely capture the relation of Y to x better than (3.1.1).

Taking the expectation with respect to M^* of the error $\hat{T}_0 - T$, we find (Exercise 3.1) that

$$E_{M^*}(\hat{T}_0 - T) = N\beta(\bar{x}_s - \bar{x}), \tag{3.1.3}$$

where $\bar{x}_s$ and $\bar{x}$ are the sample and population means of the x's, respectively. When the x's are bunched low, it will follow that $\bar{x}_s < \bar{x}$, and (3.1.3) will be negative. If we identify this expression as *the bias of* $\hat{T}_0$ (under the "true model" M^*), then we have a quantitative expression that captures our intuition that the estimate based on the expansion estimator is too low in this sample.

Suppose the sample is "representative," as in Figure 3.1(b). Here, intuitively, what we mean by representative is that each unit in the sample is close to the same number of population units, judging by the x values. In *this* sample, we would be surprised if the expansion estimator did too badly. Since nearby x's

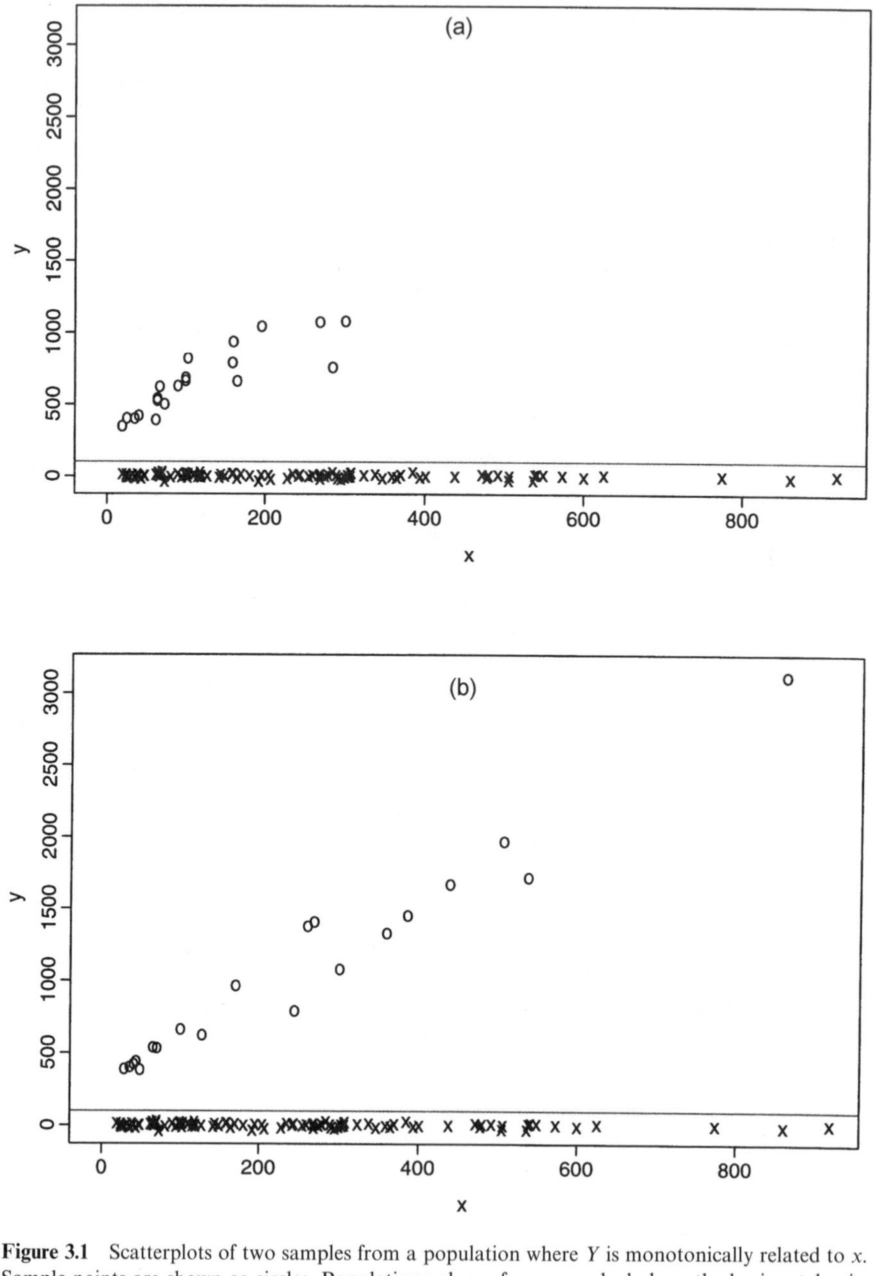

Figure 3.1 Scatterplots of two samples from a population where Y is monotonically related to x. Sample points are shown as circles. Population values of x are marked along the horizontal axis.

appear to imply nearly equal values of y, the y's in the sample should on the whole be close to those not in the sample, and multiplying each of the sample y's by N/n, the ratio of the number of population to sample points, and summing, makes sense.

Again the better model M^* clarifies what "representative sample" might mean operationally: if the sample x values are appropriately spread through the population of x values, then we would expect that $\bar{x}_s \approx \bar{x}$, and according to (3.1.3), the bias of the expansion estimator is near 0 under M^*. Such a sample is "balanced" in the sense that it has the same first moment of the x's as does the population.

This example suggests that, in the estimation of totals, we can protect against misspecification of the model by ensuring that the sample is "representative" of the population. Consideration of models *beyond* the one used to construct the estimator helps us characterize the idea of a biased estimator. In the simple example above, a key aspect of representativeness, leading to unbiasedness, is the equality of the first moments of sample and population x's. In the next section, we shall allow for greater complexity of the models, and refine our informal notion of "representative sample" into the notion of *balanced sample.*

3.2. POLYNOMIAL FRAMEWORK AND BALANCED SAMPLES

We adopt a (usually simple) model on which to base the estimator of the total T. At the same time, we suspect that other, more complicated models would fit the model better, perhaps considerably better, if we had all the data in the population to scrutinize.

Consider the general polynomial model M^*

$$Y_i = \sum_{j=0}^{J} \delta_j \beta_j x_i^j + \varepsilon_i v_i^{1/2}, \qquad (3.2.1)$$

where the errors are $\varepsilon_i \sim (0, \sigma^2)$ and uncorrelated, $\{\beta_j\}_{j=0}^{J}$ are a set of unknown parameters, and $\{\delta_j\}_{j=0}^{J}$ are 0–1 variables indicating whether the jth power term is in the model or not. We shall refer to model (3.2.1) as a Jth order polynomial model.

This model is appropriate to consider as representing the *true* underlying model when there is a single auxiliary variable x related to the variable of interest. It is well known that more general continuous functions of a single variable can be accurately approximated by polynomials. At the same time, we might be reluctant to use a very-high-order polynomial model as a basis for estimating the total T, since there is the danger of overfitting the data in the *sample*. At the extreme, if the order of the polynomial equaled the number of sample points, we would be left with an estimator claiming *zero* variance; this would be justified only in the unrealistic case that all non-sample points fit exactly on the fitted polynomial. In practice, we usually restrict ourselves to estimators based on low-order polynomials.

Thus it is critical to emphasize the distinction between the (typically unknown) *correct* underlying model, say M^*, and the *working* model M, on which we explicitly base our estimator of total. A fundamental task is to achieve a methodology that makes an estimator $\hat{T}$ derived under model M also unbiased under M^*.

Because we will be examining a variety of models and estimators in this chapter, it is convenient to have a notation that clearly links an estimator to a model. As a notational convention, let $M(\delta_0, \delta_1, \ldots, \delta_J{:}v)$ denote model (3.2.1) and $\hat{T}(\delta_0, \delta_1, \ldots, \delta_J{:}v)$ denote the *BLU* predictor under that model, following the convention in Royall and Herson (1973a). For example, $M(0, 1{:}x)$ refers to the model $Y_i = \beta_1 x_i + \varepsilon_i x_i^{1/2}$ and $\hat{T}(0, 1{:}x)$ to the ratio estimator found in Section 2.1. The notation $M(1, 0, 1{:}1)$ denotes the model $Y_i = \beta_0 + \beta_2 x_i^2 + \varepsilon_i$ and $\hat{T}(1, 0, 1{:}1)$ the corresponding *BLU* predictor.

We are now in a position to elaborate the fundamental idea of *balanced samples*, proceeding by example.

3.2.1. Expansion Estimator and Balanced Samples

We return to the expansion estimator $\hat{T}_0 = \hat{T}(1{:}1)$, the *BLU* predictor under the working model $M(1{:}1)$. If the general polynomial model M^* of (3.2.1) holds, then the bias of $\hat{T}_0$ is

$$E_M[\hat{T}(1{:}1) - T] = N \sum_{j=1}^{J} \delta_j \beta_j [\bar{x}_s^{(j)} - \bar{x}^{(j)}], \tag{3.2.2}$$

where $\bar{x}_s^{(j)} = \Sigma_s x_i^j / n$ and $\bar{x}^{(j)} = \Sigma_{i=1}^N x_i^j / N$ (see Exercise 3.3). For example, if the correct model is $M(1, 1{:}1)$, then the bias of $\hat{T}(1{:}1)$ is

$$E_M[\hat{T}(1{:}1) - T] = N\beta_1(\bar{x}_s - \bar{x}). \tag{3.2.3}$$

Let $s(J)$ denote any sample satisfying the condition

$$\bar{x}_s^{(j)} = \bar{x}^{(j)} \tag{3.2.4}$$

for $j = 1, \ldots, J$. Such a sample will be called a (*simple*) *balanced sample* (*of order* J). It is clear that, for any such sample, the bias of the expansion estimator, given by (3.2.2), equals *zero*: the expansion estimator is unbiased in balanced samples, despite the mismodeling. Thus, if we use the expansion estimator, balanced samples protect against bias.

3.2.2.* Order of the Bias of the Expansion Estimator

The expansion estimator has error variance $N^2(1 - f)\sigma^2/n$. The order of magnitude of the root of variance in terms of population and sample size is therefore $N/n^{1/2}$. For simplicity suppose the correct model is $M(1, 1{:}1)$ so that the bias is given by (3.2.3). Then, the order of the bias lies somewhere in the *range* of values $[0, N]$, depending on the order of magnitude of the difference $(\bar{x}_s - \bar{x})$, which, under one of the most common sampling schemes—simple random sampling without replacement—is $O(n^{-1/2})$ (see Theorem A.9 in the Appendix). Thus, under *srs* the bias can be of the same order of magnitude as the square root of the variance, and has the potential to be even larger in other types of samples.

This bias is eliminated in any sample that is "balanced" in the sense that $\bar{x}_s = \bar{x}$. This suggests that if we were obliged to use the estimator $\hat{T}(1{:}1)$, and had the auxiliary data x available, we should restrict our samples to those satisfying $\bar{x}_s = \bar{x}$. Selecting our sample by simple random sampling *aims at* this condition, since $E_\pi(\bar{x}_s) = \bar{x}$, but the particular sample we select might or might not meet this condition. This suggests that not all samples selected by simple random sampling are "equally good."

3.2.3. Ratio Estimator and Balanced Samples

We noted in Section 2.1 that the ratio estimator $\hat{T}_R = \hat{T}(0, 1{:}x) = N\bar{Y}_s\bar{x}/x_s$ is the *BLU* predictor under the model $M(0, 1{:}x)$. As shown in Exercise 2.8 in Chapter 2, the error variance of $\hat{T}(0, 1{:}x)$ under $M(0, 1{:}x)$ is

$$\mathrm{var}_M[\hat{T}(0, 1{:}x) - T] = N^2(1 - f)(\bar{x}_r\bar{x}/\bar{x}_s)\sigma^2/n. \tag{3.2.5}$$

This variance is minimized by making $\bar{x}_s$ as large as possible, which in turn makes $\bar{x}_r$ as small as possible. The optimum sample under $M(0, 1{:}x)$ is thus the one that contains the n units in the population with the largest values of the auxiliary x. Populations where this type of purposive selection of the largest units were optimal are described in Karmel and Jain (1987) and Kirkendall (1993). However, selecting this sample may be quite a risky procedure if the working model is wrong.

Suppose that the correct model is $M(1, 1{:}x)$, that is, the model contains an intercept. The bias in that case is

$$E_M[\hat{T}(0, 1{:}x) - T] = N\beta_0\frac{\bar{x} - \bar{x}_s}{\bar{x}_s}.$$

When β_0 is positive, the sample with the n largest x's would incur the maximal negative bias in this case. We note that the same condition that we met in the previous example, namely, $\bar{x}_s = \bar{x}$, will render the ratio estimator unbiased if in fact the correct model is $M(1, 1{:}x)$.

More generally, we can examine the bias of the ratio estimator under the polynomial model (3.2.1). A simple computation (see Exercise 3.4) shows that

$$E_M[\hat{\beta}(0, 1:x) - T] = N\bar{x} \sum_{j=0}^{J} \delta_j \beta_j \left[\frac{\bar{x}_s^{(j)}}{\bar{x}_s} - \frac{\bar{x}^{(j)}}{\bar{x}} \right]. \tag{3.2.6}$$

Note that there is no contribution from the term $j = 1$, since the model underlying the ratio estimator contains β_1. Furthermore, for any sample $s(J)$, that is, any sample with simple balance (3.2.4), the ratio estimator is unbiased under (3.2.1). In other words, the same condition—simple balance—which made the expansion estimator unbiased also makes the ratio estimator unbiased under the general polynomial model (3.2.1).

3.2.4. Bias-Robust Strategies

By a sample survey *strategy* we mean the combined selection of an estimator and a sample of a particular sort. A *bias-robust strategy* (with respect to some broad class of models) is a strategy such that the estimator is unbiased for all members of the class of models when the sample is of the particular sort. Thus the expansion estimator and $s(J)$, and also the ratio estimator and $s(J)$, are each a robust strategy with respect to (3.2.1).

Notice that for any balanced sample, the ratio estimator, $\hat{T}(0, 1:x)$, simplifies to the expansion estimator, $\hat{T}(1:1) = \hat{T}_0$ of our first example. This reduction of the *BLU* to the expansion estimator is an example of a more general phenomenon, as the following theorem indicates.

Theorem 3.2.1 (Royall and Herson, 1973a). If $s = s(J)$, then

$$\hat{T}(\delta_0, \delta_1, \ldots, \delta_J : x^j) = \hat{T}(1:1)$$

as long as $\delta_j = 1$.

In other words, the optimal estimator of the total reduces to the expansion estimator in a balanced sample as long as the correct model has the variance of Y proportional to x^j and the model includes the jth degree term $\beta_j x^j$. (Note that the ratio estimator meets this condition.) The reader is invited to prove Theorem 3.2.1 directly in Exercise 3.6. The significance of this theorem is the immediate corollary:

Corollary 3.2.1. The estimator $\hat{T}(\delta_0, \delta_1, \ldots, \delta_J : x^j)$ with $\delta_j = 1$ combined with a sample $s = s(J)$ is a bias-robust strategy with respect to the polynomial models (3.2.1).

Example 3.2.1. Linear Regression Estimator. The linear regression estimator is given by

$$\hat{T}_{LR} = \hat{T}(1, 1{:}1) = N[\bar{Y}_s + b_1(\bar{x} - x_s)],$$

where $b_1 = \Sigma_s(Y_i - \bar{Y}_s)(x_i - \bar{x}_s)/\Sigma_s(x_i - \bar{x}_s)^2$.

This estimator is *BLU* under $M(1, 1{:}1)$ with error variance

$$N^2(1-f)\sigma^2[1 + (\bar{x}_s - \bar{x})^2/\{(1-f)c_s\}]/n,$$

where $c_s = \Sigma_s(x_i - \bar{x}_s)^2/n$. It is readily seen that this estimator fits the conditions of Theorem 3.2.1. In fact, if $\bar{x}_s = \bar{x}$, then $\hat{T}_{LR}$ reduces to the expansion estimator, and so is unbiased under (3.2.1) if the sample is $s(J)$. Note that its variance reduces to that of the expansion estimator also, if the sample is $s(J)$.

The question naturally arises: if the ratio estimator (or the regression estimator, or, more generally, the estimator $\hat{T}(\delta_0, \delta_1, \ldots, \delta_J{:}x^j)$ with $\delta_j = 1$) reduces to the expansion estimator for a balanced sample $s(J)$, does it matter which estimator $\hat{T}_0 = \hat{T}(1{:}1)$ or $\hat{T}_R = \hat{T}(0, 1{:}x)$—or underlying model $M(1{:}1)$ or $M(0, 1{:}x)$—we use, given that we intend the sample to be balanced? The answer is "yes," for two reasons.

1. In practice, it may be difficult to achieve balance *exactly*, and the different estimators may be sensitive to departures from balance to a different degree. The numerical study of Section 3.2.5 illustrates this point dramatically. Almost always the ratio estimator is preferable to the expansion estimator under near-balance.
2. Variance estimators will not be the same, since the residuals under the different models are different, even at balance; see Section 5.5.1.

3.2.5. Simulation Study to Illustrate Conditional Biases and Mean Squared Errors

The Hospitals population in Appendix C can be used to illustrate the ideas from the preceding sections. Figure 3.2 is a plot of the number of discharges (Y) versus the number of beds (x) in each of the $N = 393$ hospitals in the population. A visual inspection shows that the number of discharges generally increases as the number of beds does, as does the variance of Y. Four least squares lines are plotted in the figure corresponding to models fitted using all hospitals in the population. The solid line is the fitted version of $Y_i = \beta_1 x_i + \varepsilon_i x_i^{1/2}$, the model that leads to the ratio estimator. The dotted line corresponds to $Y_i = \beta_0 + \beta_1 x_i + \varepsilon_i x_i^{1/2}$. The curved, short-dash line is fitted for the model $Y_i = \beta_1 x_i + \beta_2 x_i^2 + \varepsilon_i x_i^{1/2}$, while the long-dash line is for

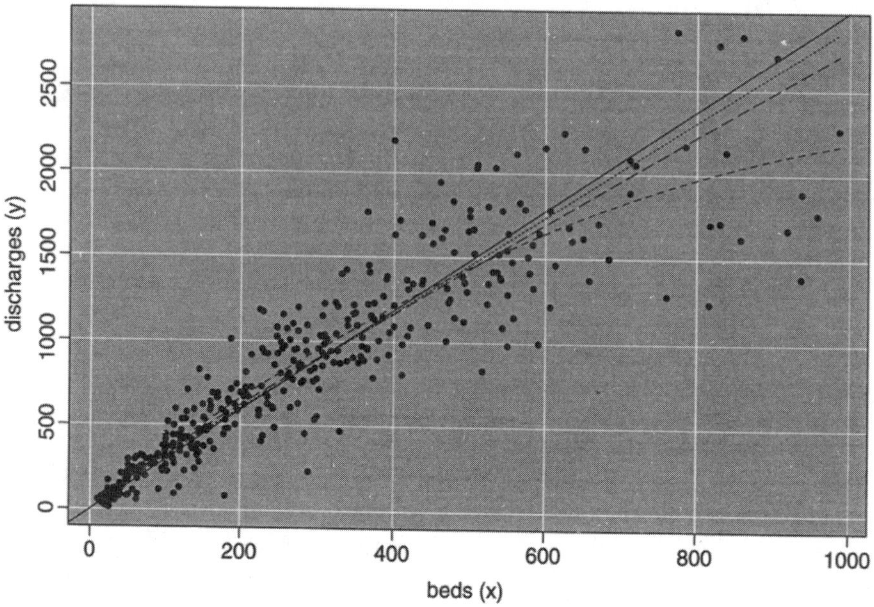

Figure 3.2 Scatterplot of Hospitals population with four fitted lines. Solid line is straight line through origin; dotted line is straight line with intercept; short-dash line is quadratic through the origin; long dash line is linear in $x^{1/2}$ and x with no intercept. Each model assumes variance of y is proportional to x.

$Y_i = \beta_{1/2}x_i^{1/2} + \beta_1 x_i + \varepsilon_i x_i^{1/2}$. The values of the parameter estimates from each model are:

Model	$\hat{\beta}_0$	$\hat{\beta}_{1/2}$	$\hat{\beta}_1$	$\hat{\beta}_2$
$Y_i = \hat{\beta}_1 x_i + \varepsilon_i x_i^{1/2}$	—	—	2.97	—
$Y_i = \beta_0 + \beta_1 x_i + \varepsilon_i x_i^{1/2}$	33.60	—	2.84	—
$Y_i = \beta_1 x_i + \beta_2 x_i^2 + \varepsilon_i x_i^{1/2}$	—	—	3.58	−0.0014
$Y_i = \beta_{1/2}x_i^{1/2} + \beta_1 x_i + \varepsilon_i x_i^{1/2}$	—	9.32	2.45	—

A model with no intercept is logical in this population because zero beds should imply zero inpatients and, hence, zero discharges. The quadratic model leads to considerably smaller predicted values for hospitals with about 600 or more beds than do the other models. The model $Y_i = \beta_{1/2}x_i^{1/2} + \beta_1 x_i + \varepsilon_i x_i^{1/2}$ may seem somewhat unusual but plays an important role in the context of weighted balance, which we shall consider below. A serious job of model fitting would naturally entail more than just estimating parameters, but for the illustration here, we need go no farther. Note also that in a real survey we will rarely have the luxury of fitting models to an entire population.

Using the Hospitals population, we conducted a simple simulation study similar to those reported in Royall and Cumberland (1981a, 1981b). A set of 2000 samples of size $n = 25$ was selected using simple random sampling without replacement. In each sample, the expansion, ratio, and linear regression estimators were calculated, along with the sample mean $\bar{x}_s$ of the auxiliary. The samples were then sorted from low to high on the basis of $\bar{x}_s$ and divided into 10 groups of 200 samples each. This method of sorting and grouping permits conditional, empirical analyses to be done. Within each group the empirical bias $\Sigma(\hat{T} - T)/S$ was calculated, where the sum is over the samples in the group, S is the number of samples in the sum, and $\hat{T}$ is one of the three estimators of the population total. The empirical square root of the mean square error (*rmse*), $\sqrt{\Sigma(\hat{T} - T)^2/S}$, was also computed for each group.

The group biases and root mean square errors are plotted in Figure 3.3. The points made theoretically are clearly illustrated here. Each estimator is unbiased in samples where $\bar{x}_s$ is near the population mean $\bar{x} = 274.7$. The expansion estimator in panel (a) has a positively sloping bias curve, similar to expectations if the model were $M(1, 1{:}v)$. In contrast, the ratio estimator in panel (b) has a downward sloping bias curve as expected under a model with a positive intercept. The linear regression estimator in panel (b) is the nearest to unbiased throughout the range of $\bar{x}_s$. Obviously, the bias becomes a substantial part of the *rmse* of the expansion and ratio estimators as $\bar{x}_s$ moves away from the balance point $\bar{x}_s = \bar{x}$. In the group of samples with $\bar{x}_s$ closest to $\bar{x}$, the *rmse*'s (in thousands of discharges) of the expansion, ratio, and linear regression estimates are very similar—17.41, 16.84, and 16.95, respectively.

Some further observations are in order. Regarding the ratio estimator, note the substantial loss that would be incurred if we selected the $n = 25$ units with the largest x's. The empirical *rmse* (in thousands) in the group of samples with the largest $\bar{x}_s$'s is 22.34, which is about 33% larger than the *rmse* of 16.84 for the group of samples nearest the balance point.

The *rmse* of the expansion estimator ranges from 17.41 thousand to over 75 thousand, depending on the value of $\bar{x}_s$. These extremes are considerably worse than those of either the ratio or regression estimator. Although $\hat{T}_0$ is design-unbiased when averaged over all simple random samples, Figure 3.3(a) shows that this is caused by negative bias when $\bar{x}_s < \bar{x}$ canceling with positive biases when $\bar{x}_s > \bar{x}$. Thus the expansion estimator cannot be recommended for general use.

3.2.6. Balance and Multiple *Y* Variables

Balanced sampling can be useful when several target Y variables are collected in a survey and the variables, though possibly dependent on the same auxiliaries, do not necessarily follow the same form of model. To illustrate, suppose we have a population where three targets are observed—Y_1, Y_2, and Y_3. The three Y variables follow the models $M(0, 1{:}x)$, $M(1, 1{:}x)$, and $M(1{:}1)$. In a population of businesses, the variables might be amount of sales, amount

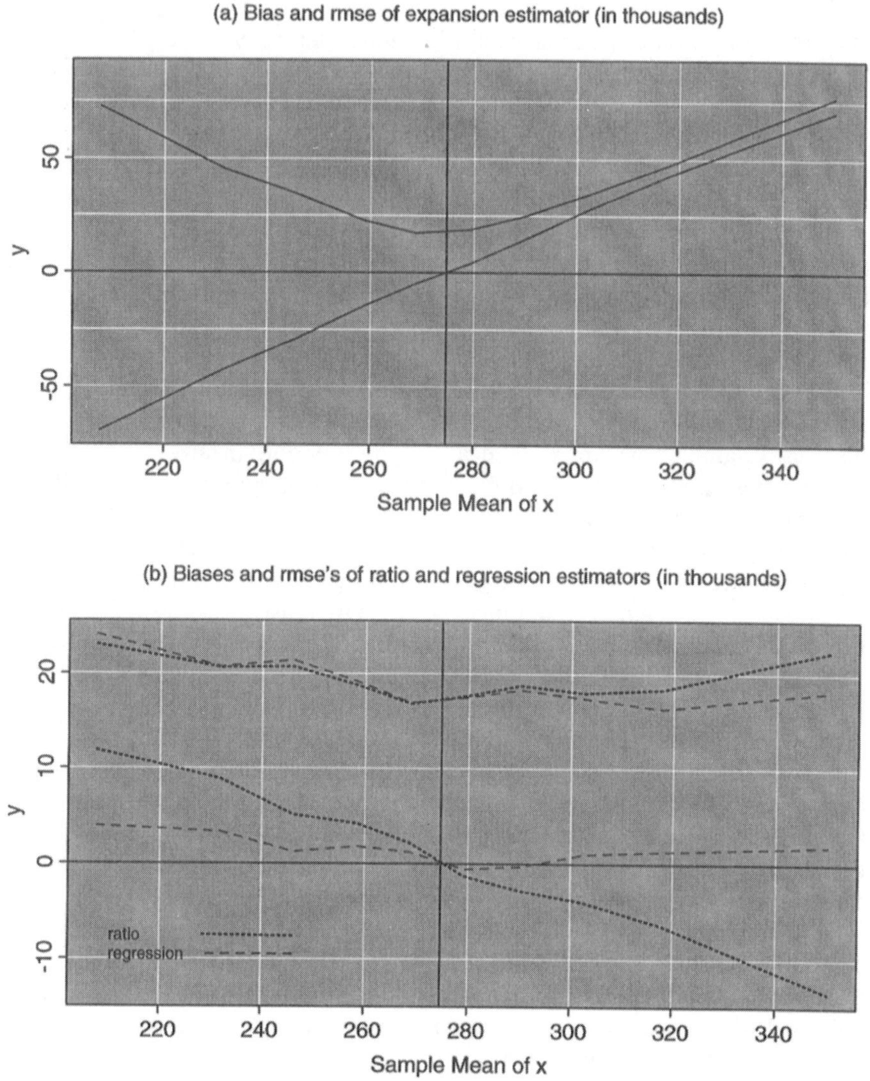

Figure 3.3 Simulation results for the expansion, ratio, and regression estimators; 2000 simple random samples, $n = 25$. Upper curves in each panel are *rmse*'s; lower curves are biases; vertical reference line drawn at population mean of x.

of capital expenditures, and net income. The auxiliary x might be the number of employees in each establishment at a certain date.

One strategy for estimating the totals would be to use the *BLU* predictor for each of the variables which would be the ratio estimator, a weighted linear regression estimator, and the expansion estimator. Using three different forms of estimator seems cumbersome and would be even more so if there were additional y's. We could select the estimator corresponding to the most inclusive of the models, in the present case, $M(1, 1:x)$, but as a general rule, this might lead to gratuitous complexity. If we select a balanced sample $s(1)$, then each of the *BLU* predictors reduces in principle to the expansion estimator, as noted in Section 3.2.5, and is also equivalent to the ratio estimator, which is less sensitive to deviation from balance. If we balance on higher moments of x, we achieve protection against bias under higher-order polynomial models besides gaining the simplicity of being able to use the ratio estimator for all of the y's.

3.3. WEIGHTED BALANCE

The notion of a simple balanced sample, introduced in Section 3.2, appears to be a helpful quantitative embodiment of the vaguer notion of a "representative sample," as discussed in Section 3.1. In such a sample, each sample unit represents (has values of the variable of interest close to) N/n population units (including itself) and so it is perhaps not surprising that the expansion estimator should turn out to be unbiased with respect to the high-order polynomial models (3.2.1) in such samples.

Does the idea of "representative sample" *require* that each sample unit represent the same number of population units? Perhaps not: if the variable of interest Y, related to an auxiliary x known over the population, is known to be more variable over part of the domain of x, then it might be sensible to design the sample to have more units over that portion of the domain, and less where predicting Y values is relatively easy. For example, we frequently find in practice that as x grows, the variance of Y conditional on x also grows (in fact, the model underlying the ratio estimator was of this sort), and in this case it might make sense to sample more heavily where x is large.

If so, then the "number of units a sample unit i represents" would properly be in some inverse relation to the variance v_i of Y_i. A generalization of the idea of simple balance, *weighted balance*, described below, clarifies this idea. For reasons of efficiency of estimation, discussed in Chapter 4, the appropriate function of the variance is the square root, that is, the number of points unit i represents should be inversely proportional to the standard deviation $v_i^{1/2}$.

Continuing to focus on the polynomial model in a single auxiliary x, we define a *weighted balanced sample* (*of order J*) as one which satisfies the following criterion:

$$\frac{\Sigma_s v_i^{-1/2} x_i^j}{n} = \frac{\Sigma_{i=1}^N x_i^j}{\Sigma_{i=1}^N v_i^{1/2}}, \qquad j = 0, 1, \ldots, J. \tag{3.3.1}$$

We also refer to such balance as *root*(v) balance or $v^{1/2}$-balance (of order J). Note that if the v_i are constant, then weighted balance reduces to the condition of unweighted balance in (3.2.4).

Why this is a desirable criterion is not, of course, self evident. One might have expected, analogous to simple balance (3.2.4), the requirement that

$$\frac{\Sigma_s v_i^{-1/2} x_i^j}{\Sigma_s v_i^{-1/2}} = \frac{\Sigma_{i=1}^N x_i^j}{N}, \qquad j = 0, 1, \ldots, J. \tag{3.3.2}$$

Notice, though, that (3.3.1) for $j = 0$ implies

$$\frac{n}{\Sigma_s v_i^{-1/2}} = \frac{\Sigma_{i=1}^N v_i^{1/2}}{N}, \tag{3.3.3}$$

that is, the *harmonic mean* of the standard deviations in the sample equals the *arithmetic mean* of the standard deviations in the population. If one solves (3.3.3) for n, and substitutes the result in (3.3.1), the result is (3.3.2). Thus (3.3.1) implies (3.3.2).

We remark additionally that, in practice, it will be easier to aim for (3.3.1) than (3.3.2), since (3.3.1) requires us to adjust only one expression in the sample values to equal a given fixed expression in the population values.

3.3.1. Elementary Estimators Unbiased Under Weighted Balance

It is elementary to see that, under weighted balance (3.3.1), the standard expansion estimator will not, in general, be unbiased if the polynomial model (3.2.1) holds; nor will be the ratio estimator or the linear regression estimator, unless v_i happens to be a constant; see Exercise 3.8.

The natural analogue to the expansion estimator is the "weighted expansion estimator,"

$$\hat{T}_0(v^{1/2}) = N \sum_s v_i^{-1/2} Y_i \Big/ \sum_s v_i^{-1/2}. \tag{3.3.4}$$

A second estimator, analogous to the Horvitz-Thompson estimator, introduced in Section 1.2, and equivalent to (3.3.4) whenever (3.3.3) holds is

$$\hat{T}_{MR}(v^{1/2}) = \sum_{i=1}^N v_i^{1/2} \sum_s v_i^{-1/2} Y_i / n, \tag{3.3.5}$$

sometimes referred to as the *mean of ratios* estimator.

Example 3.3.1. Variance Proportional to x or x^2. In the case of variance proportional to an auxiliary variable x, the corresponding weighted expansion and mean of ratio estimators are just (3.3.4) and (3.3.5), respectively, with "v" replaced by "x." In the case of variance proportional to x^2, we get $\hat{T}_0(x) = N \Sigma_s x_i^{-1} Y_i / \Sigma_s x_i^{-1}$ and the mean of ratios estimator is $\hat{T}_{MR}(x) =$

$\Sigma_{i=1}^{N} x_i \Sigma_s x_i^{-1} Y_i/n$. It is interesting to note that $\hat{T}_{MR}(x)$ is of the form $\hat{T}_{MR}(x) = \Sigma_{i=1}^{N} x_i \hat{\beta}$, for the working model $M = M(0, 1{:}x^2)$. This is *not*, in general, equal to the *BLU* under this model (Exercise 3.9), but can be expected to be close to it, for small sampling fraction n/N.

3.3.2. *BLU* Estimators and Weighted Balance

The question naturally arises whether there are any working models M and corresponding *BLU* estimators that are unbiased for the polynomial model (3.2.1) for samples meeting the weighted balance condition (3.3.1). To answer this question, we need some results, the proofs of which are deferred to Chapter 4.

Lemma 3.3.1. Suppose that in a pth order polynomial model, the variances for units in the sample are a fixed linear combination of the terms in the model, that is, $v_i = \sum_{j=0}^{p} c_j x_i^j$ for $i \in s$, for some constants c_j, $j = 0, 1, \ldots, p$. Then, the *BLU* estimator of the total is $\hat{T}(\delta_0, \delta_1, \ldots, \delta_p{:}v) = N \Sigma_{j=0}^{p} \hat{\beta}_j \bar{x}^{(j)}$, where the $\hat{\beta}_j$ are *BLU* estimators of the parameters of the model.

That is, when the (sample) variances are a linear combination of the terms in the model, the estimator can be expressed in terms of population means of the powers of x, that is, $\Sigma_s y_i + N_r \Sigma_{j=0}^{p} \hat{\beta}_j \bar{x}_r^{(j)} = N \Sigma_{j=0}^{p} \hat{\beta}_j \bar{x}^{(j)}$, with $N_r = N - n$. Models used in practice frequently have all c_j but one equal to zero. The model $M(0, 1{:}x)$ leading to the ratio estimator, and the model $M(1, 1{:}1)$ leading to the regression estimator are both of this sort. Note that although the lemma requires the condition only for the *sample* variances, this is most readily guaranteed if the condition holds over the population as a whole.

Lemma 3.3.2. Suppose that in a pth order polynomial model, the *standard deviations* are a fixed linear combination of the terms in the model, that is, $v_i^{1/2} = \Sigma_{j=0}^{p} d_j x_i^j$ for $i = 1, 2, \ldots, N$, for some constants d_j, $j = 0, 1, \ldots, p$. Then, an estimator of the total of the form $\hat{T} = N \Sigma_{j=0}^{p} \hat{\beta}_j \bar{x}^{(j)}$, where the $\hat{\beta}_j$ are *BLU* estimators of the parameters of the model, reduces to the mean of ratios estimator (3.3.5) under weighted balance of order p,

$$\frac{\Sigma_s v_i^{-1/2} x_i^j}{n} = \frac{\Sigma_{i=1}^{N} x_i^j}{\Sigma_{i=1}^{N} v_i^{1/2}}, \qquad j = 0, 1, \ldots, p.$$

Consequently, $\hat{T}$ also reduces to the weighted expansion estimator (3.3.4). Here p, the order of the working model, is typically small, and less than J, the order of the true underlying model.

The importance of this lies in the consequence that under the conditions of the lemma the estimator $\hat{T} = N \Sigma_{j=0}^{p} \hat{\beta}_j \bar{x}^{(j)}$ is unbiased under the more general model (3.2.1) under weighted balance (3.3.1). If we combine the two lemmas, we arrive at the following result on *BLU* estimators of total:

Corollary 3.3.1. Suppose that in a pth order polynomial model, we have $v_i = \Sigma_{j=0}^{p} c_j x_i^j$ and $v_i^{1/2} = \Sigma_{j=0}^{p} d_j x_i^j$ for $i = 1, 2, \ldots, N$ for some constants c_j, d_j, $j = 0, 1, \ldots, p$. Then, under weighted balance (3.3.1), the BLU estimator of the total $\hat{T}(\delta_0, \delta_1, \ldots, \delta_p{:}v) = N\,\Sigma_{j=0}^{p} \hat{\beta}_j \bar{x}^{(j)}$, reduces to the mean of ratios estimator (3.3.5), and is unbiased under the general Jth order polynomial model (3.2.1).

Example 3.3.2. Model $M(0, 1, 1{:}x^2)$. This is a quadratic model without intercept, with variance proportional to the square of x. The BLU estimator of total can be shown to be

$$\hat{T}(0, 1, 1{:}x^2) = \sum_s Y_i + \sum_r (\hat{\beta}_1 x_i + \hat{\beta}_2 x_i^2), \tag{3.3.6}$$

where

$$\hat{\beta}_1 = \frac{1}{\Sigma_s (x_i - \bar{x}_s)^2} \left\{ \frac{\Sigma_s x_i^2}{n} \sum_s \frac{Y_i}{x_i} - \frac{\Sigma_s x_i}{n} \sum_s Y_i \right\}$$

$$\hat{\beta}_2 = \frac{1}{\Sigma_s (x_i - \bar{x}_s)^2} \left\{ -\frac{\Sigma_s x_i}{n} \sum_s \frac{Y_i}{x_i} + \sum_s Y_i \right\}.$$

The BLU predictor can also be written as

$$\hat{T}(0, 1, 1{:}x^2) = \sum_s g_i Y_i \tag{3.3.7}$$

where

$$g_i = 1 + \frac{N_r}{\Sigma_s (x_i - \bar{x}_s)^2} \left\{ (\bar{x}_s^{(2)} \bar{x}_r - \bar{x}_s \bar{x}_r^{(2)}) \frac{1}{x_i} + (-\bar{x}_s \bar{x}_r + \bar{x}_r^{(2)}) \right\}.$$

Direct calculation shows that, under balance of order 2, that is,

$$\frac{\Sigma_s x_i^{-1} x_i^j}{n} = \frac{\Sigma_{i=1}^N x_i^j}{\Sigma_{i=1}^N x_i}, \qquad j = 0, 1, 2,$$

$T(0, 1, 1{:}x^2)$ reduces to the mean of ratios estimator $\hat{T}_{MR}(x) = \Sigma_{i=1}^N x_i \Sigma_s x_i^{-1} Y_i/n$, and is therefore unbiased for an underlying polynomial model (3.2.1) under the more extensive balance conditions (3.3.1). (See Exercise 3.10). Note that the balance condition holds automatically for $j = 1$, whatever the sample x values, and that, for reduction to $\hat{T}_{MR}(x)$, $j = 0$ is not actually needed. Given $v_i \propto x_i^2$, the model $M(0, 1, 1{:}x^2)$ contains the least number of distinct terms in powers of x that leads to a BLU estimator unbiased under the conditions of Corollary 3.3.1. In this sense, $M(0, 1, 1{:}x^2)$ is the smallest model with respect to the variance structure $v_i \propto x_i^2$.

Example 3.3.3. Model $M(x^{1/2}, x{:}x)$. If the variance is proportional to x (rather than x^2) then the smallest model that fulfills the conditions that the variance and standard deviation are linear combinations of the x variables is no longer polynomial:

$$Y_i = \beta_{1/2} x_i^{1/2} + \beta_1 x_i + x_i^{1/2} \varepsilon_i,$$

with independent errors $\varepsilon_i \sim (0, \sigma^2)$. It is convenient to symbolize this model by $M(x^{1/2}, x{:}x)$ and the corresponding estimator of total by $\hat{T}(x^{1/2}, x{:}x)$. $\hat{T}(x^{1/2}, x{:}x)$ is of the form (3.3.6) with x replaced by $x^{1/2}$, and under the *root*(x) balance condition $\bar{x}_s^{(1/2)} = \bar{x}/\bar{x}^{(1/2)}$, this estimator reduces to $\hat{T}_{MR}(x^{1/2}) = (N/n)(\bar{x}^{(1/2)}) \Sigma_s Y_i/x_i^{1/2}$. Under the further conditions

$$\bar{x}_s^{(j-1/2)} = \bar{x}^{(j)}/\bar{x}^{(1/2)} \quad \text{for } j = 0, 1, \ldots, J, \tag{3.3.8}$$

$\hat{T}(x^{1/2}, x{:}x)$ will be unbiased under the polynomial model (3.2.1).

We refer to $M(x^{1/2}, x{:}x)$, $M(0, 1, 1{:}x^2)$ and, in general, $M(v^{1/2}, v{:}v)$ as *minimal models* corresponding to given variance structures x, x^2, v, respectively. It is interesting to note that the model $M(x^{1/2}, x{:}x)$ is more general than the model $M_R = M(0, 1{:}x)$, reducing to M_R if $\beta_{1/2} = 0$, but the weighted balance conditions (3.3.8) are not equivalent to the simple balance conditions (3.2.4) that make the ratio estimator unbiased. This suggests that there are alternative strategies when $E_M(Y)$ is a function of x and the variance of Y is proportional to x:

Select a simple balanced sample and use the ratio estimator; *or*
Select a *root*(x) balanced sample and use the minimal estimator $\hat{T}(x^{1/2}, x{:}x)$.

This naturally raises the question as to which strategy is preferable. The simulation study of Section 3.5 suggests that the strategy of using *root*(x) balanced sample and the minimal estimator $\hat{T}(x^{1/2}, x{:}x)$ is better. The matter of efficient bias-robust estimation will be discussed in general in Chapter 4.

3.4. METHODS OF SELECTING BALANCED SAMPLES

Since balanced samples are theoretically useful, we need practical means of selecting them. Given a population, with values of the auxiliary x, and variances v, we want to find root(v) balanced samples s. In both the weighted and unweighted ($v = \text{constant}$) case, we propose a *two-step strategy* for selecting balanced samples: (1) using randomization, generate a collection of samples in such a way as to *aim* at balance with respect to v; (2) from this collection, choose a sample at or close to balance, by simple calculation of the appropriate sample moments.

For ease of presentation and for the sake of comparison, we first focus on the case of unweighted balance, although the more general case is of greater consequence in populations in which variance increases with values of an auxiliary. In Sections 3.4.1–3.4.4, we investigate four methods aiming at balanced samples of the form $s(J)$: (1) simple random sampling (*srs*), (2) systematic equal probability sampling (*sys*), (3) stratified simple random sampling (*stsrs*), and (4) restricted random sampling (*rsrs*). With the first three of these approaches, we are seeking a satisfactory *first* step, in the two-step procedure described above. With the *rsrs* method, a collection of approximately balanced samples can be generated from which one is selected at random. This method is especially useful when the sample must be clearly perceived as having been impartially selected.

Sections 3.4.5 and 3.4.6 discuss methods of selecting samples with weighted balance. Probability proportional to size (*pps*) techniques are described that use stratified and systematic sampling. We also give a restricted version of *pps* sampling, similar to *rsrs*.

Throughout this section, we assume the sample size n is predetermined and fixed. Determination of sample size is discussed in Chapter 4.

3.4.1. Simple Random Sampling

Simple unweighted balanced samples in a polynomial model satisfy $\bar{x}_s^{(j)} = \bar{x}^{(j)}$, $j = 1, \ldots, J$. One method that might be considered for attaining this type of balance is simple random sampling (*srs*), since, with respect to the probability structure of *srs*, it can be shown that $E_\pi(\bar{x}_s^{(j)}) = \bar{x}^{(j)}$ (Exercise 1.6 in Chapter 1). That is, simple random sampling gives balance *in expectation*. We may say that simple random sampling "aims at" balance.

Since simple random sampling is, for a variety of reasons, one of the most basic and popular of sampling methods, it is natural to inquire whether or not it *achieves* balance at least sufficiently closely for practical purposes. The answer to this is, unfortunately, *no*.

Consider the expansion estimator $\hat{T}(1{:}1)$, to which estimators meeting the conditions of Lemmas 3.3.1 and 3.3.2 reduce under simple balance. Under the model $M(1, 1{:}1)$, we have the bias

$$E_M[\hat{T}(1{:}1) - T] = N\beta_1(\bar{x}_s - \bar{x}). \tag{3.4.1}$$

The size of the model bias relative to the standard error of $\hat{T}(1{:}1)$ is important for deciding whether sound inference is possible. The model standard error of $\hat{T}(1{:}1)$, under $M(\delta_0, \delta_1, \ldots, \delta_J{:}1)$ is $N\sigma/\sqrt{n}$ when the sampling fraction f is negligible. The order of magnitude of the standard error is, thus, $N/\sqrt{n}$. As we observed in Section 3.2.2, in large simple random samples, $\sqrt{n}(\bar{x}_s - \bar{x})$ will be approximately normally distributed, implying that the order of the bias

$N\beta_1(\bar{x}_s - \bar{x})$ is also $N/\sqrt{n}$. In other words, as the sample size increases, the bias of $\hat{T}(1\!:\!1)$ never becomes inconsequential compared to its standard error. This undermines the performance of confidence intervals as described in the introductory section of this chapter. Royall and Cumberland (1985, 1988) discuss the inadequacies of *srs* in more detail.

This suggests, not only that *srs* does not achieve balance, but also that it does not aim at it very well: the order of $|\bar{x}_s - \bar{x}|$ is only $O_p(n^{-1/2})$; the next two sections describe standard methods that yield $|\bar{x}_s - \bar{x}| = O_p(n^{-1})$.

3.4.2. Systematic Equal Probability Sampling

A common technique in sampling is to select a systematic (*sys*) sample from a list sorted on the basis of some measure of size (see, e.g., see Brewer and Hanif, 1983; Cochran, 1946). With a single auxiliary variable x, one implementation of the method for selecting a sample of size n from N consists of four steps:

1. Sort the sampling frame from low to high based on x.
2. Compute a *skip interval* $S = N/n$.
3. Select a random start R with $0 < R \leqslant S$.
4. Calculate the sequence of selection numbers $u_i = R + (i-1)S$, $i = 1, \ldots, n$.

The ith sample unit is the first whose sequence number in the ordered list is greater than or equal to u_i. Intuitively, this procedure seems likely to be better than *srs* at selecting samples balanced on x since it spreads the x's in each sample across the range of x's in the population. Lemma 3.4.1 confirms this.

Lemma 3.4.1. (Kott, 1986). Suppose that, as $N, n \to \infty$, each x_i is bounded in the sense that $|x_i| \leqslant B < \infty$ and that a systematic sample of size n is selected as described in steps (1)–(4) above. Then, $|\bar{x}_s - \bar{x}| \leqslant 2B/n$ for all samples s.

Proof: Let $\{x_{(i)}\}_{i=1}^N$ denote the set of ordered population values of x. The minimum value that $\bar{x}_s$ can assume in the systematic selection plan is

$$\bar{x}_{s(\min)} = \sum_{i=1}^{n} x_{(1+[i-1]S)}/n$$

and the maximum is $\bar{x}_{s(\max)} = \Sigma_{i=1}^n x_{(S+[i-1]S)}/n$. Consequently,

$$\begin{aligned}\bar{x}_{s(\max)} - \bar{x}_{s(\min)} &= \frac{1}{n}(x_{(N)} - x_{(1)}) + \frac{1}{n}\sum_{i=1}^{n-1}(x_{(iS)} - x_{(iS+1)}) \\ &\leqslant 2B/n,\end{aligned}$$

since $x_{(iS)} - x_{(iS+1)} \leqslant 0$ for $i = 1, \ldots, n-1$. □

This result can also be extended to a linear combination of transformations of the x's (see Exercise 3.12). In particular, the above argument goes through, if we replace the condition $|x_i| \leqslant B < \infty$ by $|x_i^j| \leqslant B' < \infty$, so that for higher moments we have $|\bar{x}_s^{(j)} - \bar{x}^{(j)}| \leqslant 2B'/n$, and simple balance is achieved to order n^{-1}.

It may be noted that the number of distinct systematic samples from an ordered list is S—vastly lower than the number of srs samples $\binom{N}{n}$. We compare these approaches further in Section 3.4.3 below.

3.4.3. Stratification Based on the Auxiliary

Another method of selection that can also produce better balanced samples than *srs* is stratified simple random sampling (*stsrs*) with strata formed on the basis of the auxiliary variable. Stratified random sampling is discussed more generally in Chapter 6. We here focus on a particular version of stratified sampling used for the sake of obtaining balance.

As in systematic sampling, the population is first sorted based on x. Strata ($h = 1, \ldots, H$) are then formed by equalizing the number of units N_h per stratum. If n strata are formed and one sample unit selected per stratum, then we essentially have the equivalent of systematic sampling (Kott, 1986). Thus, balance will also be achieved at a faster rate under this method than under *srs*. In sampling one unit per stratum, the number of possible samples is $\left(\frac{N}{n}\right)^n$, assuming that N/n is an integer. Typically, this number of samples is much larger than the number possible under *sys* but much less than for *srs*.

Table 3.1 and Figures 3.4 and 3.5 illustrate the similarities and differences among *srs*, *sys*, and *stsrs*. Let χ^2_{df} denote the chi-square distribution with df degrees of freedom. Five hundred χ^2_6 variables were generated and 500 samples of $n = 25$ were selected. Note that for *sys* only $500/25 = 20$ samples are possible when the x's are sorted. For *stsrs*, five strata were formed with each having $N_h = 100$ and $n_h = 5$. Table 3.1 gives the means of the 500 sample

Table 3.1. Average Values of the First Three Sample Moments of an Auxiliary for Three Methods of Equal Probability Sample Selection Compared with Population Averages

Method	x	x^2	x^3
Population	6.0	52.1	629.8
srs	6.0	52.8	647.0
sys	6.0	51.7	621.7
stsrs	6.0	52.4	635.9

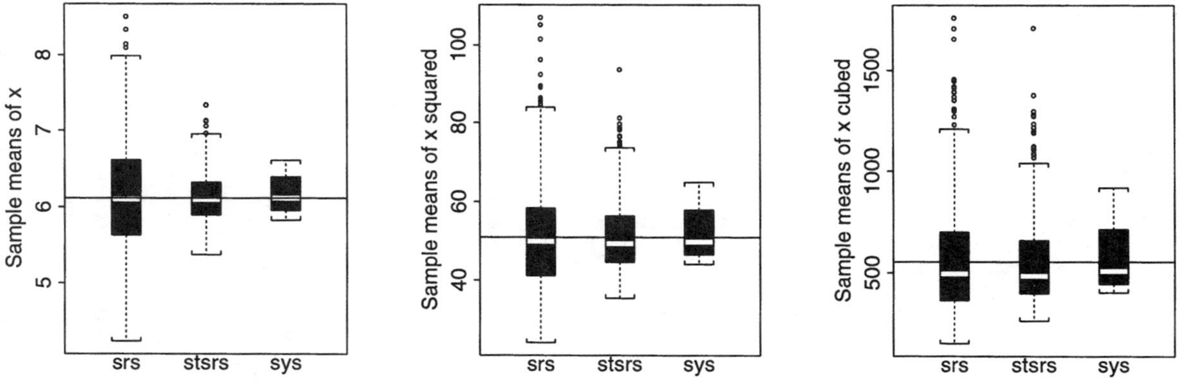

Figure 3.4 Boxplots of the distribution of the sample means of x, x^2, and x^3 for *srs*, *stsrs*, and *sys* in samples of $n = 25$.

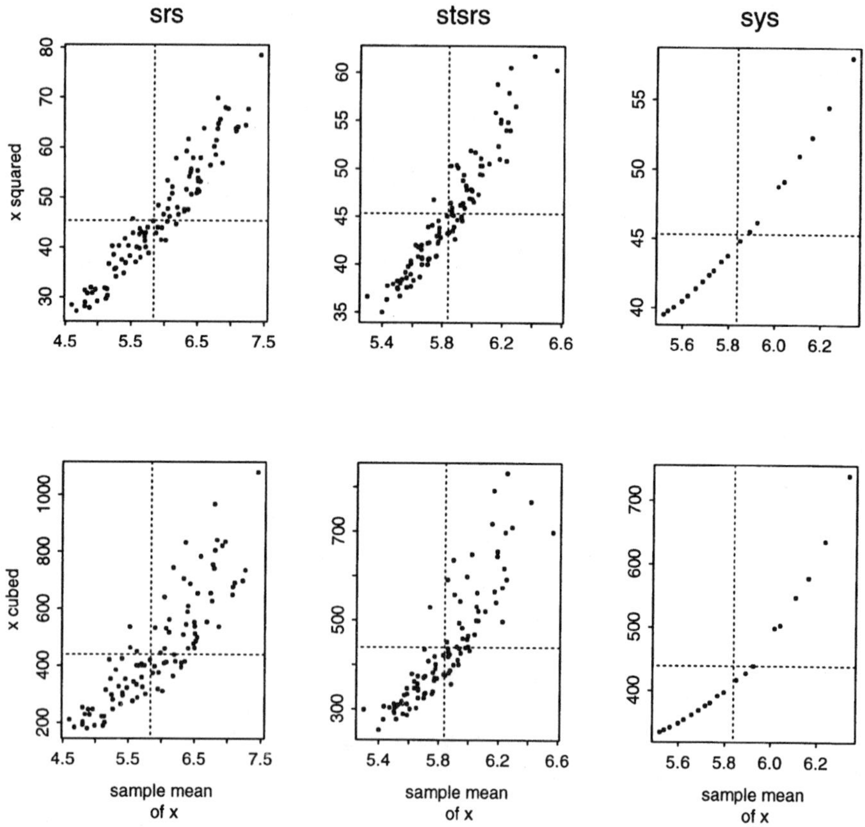

Figure 3.5 Scatterplots of $\bar{x}_s^{(2)}$ and $\bar{x}_s^{(3)}$ versus $\bar{x}_s$ for 20 samples from a population of χ_6^2 variables using *srs*, *stsrs*, and *sys* and sampling methods.

means of x, x^2, and x^3 for each method. All three methods produce nearly unbiased estimates of the first three sample moments of x.

Figure 3.4 shows boxplots of the distribution of the sample means of x, x^2, and x^3 for each method of sampling. Horizontal lines are drawn in each plot at the population means of x, x^2, and x^3. The boxes cover the interquartile ranges with a horizontal band shown in each at the median; the whiskers (brackets at the end of the dotted lines) extend to the nearest value not beyond 1.5 times the interquartile range. Sample means beyond the whiskers are shown as individual points. This analysis shows some obvious differences in selection methods. Systematic sampling leads to the shortest interquartile range for the three sample moments, followed by *stsrs*, then *srs*. Systematic sampling also clearly reduces the number of samples with extreme observations. However, in part, the narrower range is due to the fact that there are merely fewer samples generated by *sys*.

The deviation from balance on one moment also correlates with deviation from balance on another. In Figure 3.5, we consider 100 samples of size 25 from the above population, for *srs* and *stsrs*, and the 20 possible samples for *sys*, and plot the sample second and third moments against the first moment. We observe: (1) for all three methods, the higher moments are monotonic against the first moment; (2) *sys* supplies the tightest connection between the lower and higher moments, and *srs* the loosest, *stsrs* being intermediate. Thus in all cases, balance on the first moment, tends to produce balance on the higher moments, but the tendency is stronger in *stsrs* and *sys* than in *srs*, and probably strongest in *sys*. This suggests that, to achieve a balanced sample, one option is to generate a collection of samples of size n, using *sys* or *stsrs*, and then choose a sample with first moment close to the population first moment. We may well wish to check other moments as well, but there should be reasonable closeness, even on unchecked moments. We note that *srs* could be used, but that we will have to generate a greater number of samples, and compare samples more carefully on all relevant moments, before finally selecting a sample that is sufficiently balanced.

3.4.4. Restricted Random Sampling

The basic strategy of the above subsections is to generate a (possibly large) number of samples, and from among these select that sample which is best balanced up to a desired order J. An alternative approach is *restricted random sampling*, as discussed in Herson (1976). The general idea is to draw simple random samples but to reject any that are not sufficiently close to being balanced.

Suppose we seek balance on the first two moments of x. We define

$$e_1(s) = \left|\frac{\sqrt{n}(\bar{x}_s - \bar{x})}{s_x}\right| \quad \text{and} \quad e_2(s) = \left|\frac{\sqrt{n}(\bar{x}_s^{(2)} - \bar{x}^{(2)})}{s_{2x}}\right|,$$

where $s_x = [\Sigma_{i=1}^N (x_i - \bar{x})^2/(N-1)]^{1/2}$ and $s_{2x} = [\Sigma_{i=1}^N (x_i^2 - \bar{x}^{(2)})^2/(N-1)]^{1/2}$ are the population standard deviations of x and x^2. The functions $e_1(s)$ and $e_2(s)$ are standardized measures of imbalance for a given sample s.

A sample will be considered "sufficiently close to balance" if, for prescribed (small) constants E_1 and E_2,

$$e_1(s) \leqslant E_1 \qquad \text{and} \qquad e_2(s) \leqslant E_2. \tag{3.4.2}$$

A restricted random sampling (*rsrs*) plan is defined as follows:

1. Specify E_1 and E_2.
2. Select a simple random sample without replacement.
3. Retain the sample if (3.4.2) is satisfied; otherwise replace the sample into the population and repeat step (2).

The choices of E_1 and E_2 are somewhat arbitrary. In large samples $e_1(s)$ and $e_2(s)$ will be approximately normally distributed so that $E_1 = E_2 = 0.125$ will reject about 90% of samples, and provide reasonable balance (Herson, 1976; Royall and Cumberland, 1981a). Unrestricted *srs* corresponds to $E_1 = E_2 = \infty$. Notice that, as n increases, (3.4.2) becomes more stringent in terms of $|\bar{x}_s^{(j)} - \bar{x}^{(j)}|$ for $j = 1, 2$.

The method is relatively quick compared to *srs* followed by "best choice," is often easier to carry out than stratified sampling, and offers a larger array of possible samples than systematic sampling. The latter point can be important when (possibly extrastatistical) circumstances dictate the sample be clearly perceived as having been "randomly chosen."

Restricted random sampling has obvious similarities to *controlled selection* (Goodman and Kish, 1950; Kish, 1987), whose aim is to reduce the set of allowable samples by controlling the combinations of units that can occur together in a sample. Controlled selection is used, for example, in the U.S. Consumer Price Index to restrict the combination of states and areas within states that are selected (Dippo and Jacobs, 1983; Williams, Brown, and Zion, 1993).

3.4.5. Sampling for Weighted Balance

We now proceed to describe methods, analogous to the methods above, that lead to *root*(v) balance (3.3.1). For example, for a polynomial model of order J and $v_i = x_i^2\sigma^2$, we want $\bar{x}_s^{(j-1)} = \bar{x}^{(j)}/\bar{x}$, $j = 0, 1, \ldots, J$.

In general, the first step in attaining *root*(v)-balance, is to sample with probabilities proportional to the size (*pps*) measure $v^{1/2}$, that is, with inclusion probabilities $\pi_i = nv_i^{1/2}/(N\bar{v}^{(1/2)})$. This method of selection will be denoted *pp*($v^{1/2}$). In the case $v^{1/2} = x$, we use $\pi_i = nx_i/(N\bar{x})$.

As in the simple balanced case, we have *root*(v)-balance *in expectation*:

$$E_\pi\left(\sum_{i\in s} v_i^{-1/2} x_i^j/n\right) = \bar{x}^{(j)}/\bar{v}^{(1/2)}.$$

The question then becomes how to implement *pps* sampling, which is not self-evident. We discuss two methods of approximate implementation, variants of systematic sampling and stratified sampling. Each can be applied to the population with units ordered randomly, or by size of v.

The first method was studied by Hartley and Rao (1962) and is similar to that for systematic equal probability sampling:

1. Put the population in a random order.
2. Compute the cumulative sums $C_i = \Sigma_{i' \leqslant i} v_{i'}^{1/2}$, $i = 1, \ldots, N$.
3. Compute a skip interval $S = N\bar{v}^{(1/2)}/n = C_N/n$.

4. Select a random start R with $0 < R \leqslant S$.

5. Calculate the sequence $u_i = R + (i - 1)S$, $i = 1, \ldots, n$.

The ith sample unit is the first whose cumulative sum is greater than or equal to u_i. This algorithm can be characterized as "random *pps*," and in tables will be denoted by *P1:random*. If the v_i's are all equal, this method is equivalent to *srs*. Restricted *pps* selection can also be based on this procedure, as described in Section 3.4.6.

To get a version of *pps* selection analogous to systematic sampling (*sys*), the only modification in the above is in step (1):

1*. Sort the units in ascending order on $v^{1/2}$.

(2)–(5) As above.

This method gives systematic sampling, based on the size of v; it will be denoted as *P1:systematic*.

The second method was studied by Wright (1983) and discussed extensively in Särndal et al. (1992, Section 12.4). The method is similar to that for stratified equal probability sampling:

1. Put the population in a random order.

2. Compute the cumulative sums $C_i = \Sigma_{i' \leqslant i} v_{i'}^{1/2}$, $i = 1, \ldots, N$.

3. Compute the "interval length," $S = N\bar{v}^{(1/2)}/H = C_N/H$, where H is the number of strata desired.

4. let the first stratum contain the (first in order) i_1 elements with $C_{i_1} \approx S$, the second the next i_2 with $C_{i_2} \approx 2S$, etc. In short, the sum of the $v_i^{1/2}$'s are as close to S as possible within each stratum.

5. Select an equal number $n_h = n/H$ by *srs* within each stratum (or as near to n_h as possible if n_h is not an integer).

Like *P1:random*, this procedure, which may be denoted *P2:random*, yields a randomized version of *pps*, analogous to *srs*.

To get a version of *pps* selection analogous to *sys*, the only modification in the above is in step (1):

1*. Sort the units in ascending order on $v^{1/2}$.

(2)–(5) As above.

This algorithm stratifies the population strictly on the size of v; we denote it *P2:stratified*. We regard the default value of H as $H = n$; that is, take one unit per stratum; this will give results, as in the equal probability case, closest to systematic *pps*.

As in the equal probability case, the number of random-type *pps* samples (*P1:random* or *P2:random*) exceeds the number of stratified-type *pps* samples

Table 3.2. Population Values of $\bar{x}^{(j)}/\bar{x}$ and Averages Over 100 Samples of Sample Means $\bar{x}_s^{(j-1)} = \Sigma_s x_i^{j-1}/n$ ($j = 0, 2, 3$) for the Random, Stratified, and Systematic *pps* Methods from a Population of χ_6^2 Variables

		J		
		0	2	3
$\bar{x}^{(j)}/\bar{x}$	Population	0.1713	7.763	75.24
$\bar{x}_s^{(j-1)}$	*P1:random*	0.1735	7.702	73.92
$\bar{x}_s^{(j-1)}$	*P1:systematic*	0.1712	7.760	75.22
$\bar{x}_s^{(j-1)}$	*P2:stratified*	0.1720	7.761	75.31

Sample size was $n = 25$.

P2:stratified, which in turn far exceeds the number of systematic-type *pps* samples *P1:systematic*.

A simulation like the one in Section 3.4.3 compares three *pps* methods, *P1:random*, *P1: systematic*, and *P2:stratified*, for the case of $v^{1/2} = x$. It is assumed that the model is polynomial. Again, $N = 500$ χ_6^2 variables were generated and 100 samples of $n = 25$ were selected with each method. For the systematic case, enough samples were selected to guarantee 100 *distinct* samples. For *P2:stratified*, we used $H = 25$. Table 3.2 shows the population values of $\bar{x}^{(j)}/\bar{x}$ and the averages over the 100 samples of the sample means $\bar{x}_s^{(j-1)}$ ($j = 0, 2, 3$) for the three *pps* methods. All selection methods produce nearly unbiased estimates over the 100 samples. Figure 3.6 contains boxplots of the 100 samples with horizontal lines drawn at the balance points. Stratified and systematic *pps* sampling reduce the interquartile range and lead to fewer extremes.

Figure 3.7 shows the relation on a sample by sample basis between the $j = 0$ sample moments and the higher moments. We remind ourselves that $j = 0$ corresponds to balance needed for the sake of an intercept term, and corresponds to the requirement that the sample *harmonic* mean be close to the population arithmetic mean. As before, the relation between the moments is tightest for systematic sampling. Systematic *pps* seems then to be the preferred approach if we wish to generate a relatively few samples, compare them on the lowest or first few moments, and choose that sample closest to balance on these. In Section 3.5, we shall examine the consequences for bias and root mean square error of deviations from balance in two empirical studies.

3.4.6. Restricted *pps* Sampling

An alternative method of achieving root(v) balance, similar to restricted *srs* described in Section 3.4.5, is restricted probability proportional to root(v) sampling. We illustrate with restricted *pps* sampling (*rpps*), appropriate for

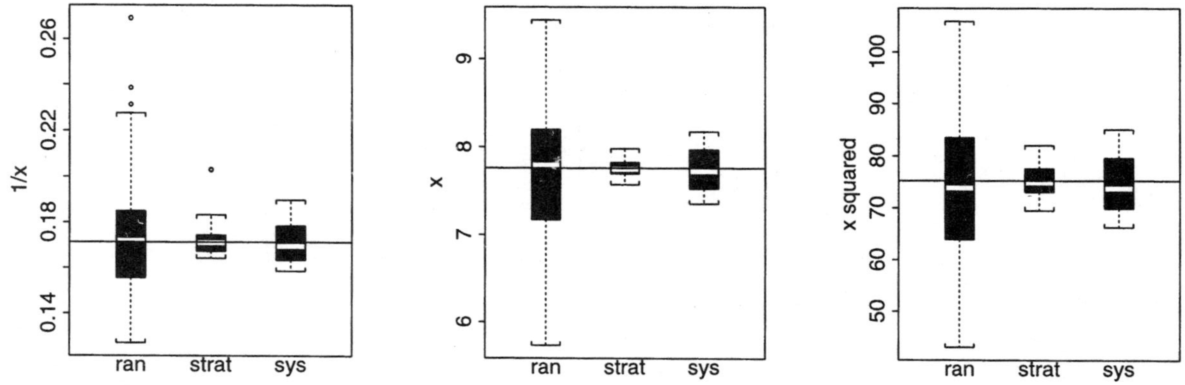

Figure 3.6 Boxplots of the distribution of the sample means of x^{-1}, x, and x^2 in random, stratified, and systematic $pp(x)$ samples of $n = 25$.

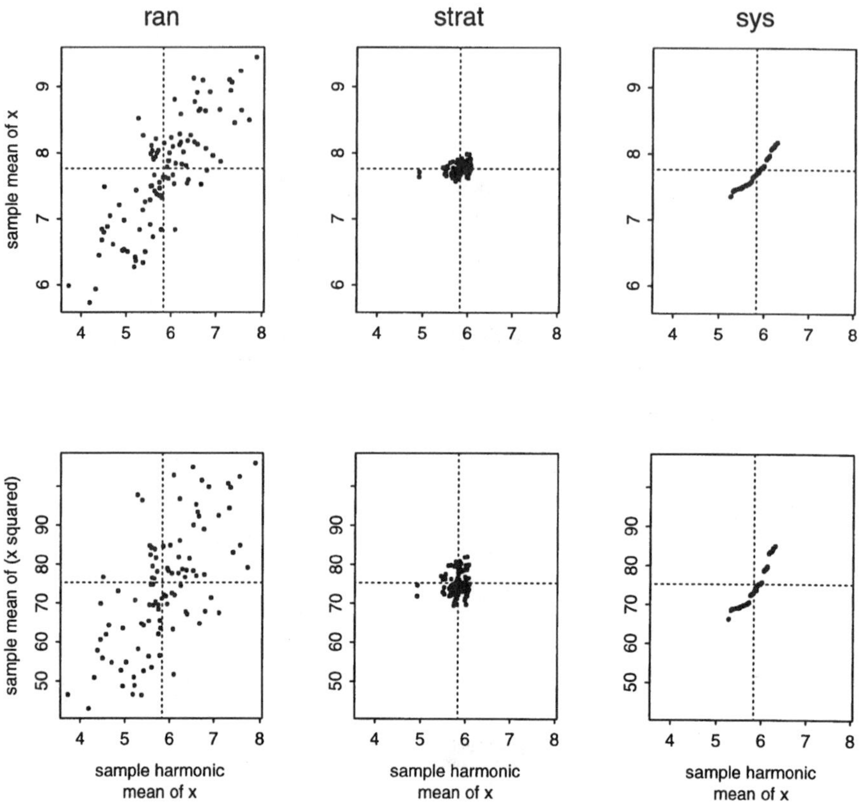

Figure 3.7 Scatterplots of sample means of x and x^2 versus sample harmonic means, $1/\bar{x}_s^{(-1)}$, for 100 samples from a population of χ_6^2 variables using random, stratified, and systematic $pp(x)$ samples of $n = 35$.

root(x^2) balance. Generalization and applications are left for the exercises. Suppose a sample is selected using the method *P1:random* or *P2:random* described in Section 3.4.5, with $v_i = x_i^2$. Define the balance constraints

$$e_{-1}(s) = \left|\frac{\sqrt{n}(\bar{x}_s^{(-1)} - 1/\bar{x})}{s_{-1,x}}\right|, \qquad e_1(s) = \left|\frac{\sqrt{n}(\bar{x}_s - \bar{x}^{(2)}/\bar{x})}{s_{1x}}\right|,$$

and

$$e_2(s) = \left|\frac{\sqrt{n}(\bar{x}_s^{(2)} - \bar{x}^{(3)}/\bar{x})}{s_{2x}}\right|,$$

where

$$s_{-1,x} = \left[\sum_{i=1}^{N} \pi_i (x_i^{-1} - 1/\bar{x})^2 \right]^{1/2}, \qquad s_{1x} = \left[\sum_{i=1}^{N} \pi_i (x_i - \bar{x}^{(2)}/\bar{x})^2 \right]^{1/2}$$

and

$$s_{2x} = \left[\sum_{i=1}^{N} \pi_i (x_i^2 - \bar{x}^{(3)}/\bar{x})^2 \right]^{1/2},$$

with $\pi_i = x_i/(N\bar{x})$. The terms $s_{ks}/\sqrt{n}$ are the design standard deviations of $\bar{x}_s^{(j)}$ in repeated *pps* sampling. Then the *rpps* procedure is

1. Select a *pps* sample using steps (1)–(5) of *P1:random* or *P2:random* in Section 3.4.5.
2. Retain the sample if $e_k(s) \leqslant E_k$, $k = -1, 1, 2$, for preset values of E_{-1}, E_1, and E_2; otherwise replace the sample and repeat (1).

As with *rsrs*, values of $E_k = 0.125$, should suffice.

3.4.7. Partial Balancing

Before concluding this section, we note a complication with weighted balance that does not arise in simple unweighted balance. When the sample is large, weighted balance is sometimes not possible, even approximately. This is a consequence of Theorem 4.2.1, in Chapter 4, and will be discussed further there. It turns out (see Exercise 4.2) that if the sampling design aiming at *root*(v) balance is possible, that is, if $\pi_i = nv_s^{(1/2)}/(N\bar{v}^{(1/2)}) \leqslant 1$ for all i, then *root*(v) balance itself is possible. Otherwise, a reasonable strategy is to pursue *partial balancing*, that is, to relegate units for which $\pi_i > 1$ to a "certainty stratum" of points estimating only themselves, and then to balance on the remainder of the population. That is, these points will not be used in estimating the model parameter β or in estimating T_r. This may seem to be throwing away possibly useful information, but including such points risks the bias that comes with lack of balance. This will be discussed further in Chapter 4. Another strategy would be to lower the rate at which the variance is supposed to increase to the point where $\pi_i \leqslant 1$; this might actually be better, but has not been explored.

3.5. SIMULATION STUDY OF WEIGHTED BALANCE

In this section, we will use simulation to study properties of some estimators arising out of the theory of weighted balance in Section 3.3. As in Section 3.2.5, the Hospitals Population serves as our testbed. Some preliminary remarks are in order.

We can expect that how different estimators will perform in a simulation depends on which model is a reasonable description for Hospitals. Indications from Figure 3.1 were that a working model with some curvature is appropriate with $M(0, 1, 1{:}x^\gamma)$ being a good choice. A quick estimate of γ can be derived as follows: Define the residual $r_i = Y_i - \mathbf{x}_i'\hat{\boldsymbol{\beta}}$ with $\mathbf{x}_i = (x_i, x_i^2)'$ and $\hat{\boldsymbol{\beta}} = (\mathbf{X}_s'\mathbf{X}_s)^{-1}\mathbf{X}_s'\mathbf{Y}_s$ being the ordinary least squares estimator of $\boldsymbol{\beta}$ in the model $M(0, 1, 1{:}1)$. Then, under some standard conditions, noted in the exercises, $E_M(r_i^2) \cong \sigma^2 x_i^\gamma[1 + o(1)]$ if the model $M(0, 1, 1{:}x^\gamma)$ is correct. Replacing $E_M(r_i^2)$ by r_i^2 and taking logs, we get the approximate relationship

$$\ln(r_i^2) \cong \ln(\sigma^2) + \gamma \ln(x_i).$$

Thus the logs of the squared residuals can be regressed on the logs of the auxiliaries x_i to get an estimate of γ. The procedure of calculating residuals and regressing $\ln(r_i^2)$ on $\ln(x_i)$ can be iterated for a more refined estimate of γ. The first iteration is completed as just described. In the tth pass, the residuals are calculated as $r_i = Y_i - \mathbf{x}_i'\hat{\boldsymbol{\beta}}^{(t)}$ with

$$\hat{\boldsymbol{\beta}}^{(t)} = [\mathbf{X}_s' \operatorname{diag}(x_i^{-\gamma^{(t-1)}})\mathbf{X}_s]^{-1}\mathbf{X}_s' \operatorname{diag}(x_i^{-\gamma^{(t-1)}})\mathbf{Y}_s,$$

where $\gamma^{(t-1)}$ is the estimate of γ from the $(t-1)^{\text{st}}$ iteration. The procedure terminates when the relative change in the estimate of γ is small between consecutive steps.

Applying this procedure to the full Hospitals population yields $\hat{\gamma} = 1.484$. Thus the value of $\hat{\gamma}$ is intermediate between the values of 1 and 2 that were examined in earlier sections. However, we seek estimators that perform well even if γ is not precisely estimated. The fact that $\hat{\gamma} = 1.484$ is bracketed by 1 and 2 suggests that estimators based on $v_i \propto x_i$ or $v_i \propto x_i^2$ should do reasonably well. We wish to explore how the estimators discussed in Section 3.3 do, combined with root(x) or root(x^2) balanced samples and, more broadly, in the context of $pp(x^{1/2})$ and $pp(x)$ sampling designs, which aim respectively at these types of balance.

3.5.1. Results Using the Hospitals Population

For the simulation, we selected two sets of 2000 samples, each of size $n = 25$. In the first set each sample was selected $pp(x^{1/2})$ with x again being the number of beds in each hospital. The version of *pps* sampling described in Section 3.4.5 was used in which the population was randomly ordered (*P1:random*). In the first set of samples we calculated the following estimates, which are motivated by the combination of a model with variance proportional to x and samples selected $pp(x^{1/2})$:

(a) $\hat{T}(0, 1, 1{:}x)$, the estimator based on a quadratic model without an intercept;
(b) $\hat{T}(1, 1, 1{:}x)$, the estimator based on a quadratic model with an intercept;
(c) $\hat{T}_{MR}(x^{1/2})$, the mean-of-ratios estimator with weights $x^{-1/2}$;
(d) $\hat{T}_0(x^{1/2}) = N\Sigma_s Y_i x_i^{-1/2}/\Sigma_s x_i^{-1/2}$, the weighted expansion estimator with weights $x^{-1/2}$; and
(e) $\hat{T}(x^{1/2}, x{:}x)$, the estimator corresponding to the minimal model.

The weighted expansion estimator, $\hat{T}_0(x^{1/2})$, is also known as a *Hájek estimator*, after Hájek (1971).

The samples were sorted by $\bar{x}_s^{(1/2)}$, divided into 10 groups of 200 samples each, and values of bias and *rmse* were computed in each group as described in Section 3.1.4. Figure 3.8 shows the results with a vertical reference line drawn at the balance point $\bar{x}_s^{(1/2)} = \bar{x}/\bar{x}^{(1/2)}$. Over much of the range of $\bar{x}_s^{(1/2)}$, $\hat{T}(0, 1, 1{:}x)$ is the best performer in terms of *rmse* by a slight margin over the minimal estimator $\hat{T}(x^{1/2}, x{:}x)$ and $\hat{T}(1, 1, 1{:}x)$. The minimal estimator does display some positive bias in the most extreme samples but in the middle of the range of $\bar{x}_s^{(1/2)}$ has little bias and competitive *rmse*. The inclusion of a possibly superfluous intercept term in $\hat{T}(1, 1, 1{:}x)$ was not particularly harmful. Notice, especially, that at balance the *rmse* of the best estimators is near its minimum, that is, a $x^{1/2}$-balanced sample is optimal or very nearly so.

The weighted expansion estimator $\hat{T}_0(x^{1/2})$, denoted Hajek(sqrt(x)) in the figure, exhibits systematic bias away from the $x^{1/2}$-balance point, $\bar{x}_s^{(1/2)} = \bar{x}/\bar{x}^{(1/2)}$, as expected when the model $M(1{:}v)$ does not hold. Though $\hat{T}_0(x^{1/2}) = \hat{T}_{MR}(x^{1/2}) = \hat{T}(x^{1/2}, x{:}x)$ if $\bar{x}_s^{(-1/2)} = 1/\bar{x}^{(1/2)}$ and $\bar{x}_s^{(1/2)} = \bar{x}/\bar{x}^{(1/2)}$, the Hájek estimator has a larger *rmse* than either of the other two in the middle of the range of $\bar{x}_s^{(1/2)}$, indicating some imbalance on $\bar{x}_s^{(-1/2)}$. The mean-of-ratios estimator $\hat{T}_{MR}(x^{1/2})$ also is biased in unbalanced samples.

The second set of 2000 samples was selected $pp(x)$, the method that aims at x-balance which is linked to models with variance proportional to x^2. The estimators were:

(a) $\hat{T}(0, 1, 1{:}x^2)$, the estimator based on a quadratic model without an intercept;
(b) $\hat{T}(1, 1, 1{:}x^2)$, the estimator based on a quadratic model with an intercept;
(c) $\hat{T}_{MR}(x)$, the mean-of-ratios estimator with weights x^{-1}; and
(d) $\hat{T}_0(x) = N\Sigma_s Y_i x_i^{-1}/\Sigma_s x_i^{-1}$, the weighted expansion estimator with weights x^{-1}.

Samples were sorted based on $\bar{x}_s$ and again divided into 10 groups for the conditional analysis pictured in Figure 3.9. The vertical reference line is now drawn at $\bar{x}_s = \bar{x}^{(2)}/\bar{x}$. The minimal estimator under a model with $\text{var}_M(Y) \propto x^2$ is $\hat{T}(0, 1, 1{:}x^2)$, which does well here, having small bias and nearly constant

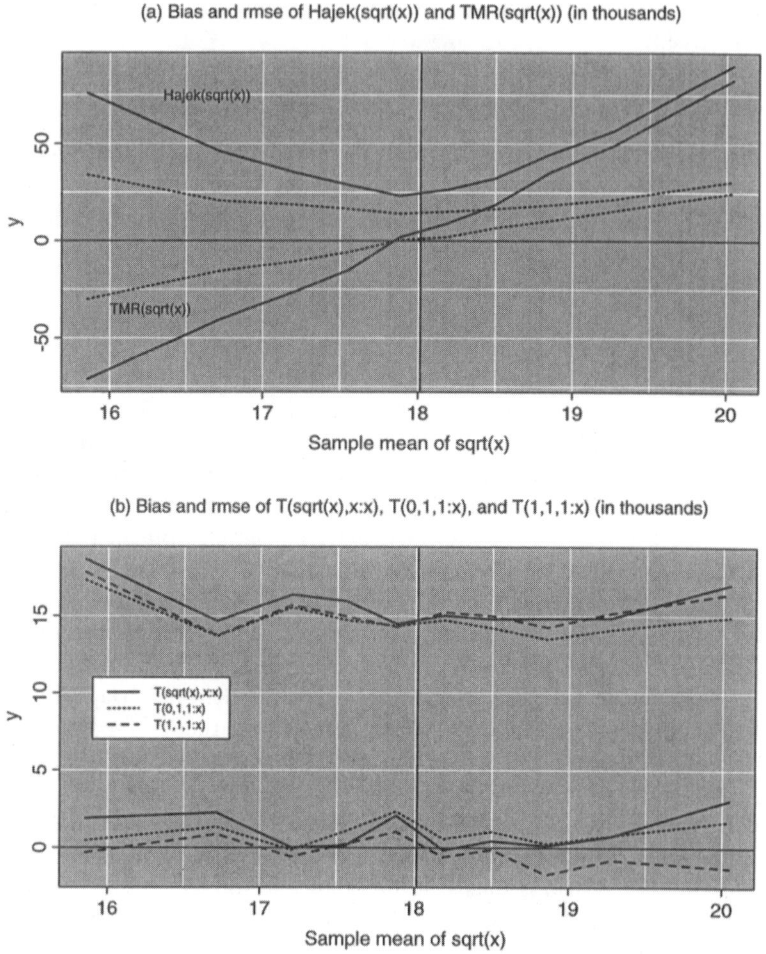

Figure 3.8 Simulation results for five estimators in 2000 $pp(x^{1/2})$ samples of size $n = 25$ from the Hospitals population. Upper curves in each panel are *rmse*'s; lower curves are biases; vertical reference line drawn at population value of $x^{1/2}$-balance, $\bar{x}/\bar{x}^{(1/2)}$.

rmse throughout the range of $\bar{x}_s$. As in Figure 3.8, a weighted balanced sample is near optimal for the best estimators. The Hájek estimator, $\hat{T}_0(x)$, and the mean-of-ratios estimator, $\hat{T}_{MR}(x)$, are poor because of substantial, conditional biases. In this case the estimator $\hat{T}(1, 1, 1:x^2)$, having what may be an unnecessary intercept, does have a noticeably larger *rmse* at balance than either $T(0, 1, 1:x^2)$ or $\hat{T}_{MR}(x)$. Since $\hat{T}(1, 1, 1:x^2) = \hat{T}(0, 1, 1:x^2) = \hat{T}_{MR}(x)$ when $\bar{x}_s = \bar{x}^{(2)}/\bar{x}$ and $\bar{x}_s^{(-1)} = 1/\bar{x}$, the reason for the larger *rmse* of $\hat{T}(1, 1, 1:x^2)$ is not immediately obvious. We will explore this more in the next section.

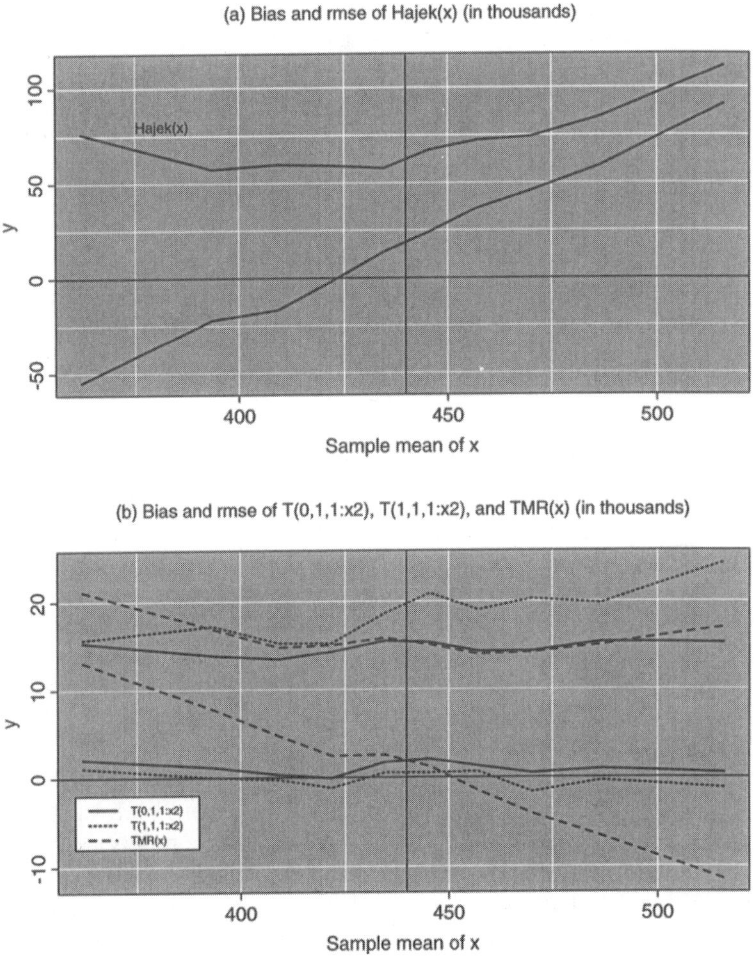

Figure 3.9 Simulation results for four estimators in 2000 $pp(x)$ samples of size $n = 25$ from the Hospitals population. Upper curves in each panel are *rmse*'s; lower curves are biases; vertical reference line drawn at population value of x-balance, $\bar{x}^{(2)}/\bar{x}$.

Table 3.3 lists the empirical *rmse*'s for the combination of the fifth and sixth groups of 200 samples in the simulations. These groups are ones that are near $x^{1/2}$-balance in the case of $pp(x^{1/2})$ sampling or near x-balance in $pp(x)$ sampling. In addition, the table includes the *rmse* for the ratio estimator from the fifth and sixth ordered groups from the simulation in Section 3.2.5. $\hat{T}(0, 1, 1{:}x)$ in $pp(x^{1/2})$ sampling is the best strategy listed, followed closely by $\hat{T}(x^{1/2}, x{:}x)$ in $pp(x^{1/2})$ and $\hat{T}(1, 1, 1{:}x)$ in $pp(x^{1/2})$.

We note especially that the estimators in $pp(x^{1/2})$ sampling—in particular, the minimal estimator $\hat{T}(x^{1/2}, x{:}x)$—are better than the ratio estimator

Table 3.3. Empirical Root Mean Square Errors of Several Estimators in the Fifth and Sixth Groups of 200 Ordered Samples of Size $n = 25$ from Hospitals that are Near $x^{1/2}$-Balance, x-Balance, or Simple Balance

Estimator	Type of Balance	*rmse*
$\hat{T}(0, 1:x)$	Simple balance	17.23
$\hat{T}(x^{1/2}, x:x)$	$x^{1/2}$-balance	14.79
$\hat{T}(0, 1, 1:x)$	$x^{1/2}$-balance	14.53
$\hat{T}(1, 1, 1:x)$	$x^{1/2}$-balance	14.75
$\hat{T}(0, 1, 1:x^2)$	x-balance	15.42
$\hat{T}(1, 1, 1:x^2)$	x-balance	19.82

$\hat{T}(0, 1:x)$ at simple balance. Further clarity on why this is so will be given in Chapter 4.

3.5.2. Interaction of Model Specification with Sample Configuration

In the simulation reported in the last section, the *rmse* of $\hat{T}(1, 1, 1:x^2)$ was substantially larger than those of $\hat{T}(0, 1, 1:x^2)$ and $\hat{T}_{MR}(x)$ in samples where $\bar{x}_s$ was near $\bar{x}^{(2)}/\bar{x}$. Since $\hat{T}(1, 1, 1:x^2)$ is approximately unbiased throughout the range of $\bar{x}_s$ in Figure 3.9, a large variance appears to be the source of the larger *rmse*.

Figure 3.10, panel (a), is a scatterplot of $\hat{T}(1, 1, 1:x^2)$ versus $\hat{T}(0, 1, 1:x^2)$ for the same set of 400 samples as in Table 3.3. The range of $\hat{T}(1, 1, 1:x^2)$ is clearly wider than that of $\hat{T}(0, 1, 1:x^2)$. Two of the more extreme points are circled and labeled sample 1 and sample 2. The particular hospitals in these samples are listed in Table 3.4. The estimates that they produce illustrate problems with $\hat{T}(1, 1, 1:x^2)$. Panels (b) and (c) are plots of discharges versus beds in Hospitals with the sample hospitals shown as circles and the nonsample units as dots. The curved lines are the least squares fitted lines for the model $M(1, 1, 1:x^2)$, whereas the dotted lines are the fitted versions of $M(0, 1, 1:x^2)$.

Judging from the earlier Figure 3.1, the intercept is near 0 in the full population, but, for the model with intercept included, the estimated intercepts are far from 0 in these two samples. Because sampling is $pp(x)$, both samples have few hospitals with small numbers of beds. There are also some high leverage points at the extremes of the sample range of x in both samples that lead to poor predictions using $M(1, 1, 1:x^2)$. In sample 1 the estimated intercept is negative, leading to an underestimate of discharges for every nonsample hospital with fewer than 200 beds. Sample 2 has the opposite problem. In contrast, the fitted lines for $M(0, 1, 1:x^2)$ are much more reasonable because the intercept is forced to be 0.

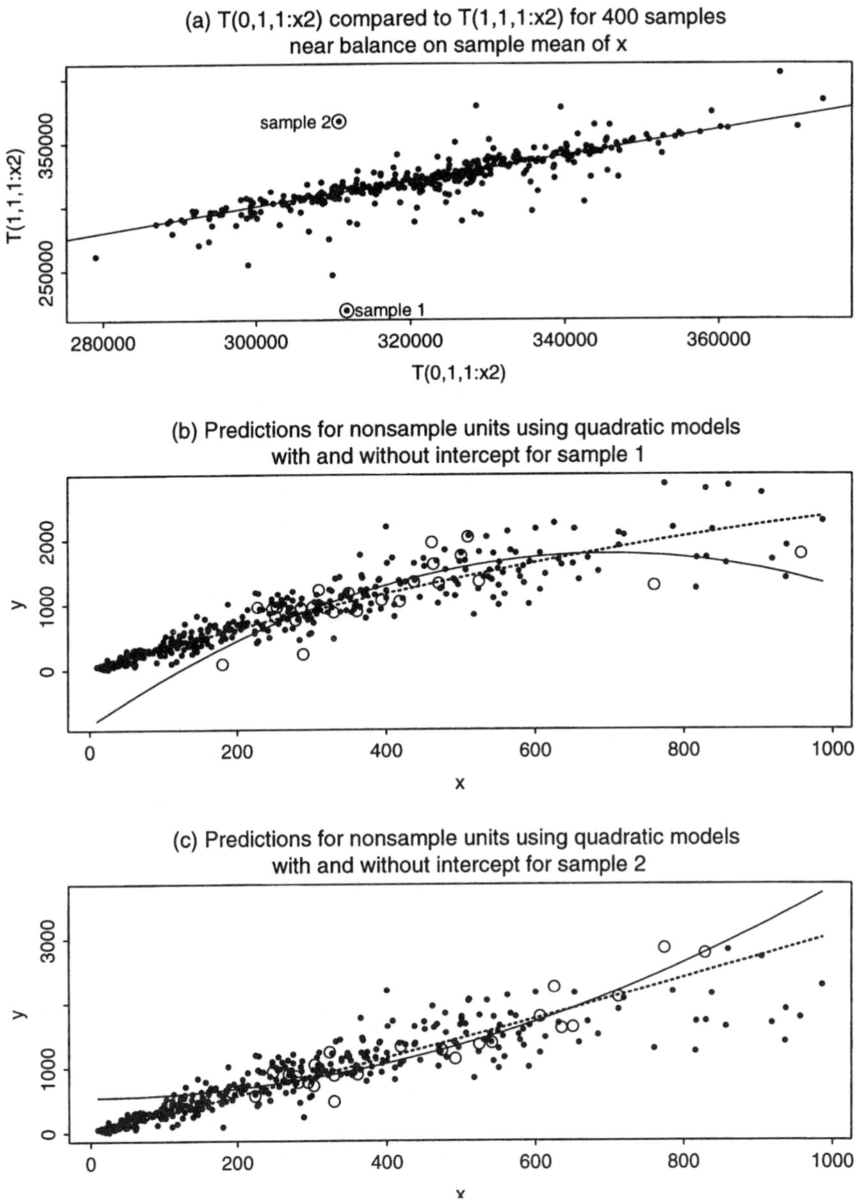

Figure 3.10 Illustrations of poor predictions by $\hat{T}(1,1,1{:}x^2)$ in two samples from the Hospitals population. In panels (b) and (c) sample points are shown as circles.

Table 3.4. Identification Numbers of the Hospitals in the Two Circled Samples in Figure 3.10

Sample 1			Sample 2		
164	253	316	173	258	364
193	256	319	189	262	367
207	263	331	207	263	369
208	272	334	223	278	370
210	278	338	231	306	376
230	296	378	232	321	379
235	305	392	239	326	383
237	310		245	338	
244	315		246	345	

Table 3.5 lists balance measures and values of the estimated totals for samples 1 and 2. Both samples are reasonably balanced on $\bar{x}_s$ but are poorly balanced on $1/\bar{x}_s^{(-1)}$. The fact that a random $pp(x)$ sample is on-balance for one moment but off-balance for another is no surprise considering the earlier simulation results shown in Figure 3.7. If stratified, systematic, or restricted $pp(x)$ sampling had been used, better balance on the one moment would lead to better balance on the other. This raises a natural question: among samples balanced on both $\bar{x}_s$ and $1/\bar{x}_s^{(-1)}$, would there be similar problems? The theoretical answer is no, since $\hat{T}(1, 1, 1{:}x^2)$ and $\hat{T}(0, 1, 1{:}x^2)$ are algebraically equal in the balanced case. Thus again we learn, in a different way, the lesson that merely aiming at balance is insufficient.

The values of the estimated totals are much different in the two samples in Table 3.4. The minimal estimator $\hat{T}(0, 1, 1{:}x^2)$ is quite close to the population total for both samples, but $\hat{T}(1, 1, 1{:}x^2)$ is a 32% underestimate for sample 1 and a 15% overestimate in sample 2. For the Hospitals we might reject $\hat{T}(1, 1, 1{:}x^2)$ purely on logical grounds given some knowledge of the population. A hospital with no beds should have no inpatient discharges and, thus, the intercept should be 0. However, for other variables in other populations, we may not have such *a priori* knowledge.

Table 3.5. Balance Measures and Values of $\hat{T}(1, 1, 1{:}x^2)$ and $\hat{T}(0, 1, 1{:}x^2)$ for Samples 1 and 2

	$1/\bar{x}_s^{(-1)}$ versus $\bar{x}$	$\bar{x}_s$ versus $\bar{x}^{(2)}/\bar{x}$	$\hat{T}(1, 1, 1{:}x^2)$ versus T	$\hat{T}(0, 1, 1{:}x^2)$ versus T
Sample 1	344.03	395.20	218,256	311,801
Sample 2	373.13	440.88	366,672	310,794
Population	274.70	439.77	320,159	320,159

The message here is clear—addition of a gratuitous parameter may add no bias but can seriously increase variance. In some unbalanced samples, high leverage points may be a problem, and fitting an extra parameter may result in completely unreasonable predictions for some nonsample units. On the other hand, under "complete" balance—balance on all relevant moments—discrimination among the models would not be a concern.

3.6. SUMMARY

To minimize bias, it is essential either (a) to have the correct model for $E_M(Y)$, an unlikely event, or (b) to have a balanced sample with a corresponding estimator, for example, the ratio estimator with simple balance or a minimal model estimator with the corresponding weighted balance. When $E_M(Y)$ is correctly specified, having the right variance structure for the estimator makes no difference in bias. On the other hand, if a balanced sample is selected, the assumed variance for the estimator *must* be the variance v on which the *root*(v) balanced sample is based. Other estimators, including simpler ones to which the minimal estimator reduces, can be risky, given that balance is typically not exact.

3.7. ROBUSTNESS AND DESIGN-BASED INFERENCE

In the statistical literature, an estimator is said to be *robust* if its efficiency—the degree to which it tends to be close to its target—is not much affected by mild changes in the mechanism (model) generating the data (see Huber 1981, Sections 1.1–2). Our concern in this chapter with *bias-robustness* is in keeping with this notion of robustness in the wider statistical literature, because, as we have argued, the squared bias, unchecked, can swamp the variance, and so be the main ingredient of an unduly large mean square error.

The notion of robustness in the design-based survey sampling literature is usually taken to be something else. Since the population data are regarded as fixed, a model of data generation is essentially unnecessary, and there is no need to worry about degradation of estimates caused by model failure. As stated succinctly by Brewer and Särndal (1983):

> Probability sampling methods are robust by definition; since they do not appeal to a model, there is no need to discuss what happens under model breakdown.

However, since "robust" tends to connote "safe," design-based proponents often characterize as robust, combinations of probability designs and estimators that give sound confidence intervals. This design-based position is stated by Hansen, Madow, and Tepping (1983) (from now on denoted by HMT):

> Robustness is usually understood to mean that inferences made from a sample are insensitive to violations of the assumptions that have been made. In principle, and ordinarily in fact, robustness is achieved in probability-sampling surveys by the use of sampling with known probabilities (i.e., randomization) and consistent estimators, and using a large enough sample that the central limit theorem applies, so that the estimates can be regarded as approximately normally distributed. (HMT, p. 786.)

> It is advantageous... and sufficient to have a 'good' estimator based on a reasonably large probability sample that provides a valid confidence interval. (HMT, p. 791.)

A brief history of the origins of this idea, arising out of the classic work of Neyman (1934), can be found in Brewer (1994). Smith (1994) observes that by this definition all randomization procedures that yield consistent, asymptotically normal estimators (with a readily estimated variance) are equally robust. Thus robustness holds regardless of any population structure relating Y to auxiliary variables, and may almost be said to be a *characteristic* of sampling carried out with regard to design-based principles.

The description of design based robustness above does include enough restrictions to exclude some pathological cases. Recall Exercise 1.13, in which the population was defined by $y_i = (1.1)^i$ for $i = 1, \ldots, 100$. In *srswor* the sample mean $\bar{Y}_s$ is design unbiased but is far from normally distributed because the population of y's contains such extreme outliers. The consequent poor coverage of confidence intervals persists, for a given sampling fraction $f = n/N$, as N gets large (see Exercise 3.17). Consequently, in this particular population $\bar{Y}_s$ would not be considered "robust," although in more well-behaved populations it would be.

Example 3.1.1, given earlier, is enough to clarify the difference between the bias-robustness we have covered earlier in this chapter and design-based robustness. If the correct model for the mean is $E_M(Y_i) = \beta_0 + \beta_1 x_i$, then the sample mean $\bar{Y}_s$ has a model bias equal to

$$E_M(\bar{Y}_s - \bar{Y}) = \beta_1(\bar{x}_s - \bar{x}).$$

Obviously, $\bar{Y}_s$ has a model bias heavily dependent on $\bar{x}_s$ like the bias we saw in Figure 3.3. However, if the sample is selected by simple random sampling without replacement, $\bar{Y}_s$ is design unbiased. If the conditions of Theorem A.9 hold, $\bar{Y}_s$ is also asymptotically normal with a variance that decreases as n increases and, hence, by Chebyshev's inequality, is design consistent. If we average the model bias over all simple random samples, then

$$\begin{aligned} E_\pi E_M(\bar{Y}_s - \bar{Y}) &= \beta_1 E_\pi(\bar{x}_s - \bar{x}) \\ &= 0. \end{aligned}$$

The positive biases when $\bar{x}_s > \bar{x}$ cancel out the negative biases when $\bar{x}_s < \bar{x}$. Thus, despite its substantial bias in some off-balance samples, the bias averages out to 0 across all simple random samples. Consequently, $\bar{Y}_s$ is "robust" in the design-based sense.

Design-based theory strives for protection against poor results through the use of large randomly selected samples and consistent estimators. Most of this chapter has been devoted to model-related solutions directed at restricting the configurations of samples that can be selected. These restricted samples are dictated by the population structure and the type of estimator implied by that structure.

The design-based approach is less formal, but has led practitioners over the years to develop techniques like stratification, equal probability systematic sampling, and sampling with probabilities proportional to size that fit well with the search for model based robustness. As elaborated in Section 3.4, these practical methods often achieve their good results by restricting the kinds of samples that randomization can produce, thereby selecting samples that are more nearly balanced.

To attempt (as this book does) to incorporate model-structure as a necessary ingredient in survey sampling inference, and to call into question (as it does) the necessity of probability based sampling designs, is to bring back into sampling a combination of ingredients which renders *nonrobustness* possible, and it seems to many, inevitable.

Indeed HMT explores the dangers of a pure model-based—or rather, as therein called, "model-*dependent*"—approach to survey sampling:

> A model-dependent 'best' estimator depends on the validity of the model and may yield confidence intervals that are seriously misleading. (HMT, p. 791.)

In a simulation study, HMT finds the model-dependent approach entailing great losses of efficiency relative to design-based estimators, as well as poor coverage of confidence intervals. An artificial population, with $N = 14{,}000$, is generated by the model $E_M(Y_i) = \beta_0 + \beta_1 x_i$ and $\mathrm{var}_M(Y_i) = \sigma^2 x_i^\gamma$, where $\beta_0 = 0.4$, $\beta_1 = 0.25$, $\sigma^2 = 0.0625$, and $\gamma = 3/2$. The intercept β_0 is small enough that if a test of $H_0{:}\beta_0 = 0$ is conducted, "such tests for a sample of 400 or less have a high probability of failing to reject the hypothesis (HMT p. 780)." The population is stratified into 10 strata using x as the size variable, and simple random samples of equal size drawn from each stratum. Their goal is to estimate the population mean $\bar{Y}$.

Estimators such as the unstratified ratio estimator, $\hat{T}_R/N$, derived on the assumption that $E_M(Y_i) = \beta_1 x_i$, are found to be biased across these stratified random samples, with consequent great loss in efficiency relative to estimators that are design unbiased or consistent. The larger the overall sample size n, the larger the relative size of the squared bias as a component of the mean square error. Also, the confidence intervals for model-dependent estimators perform poorly, giving coverage seriously below nominal, thus decisively indicating the nonrobustness of the model-dependent approach. The conclusion is that small departures from the working model can have a large effect on inferences, detrimental to the use of an estimator tied to that model.

Is this study as devastating for the model-based approach as it appears to be (and as it is generally taken to be)? Three observations can be made: (1) the first regards the relation between HMT's ideal model-dependent sampler and the paradigm we have been discussing in this chapter; (2) the second regards the test procedures used by HMT; and (3) the third relates the HMT study to the question of estimators appropriate to weighted balance. We address these three points below.

1. HMT takes the paradigm of model-dependent inference to consist of three steps:

(a) Choice of model that fits the (sample) data;

(b) Test of the model against more complex alternatives; and

(c) Subject to nonrejection of the simpler model, its use as a basis for estimation.

However, the model-based or prediction approach we have been describing proceeds by

(A) Selection of a (weighted) balanced sample, to protect against model failure;

(B) Selection of a simple model that is compatible with the particular form of balance and reflects, to the best of our knowledge, the data structure; and

(C) Estimation based on this model.

We note that testing for adequacy of model has not entered into this paradigm, and that concern with model misspecification is present from the outset, with no anticipation that we shall be able to catch small deviations from the model. The methodology is designed precisely to handle this contingency. The steps A, B, C are not the same as a, b, c. Thus the "model-dependent" paradigm is besides the point, if it is intended to apply to model-based sampling as described in the literature and (so far at least) in this book.

2. The a–b–c paradigm, however, is not so unreasonable if the survey analyst is presented with a sample that someone else has selected or was selected without regard to achieving any sort of balance. If the analyst has some knowledge of regression diagnostics and tests, and the survey situation allows the means to carry them out, the a–b–c paradigm may be exactly what is used. If the diagnostic/testing approach is used—and most students who have taken modern regression courses might be inclined to think it would be—then the failure of such tests in the relatively simple situation in HMT must be a serious worry, if they indeed do fail.

Table 3.6. Results of Testing the Hypothesis of Zero Intercept (at 5% Level)

Assumed Variance Structure in Testing	Number of Samples Rejecting Hypothesis that Intercept = 0
$v_i = \sigma^2$ (constant variance)	321
$v_i = x_i\sigma^2$ (variances proportional to x)	733
$v_i = x_i^2\sigma^2$ (variances proportional to x^2)	973

HMT's claim that the tests do fail has been readily accepted by many. However, the exact method used for testing for the presence of an intercept was not specified in HMT, nor the evidence given for their claim that "...a line through the origin is more likely than not to be accepted as a plausible model for samples of less than about 400, if one made a test before adopting the model... (HMT, p. 782)."

Examination does not bear this claim out. Results of testing will depend heavily on variance assumptions: the more weight put on points with small x, the more likely will deviations from a zero intercept be detected. We replicated HMT's population, and generated 1000 samples of size 200 (20 per stratum) according to the stratified sample design described above. Testing for a nonzero intercept yielded the results in Table 3.6, using standard testing procedures with a two-sided test at a 5% significance level, which is a strict standard. In his comments on HMT, Little (1983) recommended a 20% significance level; also, for many populations, a one-sided test for positive only intercept would be reasonable.

Our guess is that HMT carried out only the constant variance version of testing, and then relied on the assumption that testing "would depend on (*but not be sensitive to*) the conditional variances assumed in the test." (HMT, p. 782 [italics added].) However, the above results show this dependency is important. In particular, the results for variances proportional to x or x^2 clearly bring into question the claim that the intercept in the HMT model is barely detectable in large samples. It may be noted that simple residual analysis after using the simple linear or through the origin model with constant variance will, for this population, almost invariably point to variance increasing with x (see Exercise 3.18), at least at larger sample sizes. (The stratification scheme in HMT, described above, would, in fact, make most sense if the survey sampler were prepared to assume a variance increasing as x^2; see Section 6.5.2.)

There is, of course, a danger in adding too many parameters to a model, as we saw in Section 3.5.2. There, the addition of an intercept to a quadratic model led to some extremely poor predictions in a few off-balance samples. In the simple HMT case, this is less worrisome since we consider only adding an intercept to a through the origin model.

The a–b–c procedure clearly entails that the estimator ultimately used will depend on the result of the testing. It therefore is of interest to ascertain the

Table 3.7. Results for Mean Square Errors and Confidence Interval Coverage in an HMT Population, Based on 1000 Samples, $n = 200$

Estimator	Mean Square Error of Estimators of $\bar{Y}$	Percentage Coverage of 95% Confidence Intervals
Ratio estimator $T(0, 1{:}x)/N$	0.02547	71.0
amalgam, $v_i = \sigma^2$	0.01770	82.2
amalgam, $v_i = x_i\sigma^2$	0.01132	92.1
amalgam, $v_i = x_i^2\sigma^2$	0.00946	94.9
Linear estimator $T(1, 1{:}x)/N$	0.00920	95.2

sampling properties of the "amalgam estimator" which is based on the (incorrect) through the origin model when the tests fails to detect non-zero intercept and on the (proper) linear model with intercept when the test for presence of an intercept is positive. Little (1983) first noted the need for this (see also Dorfman, 1993).

Table 3.7 gives mean square errors for means and coverage of nominal 95% confidence intervals for the 1000 stratified random samples of size $n = 200$ described above. The amalgam estimators choose between the standard ratio estimator $\hat{T}_R$ and the estimator $\hat{T}(1, 1{:}v)$, depending on the results of testing. Confidence intervals were constructed using the robust variance estimator v_D described in Chapter 5. The reader may wish to refer to Tables 1 and 3 of HMT for comparison; results for the design-based estimators therein described are at about the level of the amalgam at $v_i = x_i^2\sigma^2$.

3. The stratified sampling design, with equal samples from strata defined by the size variable x can be viewed as aiming at a root(x^2) balanced sample. Using the ratio estimator with such samples is similar to using the ratio estimator with the "largest x" sample—most efficient if the through the origin model $M(0, 1{:}x)$ holds, but avoided by model-based samplers because of the vulnerability to bias under alternate models. In the light of this chapter, it is clear that such a sample calls for an estimator such as the minimal model estimator of the mean, $\hat{T}(0, 1, 1{:}x^2)/N$, to give bias robustness. At the end of Chapter 4, after the relation between efficiency and weighted balance is clarified, we will describe the improvements on efficiency provided by the combination of weighted balance and the minimal estimator over that of the estimators in HMT.

EXERCISES

3.1 Show that, under the through the origin model (3.1.2), the bias of the expansion estimator is $E_{M^*}(\hat{T}_0 - T) = N\beta(\bar{x}_s - \bar{x})$.

3.2 Suppose, for a certain population U, with $N = 234$ units, we intend to estimate a total $T = \Sigma_{j=1}^{N} Y_i$, and we take an *srs* sample of size $n = 12$ from U. Assume that no information on an auxiliary variable is available at the time of sampling. By the time we have the data in hand and are ready to estimate T, information on a variable x has become available: its population mean is $\bar{x} = 18.2$ and the sample data on x as well as Y is available:

x	3.2	4.7	6.0	9.8	10.2	11.3	13.5	17.8	22.9	28.0	32.5	34.4
Y	5.3	10.2	11.8	20.3	19.7	24.7	30.3	32.6	39.6	66.7	65.4	78.0

Find the expansion, ratio, and regression estimators for T. Which estimator do you expect to be closest to T, and why?

3.3 Suppose the general polynomial model (3.2.1) holds. Show the bias of the expansion estimator $\hat{T}_0$ is $E_M[\hat{T}(1{:}1) - T] = N\,\Sigma_{j=1}^{J}\,\delta_j\beta_j[\bar{x}_s^{(j)} - \bar{x}^{(j)}]$, where $\bar{x}_s^{(j)} = \Sigma_s x_i^j/n$ and $\bar{x}^{(j)} = \Sigma_{i=1}^{N} x_i^j/N$.

3.4 Suppose the general polynomial model (3.2.1) holds. Show the bias of the ratio estimator is given by

$$E_M[\hat{T}(0, 1{:}x) - T] = N\bar{x}\sum_{j=0}^{J}\delta_j\beta_j\left[\frac{\bar{x}_s^{(j)}}{\bar{x}_s} - \frac{\bar{x}^{(j)}}{\bar{x}}\right].$$

3.5 For a certain population, suppose the true underlying model for the variable of interest Y has $E_M(Y) = 3 + 10x - 0.25x^2$, and that $N = 1000$, $\bar{x} = 25$, $\bar{x}^{(2)} = 1125$. If a sample of size $n = 30$ is selected and found to have sample means $\bar{x}_s = 20$, $\bar{x}_s^{(2)} = 940$, what is the bias of (a) the expansion estimator of the population total of Y; (b) the ratio estimator of the total?

3.6 Prove Theorem 3.2.1.

3.7 Suppose a sample from population U of size $N = 230$ is balanced on the first three moments of an auxiliary variable x. Suppose Y_1 and Y_2 are variables of interest attaching to units of U, with Y_1 adequately modeled as linear in x, and Y_2 as quadratic in x. Suppose the sample means of the Y's are $\bar{Y}_{1s} = 10.3$ and $\bar{Y}_{2s} = 24.8$, and it is known that $\bar{x} = 25$, $\bar{x}^{(2)} = 1125$. Estimate the population totals of Y_1 and Y_2 using the *BLU* estimator for each. State any other assumptions needed in order to make the calculation.

3.8 Suppose the underlying model of variable of interest Y, in a population of size N, is quadratic, with $E(Y_i) = \beta_0 + \beta_1 x_i + \beta_2 x_i^2$, and a *root*($x$)

balanced sample of order 2 is selected, with sample size n. (a) Write out the balance conditions (3.3.1) in this case. (b) Determine the bias of the expansion estimator T_0, (c) of the ratio estimator T_R, and (d) of the linear regression estimator $T(1, 1:1)$ in this situation, that is, for this underlying model, with these balance conditions.

3.9 Suppose the working model is $M(0, 1:x^2)$. (a) Find the *BLU* estimator $\hat{\beta}$ of the slope β, to show that the mean of ratios estimator given in Example 3.3.1 can be written in the form $\hat{T}_{MR}(x) = \Sigma_{i=1}^{N} x_i \hat{\beta}$. (b) Find the *BLU* estimate of total under the working model. (c) Under what condition on the sample Y's are the *BLU* estimator and the mean of ratios estimator equal? (d) What sample size, if any, would *guarantee* their equality? (e) Calculate the value of the *BLU* estimate and of the mean of ratios estimate under the working model $M(0, 1:x^2)$, for (f) the sample of 30 units with the largest x values (beds) from the hospitals population, (g) the sample of 30 units with the smallest x values.

3.10 Suppose the working model is $M(0, 1, 1:x^2)$. (a) Show that the *BLU* estimators of the model parameters are given by

$$\hat{\beta}_1 = \frac{1}{\Sigma_s (x_i - \bar{x}_s)^2} \left\{ \frac{\Sigma_s x_i^2}{n} \sum_s \frac{Y_i}{x_i} - \frac{\Sigma_s x_i}{n} \sum_s Y_i \right\}$$

$$\hat{\beta}_2 = \frac{1}{\Sigma_s (x_i - \bar{x}_s)^2} \left\{ -\frac{\Sigma_s x_i}{n} \sum_s \frac{Y_i}{x_i} + \sum_s Y_i \right\},$$

so that expression (3.3.6) holds for the *BLU* estimator of total. (b) Verify Lemma 3.3.1 by showing that (3.3.6) can be written $\hat{T}(0, 1, 1:x^2) = N(\hat{\beta}_1 \bar{x} + \hat{\beta}_2 \bar{x}^{(2)})$. (c) Show that (3.3.7) holds. (d) Show that under the balance condition $\bar{x}_s = \Sigma_{i=1}^{N} x_i^2 / \Sigma_{i=1}^{N} x_i$,

$$\hat{T}(0, 1, 1:x^2) = \hat{T}_{MR}(x) = \sum_{i=1}^{N} x_i \sum_s x_i^{-1} Y_i / n.$$

3.11 (Generalization of Lemma 3.4.1) Suppose that, as $N, n \to \infty$, each x_j is bounded in the sense that $|x_{ji}| \leqslant B_j < \infty$ and that a systematic sample is selected as described in steps (1)–(4) of Section 3.4.2. Let $z_i = \Sigma_j c_j g_j(x_{ji})$, where c_j are constants and $g_j(\cdot)$ are monotonic functions. Then there exists a constant B_z such that $|\bar{z}_s - \bar{z}| \leqslant 2B_z/n$ for all samples s.

3.12 The Counties 70 Population in Appendix B has population size $N = 304$. The mean of the auxiliary variable is $\bar{x} = 8931$, the mean of the auxiliary squared is $\bar{x}^{(2)} = 214.5 \times 10^6$, and the total of the Y-values is $T = 11.243 \times 10^6$. Suppose T is to be estimated by a sample of size $n = 30$. Take a sample according to each of the following specifications,

(a) A simple random sample without replacement;

(b) A systematic sample;

(c) A stratified random sample, with, as nearly as possible, equal number of population units within each stratum, and one unit selected per stratum;

(d) A restricted random sample, with $e_1(s) \leqslant 0.125$ and $e_2(s) \leqslant 0.125$.

In each sample calculate (i) the sample means $\bar{x}_s$, $\bar{x}_s^{(2)}$, (ii) the expansion estimator $\hat{T}_0$, and (iii) the ratio estimator $\hat{T}_R$, and then (iv) graph $\hat{T}_0 - T$ and $\hat{T}_R - T$ versus each of $\bar{x}_s, \bar{x}_s^{(2)}$, and make whatever observations seem pertinent. (It will be of interest to compare your results to that of others who may happen to carry out the same exercise.)

3.13 Generate an artificial population using the x variable of Exercise 3.12 (i.e., the x values from Counties 70). The distribution of Y conditional on x is normal with mean $500 + 4x + (5 \times 10^{-5})x^2$ and variance $250x$. Carry out the same exercises as in 3.12.

3.14 A *best-fit* sample with respect to a particular auxiliary x has been defined (Royall and Cumberland, 1981a) to be that sample for which the empirical distribution function of x is closest to the population distribution function of x. A simple algorithm to construct such a sample (of size n) is the following. First, sort the population on x. Then, the best-fit sample $s = \{i_k : k = 1, \ldots, n\}$ is given by

$$i_k = \begin{cases} \text{either } j_k \text{ or } j_k - 1 & \text{if } j_k \text{ is an integer} \\ \text{greatest integer less than } j_k & \text{otherwise,} \end{cases}$$

where $j_k = (2k - 1)N/(2n) + 1$. Find the best-fit sample of size $n = 30$ for the Counties 70 Population, and find the quantities (i), (ii), and (iii) as in Exercise 3.12.

3.15 For the Counties 70 Population, take a sample that aims at root(x) balance, of size $n = 30$, by (a) *P1:random* sampling, (b) *P1:systematic* sampling, (c) *P2:stratified* sampling, and (d) restricted *pps* sampling that requires $e_l(s) \leqslant 0.125$, $l = -1, 1, 2$, with $e_l(s)$ defined as in Section 3.4.6, but with x replaced by the square root of x. In each case, (i) calculate $\bar{x}_s^{(-1/2)}, \bar{x}_s^{(1/2)}, \bar{x}_s$ and compare to the appropriate population ratios, (ii) calculate the mean of ratios estimator $\hat{T}_{MR}(x^{1/2})$, and (iii) the appropriate minimal model estimator.

3.16 Repeat 3.15 for the artificial population of Exercise 3.13.

3.17 As in Exercise 1.13, generate a population by $y_i = (1.1)^i$, $i = 1, 2, \ldots, N$, and take a thousand simple random samples of size n, for each of the pairs $(N, n) = (1000, 150)$, $(2000, 300)$, $(3000, 450)$. For each sample use the expansion estimator of total T and its variance estimator to construct a 95% confidence interval for T. Give your empirical estimate of coverage for each pair (N, n). Repeat, using the same population sizes, but doubling the sampling fraction. Draw conclusions.

3.18 (a) Generate an artificial population, with $N = 5{,}000$ as follows: The variable x has a gamma distribution with density

$$f(x) = 0.04 \exp(-x/5).$$

The variable Y, conditional on x, has a gamma density

$$g(y; x) = [b^c \Gamma(c)]^{-1} y^{c-1} \exp(-y/b),$$

where $b = 1.25x^{3/2}(8+5x)^{-1}$ and $c = 0.04x^{-3/2}(8+5x)^2$. (b) Show that with this structure the model for Y is $E_M(Y_i) = \beta_0 + \beta_1 x_i$ with $\mathrm{var}_M(Y_i) = \sigma^2 x_i^\gamma$ where $\beta_0 = 0.4$, $\beta_1 = 0.25$, $\sigma^2 = 0.0625$, and $\gamma = 3/2$. (c) Select simple random samples without replacement of $n = 20$, 40, 100, and 200 and fit the model $M(0, 1{:}1)$. (d) Plot the residuals from the model versus x. Is there any evidence of increasing variance as x increases? (e) Using your samples, test the hypothesis $H_0{:}\beta_0 = 0$ at the 0.05 level of significance.

CHAPTER 4

Robustness and Efficiency

4.1 INTRODUCTION

The previous chapter focused on bias-robust strategies: by selecting a balanced sample—simple or weighted—and an estimator based on a related working model, estimation of the total T is made unbiased with respect to an array of models extending beyond the working model.

In the present chapter, we extend the results on balance, addressing mainly the question of what sort of balance gives the greatest efficiency, that is, reduces the error variance as much as possible, under specified conditions. In the simulation study of Section 3.5, we have already had a clue that not all weights serve equally well in seeking a weighted balanced sample. There we saw that the minimal estimator for root(x) balance gave lower root mean square error than did the ratio estimator under simple unweighted balance. This result is not entirely surprising, given that the root(x) balanced sample has more units where the variation in the Y variable is greater.

Consider again the ratio estimator $\hat{T}_R = \hat{T}(0, 1\colon x) = N\bar{Y}_s\bar{x}/\bar{x}_s$, the *BLU* predictor under the model $M(0, 1\colon x)$. As shown in Exercise 2.8, the error variance of $\hat{T}(0, 1\colon x)$ under $M(0, 1\colon x)$ is

$$\text{var}_M[\hat{T}(0, 1\colon x) - T] = N^2(1 - f)(\bar{x}_r\bar{x}/\bar{x}_s)\sigma^2/n, \tag{4.1.1a}$$

which, under simple balance, $s \in s(1)$, is

$$\text{var}_M[\hat{T}(0, 1\colon x) - T] = N^2(1 - f)\bar{x}\sigma^2/n. \tag{4.1.1b}$$

Consider the minimal model estimator $T(x^{1/2}, x\colon x)$ or, more simply, the "weighted expansion estimator" $\hat{T}_0(x^{1/2}) = N\Sigma_s x_i^{-1/2}Y_i/\Sigma_s x_i^{-1/2}$, to which the minimal model estimator reduces under the weighted balance conditions

$$\frac{n}{\Sigma_s x_i^{-1/2}} = \bar{x}^{(1/2)}, \tag{4.1.2}$$

$$\frac{\Sigma_s x_i^{-1/2}x_i}{\Sigma_s x_i^{-1/2}} = \bar{x}. \tag{4.1.3}$$

It is readily checked that the estimator $\hat{T}_0(x^{1/2})$ is unbiased under $M(0, 1\colon x)$ when (4.1.3) holds.

Suppose $M(0, 1\colon x)$ is the true model, leading as we have seen to the ratio estimator as *BLU* estimator of the total. Then, under the weighted balance conditions, (4.1.2) and (4.1.3), $\hat{T}_0(x^{1/2})$ has smaller variance than the ratio estimator itself has under simple balance. In particular, under these conditions,

$$\begin{aligned}\mathrm{var}_M(\hat{T}_0(x^{1/2}) - T) &= \left(\frac{N^2}{n}[\bar{x}^{(1/2)}]^2 - N\bar{x}\right)\sigma^2 \leqslant \left(\frac{N^2}{n}\left(1 - \frac{n}{N}\right)\right)\bar{x}\sigma^2 \\ &= \mathrm{var}_M(\hat{T}(0, 1\colon x) - T).\end{aligned}$$

Thus, $\hat{T}_0(x^{1/2})$ under root(x) balance is actually more efficient than the ratio estimator under simple balance, when the model justifying the ratio estimator holds.

This has two important implications: (1) The ratio estimator, which has a longstanding place in the literature, and appeals to us by its commonsense simplicity, less easily combines bias-robustness and efficiency than the slightly more general minimal estimator $T(x^{1/2}, x\colon x)$. One can certainly make the ratio estimator a good deal more efficient than it is in a balanced sample, *if* the through the origin model $M(0, 1\colon x)$ holds, by taking a sample that is *not* balanced. In fact, the error variance (4.1.1a) is easily seen to be minimized when the sample consists of the units with the largest n x's in the population. However, this sort of extreme sample sacrifices the safety that balance provides if the through the origin model fails. The strategy of minimal model and root(x) balance maintains bias-robustness, and achieves greater efficiency than the ratio estimator under simple balance. (2) It suggests the general principle that careful choice of the weights used to construct a balanced sample is important for attaining efficiency.

The outline of the remainder of this chapter is as follows: Section 4.2 discusses weighted balance in the context of the general linear model with diagonal covariance matrix, proving an important theorem on the achievement of bias-robustness and maximal efficiency. Section 4.3 elaborates these ideas in an extensive empirical study of biases and standard errors using different samples and estimators, under a known underlying population model. Section 4 applies results on the variance of estimators under balance to the determination of sample size. Section 5 sums up, and makes some remarks regarding standard procedures in design-based sampling.

4.2 GENERAL LINEAR MODEL

We return to the general linear model with a diagonal covariance matrix:

$$E_M(\mathbf{Y}) = \mathbf{X}\boldsymbol{\beta}, \ \mathrm{var}_M(\mathbf{Y}) = \mathbf{V}\sigma^2, \tag{4.2.1}$$

which will be referred to as $M(\mathbf{X}: \mathbf{V})$. As in chapter 2, $\boldsymbol{\beta}$ is $p \times 1$, $\mathbf{X}$ is $N \times p$, and $\mathbf{V} = diag(v_1, v_2, \ldots, v_N)$ is $N \times N$. The *BLU* predictor under this model is $\hat{T}(\mathbf{X}: \mathbf{V}) = \mathbf{1}_s'\mathbf{Y}_s + \mathbf{1}_r'\mathbf{X}_r\hat{\boldsymbol{\beta}}$, where $\mathbf{1}_s$ and $\mathbf{1}_r$ are vectors of n 1's and N-n 1's, and $\hat{\boldsymbol{\beta}} = \mathbf{A}_s^{-1}\mathbf{X}_s'\mathbf{V}_{ss}^{-1}\mathbf{Y}_s$ with $\mathbf{A}_s = \mathbf{X}_s'\mathbf{V}_{ss}^{-1}\mathbf{X}_s$. Let $\mathcal{M}(\mathbf{X})$ denote the linear manifold generated by the columns of $\mathbf{X}$, that is, the vector space spanned by all linear combinations of the columns of $\mathbf{X}$. We will also need 1_N, an N-vector of 1's.

Lemma 4.2.1 (Royall, 1992). If $\mathbf{V1}_N \in \mathcal{M}(\mathbf{X})$, then the *BLU* predictor and its error variance are

$$\hat{T}(\mathbf{X}: \mathbf{V}) = \mathbf{1}_N'\mathbf{X}\hat{\boldsymbol{\beta}} \tag{4.2.2}$$

$$\text{var}_M[\hat{T}(\mathbf{X}: \mathbf{V}) - T] = (\mathbf{1}_N'\mathbf{X}\mathbf{A}_s^{-1}\mathbf{X}'\mathbf{1}_N - \mathbf{1}_N'\mathbf{V}\mathbf{1}_N)\sigma^2. \tag{4.2.3}$$

Proof: The assumption that $\mathbf{V1}_N \in \mathcal{M}(\mathbf{X})$ implies that $\mathbf{V1}_N = \mathbf{Xc}$ for some p-vector $\mathbf{c}$. It follows that $\mathbf{c}'\mathbf{X}_s'\mathbf{V}_{ss}^{-1} = \mathbf{1}_s$ or $\mathbf{c}'\mathbf{X}_s'\mathbf{V}_{ss}^{-1}\mathbf{Y}_s = \mathbf{1}_s'\mathbf{X}_s\hat{\boldsymbol{\beta}}$. Substituting $\mathbf{c}'\mathbf{X}_s' = \mathbf{1}_s'\mathbf{V}_{ss}$ gives $\mathbf{1}_s'\mathbf{Y}_s = \mathbf{1}_s'\mathbf{X}_s\hat{\boldsymbol{\beta}}$. The *BLU* predictor, thus, becomes $\hat{T}(\mathbf{X}: \mathbf{V}) = \mathbf{1}_s'\mathbf{X}_s\hat{\boldsymbol{\beta}} + \mathbf{1}_r'\mathbf{X}_r\hat{\boldsymbol{\beta}} = \mathbf{1}_N'\mathbf{X}\hat{\boldsymbol{\beta}}$. The error variance is

$$\begin{aligned}\text{var}_M[\hat{T}(\mathbf{X}: \mathbf{V}) - T] &= \text{var}_M(\mathbf{1}_N'\mathbf{X}\hat{\boldsymbol{\beta}}) + \text{var}_M(\mathbf{1}_s'\mathbf{Y}_s) + \text{var}_M(\mathbf{1}_r'\mathbf{Y}_r)\\ &\quad - 2\,\text{cov}_M(\mathbf{1}_N'\mathbf{X}\hat{\boldsymbol{\beta}}, \mathbf{1}_s'\mathbf{Y}_s).\end{aligned}$$

We have $\text{var}_M(\mathbf{1}_N'\mathbf{X}\hat{\boldsymbol{\beta}}) = \mathbf{1}_N'\mathbf{X}\mathbf{A}_s^{-1}\mathbf{X}'\mathbf{1}_N$, $\text{var}_M(\mathbf{1}_s'\mathbf{Y}_s) + \text{var}_M(\mathbf{1}_r'\mathbf{Y}_r) = \mathbf{1}_N'\mathbf{V}\mathbf{1}_N'$, and $\text{cov}_M(\mathbf{1}_N'\mathbf{X}\hat{\boldsymbol{\beta}}, \mathbf{1}_s'\mathbf{Y}_s) = \mathbf{1}_N'\mathbf{X}\mathbf{A}_s^{-1}\mathbf{X}_s'\mathbf{1}_s$. Because $\mathbf{1}_s = \mathbf{V}_{ss}^{-1}\mathbf{X}_s\mathbf{c}$, we have $\text{cov}_M(\mathbf{1}_N'\mathbf{X}\hat{\boldsymbol{\beta}}, \mathbf{1}_s'\mathbf{Y}_s) = \mathbf{1}_N'\mathbf{Xc} = \mathbf{1}_N'\mathbf{V}\mathbf{1}_N'$. Substituting the various components produces (4.2.3) above. □

The condition $\mathbf{V1}_N \in \mathcal{M}(\mathbf{X})$ is a generalization of the condition on the variances in Lemma 3.3.1, and the above proof of (4.2.2) suffices to prove that Lemma.

4.2.1. *BLU* Predictor Under the General Linear Model with Diagonal Variance Matrix

In this section we generalize the idea of balanced samples and show that *BLU* estimators of total based on models fulfilling certain weak conditions are bias-robust in weighted balanced samples. Furthermore, there exists a lower bound on the error variance of the *BLU* predictor under these conditions, and this bound is only achieved if the sample is balanced.

The collection of samples that satisfy

$$\frac{1}{n}\mathbf{1}_s'\mathbf{W}_s^{-1/2}\mathbf{X}_s = \frac{\mathbf{1}_N'X}{\mathbf{1}_N'\mathbf{W}^{1/2}\mathbf{1}_N} \tag{4.2.4}$$

will be denoted $B(\mathbf{X}:\mathbf{W})$, when the working model is (4.2.1), and said to be *balanced with respect to the weights root*$(\mathbf{W})$ or to be *root*(W) *balanced*. Here $\mathbf{W}$ is an $N \times N$ matrix and $\mathbf{W}_s$ is the $n \times n$ submatrix for the sample units.

As in the discussion of balance for polynomial models in Section 3.3, one might have anticipated a different definition of weighted balance, namely, that

$$\frac{\mathbf{1}_s'\mathbf{W}_s^{-1/2}\mathbf{X}_s}{\mathbf{1}_s'\mathbf{W}_s^{-1/2}\mathbf{1}_s} = \frac{1}{N}\mathbf{1}_N'\mathbf{X}.$$

It is readily seen that the two definitions are equivalent if

$$\frac{n}{\mathbf{1}_s'\mathbf{W}_s^{-1/2}\mathbf{1}_s} = \frac{1}{N}\mathbf{1}_N' W^{1/2}\mathbf{1}_N, \tag{4.2.5}$$

that is, if the harmonic mean of the sample weights equals the mean of the population weights. This condition is implied by equation (4.2.4), if the model contains an intercept, that is, if $\mathbf{X}$ contains a column of ones. Also, to achieve (4.2.4) it is necessary only to focus on a single weighted sample *sum*, whereas the alternative requires adjusting the ratio of sample sums. Thus (4.2.4) is preferred.

When $\mathbf{W} = \mathbf{V}$, and the columns of $\mathbf{X}$ are integral powers of a single scalar x, through the Jth power, then $B(\mathbf{X}:\mathbf{W})$ is the weighted balance condition (3.3.1), considered earlier for the polynomial model.

The key point is that, when the working model is (4.2.1), a balanced sample with $\mathbf{W} = \mathbf{V}$ yields maximal efficiency under that model, as the following theorem shows:

Theorem 4.2.1 (Royall, 1992). Under $M(\mathbf{X}:\mathbf{V})$ if both $\mathbf{V}\mathbf{1}_N$ and $\mathbf{V}^{1/2}\mathbf{1}_N \in \mathcal{M}(\mathbf{X})$, then

$$\operatorname{var}_M[\hat{T}(\mathbf{X}:\mathbf{V}) - T] \geqslant [n^{-1}(\mathbf{1}_N'\mathbf{V}^{1/2}\mathbf{1}_N)^2 - \mathbf{1}_N'\mathbf{V}\mathbf{1}_N]\sigma^2.$$

The bound is achieved if and only if $s \in B(\mathbf{X}:\mathbf{V})$, in which case

$$\hat{T}(\mathbf{X}:\mathbf{V}) = n^{-1}(\mathbf{1}_N'\mathbf{V}^{1/2}\mathbf{1}_N)(\mathbf{1}_s'\mathbf{V}_{ss}^{-1/2}\mathbf{Y}_s).$$

Proof: Lemma 4.2.1 applies so that the variance in (4.2.5) is minimized when $\mathbf{a}'\mathbf{A}_s^{-1}\mathbf{a}$ is minimized, where $\mathbf{a} = \mathbf{X}'\mathbf{1}_N$. The assumption that $\mathbf{V}^{1/2}\mathbf{1}_N \in \mathcal{M}(\mathbf{X})$ means that there is a p-vector $\mathbf{c}_1$ such that $\mathbf{V}^{1/2}\mathbf{1}_N = \mathbf{X}\mathbf{c}_1$ and, since $\mathbf{V}$ is diagonal, this ensures that $\mathbf{V}_{ss}^{1/2}\mathbf{1}_s = \mathbf{X}_s\mathbf{c}_1$ for every sample s. From this it follows that $\mathbf{1}_s'\mathbf{V}_{ss}^{1/2}\mathbf{V}_{ss}^{-1}\mathbf{V}_{ss}^{1/2}\mathbf{1}_s = \mathbf{c}_1'\mathbf{X}_s'\mathbf{V}_{ss}^{-1}\mathbf{X}_s\mathbf{c}_1$ or $n = \mathbf{c}_1'\mathbf{A}_s\mathbf{c}_1$. Because $\mathbf{A}_s$ is invertible, both $\mathbf{A}_s$ and $\mathbf{A}_s^{-1}$ have symmetric square roots, that is, $\mathbf{A}_s^{-1} = \mathbf{A}_s^{-1/2}\mathbf{A}_s^{-1/2}$ and $\mathbf{A}_s = \mathbf{A}_s^{1/2}\mathbf{A}_s^{1/2}$. The desired inequality then follows from Schwarz's inequality (see Appendix A):

$$\begin{aligned} n(\mathbf{a}'\mathbf{A}_s^{-1}\mathbf{a}) &= (\mathbf{a}'\mathbf{A}_s^{-1}\mathbf{a})(\mathbf{c}_1'\mathbf{A}_s\mathbf{c}_1) \\ &= (\mathbf{a}'\mathbf{A}_s^{-1/2}\mathbf{A}_s^{-1/2}\mathbf{a})(\mathbf{c}_1'\mathbf{A}_s^{1/2}\mathbf{A}_s^{1/2}\mathbf{c}_1) \\ &\geqslant [(\mathbf{a}'\mathbf{A}_s^{-1/2})(\mathbf{A}_s^{1/2}\mathbf{c}_1)]^2 \\ &= (\mathbf{a}'\mathbf{c}_1)^2 \\ &= (\mathbf{1}_N'\mathbf{V}^{1/2}\mathbf{1}_N)^2. \end{aligned}$$

Equality occurs above if and only if $\mathbf{a}' = k\mathbf{c}_1'\mathbf{A}_s$ where $k = n^{-1}(\mathbf{1}_N'\mathbf{V}^{1/2}\mathbf{1}_N)$. This is equivalent to $s \in B(\mathbf{X}:\mathbf{V})$ because $\mathbf{c}_1'\mathbf{A}_s = \mathbf{1}_s'\mathbf{V}_{ss}^{-1/2}\mathbf{X}_s$. To obtain the reduced form of $\hat{T}$, note that $\hat{T}(\mathbf{X}:\mathbf{V}) = \mathbf{1}_N'\mathbf{X}\hat{\boldsymbol{\beta}}$ from Lemma 3.2.1, $\mathbf{1}_N'\mathbf{X} = (\mathbf{1}_N'\mathbf{V}^{1/2}\mathbf{1}_N)(\mathbf{1}_s'\mathbf{V}_{ss}^{-1/2}\mathbf{X}_s)/n$ since $s \in B(\mathbf{X}:\mathbf{V})$, $\mathbf{1}_s'\mathbf{V}_{ss}^{-1/2} = \mathbf{c}_2'\mathbf{X}_s'\mathbf{V}_{ss}^{-1}$ and $\mathbf{c}_2'\mathbf{X}_s' = \mathbf{V}_{ss}^{1/2}\mathbf{1}_s$ since $\mathbf{V}^{1/2}\mathbf{1}_N = \mathbf{X}\mathbf{c}_2$ for some p-vector $\mathbf{c}_2$. Making these substitutions yields

$$\begin{aligned} \hat{T}(\mathbf{X}:\mathbf{V}) &= \mathbf{1}_N'\mathbf{X}\hat{\boldsymbol{\beta}} \\ &= \frac{(\mathbf{1}_N'\mathbf{V}^{1/2}\mathbf{1}_N)}{n}\mathbf{1}_s'\mathbf{V}_{ss}^{-1/2}\mathbf{X}_s\mathbf{A}_s^{-1}\mathbf{X}_s'\mathbf{V}_{ss}^{-1}\mathbf{Y}_s \\ &= \frac{(\mathbf{1}_N'\mathbf{V}^{1/2}\mathbf{1}_N)}{n}(\mathbf{1}_s'\mathbf{V}_{ss}^{-1/2}\mathbf{Y}_s). \end{aligned}$$

The form of the error variance follows from the fact $\mathbf{a}'\mathbf{A}_s^{-1}\mathbf{a} = (\mathbf{1}_N'\mathbf{V}^{1/2}\mathbf{1}_N)^2/n$ at balance. □

We can also rewrite the reduced form of the *BLU* predictor under weighted balance in the simple form

$$\hat{T}(\mathbf{X}:\mathbf{V}) = \frac{N\bar{v}^{(1/2)}}{n}\sum_{i \in s}\frac{Y_i}{v_i^{1/2}}.$$

where $\bar{v}^{(1/2)} = \Sigma_{i=1}^N v_i^{1/2}/N$. This is just the mean-of-ratios estimator $\hat{T}_{MR}(v^{1/2})$ given in equation (3.3.5). Note also that Corollary 3.3.1 follows directly from Theorem 4.2.1.

Several other observations are in order:

1. The estimator $\hat{T}(\mathbf{X}:\mathbf{V}) = n^{-1}(\mathbf{1}_N'\mathbf{V}^{1/2}\mathbf{1}_N)(\mathbf{1}_s'\mathbf{V}_{ss}^{-1/2}\mathbf{Y}_s)$, which the *BLU* predictor reduces to under $\mathbf{B}(\mathbf{X}:\mathbf{V})$, does not depend on the particular $\mathbf{X}$ matrix. This implies that the *BLU* predictor for a particular model $M(\mathbf{X}:\mathbf{V})$ will remain *BLU* for an extended model $M(\mathbf{X}^*:\mathbf{V})$, $\mathbf{X}^* \supset \mathbf{X}$, if balance for the extended model holds, that is, if $s \in B(\mathbf{X}^*:\mathbf{V})$.
2. The variance bound $[n^{-1}(\mathbf{1}_N'\mathbf{V}^{1/2}\mathbf{1}_N)^2 - \mathbf{1}_N'\mathbf{V}\mathbf{1}_N]\sigma^2$, achieved under $B(\mathbf{X}:\mathbf{V})$, depends neither on the particular $\mathbf{X}$ matrix nor on the particular

sample. This implies that nothing is lost in efficiency by adding columns to $\mathbf{X}$, once the conditions of the theorem are met.

3. We note that, in the case of unweighted balance, somewhat stronger results on the reduction of $\hat{T}(\mathbf{X}:\mathbf{V})$ are possible:

Theorem 4.2.2 (Tallis, 1978). Under model $M(\mathbf{X}:\mathbf{V})$ [$\mathbf{V}$ not necessarily diagonal], $\hat{T}(\mathbf{X}:\mathbf{V}) = \hat{T}(1{:}1)$ if and only if $\mathbf{V}_{ss}\mathbf{1}_s \in \mathscr{M}(\mathbf{X}_s)$ and $s \in B(\mathbf{X}:\mathbf{I})$. The proof is left to Exercise 4.5.

4. The formulas for the *BLU* predictor and its lower bound are also found in design-based theory. Let the sample size be fixed at n. If the inclusion probability of unit i is $\pi_i \propto v_i^{1/2}$, that is, $\pi_i = nv_i^{1/2}/\mathbf{1}'_N\mathbf{V}^{1/2}\mathbf{1}_N$, then the *BLU* predictor $\hat{T}(\mathbf{X}:\mathbf{V}) = n^{-1}(\mathbf{1}'_N\mathbf{V}^{1/2}\mathbf{1}_N)(\mathbf{1}'_s\mathbf{V}_{ss}^{-1/2}\mathbf{Y}_s)$ is the Horvitz-Thompson estimator $\Sigma_s y_i/\pi_i$. The variance bound is the one established by Godambe and Joshi (1965, Theorem 6.1) for the model-based expectation of the design-based variance of the Horvitz-Thompson estimator. Isaki and Fuller (1982) show that this bound is approached asymptotically by a estimator based on a regression model that includes the standard deviations and variances (of the Y's given the x's) in the column space of the $\mathbf{X}$ matrix, when the inclusion probabilities are proportional to the standard deviations.
5. It should be noted that when $\mathbf{V} \neq \mathbf{I}$, balance is not necessarily achievable (even approximately). This follows from the expression for the variance at balance, which implies that, at balance, $n^{-1}(\mathbf{1}'_N\mathbf{V}^{1/2}\mathbf{1}_N)^2 - \mathbf{1}'_N\mathbf{V}\mathbf{1}_N \geqslant 0$ or, equivalently, $n \leqslant N\{\bar{v}^{(1/2)}\}^2/\bar{v}$. This upper bound is less than N, unless the v's are constant. Thus for n large enough, balance will be unattainable. As noted in Section 3.4.7, the basic strategy in such circumstances will be to "stratify off" population elements corresponding to the largest values of $\mathbf{V}$, and balance on the remainder. The condition $\pi_i = nv_i^{1/2}/(N\bar{v}^{(1/2)}) \leqslant 1$ for all i implies that $n \leqslant N\{\bar{v}^{(1/2)}\}^2/\bar{v}$ (see Exercise 4.2), so that we know that root(v) balance is possible whenever *aiming* at it through *pps* sampling is possible.
6. A natural consequence of Theorem 4.2.1 is the idea of *minimal models for balance for given variance matrix* $\mathbf{V}$ or just *minimal models* (see Section 3.3). Given (suspicion of) a particular variance structure, we want to guarantee that the conditions of the theorem—both $\mathbf{V}\mathbf{1}_N$ and $\mathbf{V}^{1/2}\mathbf{1}_N \in \mathscr{M}(\mathbf{X})$—are met. The *minimal model* to accomplish this has the matrix of auxiliaries $\mathbf{X}_v = (\mathbf{V}^{1/2}\mathbf{1}_N, \mathbf{V}\mathbf{1}_N)$. The sample is taken from the class of *root*($\mathbf{V}$) balanced samples, $B(\mathbf{X}:\mathbf{V})$. Some elementary examples of minimal models were already discussed in Section 3.3.2. It should be noted that a minimal model is just that—*minimal*; we may well want to include additional columns that we suspect are appropriate, for example, a column corresponding to an intercept or to an additional auxiliary variable.

7. If now we reconsider the "multiple Y" situation described in Section 3.2.6, we realize that there is, as it were, a conflict of interest among the variables. Totals for the Y's with variance proportional to x, are best estimated having taken a *root*(x) balanced sample, while for the constant variance Y, we would prefer to use simple balance. Resolution of the conflict will depend on circumstance. If one of the variables is of greater import than the others, we might use the type of balance suited to *its* likely variance structure and sacrifice efficiency on the others; if the variables are of equal importance, then an intermediate sort of balance such as root($v^{1/2}$) might be appropriate, to minimize efficiency loss overall. This is a largely unexplored area.
8. Simple balance between sample and population means implies balance between *sample* and *nonsample* means; that is, $\bar{x}_s^{(j)} = \bar{x}^{(j)} \Rightarrow \bar{x}^{(j)} = \bar{x}_r^{(j)}$; this is not true in general, and an alternative is to work with a type of non-sample balance $B_r(\mathbf{X}: \mathbf{V})$ defined by $\Sigma_s v_i^{-1/2} x_i^j / \Sigma_s v_i^{-1/2} = N_r^{-1} \Sigma_r x_i^j$. A type of balance involving nonsample moments was studied by Scott, Brewer, and Ho (1978). The theorem shows that, for purposes of efficiency, weighted balance with respect to the *population* is preferable when the variance is correctly specified. Given the likelihood of misspecification of the variance, it becomes less clear which will be better in specific situations (see Exercise 4.6).
9. For Y variables that follow nonlinear models (see Section 11.2) , a theory of balanced samples has yet to be developed. Balance conditions for zero-one Y variables that follow logistic or similar models may be much different from (4.2.5) or may be impossible to derive at all.
10. For goals beyond the total or mean of Y, for example quantiles, balance may not suffice to guarantee bias-robustness. See section 11.4.

4.2.2. Examples of Minimal Models

For completeness, we consider again the examples of minimal models discussed earlier in Chapter 3. We extend earlier notation to allow $\bar{x}_s^{(j)}$ to cover non-integer values of j. $M(x^j, x^k : x^l)$ will denote a model with $E_M(Y) = \beta_j x^j + \beta_k x^k$ and $\text{var}_M(Y) = \sigma^2 x^l$, where j, k, and l may not be integers. The BLU under $M(x^j, x^k : x^l)$ will be $\hat{T}(x^j, x^k, : x^l)$.

Example 4.2.1 Variance proportional to x^2

If the variance is proportional to x^2, the minimal working model is $M = M(0, 1, 1 : x^2)$, with corresponding estimator $\hat{T}(0, 1, 1 : x^2)$. The condition $\mathbf{V1}_N$ and $\mathbf{V}^{1/2}\mathbf{1}_N \in \mathscr{M}(\mathbf{X})$ are met since *ipso facto* both x and x^2 are in the model for $E_M(Y)$. The lower bound on the variance is

$$\left[\frac{(N\bar{x})^2}{n} - \sum_{i=1}^{N} x_i^2\right]\sigma^2. \tag{4.2.6}$$

This bound is achieved under $M(0, 1, 1: x^2)$ in any balanced sample with $\bar{x}_s = \bar{x}^{(2)}/\bar{x}$. Bias protection against more general polynomial models is obtained at no cost in efficiency under the working model by balancing on additional powers:

$$\bar{x}_s^{(j-1)} = \bar{x}^{(j)}/\bar{x} \text{ for } j = 0, 2, 3, \ldots, J. \tag{4.2.7}$$

We refer to such balance as *root*(x^2) balance or just x-balance (of order J). It is also known in the literature as π-balance (Cumberland and Royall, 1981).

With these balance constraints, the *BLU* predictor, using the formula in Theorem 4.2.1, reduces to $\hat{T}_{MR}(x) = N\bar{x}\Sigma_s Y_i/(nx_i)$, the mean-of-ratios estimator with weights x^{-1}—a result first derived by Kott (1984). This estimator is in common use: it is the Horvitz-Thompson estimator for $\pi_i \propto x_i$; it approximates $\hat{T}(0, 1: x^2) = n\bar{Y}_s + (N - n)\bar{x}_r\Sigma_s Y_i/(nx_i)$ when the sampling fraction is small. However, the estimator $\hat{T}(0, 1, 1: x^2)$, given in (3.3.6) or (3.3.7), is preferable, because in practice *exact* x-balance may be hard to achieve, and the more complicated version will be less sensitive to deviations from balance. (Software for S-PLUS™ to calculate $\hat{T}(\mathbf{X}: \mathbf{V})$ for general $\mathbf{X}$ and diagonal $\mathbf{V}$ is included in Appendix C.)

Example 4.2.2 Variance Proportional to x

If the variance is proportional to x, the minimal working model corresponds to $M(x^{1/2}, x: x)$ with *BLU* estimator $\hat{T}(x^{1/2}, x: x)$. The lower bound on the variance is

$$\left[\frac{(N\bar{x}^{(1/2)})^2}{n} - \sum_{i=1}^{N} x_i\right]\sigma^2. \tag{4.2.8}$$

This bound is achieved under $M(x^{1/2}, x: x)$ in any balanced sample with $\bar{x}_s^{(1/2)} = \bar{x}/\bar{x}^{(1/2)}$. Bias protection against more general polynomial models is obtained by balancing on additional powers:

$$\bar{x}_s^{(j-1/2)} = \bar{x}^{(j)}/\bar{x}^{(1/2)} \text{ for } j = 0, 1, 2, \ldots, J. \tag{4.2.9}$$

As in Section 3.3, we refer to such balance as *root*(x) balance (of order J) or $x^{1/2}$-balance. Under balance, the estimator $\hat{T}(x^{1/2}, x: x)$ reduces to $\hat{T} = (N/n)\bar{x}^{(1/2)}\Sigma_s Y_i/x_i^{1/2}$. It has already been shown that this estimator surpasses in efficiency the ratio estimator $\hat{T}_R = \hat{T}(0, 1: x) = N\bar{Y}_s\bar{x}/\bar{x}_s$ most often associated with estimation under the variance structure $v_i = x_i\sigma^2$.

4.3 COMPARISONS USING AN ARTIFICIAL POPULATION

The theoretical results of Section 4.2 raise a variety of questions. The "minimal model" may require the addition of a gratuitous term. For example, if one is reasonably sure that the model $M(0, 1{:}x)$ underlying the ratio estimator is correct, but still wants to both protect against bias and achieve reasonable efficiency, then Theorem 4.2.1 leads to adding an $x^{1/2}$ term to the model (and using a $x^{1/2}$-balanced sample). How much efficiency is lost by adding this term to the original working model? In general, how much is lost in efficiency by overparameterizing? What is the loss if a balanced sample is chosen based on the preconception that the variance is v, but v^* is the actual variance? If evidence of data in the sample is for variance structure v^*, but the sample is *root*(v) balanced, should we then switch to an estimator based on v^* or stick with v? Since a typical population has variance v monotonically increasing in x, is seeking unweighted balance (and using the well-known ratio estimator) ever warranted as a typical procedure? What should we do if sample size is sufficiently large that weighted balance with respect to v is impossible? What if, as often happens in practice, the sample has *already* been chosen, is not among a class of balanced samples, and we are required to draw inference?

Toward the end of understanding the interplay among a working model, an underlying model, and a given sample, we undertake a numerical exploration. Its purpose is twofold: (1) to give suggested answers to the above questions, and (2) to serve as a paradigm for the kind of investigation someone planning a survey might wish to carry out, based on knowledge of a given population's auxiliary variable and on reasonable suspicions about underlying models and variance structure. Nothing is calculated that requires data from the sample. Thus we have the opportunity to test beforehand a range of samples, and accompanying estimators, under a variety of reasonable assumptions regarding the working and background models. In what follows we will limit ourselves to one mildly complex underlying model, a cubic.

As the x-variate, we take the x variable ("number of beds") for the Hospitals population that we have already considered several times. The intent is to have a spread of x-values such as we might find in practice. To fix ideas, assume we have the model $M(1, 1, 0, 1{:}x^2)$ that lacks a quadratic term. Specifically, suppose that the dependent variable follows the law

$$Y_i = 50 + x_i + 10^{-6}x_i^3 + 0.2x_i\varepsilon_i \qquad (4.3.1)$$

with $\varepsilon_i \sim (0, 1)$ and independent. Thus, the ith row of $\mathbf{Z}$ is $(1, x_i, x_i^3)$, the true underlying parameter is $\boldsymbol{\beta} = (50, 1, 10^{-6})$, and the true variance is $\mathbf{W} = 0.04diag(x^2)$. It may verified that the median coefficient of variation, defined as $\text{median}(w_i^{1/2})/\text{median}[E_M(Y_i)]$, is about 0.16, in the midrange of what we would expect in an actual population (see Carroll and Ruppert, 1988, for a discussion of this measure). Figure 4.1 is a scatterplot of a population generated from model (4.3.1) using $N(0, 1)$ errors. However, note that for the

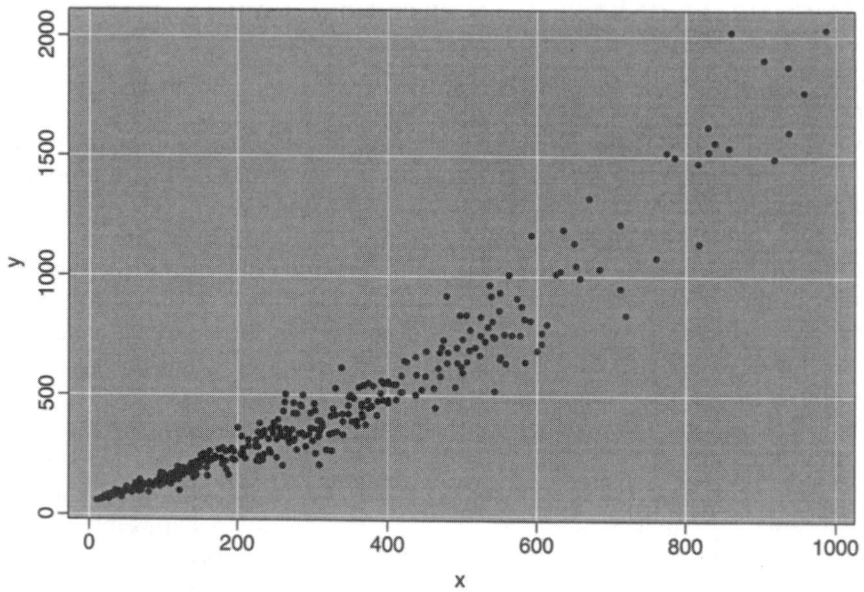

Figure 4.1 Example of a population generated from model (4.3.1) using x's from the Hospitals population and normally distributed errors.

analytic calculations of biases and variances below, we need only for the errors to have mean 0 and variance 1; normality is unnecessary.

For a given working model $M(\mathbf{X}:\mathbf{V})$, the estimator $T(\mathbf{X}:\mathbf{V})$, being linear, can be written in the form $\hat{T} = \mathbf{1}'_s Y_s + \mathbf{a}'Y_s$ with $\mathbf{a}$ dependent on $\mathbf{X}$ and $\mathbf{V}$. The bias will be $(\mathbf{a}'\mathbf{Z}_s - \mathbf{1}'_r\mathbf{Z}_r)\boldsymbol{\beta}$ and the error variance $\mathbf{a}'\mathbf{W}_s\mathbf{a} + \mathbf{1}'_r\mathbf{W}_r\mathbf{1}_r$ for an underlying model $M(\mathbf{Z}:\mathbf{W})$. We can thus analytically ascertain bias, standard deviation, and root mean square error of a given estimator on a given sample under a hypothesized true model without recourse to simulations. Thus, we have a relatively quick, cheap tool of exploration.

Sample size is fixed at $n = 40$ in what follows. Three broad scenarios are imagined, corresponding to sampling aimed at (1) simple unweighted balance, (2) $x^{1/2}$-balance, and (3) x-balance, respectively. The last would have been the correct choice for model (4.3.1), given knowledge of $\mathbf{W}$ (up to a constant multiplier.) Thus we generated 100 samples each for the systematic equal probability, systematic $pp(x)$, and systematic $pp(x^{1/2})$ sampling schemes, as described in Section 3.4. Samples were ordered by their corresponding value of $\bar{x}_s$ in the case of the scenario aiming at simple balance, and by the moments $1/\bar{x}_s^{(-1/2)}$ and $1/\bar{x}_s^{(-1)}$ in the latter two scenarios, respectively. (Recall that for a model containing an intercept term, $x^{1/2}$-balance requires $1/\bar{x}_s^{(-1/2)} = \bar{x}^{(1/2)}$, and likewise x-balance requires $1/\bar{x}_s^{(-1)} = \bar{x}$).

From the 100 samples, we chose a few samples, representing various degrees of achievement of the sort of balance sought. In particular, we chose the

Table 4.1a. Sample Moments for Five Equal Probability Systematic Samples ($n = 40$) from Hospitals Population, with $\bar{x} = 275$, $\bar{x}^{(2)} = 120{,}805$, and $\bar{x}^{(3)} = 67{,}736{,}501$

Quantile of $\bar{x}_s$	$\bar{x}_s$	$\bar{x}_s^{(2)}/10^3$	$\bar{x}_s^{(3)}/10^6$
10%	264.4	110.8	57.9
25%	269.4	114.9	61.6
50%	275.0	121.4	68.5
75%	281.1	127.2	74.0
90%	282.8	129.0	75.9

samples corresponding to the 10th, 25th, 50th, 75th, and 90th quantiles of $\bar{x}_s$, $1/\bar{x}_s^{(-1/2)}$, and $1/\bar{x}_s^{(-1)}$ adding a sixth balanced sample, in the case of $x^{1/2}$-balance and x-balance. The median sample sufficed for unweighted balance. Tables 4.1a–c give the first several moments for each of the samples in the several scenarios, and the corresponding population quantities. In Table 4.1a, for example, the 25th percentile of $\bar{x}_s$ in the 100 equal probability samples was 269.4. For the sample having this value of $\bar{x}_s$, other sample moments were $\bar{x}_s^{(2)} = 114.9 \times 10^3$ and $\bar{x}_s^{(3)} = 61.6 \times 10^6$. Tables 4.1b and 4.1c are read similarly.

4.3.1 Results for Probability Proportional to x Sampling and x-Balance

For each scenario, an array of estimators, of varying degrees of appropriateness, were tried. Since x-balance would have been the best choice here (the variance being proportional to x^2), we begin with this set of results. The estimators tried were:

Table 4.1b. Sample Moments for Six Probabilility Proportional to $x^{1/2}$ Systematic Samples ($n = 40$) from Hospitals Population, with $1/\bar{x}^{(1/2)} = .06558$, $\bar{x}/\bar{x}^{(1/2)} = 18.016$, $\bar{x}^{(2)}/\bar{x}^{(1/2)}) = 7{,}922.9$, $\bar{x}^{(3)}/\bar{x}^{(1/2)} = 4{,}442{,}453$.

Quantile of $\bar{x}_s^{(-1/2)}$	$\bar{x}_s^{(-1/2)}$	$\bar{x}_s^{(1/2)}$	$\bar{x}_s^{(3/2)}$	$\bar{x}_s^{(5/2)}/10^3$
10%	0.0676	17.76	7628	4152
25%	0.0672	17.82	7711	4240
42%	0.0656	17.98	7881	4407
50%	0.0654	18.02	7935	4450
75%	0.0641	18.21	8151	4686
90%	0.0633	18.30	8256	4789

The 42th percentile sample is one for which $\bar{x}_s^{(-1/2)} \approx 1/\bar{x}^{(1/2)}$.

Table 4.1c. Sample Moments for Six Probability Proportional to x Systematic Samples ($n = 40$) from Hospitals Population, with $1/\bar{x} = .00364$, $\bar{x}^{(2)}/\bar{x} = 439.77$, $\bar{x}^{(3)}/\bar{x} = 246{,}586$

Quantile of $\bar{x}_s^{(-1)}$	$\bar{x}_s^{(-1)}$	$\bar{x}_s$	$\bar{x}_s^{(2)}/10^3$
10%	0.00443	429.2	236.4
25%	0.00390	433.3	239.8
38%	0.00364	437.4	244.3
50%	0.00348	439.6	245.9
75%	0.00333	444.7	251.1
90%	0.00327	449.5	256.4

The 38th percentile sample is one for which $x_s^{(-1)} \approx 1/\bar{x}$.

(a) $\hat{T}(1, 1, 0, 1\!:\!x^2)$ the *BLU* estimator corresponding to the true underlying model;

(b) $\hat{T}(0, 1, 1\!:\!x^2)$ the *BLU* estimator corresponding to the minimal model for variance proportional to x^2;

(c) $\hat{T}_{MR}(x) = N\bar{x}\Sigma_s Y_i x_i^{-1}/n$ the estimator to which the minimal estimator reduces at x-balance for $j = 0, 2$;

(d) The weighted expansion estimator $\hat{T}_0(x) = N\Sigma_s Y_i x_i^{-1}/\Sigma_s x_i^{-1}$. At x-balance this estimator and the minimal estimator are also equal. When the sampling fraction is small, this estimator closely approximates, but is not identical to, $\hat{T}(1\!:\!x)$ (cf. Remark (9) in Section 4.2.1). This estimator is also known as a Hájek estimator (Hájek, 1971).

(e) $\hat{T}(1, 1, 1, 1\!:\!x^2)$, which includes the true underlying model as a special case, the quadratic term being gratuitous; and

(f) $\hat{T}(1, 1, 0, 1\!:\!x)$, which matches the underlying model, except that the variance is misspecified.

Table 4.2 gives the bias, standard deviation, and root mean square error (*rmse*) for the minimal estimator (b) obtained by evaluating the analytic formulas for the six samples. Two facts are at once patent: (1) the standard deviation remains virtually constant across the samples; (2) the bias is smallest in magnitude at balance (the third sample) and grows quite sizable as the samples deviate from balance, to the point where the *rmse* is 44% higher at the lowest quantile than at balance. It is thus clear that, using the minimal estimator, we cannot rest content with *aiming* at balance through an appropriate sampling design, but want our sample to *achieve* balance.

Figure 4.2 gives the absolute value of the bias, standard deviation, and *rmse* for the 6 estimators (a)–(f), for each of the six $pp(x)$ samples. Results are listed down the left of the figure by estimator and sample. The bars in the left panel

Table 4.2. Standard Deviation, and Root Mean Square Error for Minimal Model Estimator $T(0, 1, 1 : x^2)$ Under Cubic Model, for Six Probability Proportional to x Samples

	10%	25%	$\bar{x}_s^{(-1)} \approx 1/\bar{x}$	50%	75%	100%
Bias	3258	864	−151	−910	−1494	−1525
Standard deviation	3127	3125	3124	3123	3124	3127
rmse	4516	3242	3127	3253	3463	3479

show the absolute value of the bias. The middle panel represents the standard deviation of each estimator/sample combination. The right-hand panel shows the *rmse*'s. Note that the second cluster depicts the same data as Table 4.2 for the minimal estimator.

One overall observation is that there is remarkable stability in the standard deviation across all estimators and all samples, with the exception of the weighted expansion estimator (d). Thus differences in *rmse*, which can be severe, are solely due to differences in bias. Very little difference from the optimal estimator (a) can be discerned for estimators (e) and (f); that is, adding an extra parameter to the model, or getting the variance structure alone wrong, does not seem really much to matter, so long as the bias is limited. The differences between the minimal estimator (b) and the mean of ratios estimator (c) are minor, suggesting that adherence to the traditional, simpler estimator would be all right in this instance, so long as the sample remains close to x-balance. However, the weighted expansion estimator (d), labeled Hájek(x) in Figure 4.2, shows egregious biases even in samples a small way removed from balance, and should be *avoided*.

It should perhaps be noted at this point, that the relative size of the bias and standard error is dependent on the particular values of the parameters $\boldsymbol{\beta} = (50, 1, 10^{-6})$, and $\sigma^2 = 0.04$ that we happen to have selected; for example, if, other things remaining the same, σ^2 were quadrupled, then the standard deviation of the estimators of T would double, and the relative impact of the bias would be reduced. On the other hand, other things being equal, the relative size of the bias will increase if the sample size is increased. In the given parametrization, the (hypothetical) population total would be 127,606; thus the worst bias in Table 4.2, for example, would represent a relative bias of only 2.6%; nonetheless, as a large component of the *rmse*, it would have to be considered worrisome.

4.3.2 Results for Probability Proportional to $x^{1/2}$ Sampling and $x^{1/2}$-Balance

Figure 4.3 gives the analytic results for the samples obtained from sampling proportional to $x^{1/2}$, that is, aiming at $x^{1/2}$-balance. The estimators in this situation were:

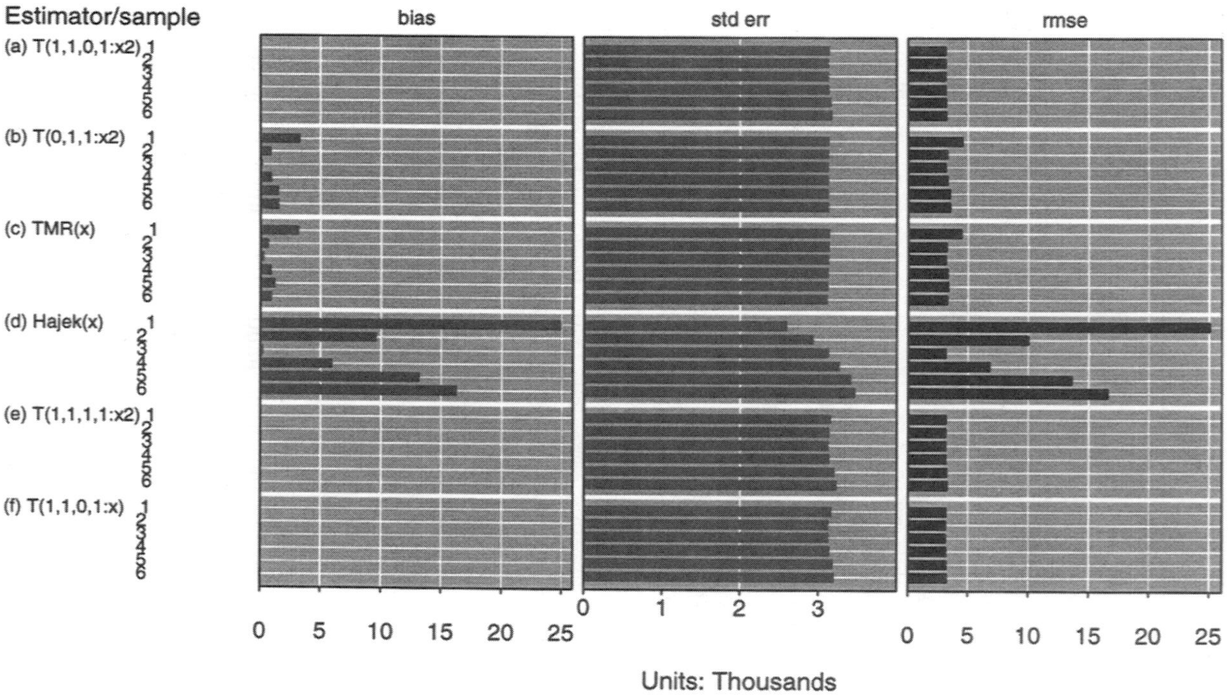

Figure 4.2 Comparisons of the absolute value of bias, standard error, and *rmse* of six estimators for six *ppx* samples with different degrees of $root(x^2)$ balance.

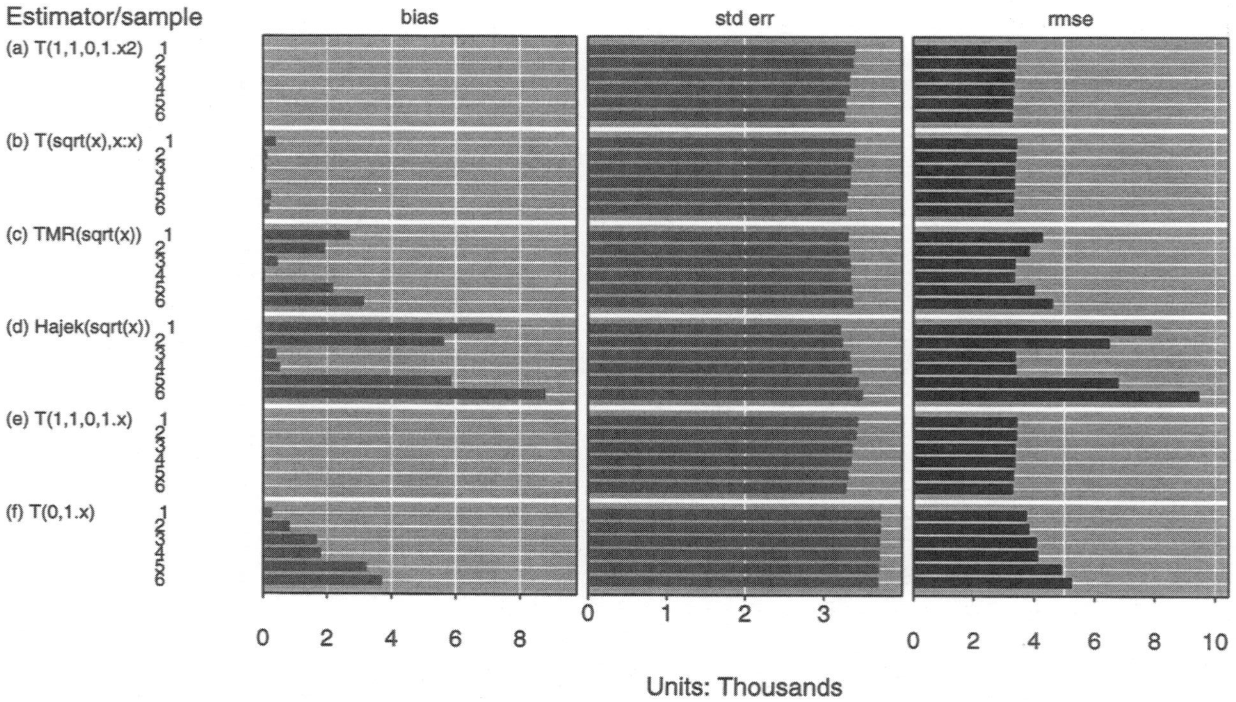

Figure 4.3 Comparisons of the absolute value of bias, standard error, and *rmse* of six estimators for six $pp(x^{1/2})$ samples with different degrees of $root(x)$ balance.

(a) $\hat{T}(1, 1, 0, 1 : x^2)$, the *BLU* estimator corresponding to the true underlying model.

(b) $\hat{T}(x^{1/2}, x : x)$, the *BLU* estimator corresponding to the minimal model on assumption of variance proportional to x.

(c) $\hat{T}_{MR}(x^{1/2}) = N\bar{x}^{(1/2)}\Sigma_s Y_i x_i^{-1/2}/n$, the estimator to which the minimal estimator reduces at $x^{1/2}$-balance according to Theorem 4.2.1; also another type of mean-of-ratios (or Horvitz-Thompson) estimator similar to $\hat{T}_{MR}(x)$.

(d) The weighted expansion estimator $\hat{T}_0(x^{1/2}) = N\Sigma_s x_i^{-1/2} Y_i / \Sigma_s x_i^{-1/2}$; this estimator, with the balance condition defined by (4.1.2-3) for $j = 1$, reduces to the same value as the minimal estimator at $x^{1/2}$-balance; $\hat{T}_0(x^{1/2})$ is also approximated by $\hat{T}(1 : x^{1/2})$ when the sampling fraction is small and is the Hájek estimator in $pp(x^{1/2})$ sampling.

(e) $\hat{T}(1, 1, 0, 1 : x)$, the estimator based on correct model for $E_M(Y)$, with (incorrect) variance that was assumed for sampling.

(f) $\hat{T}(0, 1 : x)$, the ratio estimator, which has variance in keeping with the sampling choice, but is otherwise inappropriate, being based neither on the correct model, nor on considerations of balance in this scenario.

We note again a relative constancy in Figure 4.3 across samples and estimators of the standard deviation, which roughly hovers between 3300 and 3400, about a 5% loss compared to that in the probability proportional to x samples—not very severe. There is very little difference in *rmse* between the completely correct estimator in (a) and the estimator (e) which uses the correct model, but with misspecified variance structure. The minimal estimator (b) behaves well at balance, and is not as affected by deviation from balance as was the minimal estimator in the x-sampling case. This suggests that a routine of sampling proportional to $x^{1/2}$, and using the minimal estimator $\hat{T}(x^{1/2}, x : x)$ might be a reasonable all-purpose tool. The two estimators (c) and (d) to which the minimal estimator reduces are extremely sensitive to deviation from balance, especially the weighted expansion estimator denoted Hájek(sqrt(x)) in Figure 4.3, and are, again, to be *avoided*. The ratio estimator has, as we would expect, large biases here, and reveals itself as not being an all purpose tool.

4.3.3 Results for Equal Probability Sampling and Simple Balance

Finally, at simple unweighted balance and its "nearby samples," we used the estimators:

(a) $\hat{T}(1, 1, 0, 1 : x^2)$, the correct estimator;

(b) $\hat{T}(0, 1 : x)$, the ratio estimator, suitable under simple balance;

(c) $\hat{T}(1 : 1)$, the expansion estimator to which the ratio estimator reduces at simple balance;

(d) $\hat{T}(1, 1, 0, 1{:}1)$, the estimator with the correct model for $E_M(Y)$, but constant variance; and

(e) $\hat{T}_{MR}(x)$, the mean of ratios estimator for which no rationale exists under unweighted balance, but which someone, convinced of the correct variance structure, might be tempted to use as analogous to the ratio estimator.

Figure 4.4 shows the analytic results for the five samples in Table 4.1a. We here note a certain amount of variation in the standard deviation for the estimators (a) and (d) that use the correct model for $E_M(Y)$: as the sample gets farther from unweighted balance in the direction of x-balance, the standard deviation comes down. This may be taken as a measure of the distance unweighted balance is from desirable in the variance proportional to x^2 case. As we might anticipate, the ratio estimator (b), and its equivalent-at-balance, the ordinary expansion estimator (c) do well at balance, having low bias, but show marked bias, especially severe for the expansion estimator, in samples that deviate from balance even to a minor degree. The "off-sample" mean of ratios estimator (e) shows severe biases for all samples.

Before ending this section, we consider the case where weighted balance becomes impossible because of large sample size n. We noted in Section 3.4.7 and in comment (5) following Theorem 4.2.1—that if a probability proportional to $v^{1/2}$ sampling scheme is possible, then a $v^{1/2}$-balanced sample is also possible. This suggested the preliminary step, in aiming at $v^{1/2}$-balance, of checking the values of $\pi_i = nv_i^{1/2}/\Sigma_{i=1}^{N} v_i^{1/2}$. Units i with $\pi_i > 1$ should be included in the sample as "certainties"; they contribute to the T_s component of $\hat{T}$ but not to $\hat{T}_r$. Say there are m such units. This leaves a reduced population P', of size $N' = N - m$, on which to balance and a reduced sample size of $n' = n - m$. An S-PLUS™ function, *weed.high* (see Appendix C) has been included in the software accompanying this book to identify such points. Note that iteration may be necessary, since the denominator of $\pi_i = n'v^{1/2}/\Sigma_{i=1}^{N'} v^{1/2}$ shrinks for the reduced population, and additional values of i can yield $\pi_i > 1$.

Having "shrunk" the population to P', one then chooses a balanced sample of size $n' = n - m$ from P'. A natural question is how much is lost by not including the certainties in estimating parameters of the model. Results for the x-balance scenario using the same underlying cubic model as above, with a sample size $n = 150$ (recall $N = 393$ for the Hospitals population), suggest, that little is lost in *not* including the certainties. The 19 largest x's are certainties, leaving a population of size 374, on which to balance. Table 4.3 gives the x-moments of the population, and the corresponding sample moments for a selected sample s' of size 131. By contrast the harmonic mean of the whole ($n = 150$) sample is 269 < 274, the mean of the entire population, and likewise the higher sample moments are less than the corresponding population expressions.

Table 4.4 gives the bias and *rmse* of $\hat{T}(1, 1, 0, 1{:}x^2)$, that is, the estimator based on the correct model, and $\hat{T}(0, 1, 1{:}x^2)$, the minimal estimator. Rows 1

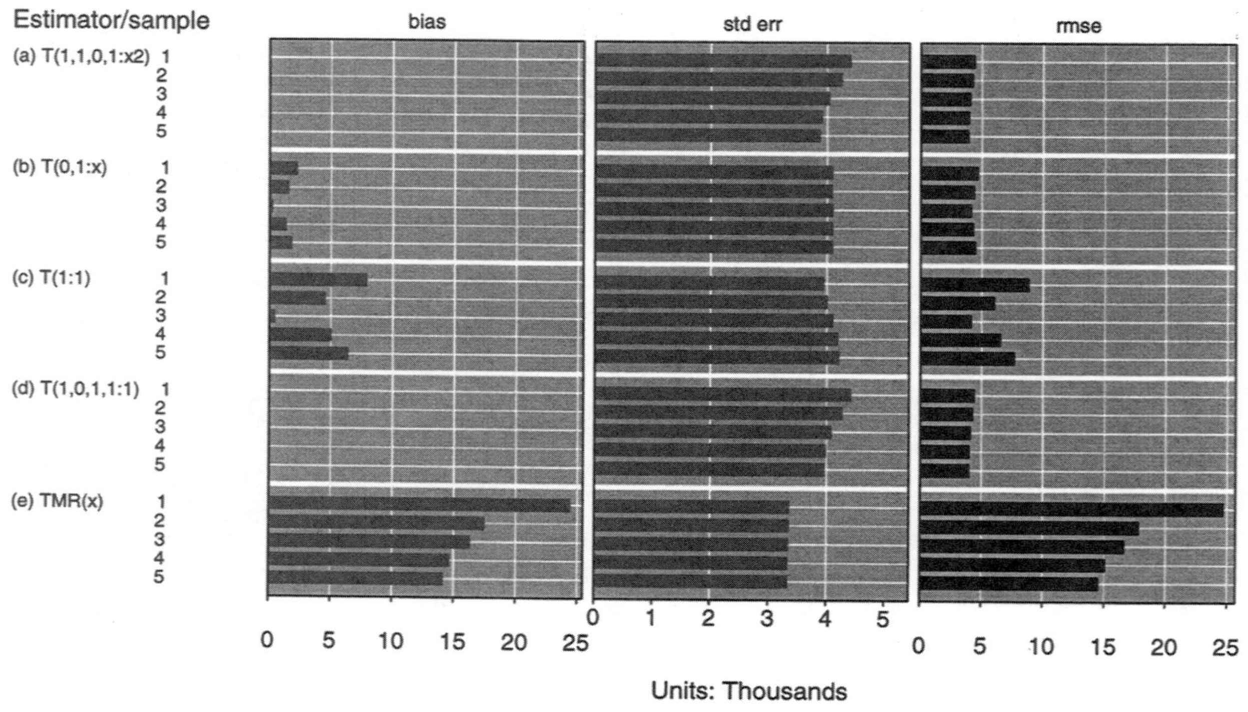

Figure 4.4 Comparisons of the absolute value of bias, standard error, and *rmse* of six estimators for five systematic, equal probability samples with different degrees of simple balance.

Table 4.3. Moments of Partial Sample s' Compared to Partial Population P' Quantities

	$\bar{x}'$ versus $1/\bar{x}_s^{\prime(-1)}$	$\bar{x}'^{(2)}/\bar{x}'$ versus $\bar{x}'_s$	$\bar{x}'^{(3)}/\bar{x}'$ versus $\bar{x}_s^{\prime(2)}$
P'-value	246.0	369.1	163,747
s'-value	245.3	368.2	163,135

and 2, labeled "full sample," give results for estimates in which all the data in sample s is used to estimate the parameters of the model; in rows 3 and 4, labeled "partial sample," only the $n' = 131$ noncertainty points are used to make estimates.

It is readily seen that there is little difference between results for the full and partial sample estimator $\hat{T}(1, 1, 0, 1 : x^2)$ that uses the correct model. On reflection, this is not too surprising, since the points omitted in the partial estimator, are precisely those which will, under a working model with variance proportional to x, be severely downweighted in calculating the full estimator. Since, at balance on P', the minimal estimator can be expected to approximate the correct model estimator using partial data, little is lost by using the partial sample based $\hat{T}(0, 1, 1 : x^2)$. But it is important to note that, since the full sample s is not balanced with respect to P, this estimator is no longer bias-robust if *all* the sample data are used in determining $\hat{T}_r$. A certainty stratum under a given working model provides protection against the model's failing, when balance on the whole cannot be achieved.

Table 4.4. Bias and Root Mean Square Error (*rmse*) of Two Estimators for Partially Balanced Sample in Cubic Model with Hospitals x-Variable for a Sample of $n = 150$, "Partially Balanced" on Subsample of $n' = 131$

	Bias	*rmse*
(a) *Estimator based on full sample of* $n = 150$		
$\hat{T}(1, 1, 0, 1 : x^2)$	0	1096.2
$\hat{T}(0, 1, 1 : x^2)$	−657.8	1284.7
(b) *Estimator based on partial sample excluding 19 certainties*		
$\hat{T}(1, 1, 0, 1 : x^2)$	0	1106.4
$\hat{T}(0, 1, 1 : x^2)$	−15.4	1107.4

4.4 SAMPLE SIZE DETERMINATION

To this point, Chapter 4 has dealt with the question of efficiency and the idea of protection against bias through balancing. Comparisons between estimators or sample designs assumed a given sample size. We now address an important question which arises when designing any sample: how big should the sample be?

The answer depends on many factors, particularly the goals of the survey and the budget. A survey rarely has a single goal or estimate that is of primary importance. For example, in a survey to study the health status of the population in a geographic area, data may be collected to estimate the percentage of the population having one or more chronic health problems like rheumatoid arthritis, lung disease, or osteoporosis. Blood pressure and serum cholesterol levels may also be measured on sample persons. Estimates may be desired for the full population but also for subgroups or domains defined by age, ethnicity, and sex.

One way of proceeding is to decide what level of precision, as measured by the standard error, is needed to permit analyses or decisions to be made using the sample data. Calculations can then be made to find the sample size that will satisfy the estimation goals for the full population and the domains. When more than one target variable is involved, the calculations may lead to conflicting values of n—one for each variable. The total sample size may also conflict with the amount of money available and some realignment of goals and budget will be required. Chapter 6 will cover the situation in which survey costs differ among strata. Cochran (1977) and Groves (1989) also discuss these issues at length.

We here consider the case of a single Y variable. A useful statistic in sample size calculations is the *coefficient of variation* (cv), defined as the standard error of an estimator divided by the expected value of the quantity being estimated. In the case of an estimator $\hat{T}$ of the total, the cv is

$$cv(\hat{T}) = \frac{\sqrt{\mathrm{var}_M(\hat{T} - T)}}{E_M(T)}. \tag{4.4.1}$$

The square of the cv is known as the relative variance or *relvariance*. Expression (4.4.1) is also often multiplied by 100 and expressed as a percentage. The cv is a unitless measure because the units of the numerator and denominator cancel out. If, for example, Y is in dollars, both $\sqrt{\mathrm{var}_M \hat{T} - T)}$ and $E_M(T)$ are also. Because of its lack of units, the cv is a good criterion for establishing precision goals in different situations. An estimator whose cv is 50% or 100% is obviously fairly useless in pinning down the actual value. Coefficients of variation of 10% or less are usually considered to be reasonably precise, but how small an acceptable value is depends on the uses of the survey data.

One method of determining the required sample size is to set the cv to a specified value k and then solve for the value of n that will produce $cv(\hat{T}) = k$.

Example 4.4.1 Homoscedastic Case. Suppose that the working model is $M(1:1)$, we use the expansion estimator $\hat{T}_0 = \hat{T}(1:1)$ and, to protect against bias under the model $M(\delta_0, \delta_1, \ldots, \delta_J : v)$, we select a balanced sample $s(J)$. Under the working model $M(:1)$,

$$\text{var}_M(\hat{T}_0 - T) = N\left(\frac{N}{n} - 1\right)\sigma^2$$

and

$$[cv(\hat{T}_0)]^2 = \left(\frac{N}{n} - 1\right)\frac{\sigma^2}{N\mu^2}. \tag{4.4.2}$$

Setting (4.4.2) to a desired value k^2 and solving for n gives

$$n = \left[k^2\frac{\mu^2}{\sigma^2} + \frac{1}{N}\right]^{-1}. \tag{4.4.3}$$

Evaluation of (4.4.3) requires several quantities: the desired *cv* k, the population mean μ and variance σ^2, and the population size N. The *cv* k we set based on the survey goals while the others require some knowledge either from previous surveys or work with similar populations. Sometimes a small preliminary or "test" survey will be in order. If a complete frame is available, N will be known. If an earlier survey has been conducted, for the parameters μ and variance σ^2, we can use that survey's estimates $\hat{\mu} = \bar{Y}_s$ and $\hat{\sigma}^2 = \Sigma_s(Y_i - \bar{Y}_s)^2/(n - 1)$. The quantity σ^2/μ^2 is sometimes called the *unit relvariance* and is a measure of the spread of y in the population as a whole.

Table 4.5 lists the sample sizes needed to achieve *cv*'s of 0.05 and .10 in several sizes of population, all having a unit relvariance of 1. For the smaller population sizes in the table, the value of N does effect how large n should be. To achieve a *cv* of 0.05, for example, a sample of 44 is necessary when $N = 50$ but $n = 286$ is required when $N = 1{,}000$. For extremely large populations, though, the particular size of the population has no effect. As $N \to \infty$, $n = \sigma^2/(k^2\mu^2)$, which is $n = 400$ for a *cv* of 0.05 and a unit relvariance of 1.

Example 4.4.2 Variance Proportional to x: Ratio Estimator. Consider first the ratio estimator whose variance under the working model $M(0, 1:x)$ depends on the size of the x's in the sample (see equation 4.1.1). As has been indicated in several places, this is *not* the preferred model and estimator; however, it is useful to consider not only for purposes of comparison, but also as one means to get a preliminary idea of the unknowns needed for sample size determination. If a simple balanced sample were selected to guard against bias under a polynomial model like $M(\delta_0, \delta_1, \ldots, \delta_J : v)$, then the relvariance under the working model $M(0, 1:x)$ is

$$[cv(\hat{T}_R)]^2 = \left(\frac{1}{n} - \frac{1}{N}\right)\frac{\sigma^2}{\beta^2\bar{x}}.$$

Table 4.5. Sample Sizes Needed to Obtain Specified Levels of Coefficient of Variation in Populations with Unit Relvariance Equal to 1

k (specified cv)	N	n
0.05	50	44
	100	80
	1,000	286
	1,000,000	400
0.10	50	33
	100	50
	1,000	91
	1,000,000	100

Setting the cv equal to a desired value k and solving for n yields

$$n = \left[k^2 \bar{x} \frac{\beta^2}{\sigma^2} + \frac{1}{N} \right]^{-1}. \tag{4.4.4}$$

Estimates or population values are required for $\bar{x}$, β, and σ^2. When a complete frame is available, the mean $\bar{x}$ of the auxiliary x should be known. Under the working model $M(0, 1\!:\!x)$, unbiased estimators of the other two parameters are $\hat{\beta} = \bar{Y}_s/\bar{x}_s$ and $\hat{\sigma}^2 = \Sigma_s x_i^{-1}(Y_i - \hat{\beta}x_i)^2/(n-1)$. Common procedures are to select a preliminary sample or to use survey results from an earlier time period in order to estimate the parameters, and then to substitute the estimates into (4.4.4) to arrive at a sample size.

A sample of $n = 10$ from the Hospitals population that is approximately balanced on the first moment of x is the following:

Hospital	x	y
63	65	239
124	126	427
153	160	637
155	160	402
210	254	956
229	276	852
249	307	935
256	318	944
266	339	885
371	652	2150

The parameter estimates from this sample are $\hat{\beta} = 3.172$ and $\hat{\sigma}^2 = 47.000$, while $\bar{x} = 274.697$. To achieve a *cv* of 0.05 for the ratio estimator under the model $M(0, 1{:}x)$ the estimated sample size is

$$\left[(.05)^2 \frac{3.17^2}{47}(274.697) + \frac{1}{393}\right]^{-1} \cong 7, \tag{4.4.5}$$

which is quite small. If we require a more precise estimator with a *cv* of 0.01, then replacement of 0.05 by 0.01 in (4.4.5), estimates the sample needed as $n = 119$.

Example 4.4.3 Variance Proportional to x: Root(x) Balance. Consider now the estimator $\hat{T}(x^{1/2}, x{:}x)$. Using this estimator with *root*(x) balance will be more efficient than the ratio estimator with unweighted balance.

Using the expression (4.2.8) for the variance under *root*(x) balance, and solving for n, we find

$$n = \frac{(\Sigma_{i=1}^N x_i^{1/2})^2}{\Sigma_{i=1}^N x_i + k^2\tilde{T}^2/\tilde{\sigma}^2},$$

where $\tilde{T}$ is a preliminary guess at T or $E_M(T)$. Using the results of the prelimary study discussed in the preceding example, we can take $\tilde{T} = \hat{\beta}N\bar{x} = (3.172)(393)(274.697) \approx 342{,}436$ and $\tilde{\sigma}^2 = 47$. From the population of x values we can determine that $\Sigma_{i=1}^N x_i^{1/2} \approx 5992$. Hence for a *cv* of 0.01, the estimated sample size is

$$n = \frac{(5992)^2}{393(274.697) + (.01)^2 342436^2/47} \approx 100. \tag{4.4.6}$$

This is about 5/6th of the sample size of 119 for simple balanced samples in Example 4.4.2.

In general, so long as we anticipate taking a *root*(v)-balanced sample, we can calculate the desired sample size by

$$n = \frac{(\Sigma_{i=1}^N v_i^{1/2})^2}{\Sigma_{i=1}^N v_i + k^2\tilde{T}^2/\tilde{\sigma}^2}. \tag{4.4.7}$$

Conservatism and the likelihood of losses from the initial sample due to nonresponse and other problems will often impel samplers to use larger sample sizes than those computed from (4.4.7). If, in the last example, we suspect that data will not be collected for 20% of the initial sample, and wished for a *cv* of 0.01, we might select $100/0.8 \cong 125$ units. Note, however, that not all sample units are equal here, and more damage will be done by losing units with large variance than those with small. Also, since the *planned* sample was $x^{1/2}$-balanced, we shall have to worry whether the final sample is still balanced, and take appropriate measures if not.

4.5 SUMMARY AND PERSPECTIVE

In this chapter, we have continued the discussion, begun in Chapter 3, on bias-robustness through balanced samples, but emphasizing the question of efficiency under the working model. The major points are:

(1) Bias-robustness of estimators of total (or mean) can be achieved through selecting a balanced sample $B(\mathbf{X}^* : \mathbf{V})$ with an estimator $\hat{T}(\mathbf{X}:\mathbf{V})$, $\mathbf{X} \subset \mathbf{X}^*$ that meets certain conditions relating $\mathbf{V}$ to $\mathbf{X}$ (see Theorem 4.2.1).

(2) If a balanced sample is selected, the assumed variance for the estimator *must* be the variance v on which the *root*(v) balanced sample is based. This is true even if the sample presents some evidence that a different variance structure might fit the population. More specifically, if the sample balance is based on V but the sample data suggest $\tilde{V}$ is more appropriate, the estimator should still be $\hat{T}(\mathbf{X}:\mathbf{V})$, not $\hat{T}(\mathbf{X}:\tilde{\mathbf{V}})$. Otherwise the bias robustness conveyed by $s \in B(X:V)$ can be lost. Using estimators simpler than the minimal estimator can be risky, given that balance is typically not exact.

(3) To minimize the error variance, use *root*(v) balance, if there is reason to think that the v reflects the correct variance structure. Previous experience with similar populations will be helpful here. In continuing surveys that are periodically repeated, extensive data analysis of different survey variables can be done. In the absence of prior information, a reasonable strategy, in the case of a single auxiliary x, is to assume $v \propto x$, and aim for *root*(x) balance; this should put us reasonably close to the optimal sample, without too much loss of precision.

(4) For balanced samples in $B(\mathbf{X}^*:\mathbf{V})$ and an appropriate estimator $\hat{T}(\mathbf{X}:\mathbf{V})$ as in (1) above, the matter of determining the proper sample size is straightforward, since the variance reduces to the simple form in Theorem 4.2.1 that depends neither on the specifics of the sample nor of the estimator.

(5) If inference must be made from an unbalanced sample, careful modeling is important. Parsimonious models with few parameters are appealing, but, as noted in Sections 3.5 and 3.7, care must be exercised in testing whether a parameter should be included or not, particularly in the choice of variance specification used. Another possibility for obtaining bias-robustness, when a single auxiliary is present, is nonparametric regression, discussed later in Section 11.3.

(6) We have not considered what needs to be done for targets other than the total or mean, but will address estimation of cumulative distribution functions and quantiles in Section 11.4.

4.6* REMARKS ON DESIGN-BASED INFERENCE

We add here a few brief remarks on the perspective the results of this and the previous chapter shed on a good deal of design-based practice. The use of *pps* sampling combined with a Hájek, Horvitz-Thompson, or model-assisted regression estimator, can be justified, to an extent, by the results on bias-robustness and efficiency shown in this and the preceding chapter. The qualification "to an extent" is important. Use of *pps* and of these estimators is usually justified on the grounds of efficiency only, where efficiency is measured in terms of a model-based expectation of the design variance. The Godambe-Joshi result noted in observation (4) in Section 4.2.1, links the present results to those design-based concerns with efficiency. What we now see more clearly is:

(a) What really makes those estimators "work", to the extent that they do work, is the achievement of approximate weighted balance by the *pps* sampling and the estimators being unbiased, with respect to possible underlying *models*, under such weighted balance. If the results of this and the preceding chapter had turned out differently, we would have had serious doubts about the use of those estimators.

(b) But this chapter has raised some serious caveats about the use of the Hájek and the Horvitz-Thompson based on a continuous auxiliary variable, since balance is typically *not* achieved to a satisfactory tolerance by the standard ways of doing *pps* sampling.

(c) Model-assisted estimators (GREG) are better supported, but we would qualify the usual claims for them by two observations: (i) unless the model selected is very dependable, a strict (weighted) balanced sample gives reassurance that *pps* sampling, merely aiming at that balance, cannot give. (ii) the texts often refer to modification of the basic estimator in accordance with a suspected variance structure, in effect using the weights $\pi_i^{-1} v_i^{-1}$ instead of π_i^{-1} in estimating the regression coefficients of the model. This is probably not much done in practice, which is fortunate, since, as remark (2) in Section 4.5 indicates, there needs to be a tight connection between the form of the weighted balance achieved (or at least aimed at) and the form of the estimator. If π_i is proportional to $v_i^{1/2}$, then $\pi_i^{-1} v_i^{-1} \propto v_i^{-3/2}$, which can be an inefficient choice, as we also noted in Section 2.8.

We observed in Section 3.7 that randomization procedures that yield consistent, asymptotically normal estimators are equally robust with the design-based definition of robustness. To attain distinctions of efficiency, models are invoked. For a given probability design and a chosen model, Isaki and Fuller (1982) term the model expectation of the design variance the *anticipated variance*:

$$AV(\hat{T} - T) = E_M\{E_\pi[(\hat{T} - T)^2]\} - E_M\{E_\pi[\hat{T} - T]\}^2,$$

and define efficiency in terms of the AV. When $\hat{T}$ is design-unbiased, the AV

reduces to $AV(\hat{T}-T) = E_M\{E_\pi[(\hat{T}-T)^2]\}$. The anticipated variance can be used to clarify the relation between design efficiency and design robustness.

We give a simple example. Suppose that the correct model is $M(1, 1{:}1)$, that is, $Y_i = \beta_0 + \beta_1 x_i + \varepsilon_i$ with the errors being independent and $\varepsilon_i \sim (0, \sigma^2)$. Assume an *srswor* of some size n is selected. If the sample mean $\bar{Y}_s$ is used, it has a model bias equal to

$$E_M(\bar{Y}_s - \bar{Y}) = \beta_1(\bar{x}_s - \bar{x}). \tag{4.6.1}$$

Obviously, $\bar{Y}_s$ has a model bias heavily dependent on $\bar{x}_s$ like the one we saw in Figure 3.2. However, $\bar{Y}_s$ is design unbiased under simple random sampling without replacement. If the conditions of Theorem A.9 hold, $\bar{Y}_s$ is also asymptotically normal with a variance that decreases as n increases and, hence, by Chebyshev's inequality, is design consistent. If we average the model bias over all simple random samples, then

$$\begin{aligned} E_\pi E_M(\bar{Y}_s - \bar{Y}) &= \beta_1 E_\pi(\bar{x}_s - \bar{x}) \\ &= 0. \end{aligned}$$

The positive biases when $\bar{x}_s > \bar{x}$ cancel out the negative biases when $\bar{x}_s < \bar{x}$. Thus, despite its substantial bias in some off-balance samples, the bias averages out to 0 across all simple random samples. Consequently, $\bar{Y}_s$ is "robust" in the design-based sense.

The conditional bias of $\bar{Y}_s$ is indirectly accounted for in a design-based analysis because $\bar{Y}_s$ proves to be extremely inefficient if $Y_i = \beta_0 + \beta_1 x_i + \varepsilon_i$ is the right model. To see this, compute the anticipated variance in the form

$$E_\pi E_M(\bar{Y}_s - \bar{Y})^2 = E_\pi\{\text{var}_M(\bar{Y}_s - \bar{Y}) + [E_M(\bar{Y}_s - \bar{Y})]^2\}.$$

The model variance is easily calculated as $\text{var}_M(\bar{Y}_s - \bar{Y}) = \sigma^2(1-f)/n$. Combining this with the model bias, the anticipated variance is

$$E_\pi E_M(\bar{Y}_s - \bar{Y})^2 = \frac{1-f}{n}(\sigma^2 + \beta_1^2 S_x^2), \tag{4.6.2}$$

where $S_x^2 = \Sigma_{i=1}^N (x_i - \bar{x})^2/(N-1)$. The model bias of a sample with a particular value of $\bar{x}_s$ becomes a component of variance when the model error variance is averaged over all possible samples.

If the model were $Y_i = \beta_0 + \varepsilon_i$ so that $\bar{Y}_s$ is model unbiased, then the term $\beta_1^2 S_x^2$ disappears from (4.6.2) but otherwise it can make a substantial contribution to the *mse*. On the other hand, suppose we use the linear regression estimator of the mean, $\hat{\bar{Y}}_{LR} = \bar{Y}_s + b(\bar{x} - \bar{x}_s)$ with $b = \Sigma_s(Y_i - \bar{Y}_s)(x_i - \bar{x}_s)/\Sigma_s(x_i - \bar{x}_s)^2$. Then $\hat{\bar{Y}}_{LR}$ is approximately design unbiased, asymptotically normal in repeated *srswor*'s (under appropriate conditions) and thus is also robust by design-based standards. Using the result in Example 2.3.2, its error variance is $(1-f)\sigma^2[1+(\bar{x}_s-\bar{x})^2/\{(1-f)c_s\}]/n$ with $c_s=\Sigma_s(x_i-\bar{x}_s)^2/n$. Thus, in large simple random samples, if the sampling fraction converges to a constant, the anticipated variance is approximately

$$E_\pi E_M(\hat{\bar{Y}}_{LR}-\bar{Y})^2 \approx \frac{(1-f)}{n}\sigma^2\left[1+\frac{E_\pi(\bar{x}_s-\bar{x})^2}{(1-f)E_\pi(c_s)}\right]$$

$$\approx \frac{(1-f)}{n}\sigma^2,$$

where we use $E_\pi(\bar{x}_s-\bar{x})^2=(1-f)S_x^2/n$ and $E_\pi(c_s)=(n-1)S_x^2/n$. As a result, $\hat{\bar{Y}}_{LR}$ eliminates the term $\beta_1^2S_x^2$ from the *mse* and is much more efficient than $\bar{Y}_s$. Notice that, if we take the design-based approach, the issue is not whether one estimator is less biased than the other. In fact, $\bar{Y}_s$ is design unbiased whereas $\hat{\bar{Y}}_{LR}$ is only approximately so. So, considering only the design bias, we might choose $\bar{Y}_s$. The linear regression estimator would only be selected over the sample mean because it has a smaller anticipated variance. Thus it is hard to see design unbiasedness or consistency *per se* as a means to choose among estimators.

The model-based arguments of this and the preceding chapter distinguish (model) bias and variance, shedding a clear, simple light on when and whether an estimator/sample combination is good. We conclude with a return to the HMT example of Section 3.7. We saw there that HMT's fears about the inadequacy of the model-based approach would be allayed by a better application of regression diagnostics. We add here that balanced sampling, properly carried out, actually achieves efficiencies compared to the results they had.

We noted in Section 3.7 that the stratified simple random sampling plan used by HMT with strata created by equalizing the cumulative values of x aims at root(x^2) balanced sampling. The method is an example of one of the methods of approximating sampling $pp(x)$ described in Section 3.4.5. The minimal estimator when variances are proportional to x^2 is $\hat{T}(0,1,1:x^2)$. Recall from Example 4.2.1 that the minimal estimator achieves its smallest variance under $M(0,1,1:x^2)$ in any balanced sample with $\bar{x}_s=\bar{x}^{(2)}/\bar{x}$.

We can protect the minimal estimator against bias if there is an intercept (as there was in the HMT population) by the balancing condition $\bar{x}_s^{(-1)}=1/\bar{x}$. The mean of the x variable in our version of their study was $\bar{x}=10.03$. As in Section 3.7, we selected 1,000 *stsrs*'s with 10 strata formed by size, as noted above, and 20 units selected per stratum. The range of values of the sample harmonic mean $1/\bar{x}_s^{(-1)}$ across the thousand *stsrs*'s was from 1.78 to 10.58, with an inter-quartile range of 8.466 to 9.48 (suggesting that a 20 unit/stratum design is not ideal for aiming at balance). Recall that the mean square error (*mse*) across all samples tended at best to be about 0.009 to 0.010 both in the original study and in our replication.

Table 4.6. Mean Square Error of Minimal Estimator for HMT Population

Ranges of Values of Sample Harmonic Mean	Number of Samples	Mean Square Error
1.78–10.58	1000	0.0282
8.47–9.48	500	0.0118
9.70–10.30	140	0.0101
9.80–10.20	80	0.0082
9.90–10.10	39	0.0065

Stratified simple random sampling, 10 strata, 20 units per stratum

Table 4.6 gives mean square error for the minimal estimator, $\hat{T}(0, 1, 1: x^2)$, under various degrees of restricted sampling. The first line in the table gives the *mse* across all 1,000 samples as 0.0282. The following four lines give subsets of the 1,000 samples based on how close the harmonic mean, $1/\bar{x}_s^{(-1)}$, is to the balance point, $\bar{x} = 10.03$. As we observed in Figure 3.7, $pp(x)$ samples that have $1/\bar{x}_s^{(-1)} \cong \bar{x}$ also tend to have $\bar{x}_s \cong \bar{x}^{(2)}/\bar{x}$, which would be optimal for the minimal estimator under $M(0, 1, 1: x^2)$. It is readily seen that, on the one hand, the minimal estimator in this population is rather vulnerable, if the sample is far from x-balance. On the other hand, there are appreciable gains to be achieved by restricting the sample to near balance.

EXERCISES

4.1 Show that the estimator $\hat{T}_0(x^{1/2})$ is unbiased under $M(0, 1: x)$ under (4.1.3).

4.2 Prove that the condition $\pi_i = nv_i^{1/2}/(N\bar{v}^{(1/2)}) \leqslant 1$, $i = 1, 2, \ldots, N$ implies that $n \leqslant N\{\bar{v}^{(1/2)}\}^2/\bar{v}$. What are the implications of this for balanced $pp(v^{1/2})$ sampling?

4.3 The lower bound for the variance given in Theorem 4.2.1 holds in any sample for a model in which the conditions of the theorem hold, as well as being the Godambe-Joshi lower bound of the anticipated variance. Show by example that there exists a model (in which the conditions in the theorem do not hold), a population, and a sample in which the error variance for the *BLU* estimator of the total is less than this bound.

4.4 Using the x variables (beds) from the Hospitals population and sample size $n = 30$, and assuming that the correct model is $M(0, 1: x)$, calculate the variances of (a) the ratio estimator for a balanced sample, (b) the ratio estimator for the 30 units with largest x-values, and (c) the minimal estimator $T(x^{1/2}, x:x)$ for a sample with the weighted balance conditions (4.1.2–3). Write the variances in the form $K\sigma^2$ where K is determined separately for parts (a), (b), and (c).

4.5 Let the population of x-values be $1, 2, \ldots, 100$. Find the minimal variance of a *BLU* estimator of total that is achievable by a sample of 10, for the models

(a) $Y_i = \Sigma_{j=0}^{14} \beta_j x_i^j + x_i^4 \varepsilon_i$, $\varepsilon_i \sim (0, \sigma^2)$ independently $i = 1, 2, \ldots, N$

(b) $Y_i = \Sigma_{j=0}^{1} \beta_j x_i^j + x_i^{3/2} \varepsilon_i$, $\varepsilon_i \sim (0, \sigma^2)$ independently $i = 1, 2, \ldots, N$.

4.6 Prove Theorem 4.2.2 (Tallis' Theorem).

4.7 A sample is said to be *overbalanced* (of order J) if $\Sigma_s x_i^{j-1}/n = \Sigma_r x_i^j / \Sigma_r x_i$, $j = 1, 2, \ldots, J$ (Scott, Brewer, and Ho, 1978).

(a) Let $\hat{\beta} = (\Sigma_s Y_i/x_i)/n$. Given the sample s is overbalanced, show that $\hat{T}(0, 1:x^2) = \Sigma_s Y_i + \hat{\beta}\Sigma_r x_i$ is unbiased under the general polynomial model (3.2.1).

(b) Suppose the variances are $v_i = \sigma_1^2 x_i + \sigma_2^2 x_i^2$. Find the variances of

(i) $\hat{T}(0, 1:x^2)$ under overbalance;

(ii) The minimal estimator $\hat{T}(0, 1, 1:x^2)$ under root(x^2) balance, and

(iii) The minimal estimator $\hat{T}(x^{1/2}, x:x)$ under root(x) balance.

(c) Show explicitly that if $\sigma_1^2 = 0$, then variance under (b)(ii) is smallest and (b)(iii) is largest, and that if $\sigma_2^2 = 0$, then variance under (b)(iii) is smallest and (b)(ii) is largest.

4.8 For the ratio estimator, find an expression for the relvariance, that is, the coefficient of variation squared, for an arbitrary (i.e., not necessarily balanced) sample under the model $M(0, 1:x)$. For given target coefficient of variation, what happens to the required sample size, as $\bar{x}_s$ more and more exceeds $\bar{x}$?

4.9 Consider the Counties70 Population. Suppose you suspect variances increase as the x variable, and you plan to take a root(x) balanced sample. Take a preliminary sample of size $n = 10$, and estimate sample size required to achieve a coefficient of variation of 0.03 under the minimal working model $M(x^{1/2}, x:x)$. Take a balanced sample of the required size, and estimate the population total using the minimal estimator. How far is it, in fact, from the total of the Y's?

4.10 For the Counties70 Population, how large does the sample need to be, before a root(x^2) balanced sample is clearly impossible? Call the resulting sample size n^*. Assume the variances are proportional to x^2, and take a sample of size $n^* + 8$, that is root(x^2) balanced on as large a sub-sample as possible. Estimate the total T, making use of the minimal estimator. Find the largest value of γ such that, for $n = n^* + 8$, a root(x^γ) sample *is* possible, and get the corresponding minimal estimator of T. Which estimator is, in fact, closer to T?

4.11 Suppose we have a population of size N, for which full information is available on an auxiliary variable x, and having two variables of interest Y_1 and Y_2. Suppose further a reasonable working model for Y_1 is $Y_{1i} = \Sigma_{j=1}^{2} \beta_j x_i^j + \varepsilon_i$ with $\varepsilon_i \sim (0, \sigma_1^2)$, $i = 1, 2, \ldots, N$, and for Y_2, $Y_{2i} = \Sigma_{j=1}^{3} \gamma_j x_i^j + x_i \eta_i$ with $\eta_i \sim (0, \sigma_2^2)$, $i = 1, 2, \ldots, N$. In addition, preliminary estimates $\tilde{T}_1$, $\tilde{T}_2$, $\tilde{\sigma}_1$, $\tilde{\sigma}_2$ are available for the population totals T_1, T_2 of Y_1 and Y_2, and the standard deviations σ_1, σ_2, and a desired coefficient of variation k is set. One wishes to take a single balanced sample to protect both *BLU* estimators of total from bias under higher degree polynomial models, and the question is whether to use unweighted or root(x^2) balance, with the *BLU* estimators $\hat{T}_1(1, 1, 1{:}1)$, $\hat{T}_2(1, 1, 1, 1{:}1)$ or the *BLU* estimators $\hat{T}_1(1, 1, 1{:}x^2)$, $\hat{T}_2(1, 1, 1, 1{:}x^2)$ respectively.

(a) Determine expressions for minimal sample size for each of these estimators, in terms of k, N, appropriate population moments of x, and $\tilde{T}_1$, $\tilde{\sigma}_1$ or $\tilde{T}_2$, $\tilde{\sigma}_2$. Assume balance extends to negative powers of x, where necessary.

(b) Suppose x coincides with the $N = 393$ x-values of the Hospitals Population, that the preliminary estimates are $\tilde{T}_1 = 300{,}000$, $\tilde{T}_2 = 600{,}000$, $\tilde{\sigma}_1 = 240$, $\tilde{\sigma}_2 = 2$. Determine the sample sizes each estimator requires to achieve $k = 0.03$. What is the best strategy?

(c) Repeat (b), but with the values $\tilde{T}_1 = 300{,}000$, $\tilde{T}_2 = 600{,}000$, $\tilde{\sigma}_1 = 120$, $\tilde{\sigma}_2 = 8$.

CHAPTER 5

Variance Estimation

The preceding chapters have been concerned with predicting the population total $T = \Sigma_{i=1}^{N} y_i$ of a variate of interest y. We have focused on this particular function of population values both for its common interest, and for its relative simplicity. We have been concerned with the frequently met situation where an auxiliary variable x is available on the population and it is reasonable to assume that the y_i's are realizations of random variables Y_i with the properties that (1) Y_i is linearly dependent on x_i and (2) the Y_i's are mutually independent, given the values x_i. In later chapters we will loosen these goals and assumptions.

In the present chapter, we attend to the question of estimating the precision of an estimator $\hat{T}$ of T. The precision, as measured by the error variance, is used to assess how close our estimate is to the unknown population total T.

Keeping in mind that T itself is the realization of a random variable, we want to estimate the mean square error or error variance $E_M(\hat{T} - T)^2$. If $\hat{T}$ is unbiased, the error variance is just the variance $v \equiv \text{var}_M(\hat{T} - T)$. Once we have an estimate $\hat{v}$ of v, we then typically construct a confidence interval for T of the form $\hat{T} \pm z\hat{v}^{1/2}$, where z is the appropriate quantile of the standard normal distribution. The bounds of the confidence interval are intended to assess how close the estimate $\hat{T}$ is to T. This, of course, is in keeping with much of what we do in the broader, general context of statistical inference.

Just as we were concerned in the last chapter about making our estimator $\hat{T}$ robust to model failure, that is, to deviations of the working model from models which might, in fact, have generated the population values, so, too, in estimating the variance v we need to be concerned with robustness to model failure. At least initially, this means an emphasis on the question of what happens if the variance structure we have assumed in our working model is off base. Under this circumstance, which is surely very common, and not always easy to detect, the standard least squares variance estimator leaves us vulnerable, and we have recourse to "robust" variance estimators. Robust variance estimation, its variations and consequences, strengths and weaknesses, is the theme of this chapter.

The estimation error of any estimator of the total can be written as $\hat{T} - T = \hat{T}_r - T_r$, where T_r is the total for the nonsample units and $\hat{T}_r$ is an estimator of T_r based on the sample units. When the nonsample units are independent of those in the sample, we can write $\mathrm{var}_M(\hat{T} - T) = \mathrm{var}_M(\hat{T}_r - T_r) = \mathrm{var}_M(\hat{T}_r) + \mathrm{var}_M(T_r)$. Under typical sampling conditions, the first component, $\mathrm{var}_M(\hat{T}_r)$, dominates (see Section 5.5 below). When the sample and nonsample units are independent, $\hat{T}_r = \mathbf{1}_r'\mathbf{X}_r\hat{\boldsymbol{\beta}}$, with $\hat{\boldsymbol{\beta}}$ a weighted least squares estimator of an underlying parameter $\boldsymbol{\beta}$. In preparation for studying the finite population problem, we consider variance estimation for estimators of the general form $\hat{\Theta} = \mathbf{q}'\hat{\boldsymbol{\beta}}$.

The basic idea of robust variance estimation is a simple one. Since $\hat{\boldsymbol{\beta}}$ is a linear function of the Y's, we can write $\hat{\Theta} = \mathbf{a}'\mathbf{Y} = \Sigma_s a_i Y_i$, where the a's are given in terms of the known x's and $\mathbf{q}$. Let $v = \mathrm{var}_M(\hat{\Theta})$. If $\mathrm{var}_M(\mathbf{Y}_s) = \boldsymbol{\Psi} = diag(\psi_i)$, then $v = \mathbf{a}'\boldsymbol{\Psi}\mathbf{a} = \Sigma_s a_i^2 \psi_i$. Then a simple plug-in estimator that replaces the unknown $\boldsymbol{\Psi}$ is just $\hat{v} = \mathbf{a}'\hat{\boldsymbol{\Psi}}\mathbf{a} \equiv \mathbf{a}' diag(r_i^2)\mathbf{a} = \Sigma_s a_i^2 r_i^2$, where r_i's are the *residuals* gotten by fitting the working regression model to the sample data. Because of its matrix form, this estimator has come to be called the *Sandwich Estimator*: the known vector $\mathbf{a}$ is the "bread", and the matrix $\hat{\boldsymbol{\Psi}} = diag(r_i^2)$ is the "filling." Improvements on this estimator (the "condiments") are achieved using expressions involving the *leverages*, well known from regression diagnostics (Belsley, Kuh, and Welsch, 1980). For finite population estimation, these methods were formulated for specific cases in Royall and Eberhardt (1975) and more generally by Royall and Cumberland (1978).

The outline of this chapter is as follows: In the first section, Section 5.1, we give two examples to illustrate procedures in a nutshell. In Section 5.2 we define some terms, including, for the sake of completeness, the standard least squares variance estimator. This variance estimator, well known from regression textbooks, has a potential for bias that can undermine inference when the working model misspecifies the variance structure. In Section 5.3 and 5.4, we consider methods of robust variance estimation for $\hat{\Theta} = \mathbf{q}'\hat{\boldsymbol{\beta}}$, where "robust" means primarily robust against misspecification of the variance structure. In Section 5.5, we apply these results to estimation of $\mathrm{var}_M(\hat{T}_r)$, and discuss options for taking into account the secondary component $\mathrm{var}_M(T_r)$.

The first five sections of this chapter regard the regression component $E_M(Y_i) = \mathbf{x}_i'\boldsymbol{\beta}$ of the model for Y_i as correct. In Section 5.6, we consider conditions where the regression component itself is incorrect, that is, where $\hat{T}$ may be biased under the correct model. This can have serious consequences for confidence intervals, unless, by balanced sampling, earlier discussed in Chapters 3 and 4, we have managed to achieve effective unbiasedness despite model failure. In this case robust variance estimation actually yields conservative coverage of confidence intervals. A sub-theme is that many standard sampling designs, in particular, simple random sampling, do not supply balance sufficiently often to achieve sound inference.

5.1 EXAMPLES OF ROBUST VARIANCE ESTIMATION

In this section, we consider two examples of variance estimation for the simple through the origin model characterized by $E_M(Y) = \beta x$. In the first example, the working model assumes equal variances ("homoscedasticity"), and we digress from the finite population context to focus on variance estimation for estimating the unknown parameter β. A great many of the points to be made about robust variance estimation hold not just for sampling with the aim of estimating finite population quantities, but in the regression context generally. The second example will assume variances proportional to the auxiliary x, and focus on estimating the variance of the ratio estimator of the population total. Between these two examples, the main points of robust variance estimation, generalizable to more complex situations, will become clear.

5.1.1 Homoscedastic Through the Origin Model

Consider the simple model $M(0, 1{:}1)$, that is, the through the origin model $E_M(Y) = \beta x$, having constant variances, $\text{var}_M(Y) = \sigma^2$. This is the simplest of models considered in standard regression texts. The *BLU* estimator of the slope β is $\hat{\beta} = \Sigma_s x_i Y_i / \Sigma_s x_i^2$, which can be written in the linear form $\hat{\beta} = \Sigma_s a_i Y_i$, with $a_i = x_i / \Sigma_s x_k^2$. The factors a_i will be termed the *estimation components*; they depend on the particular model and on what is being estimated, but are always *known* to the analyst.

We remind ourselves of some standard regression terminology:

The *fitted values* $\hat{Y}_i = \hat{\beta} x_i$ estimate $E_M(Y_i) = \beta x_i$ and can be written $\hat{Y}_i = \Sigma_j h_{ij} Y_j$, where (in the present case) $h_{ij} = x_i x_j / \Sigma_s x_k^2$. The value h_{ii} measures the impact of Y_i on its own fitted value, and is the ith *leverage*. Note that (in the present case) the sum of the leverages is $\Sigma_s h_{ii} = 1$, so their average value is n^{-1}. The matrix $\mathbf{H} = (h_{ij})$ is referred to as the *Hat Matrix* (Belsley, Kuh, and Welsch, 1980, p.16; Hoaglin and Welsch, 1978), because "it puts the hat on $\mathbf{Y}$": $\hat{\mathbf{Y}}_s = \mathbf{H}\mathbf{Y}_s$. The leverages are the diagonal elements of $\mathbf{H}$.

The *residuals* are $r_i = Y_i - \hat{Y}_i$, the difference between the ith data value and the corresponding fitted value. The major ingredient of the variance estimators we shall consider is the set of *squared residuals* r_i^2.

Under the (working) model $M(0, 1{:}1)$, the variance of $\hat{\beta}$ is readily found to be $\text{var}_M(\hat{\beta}) = \sigma^2 / \Sigma_s x_i^2$, where σ^2 is the unknown variance, assumed constant, of the errors. Thus, so long as we are not concerned with violations of the model, all that is necessary is an estimate of σ^2. An efficient (under normality) unbiased estimator of σ^2 is the ordinary least squares variance estimator $s^2 = \Sigma_s r_i^2 / (n-1)$. This gives the standard least squares variance estimator $v_L(\hat{\beta}) = s^2 / \Sigma_s x_i^2$.

What happens if the variance as modeled in the working model is incorrect? To achieve some generality, suppose $\operatorname{var}_M(Y_i) = \psi_i$. Then

$$\operatorname{var}_M(\hat{\beta}) = \sum_s a_i^2 \psi_i = \sum_s x_i^2 \psi_i \bigg/ \left(\sum_s x_i^2\right)^2. \tag{5.1.1}$$

We wish to determine the expected value of v_L, under this general variance condition.

From $E_M(r_i) = 0$ (the regression component of the working model still being assumed correct), it follows that

$$E_M(r_i^2) = \operatorname{var}_M(r_i) = \operatorname{var}_M(Y_i - \hat{Y}_i) = \operatorname{var}_M\left(Y_i[1 - h_{ii}] - \sum_{j \neq i} h_{ij} Y_j\right)$$
$$= \psi_i(1 - h_{ii})^2 + \sum_{j \neq i} h_{ij}^2 \psi_j. \tag{5.1.2}$$

When the working model is $M(0, 1{:}1)$, this is equal to

$$E_M(r_i^2) = \left(1 - \frac{2x_i^2}{\Sigma_s x_k^2}\right)\psi_i + \frac{x_i^2}{(\Sigma_s x_k^2)^2} \sum_s x_k^2 \psi_k, \tag{5.1.3}$$

whence it follows that

$$E_M(v_L) = \frac{1}{\Sigma_s x_i^2} \frac{1}{n-1}\left(\sum_s \psi_i - \frac{\Sigma_s x_i^2 \psi_i}{\Sigma_s x_i^2}\right). \tag{5.1.4}$$

If the variance elements ψ_i are constant, then (5.1.4) reduces to (5.1.1), as we would anticipate, but, in general, the two expressions are quite different. This can be seen using an interesting result of Pathak (1966) (see also Mukhopadhyay, 1977):

Lemma 5.1.1 Let $a_1 \leqslant a_2 \leqslant \cdots \leqslant a_k$ and $b_1 \leqslant b_2 \leqslant \cdots \leqslant b_k$ be two monotonic sequences of positive numbers, then $\Sigma_{i=1}^k a_i \Sigma_{i=1}^k b_i \leqslant k \Sigma_{i=1}^k a_i b_i$. If the inequalities are reversed on one of the sequences, then $\Sigma_{i=1}^k a_i \Sigma_{i=1}^k b_i \geqslant k \Sigma_{i=1}^k a_i b_i$. Equality holds only if $a_i = a$ or $b_i = b$ for all i.

Proof is by induction (see Exercise 5.1).

The ratio of (5.1.4) to (5.1.1) is less than or equal to 1, whenever $\Sigma_s \psi_i \Sigma_s x_i^2 \leqslant n \Sigma_s x_i^2 \psi_i$. By the above lemma, this condition holds whenever ψ increases with x. Thus, in this very common circumstance, the standard ordinary least squares variance estimator will tend to underestimate the variance.

An alternative variance estimator begins with the expression (5.1.1) for the variance, and seeks an appropriate surrogate for *each* of the unknowns ψ_i. Now, in (5.1.3) we note that, if the ψ_i and x_i are reasonably bounded, then $E_M(r_i^2) = \psi_i + O(n^{-1})$. More generally, if the ψ_i are bounded in (5.1.2) and $h_{ij} = O(n^{-1})$, then r_i^2 is approximately unbiased for ψ_i. This suggests that a variance estimator of the form $v_R = \Sigma_s a_i^2 r_i^2$ will be approximately unbiased for the variance (5.1.1), *whatever* the actual variance structure. In Section 5.3, we argue that, under reasonable conditions, v_R is consistent for the variance. This idea for robust variance estimation has a long history, being first suggested by Eicker (1963), and later rediscovered and popularized by White (1982). Other authors who have extended the study of the technique include Binder (1983), Liang and Zeger (1986), and Royall (1986). The estimator also has actually found its way into statistical software like SAS (SAS Institute 1990, p.1431), STATA (STATA Corporation 1997, Section 26.10), and SUDAAN (Shah, Barnwell, and Bieler 1997, p.9-7).

Although v_R is approximately unbiased, what bias it has tends to be predictably *negative*. To see this, suppose that the variance elements are actually constant and equal to σ^2, that is, suppose that the working model was actually correct. Then (5.1.3) reduces to

$$E_M(r_i^2) = \left(1 - \frac{x_i^2}{\Sigma_s x_i^2}\right)\sigma^2 = (1 - h_{ii})\sigma^2.$$

In this circumstance, $E_M(v_R) = \Sigma_s a_i^2(1 - h_{ii})\sigma^2$, which is less than $\Sigma_s a_i^2 \sigma^2$, the variance (5.1.1) when the variance elements are equal to a constant σ^2. If some values of the leverages are high, this bias can be appreciable.

There are two methods to rectify this bias at the working model M. The first, an "external" adjustment to v_R, yields an estimator referred to in the literature as v_H. The idea is to multiply v_R by $\text{var}_M(\hat{\beta})/E_M(v_R)$, the ratio under the working model M of the actual variance of the estimator to the expected value of v_R, obtaining

$$v_H = \frac{\text{var}_M(\hat{\beta})}{E_M(v_R)} v_R. \tag{5.1.5}$$

In the present case, we have

$$\frac{\text{var}_M(\hat{\beta})}{E_M(v_R)} = \frac{\Sigma_s a_i^2 \sigma^2}{\Sigma_s a_i^2 (1 - h_{ii})\sigma^2} = \frac{\Sigma_s a_i^2}{\Sigma_s a_i^2(1 - h_{ii})},$$

so that

$$v_H = \frac{\Sigma_s a_i^2}{\Sigma_s a_i^2(1 - h_{ii})} \sum_s a_i^2 r_i^2 .$$

The result is that v_H is unbiased for the variance under the working model.

The second, an "internal" adjustment, is given by

$$v_D = \sum_s a_i^2 \tilde{\psi}_i, \tag{5.1.6}$$

where $\tilde{\psi}_i = r_i^2/(1 - h_{ii})$, for $i \in s$. One readily sees that $E_M(v_D) = \text{var}_M(\hat{\beta})$.

Either of these estimators will be unbiased at the working model, and shares the property with v_R of being consistent for the variance (5.1.1) under very general conditions. In most circumstances, there is little to choose between v_H and v_D, although in the case of high leverage, that is, where one of the $h_{ii} \gg n^{-1}$, v_H will be more stable.

The fact that v_H and v_D are unbiased *at* the working model does not imply, of course, unbiasedness when the variance structure deviates from that of the working model. As the leverages go to zero, as will tend to happen, under reasonable conditions, as n gets large, this will not much matter. On the other hand, in the presence of large leverages, some additional precaution may be prudent. It follows from equation (5.1.2) that, if $\tilde{\psi}_i = r_i^2/(1 - h_{ii})^2$, then, *whatever* the underlying variance structure, we have $E(\tilde{\psi}_i) \geqslant \psi_i$, for $i \in s$ (see Exercise 5.2). This gives us a guaranteed positively biased variance estimator

$$v_{J*} = \sum_s a_i^2 \tilde{\psi}_i. \tag{5.1.7}$$

This variance estimator is a close approximation to the well-known *jackknife variance estimator* (see Exercise 5.3 and Section 5.4); hence the notation. The leverage adjustments in v_D and v_J^* are also discussed in Davidson and MacKinnon (1993) and MacKinnon and White (1985).

In constructing confidence intervals, we may prefer to use the variance estimator v_{J*} that is biased upward, to estimators that are unbiased under the working model, in order to lessen the risk of low coverage.

5.1.2 Variance Estimators for the Ratio Estimator

Consider the working model $M(0, 1{:}x)$, that is,

$$Y_i = \beta x_i + e_i, \qquad e_i \sim (0, v_i) \qquad i = 1, 2, \ldots, N, \tag{5.1.8}$$

with $v_i = x_i \sigma^2$. Then, as we have seen, the weighted least squares estimator of β is $\hat{\beta} = \Sigma_s Y_i / \Sigma_s x_i = \bar{Y}_s / \bar{x}_s$, the ratio of the sample sum of Y's to the sample sum of x's. The best linear unbiased predictor of $T = \Sigma_{i=1}^N Y_i = \Sigma_{i \in s} Y_i + \Sigma_{i \in r} Y_i$ is the ratio estimator

$$\hat{T} = \sum_{i \in s} Y_i + \sum_{i \in r} \hat{Y}_i = \sum_{i \in s} Y_i + \sum_{i \in r} \hat{\beta} x_i = \frac{\Sigma_{i=1}^N x_i \Sigma_{i \in s} Y_i}{\Sigma_{i \in s} x_i} = N\bar{x}\frac{\bar{Y}_s}{\bar{x}_s}$$

In the present case, the fitted value is the sum of constants times the sample Y's, that is, $\hat{Y}_i = \Sigma_{k\in s}(x_i/\Sigma_{i'\in s}x_{i'})\,Y_k$ so that the ith leverage is $h_{ii} = x_i/\Sigma_{i'\in s}x_{i'}$. The error variance is

$$\text{var}_M(\hat{T} - T) = \text{var}_M\left(\sum_{i\in r}\hat{Y}_i - \sum_{i\in s}Y_i\right) = \text{var}_M\left(\sum_{i\in r}\hat{Y}_i\right) + \text{var}_M\left(\sum_{i\in r}Y_i\right), \tag{5.1.9}$$

since the fitted values, dependent on sample Y values, are independent of the nonsample Y_i's. Thus, in the case of predicting a total (and in prediction problems in general when the Y's are independent) there are *two components of the error variance.* The first term is the variance of an expression in the *sample* Y_i's; the second, of the *nonsample* Y_i's. The task of estimating the variance is then twofold:

(a) Estimate the first sample part, $\text{var}_M(\Sigma_{i\in r}\hat{Y}_i)$ and

(b) Estimate the nonsample part $\text{var}_M(\Sigma_{i\in r}Y_i)$.

In typical sampling circumstances, (a) is much bigger than (b), and therefore much more important. As we shall see below, it turns out that the sample component (a) is of order N_r^2/n, where $N_r = N - n$, while (b) is of order N_r. Often the sampling fraction, n/N, is small and $n \ll N_r$, so that (a) is dominant. Procedures are largely unexplored for obtaining robust estimators of (b) when $n \ll N_r$ does *not* hold. We shall discuss this briefly in Section 5.5.1 below.

We first focus on estimating the (a) component, $\text{var}_M(\hat{T}_r) = \text{var}(\Sigma_{i\in r}\hat{Y}_i)$, of the error variance. We have $\Sigma_{i\in r}\hat{Y}_i = \hat{\beta}\Sigma_r x_i = \Sigma_s a_i Y_i$, with $a_i = \Sigma_r x_i/\Sigma_s x_i$, $i\in s$ (in this case the estimation coefficients are all equal). Then

$$v = \text{var}_M\left(\sum_{i\in r}\hat{Y}_i\right) = \sum_s a_i^2\psi_i = \left(\frac{\Sigma_r x_i}{\Sigma_s x_i}\right)^2\left(\sum_s \psi_i\right).$$

Thus the sandwich variance estimator for the (a) component of (5.1.9) is just

$$v_R(\hat{T}_r) = \left(\frac{\Sigma_r x_i}{\Sigma_s x_i}\right)^2\left(\sum_s r_i^2\right), \tag{5.1.10}$$

where $r_i = Y_i - \hat{\beta}x_i$. Using (5.1.2), the expectation of the squared residual under the working model (5.1.8) is $E_M(r_i^2) = \sigma^2 x_i(1 - x_i/\Sigma_s x_k)$. Based on this expression, we can make an overall adjustment to $v_R(\hat{T}_r)$, analogous to (5.1.5), to obtain

$$\begin{aligned} v_H(\hat{T}_r) &= \frac{\text{var}_M(\hat{T}_r)}{E_M[v_R(\hat{T}_r)]}\,v_R(\hat{T}_r) \\ &= \frac{1}{1 - \Sigma_s(x_i/\Sigma_s x_i)^2}\,v_R(\hat{T}_r) \\ &= \frac{1}{1 - \Sigma_s h_{ii}^2}\,v_R(\hat{T}_r). \end{aligned} \tag{5.1.11}$$

The adjusted estimators v_D and v_{j*} are

$$v_D(\hat{T}_r) = \left(\frac{\Sigma_r x_i}{\Sigma_s x_i}\right)^2 \left(\sum_s \frac{r_i^2}{1-h_{ii}}\right), \tag{5.1.12}$$

and

$$v_{J*}(\hat{T}_r) = \left(\frac{\Sigma_r x_i}{\Sigma_s x_i}\right)^2 \left(\sum_s \left(\frac{r_i}{1-h_{ii}}\right)^2\right). \tag{5.1.13}$$

The estimators v_D and v_H are unbiased for $\text{var}_M(\hat{T}_r)$ under model (5.1.8), while v_R and v_{J*} are approximately so in large samples.

We still want to estimate the (b) component, $\text{var}(\Sigma_{i\in r} Y_i) = \Sigma_r \psi_i$, even though, as already noted, it will usually be far smaller than (a). We do not have residuals for the nonsample units, since Y_i is not available, and so a straight "plug-in" estimator is not possible. A reasonable strategy takes a "plug-in" estimator of $\Sigma_s \psi_i$ under the working model, and "blows it up" to nonsample size, using a multiplier $\text{var}_M(\Sigma_r Y_i)/\text{var}_M(\Sigma_s Y_i) = \Sigma_r x_i/\Sigma_s x_i$. In this way we get

$$v_R(T_r) = \left(\frac{\Sigma_r x_i}{\Sigma_s x_i}\right)\left(\sum_s r_i^2\right)$$

$$v_H(T_r) = \left(\frac{\Sigma_r x_i}{\Sigma_s x_i}\right)\frac{1}{1-\Sigma_s(x_i/\Sigma_s x_i)^2}\left(\sum_s r_i^2\right)$$

$$v_D(T_r) = \left(\frac{\Sigma_r x_i}{\Sigma_s x_i}\right)\left(\sum_s \frac{r_i^2}{1-h_{ii}}\right),$$

and

$$v_{J*}(T_r) = \left(\frac{\Sigma_r x_i}{\Sigma_s x_i}\right)\left(\sum_s \left(\frac{r_i}{1-h_{ii}}\right)^2\right).$$

Combining these estimators of the (b) component with those of the (a) component, one gets

$$v_R(\hat{T}_r - T_r) = \left(\frac{\Sigma_r x_i}{\Sigma_s x_i}\right)^2\left(\sum_s r_i^2\right) + \left(\frac{\Sigma_r x_i}{\Sigma_s x_i}\right)\left(\sum_s r_i^2\right) = \frac{\Sigma_1^N x_i \Sigma_r x_i}{(\Sigma_s x_i)^2}\left(\sum_s r_i^2\right),$$

or, equivalently,

$$v_R(\hat{T}_r - T_r) = \frac{N^2}{n}\left(1-\frac{n}{N}\right)\frac{\bar{x}\bar{x}_r}{\bar{x}_s^2}\left(\frac{1}{n}\sum_s r_i^2\right),$$

where $\bar{x}_r = \Sigma_r x_i/N_r$. The latter form is instructive since it makes clear the role of the sampling fraction n/N. Likewise, we have

$$v_H(\hat{T}_r - T_r) = \frac{\Sigma_1^N x_i \Sigma_r x_i}{(\Sigma_s x_i)^2} \frac{1}{1 - \Sigma_s (x_i / \Sigma_s x_i)^2} \left(\sum_s r_i^2 \right)$$

$$= \frac{N^2}{n} \left(1 - \frac{n}{N} \right) \frac{\bar{x}\bar{x}_r}{\bar{x}_s^2} \left(\frac{1}{n} \frac{\Sigma_s r_i^2}{1 - \Sigma_s [x_i / \Sigma_s x_k]^2} \right), \qquad (5.1.14)$$

$$v_D(\hat{T}_r - T_r) = \frac{\Sigma_1^N x_i \Sigma_r x_i}{(\Sigma_s x_i)^2} \left(\sum_s \frac{r_i^2}{1 - h_{ii}} \right)$$

$$= \frac{N^2}{n} \left(1 - \frac{n}{N} \right) \frac{\bar{x}\bar{x}_r}{\bar{x}_s^2} \left(\frac{1}{n} \sum_s \frac{r_i^2}{1 - h_{ii}} \right), \qquad (5.1.15)$$

and

$$v_{J*}(\hat{T}_r - T_r) = \frac{\Sigma_1^N x_i \Sigma_r x_i}{(\Sigma_s x_i)^2} \left(\sum_s \left(\frac{r_i}{1 - h_{ii}} \right)^2 \right)$$

$$= \frac{N^2}{n} \left(1 - \frac{n}{N} \right) \frac{\bar{x}\bar{x}_r}{\bar{x}_s^2} \left[\frac{1}{n} \sum_s \left(\frac{r_i}{1 - h_{ii}} \right)^2 \right]. \qquad (5.1.16)$$

Here is a small numerical illustration.

Example 5.1.1. An Artificial Population with $N = 7$ and $n = 3$. The dashes denote nonsample cases.

x	2.1	3.8	8.1	1.0	3.3	4.2	5.5	7.0	9.0	10.0
Y	5.9	12.4	21.5	—	—	—	—	—	—	—

We have

$$\hat{T} = \frac{5.9 + 12.4 + 21.5}{2.1 + 3.8 + 8.1} (2.1 + 3.8 + 8.1 + 1 + 3.3 + 4.2 + 5.5 + 7 + 9 + 10)$$

$$= \frac{39.8}{14.0}(54) \approx 153.5$$

with

$$\sum_1^N x_i = 54, \sum_s x_i = 14 \text{ and } \sum_r x_i = 40.$$

The residuals are

$$r_1 = 5.9 - (39.8/14)\,2.1 = -.07 \Rightarrow r_1^2 = 0.0049$$
$$r_2 = 12.4 - (39.8/14)\,3.8 = 1.60 \Rightarrow r_2^2 = 2.56$$
$$r_3 = 21.5 - (39.8/14)\,8.1 = -1.53 \Rightarrow r_3^2 = 2.34$$

The corresponding leverages h_{ii} will be:

$$h_{11} = 2.1/14 = 0.15 \quad h_{22} = 3.8/14 = 0.27 \quad h_{33} = 8.1/14 = 0.58$$

Putting these ingredients together, we have

$$v_D = \frac{(54)(40)}{(14)^2}(0.0049/.85 + 2.56/.73 + 2.34/.42) = \frac{(54)(40)}{(14)^2}9.08 = 100.1.$$

In this case, the sample is very small, the leverages, especially the third, quite large and there is a major difference between v_D and the sandwich estimator v_R based on the straight squared residuals:

$$v_R = \frac{(54)(40)}{(14)^2}(0.0049 + 2.56 + 2.34) = \frac{(54)(40)}{(14)^2}4.90 = 54.05.$$

The reader is invited to calculate v_H and v_{J*} for this simple example.

5.2 VARIANCE ESTIMATION OF A LINEAR FUNCTION OF THE PARAMETER

We consider estimation of $\text{var}_M(\hat{\Theta})$ for an estimator $\hat{\Theta} = \mathbf{q}'\hat{\boldsymbol{\beta}} = \mathbf{q}'(\mathbf{X}'\mathbf{W}\mathbf{X})^{-1} \times \mathbf{X}'\mathbf{W}\mathbf{Y}$ of $\hat{\Theta} = \mathbf{q}'\boldsymbol{\beta}$. Here $\mathbf{q}$ is an arbitrary p-vector, possibly depending on the sample size n, and $\hat{\boldsymbol{\beta}}$ is the standard weighted least squares estimator based on a working model $M = M(\mathbf{X}:\mathbf{V})$, with the weights $\mathbf{W} = diag(w_i) = \mathbf{V}^{-1}$. The subscript s, indicating the data are based on a sample, is temporarily omitted.

The variance of $\hat{\Theta}$ under M is $\text{var}_M(\hat{\Theta}) = c\sigma^2$, where $c = \mathbf{q}'(\mathbf{X}'\mathbf{W}\mathbf{X})^{-1}\mathbf{q}$. Let $\mathbf{r}$ be the vector of residuals from regressing Y on $\mathbf{X}$, given by $\mathbf{r} = (\mathbf{I} - \mathbf{H})\mathbf{Y}$, where $\mathbf{H} = \mathbf{X}(\mathbf{X}'\mathbf{W}\mathbf{X})^{-1}\mathbf{X}'\mathbf{W} \equiv (h_{ij})$ is the hat matrix. Its diagonal elements $h_{ii} = \mathbf{x}_i'(\mathbf{X}'\mathbf{W}\mathbf{X})^{-1}\mathbf{x}_i w_i$ are known as *leverages* and are the generalizations of the scalar version given in Section 5.1.1.

Under the working model M, an unbiased variance estimator, most efficient among variance estimators if the variates Y are normally distributed, is $v_L = c\mathbf{r}'\mathbf{W}\mathbf{r}/(n-p)$. This is the standard, weighted least squares variance estimator.

However, suppose the actual model is $\tilde{M} = M(\mathbf{X}:\Psi)$; that is, the regression component of the working model is correct, but the variance structure is possibly misspecified: $\Psi = diag(\psi_1, \ldots, \psi_n)$ may or may not be the same as $\mathbf{V}$. This is a likely circumstance in practice, and it is not always clear that regression diagnostics, even when there is sufficient time and expertise to use them, can distinguish M and $\tilde{M}$. The estimator $\hat{\Theta}$ is unbiased under either model, but v_L is no longer unbiased or consistent for the variance

$$v = \text{var}_{\tilde{M}}(\hat{\Theta}) = \mathbf{q}'(\mathbf{X}'\mathbf{W}\mathbf{X})^{-1}\mathbf{X}'\mathbf{W}\Psi\mathbf{W}\mathbf{X}(\mathbf{X}'\mathbf{W}\mathbf{X})^{-1}\mathbf{q}. \tag{5.2.1}$$

As in Section 5.1.1 the squared residual has expectation

$$E_{\tilde{M}}(r_i^2) = \psi_i(1 - h_{ii})^2 + \sum_{j \neq i} h_{ij}^2 \psi_j, \tag{5.2.2}$$

and from this it follows that $E_{\tilde{M}}(v_L) = c(n-p)^{-1}\Sigma_s w_i\{\psi_i(1-h_{ii})^2 + \Sigma_{j\neq i} h_{ij}^2 \psi_j\}$, which is not equal to v, in general. The first example of Section 5.1 above, in which the working model was $M(0, 1{:}1)$, shows how biased v_L can be.

5.3 SANDWICH ESTIMATOR OF VARIANCE

In this section, we describe a simple robust estimator of the variance of estimators of an arbitrary linear function of the model parameter β. Suppose again the working model is $M = M(\mathbf{X}:\mathbf{V})$ and the underlying model is $\tilde{M} = M(\mathbf{X}:\Psi)$. Define the "estimation coefficients" $\mathbf{a}' = (a_1, \ldots, a_n) \equiv \mathbf{q}'(\mathbf{X}'\mathbf{W}\mathbf{X})^{-1}\mathbf{X}'\mathbf{W}$. We are interested in consistently estimating the variance of the linear expression $\hat{\Theta} = \Sigma_s a_i Y_i$. Such an estimator should yield valid confidence intervals for $\hat{\Theta} = \mathbf{q}'\boldsymbol{\beta}$, where $\boldsymbol{\beta}$ is the unknown parameter implied by M or $\tilde{M}$. The unknown variance can be written as $v \equiv \text{var}_{\tilde{M}}(\hat{\Theta}) = \mathbf{a}'\Psi\mathbf{a} = \Sigma_s a_i^2 \psi_i$.

The basic idea of robust variance estimation is to replace the unknown variance elements ψ_i by the squares of the corresponding residuals r_i^2 based on the regression fit, to arrive at the variance estimator

$$\begin{aligned} v_R &= \mathbf{a}'diag(r_i^2)\mathbf{a} \\ &= \sum_s a_i^2 r_i^2. \end{aligned} \tag{5.3.1}$$

In expression (5.2.2), under certain regularity conditions, the second term is typically of lower order than the first. The basic idea of this was illustrated in Section 5.1, and the claim is more generally implied by Theorem 5.3.1 below. Thus, asymptotically, $E_{\tilde{M}}(r_i^2) \approx \psi_i$ and so $E_{\tilde{M}}(v_R) \approx v$. This suggests that v_R is robust against deviations from the assumed variance structure. The mathematics underlying this idea is given in the next subsection.

5.3.1* Consistency of v_R

To show the consistency of v_R requires three results: a lemma gathering together several results relative to the hat matrix and estimation coefficients, and two fundamental theorems in limit theory found in Appendix A.8. The development we follow is largely inspired by Eicker (1963, 1967).

Lemma 5.3.1. Let $\mathbf{W}$, $\mathbf{H}$, h_{ij}, h_{ii}, a_i, $c = \mathbf{q}'(\mathbf{X}'\mathbf{W}\mathbf{X})^{-1}\mathbf{q}$, $r_i = Y_i - \mathbf{x}_i'\hat{\boldsymbol{\beta}}$, ψ_i, $\varepsilon_i = Y_i - \mathbf{x}_i'\boldsymbol{\beta}$ be defined as above under the working model $M(\mathbf{X}:\mathbf{V})$ and the underlying, true model $M(\mathbf{X}:\boldsymbol{\Psi})$. Then the following relationships hold:

(a) $h_{ii} = \mathbf{X}_i'(\mathbf{X}'\mathbf{W}\mathbf{X})^{-1}\mathbf{x}_i w_i$,

(b) $\mathbf{H}\mathbf{X} = \mathbf{X}$,

(c) $\mathbf{X}'\mathbf{W}(\mathbf{I} - \mathbf{H}) = \mathbf{0}$

(d) $0 \leqslant h_{ii} \leqslant 1$,

(e) $\sum_{i \in s} h_{ii} = p$,

(f) $h_{ij}h_{ji} \leqslant h_{ii}h_{jj}$,

(g) $h_{ij}^2 \leqslant h_{ii}h_{jj}w_i^{-1}w_j$,

(h) $\sum_j h_{ij}^2 = h_{ii}w_i^{-1}w_j$,

(i) $a_i^2 \leqslant c h_{ii} w_i$,

(j) $c \min\{w_i\} \leqslant \mathbf{a}'\mathbf{a} \leqslant c \max\{w_i\}$,

(k) $r_i = \varepsilon_i - \sum_j h_{ij}\varepsilon_j$,

(l) $\mathrm{var}(r_i) = \psi_i(1 - h_{ii})^2 + \sum_{j \neq i} h_{ij}^2 \psi_j$,

(m) $\mathrm{cov}(r_i, r_{i'}) = -h_{ii'}\psi_i - h_{i'i}\psi_{i'} + \sum_j h_{ij}h_{i'j}\psi_j$, and

(n) $\sum_i a_i r_i = 0$.

Proof: We will prove parts (f) and (j). The others are left to Exercise 5.5.

Let $\mathbf{A} = \mathbf{X}'\mathbf{W}\mathbf{X}$. To prove (f), note that $h_{ij}h_{ji} = \mathbf{x}_i'\mathbf{A}^{-1}\mathbf{x}_j w_j \mathbf{x}_j'\mathbf{A}^{-1}\mathbf{x}_i w_i$. Factoring $\mathbf{A}$ as $\mathbf{A} = \mathbf{A}^{1/2}\mathbf{A}^{1/2}$, rearranging, and applying the Cauchy-Schwarz inequality, we have

$$\begin{aligned} h_{ij}h_{ji} &= w_i w_j (\mathbf{x}_i'\mathbf{A}^{-1/2}\mathbf{A}^{-1/2}\mathbf{x}_j)^2 \\ &\leqslant w_i w_j (\mathbf{x}_i'\mathbf{A}^{-1}\mathbf{x}_i)(\mathbf{x}_j'\mathbf{A}^{-1}\mathbf{x}_j) = h_{ii}h_{jj}. \end{aligned}$$

For part (j), use $\mathbf{a}' = \mathbf{q}'\mathbf{A}^{-1}\mathbf{X}'\mathbf{W}$ to write

$$\begin{aligned}\mathbf{a}'\mathbf{a} &\leqslant \mathbf{q}'\mathbf{A}^{-1}\mathbf{X}'\mathbf{W}\, diag[\max(w_i)]\,\mathbf{X}'\mathbf{A}^{-1}\mathbf{q}\\ &= c\max(w_i).\end{aligned}$$

The "less than" direction of (j) is similar. □

We now state a central result on robust variance estimation, giving conditions under which the sandwich estimator is consistent. The theorem uses two theoretical results, given in Appendix A.8, on the large sample behavior of independent, non-identically distributed random variables.

Theorem 5.3.1 (Dorfman 1991). Let $t = (\hat{\Theta} - \Theta)/v_R^{1/2} = \Sigma_{i=1}^n a_i \varepsilon_i / (\Sigma_{i=1}^n a_i^2 r_i^2)^{1/2}$, where the errors $\{\varepsilon_i\}$, having mean 0 and variance ψ_i are defined by $Y_i = \mathbf{x}_i'\boldsymbol{\beta} + \varepsilon_i$, $i = 1, 2, \ldots$. Assume that $\varepsilon_1, \varepsilon_2, \ldots$ are independent with distribution functions $F_1, F_2, \ldots$ respectively. Let z represent the standard normal random variable. Suppose for positive constants b_ψ, B_ψ, b_W, B_W that the following conditions hold:

(i) $\displaystyle \sup_{1 \leqslant i \leqslant n; n = p+1, \ldots} \left(\int_{|u| > d} u^2 \, dF_i \right) \to 0, \qquad \text{as } d \to \infty$

(ii) $\displaystyle b_\psi \leqslant \int u^2 \, dF_i \leqslant B_\psi, \qquad i = 1, 2, \ldots$

(iii) $\displaystyle b_W \leqslant \inf_{n=p+1,\ldots} \min_{1 \leqslant i \leqslant n} \{w_i\} \leqslant \sup_{n=p+1,\ldots} \max_{1 \leqslant i \leqslant n} \{w_i\} \leqslant B_W$

(iv) $\displaystyle \max_{1 \leqslant i \leqslant n} \{h_{ii}\} \to 0, \qquad \text{as } n \to \infty.$

Then (a) $v_R/v \to 1$ in probability, and (b) $t \to z$ in distribution, as $n \to \infty$.

Proof. Let

$$z^{(n)} = \sum_{i=1}^n a_i \varepsilon_i \Bigg/ \left(\sum_{i=1}^n a_i^2 \psi_i \right)^{1/2} \quad \text{and} \quad s^{(n)} = \sum_{i=1}^n a_i \varepsilon_i \Bigg/ \left(\sum_{i=1}^n a_i^2 \varepsilon_i^2 \right)^{1/2}.$$

By Theorem A.8.1, $z^{(n)} \to z$ in distribution, with summands going to zero, if and only if

$$\max_{1 \leqslant i \leqslant n} \left(a_i^2 \Bigg/ \sum_{l=1}^n a_l^2 \right) \to 0, \text{ as } n \to \infty.$$

Conditions (iii) and (iv) imply this, through inequalities (g) and (h) of Lemma 5.3.1. Then, since $z^{(n)} \to z$ in distribution, we have by Theorem A.8.2 (the "only if" part), that

$$\sum a_i^2 \varepsilon_i^2 \Big/ \sum a_i^2 \psi_i \to 1 \text{ in probability, as } n \to \infty.$$

It follows that $s^{(n)} \to z$ in distribution as $n \to \infty$.

We can write the sandwich estimator as

$$\sum a_i^2 r_i^2 = \sum a_i^2 \varepsilon_i^2 - 2 \sum a_i^2 \varepsilon_i \sum h_{ij} \varepsilon_i + \sum a_i^2 \left(\sum h_{ij} \varepsilon_i \right)^2,$$

so that it is enough, for both (i) and (ii) to hold, to show that

$$u^{(n)} \equiv \sum_{k=1}^{n} a_k^2 \left(\sum_{j=1}^{n} h_{kj} \varepsilon_k \right)^2 \Big/ \sum_{i=1}^{n} a_i \psi_i \to 0 \text{ in probability as } n \to \infty.$$

But

$$E_{\tilde{M}} \left\{ \left(\sum_{j=1}^{n} h_{kj} \varepsilon_k \right)^2 \right\} = \sum_l h_{lk} h_{kl} (w_l / w_k) \psi_k \leqslant b_w^{-1} B_W B_\psi \max_{1 \leqslant i \leqslant n} \{h_{ii}\},$$

so that $E_{\tilde{M}}(u^{(n)}) \to 0$, as $n \to \infty$. But a positive random variable going to zero in expectation, must go to zero in probability (see exercise 5.6). The result follows. □

5.3.2 Some Comments on the Requirements for Consistency of the Sandwich Estimator

Some further commentary on Theorem 5.3.1 and its implications are given below.

(1) The condition (iii) can be replaced by the slightly more general

$$\text{(iii}'\text{)} \quad \sup_{n=p+1} \max_{1 \leqslant i \leqslant n} \{w_i\} / \min_{1 \leqslant i \leqslant n} \{w_i\} \leqslant B_W^*.$$

(2) Conditions (i) and (ii) refer, as in Theorem A.8.1, to conditions on the underlying random structure which are not open to the inspection of the analyst. One here can only go by indirection: previous experience, common sense, and perhaps some indirect information from diagnostics.

(3) If the fourth moments of the errors are bounded, then condition (i) is met.

(4) Conditions (iii) and (iv) translate into conditions in finite samples that may be more useful to the analyst: the range of w_i weights used in the regression should not be extreme. In particular, we should worry if there are either outliers among them on the high side, or near zero, and likewise, the leverages should be small. Some experience suggests "small" might be taken to mean less than 2 or 3 times their average p/n. If either of these conditions is violated, we should be cautious in our conclusions, even if we are using robust variance estimation.

(5) As noted in the proof of Theorem 5.3.1, conditions (iii) and (iv) also imply that no squared coefficient, a_i^2, should dominate the others. Recall from Section 2.5 that the terms $g_i = a_i + 1$ are often called the "survey weights"—especially in the context of estimating finite population means and totals. Practitioners often use *ad hoc* weight-trimming methods to cut back extreme g_i's. The theorem supports the practice, at least in a general way, as a method of ensuring asymptotic normality and well-behaved variance estimates.

(6) We should not forget that these are *asymptotic* results. In particular, if the distributions of the errors are *skewed* then it may only be with quite large n that the normality results are valid.

5.4 VARIANTS ON THE BASIC ROBUST VARIANCE ESTIMATOR

In this section, we offer some alternatives to the fundamental variance estimator, which are asymptotically equivalent to it, but in finite samples have an advantage with regard to bias. These are (a) an almost unbiased (robust) variance estimator of "internal type", labeled, for historic reasons v_D, (b) an almost unbiased (robust) variance estimator of "external type", labeled v_H, and (c) a guaranteed positively biased variance estimator v_{J*}. We also discuss (d) the classical jackknife variance estimator v_J, which, it turns out, is virtually the same as v_{J*}. In Section 5.1 we introduced some special cases of these alternative estimators. Now, we describe the forms they take for more general linear models.

The terminology "almost unbiased variance estimator" is after Horn, Horn, and Duncan (1975), and means that if the working model $M = M(\mathbf{X}:\mathbf{V})$ is in some sense close to the actual model $\tilde{M} = M(\mathbf{X}:\Psi)$, then the variance estimator should be close to being unbiased. In particular, it requires that if the working model M *coincides* with the actual model $\tilde{M}$, the variance estimator should be strictly unbiased for the variance.

5.4.1. Internal and External Adjustments to the Sandwich Estimator

From equation (5.1.2), it follows that if $M = \tilde{M}$, then $E_{\tilde{M}}(r_i^2) = (1 - h_{ii})\psi_i$, for $E_{\tilde{M}}(\varepsilon_i^2) = \psi_i$ (see Exercise 5.8). Hence

$$E_{\tilde{M}}(v_R) = E_{\tilde{M}}\left(\sum_s a_i^2 r_i^2\right) = \sum_s a_i^2(1 - h_{ii})\psi_i < v,$$

if the working model is correct. There are several possible strategies to correct for this downward bias:

(a) The almost unbiased estimator of *internal type* is defined by

$$v_D = \sum_s a_i^2 \hat{\psi}_i, \qquad \text{with } \hat{\psi}_i = (1 - h_{ii})^{-1} r_i^2. \tag{5.4.1}$$

This estimator adjusts the squared residuals individually to make each an unbiased estimator of ψ_i under the working model. It is immediate that, for $M = \tilde{M}$,

$$E_{\tilde{M}}(v_D) = \sum_s a_i^2 \psi_i = v.$$

Under condition (iv) of Theorem 5.3.1, we have $1 \leqslant v_D/v_R \leqslant 1 + R_n$, with $R_n \to 0$, as $n \to \infty$, so that in any case, under the conditions of that theorem, v_D is robust against unknown heteroscedasticity; that is, $v_D/v \to 1$ in probability, and $(\hat{\Theta} - \Theta)/v_D^{1/2}$ is asymptotically distributed as a standard normal variate.

(b) The almost unbiased estimator of *external type* is defined by

$$v_H = v_R \frac{\Sigma_s a_i^2 v_i}{\Sigma_s a_i^2 v_i (1 - h_{ii})}, \tag{5.4.2}$$

for v_i equal or proportional to the variances under the working model M. Note that we can also write v_H in the same general form as v_D:

$$v_H = \sum_s a_i^2 \hat{\psi}_i,$$

where

$$\hat{\psi}_i = r_i^2 \frac{\Sigma_s a_i^2 v_i}{\Sigma_s a_i^2 v_i (1 - h_{ii})}.$$

The estimator v_H makes an overall adjustment to v_R. Again, it is readily shown that $E_M(v_H) = v$ for $M = \tilde{M}$, and that under condition (iv) of Theorem 5.3.1, $1 \leqslant v_H/v_R \leqslant 1 + R_n$, with $R_n \to 0$, as $n \to \infty$.

The symbols v_D and v_H are adopted from the literature, and these variance estimators are sometimes referred to as being of "*D*-type" and of "*H*-type." One mnemonic for distinguishing them, is that the *D*-type has the bias adjustment *d*own in the point estimate $\hat{\psi}_i$ of ψ_i, while the adjustment for the *H*-type variance estimator v_H *h*overs outside of v_R. Both of these estimators are geared to the idea of forcing the variance estimator to be unbiased under the working model.

(c) A more cautious approach is to guarantee that the variance estimator is not biased downwards, whatever the actual variance structure. This should lead to valid or conservative confidence intervals. Such a bias-positive variance estimator is surprisingly easy to achieve. We define

$$v_{J*} = \sum_s a_i^2 \hat{\psi}_i \qquad \text{with} \quad \hat{\psi}_i = (1 - h_{ii})^{-2} r_i^2. \tag{5.4.3}$$

Then using (5.2.2) one can show that $E_{\tilde{M}}(v_{J*}) \geqslant v$, whatever the variance structure of $\tilde{M}$ (see Exercise 5.9). Note that the only difference between this variance estimator and v_D is in the exponent of the factor $(1 - h_{ii})$. As for v_D, one can show that under condition (iv) of Theorem 5.3.1, this variance estimator is asymptotically close to v_R, with consequent consistency for the variance, and asymptotic standard normality of the t-statistic standardized by the square root of v_{J*}. In addition, we can show that, under the standard conditions of Theorem 5.3.1, v_{J*} is virtually identical to the standard jackknife variance estimator, except that it is simpler and easier to calculate. We may therefore refer to v_{J*} as the *quasi-jackknife variance estimator*.

5.4.2 Jackknife Variance Estimator

The jackknife is a standard tool in the sampler's armament of variance estimators (Wolter, 1985, Chapter 4). The general idea in the delete-one jackknife is to omit one of the sample units, compute a point estimate from the remaining $n - 1$ units, cycle through all n units this way, and then compute the variance among the n delete-one estimates. There are versions of the jackknife based on deleting more than one unit at a time (Shao and Wu, 1989; Shao and Tu, 1995), but we shall not cover these here. In the context of the general estimation of $\Theta = \mathbf{q}'\boldsymbol{\beta}$, we define $v_J = [(n-1)/n]\Sigma_{i=1}^n(\hat{\Theta}_{(i)} - \hat{\Theta}_{(\cdot)})^2$, where $\hat{\Theta}_{(i)} = \mathbf{q}'\hat{\boldsymbol{\beta}}_{(i)}$, with $\hat{\boldsymbol{\beta}}_{(i)} = (\mathbf{X}'_{(i)}\mathbf{W}_{(i)}\mathbf{X}_{(i)})^{-1}\mathbf{X}'_{(i)}\mathbf{W}_{(i)}\mathbf{Y}_{(i)}$ being the best linear unbiased estimator of $\boldsymbol{\beta}$ under the working model M, omitting the ith point, and $\hat{\Theta}_{(\cdot)} = n^{-1}\Sigma_{i=1}^n\hat{\Theta}_{(i)}$. In what follows we ignore the minor adjustment $(n-1)/n$.

To connect v_J with the other robust variance estimators of this section, particularly v_{J*}, we first note the next lemma.

Lemma 5.4.1 Let $\mathbf{A}$ be an $n \times n$ nonsingular matrix and $\mathbf{u}$, $\mathbf{v}$ be vectors of length n. Then

$$(\mathbf{A} - \mathbf{u}\mathbf{v}')^{-1} = \mathbf{A}^{-1} + \frac{\mathbf{A}^{-1}\mathbf{u}\mathbf{v}'\mathbf{A}^{-1}}{1 - \mathbf{v}'\mathbf{A}^{-1}\mathbf{u}}.$$

Proof: This standard result is readily confirmed by direct multiplication. □

Theorem 5.4.1. Under the conditions of Theorem 5.3.1, $v_J/v_{J*} \to 1$ in probability.

Proof: We note that $\mathbf{X}'_{(i)}\mathbf{W}_{(i)}\mathbf{X}_{(i)} = \mathbf{X}'\mathbf{W}\mathbf{X} - w_i\mathbf{x}_i\mathbf{x}_i'$, where $\mathbf{x}_i$ represents the

ith row of $\mathbf{X}$, and $\mathbf{X}'_{(i)}\mathbf{W}_{(i)}\mathbf{Y}_{(i)} = \mathbf{X}'\mathbf{W}\mathbf{Y} - w_i\mathbf{x}_i Y_i$. Then application of Lemma 5.4.1 yields $\hat{\Theta}_{(i)} - \hat{\Theta} = -\mathbf{q}'(\mathbf{X}'\mathbf{W}\mathbf{X})^{-1}\mathbf{x}_i w_i r_i/(1-h_{ii}) = -a_i r_i/(1-h_{ii})$. Using this relationship, we can rewrite the jackknife variance estimator as $v_J = \Sigma a_i^2 r_i^2/(1-h_{ii})^2 - n^{-1}\{\Sigma a_i r_i/(1-h_{ii})\}^2$. The first term is just v_{J*}.

We now argue that, under the conditions of Theorem 5.3.1, the second term is negligible relative to the first. Since Theorem 5.3.1 (iv) implies that $v_{J*}/\Sigma a_i^2\psi_i \to 1$ in probability, it suffices to show that $\{\Sigma a_i r_i/(1-h_{ii})\}/\{n\Sigma a_i^2\psi_i\}^{1/2} \to 0$ in probability. Since $\Sigma a_i r_i = 0$, we have $\Sigma a_i r_i/(1-h_{ii}) = \Sigma h_{ii} a_i r_i/(1-h_{ii})$. The expectation of this expression is zero, so that by Chebyshev's Inequality (Appendix A.4) the proof reduces to the claim that

$$\operatorname{var}_{\tilde{M}}\left\{\sum h_{ii}a_i r_i/(1-h_{ii})\right\}\bigg/\left\{n\sum a_i^2\psi_i\right\} \to 0 \tag{5.4.4}$$

From Lemma 5.3.1(g) and (l) and Theorem 5.3.1(ii), it follows that

$$\begin{aligned}
\operatorname{var}(r_i) &= \psi_i(1-h_{ii})^2 + \sum_{j\neq i} h_{ij}^2\psi_j \\
&\leqslant B_\psi\left[1 - 2h_{ii} + \sum_j h_{ij}^2\right] \qquad \text{by Theorem 5.3.1(ii)} \\
&\leqslant B_\psi[1 + h_{ii}w_j/w_i] \qquad \text{by Lemma 5.3.1(h) and } h_{ii} \geqslant 0 \\
&\leqslant B_\psi[1 + b_w^{-1}B_W \max(h_{ii})] \qquad \text{by Theorem 5.3.1(iii)}
\end{aligned}$$

Lemma 5.3.1(m) and (g) imply that

$$|\operatorname{cov}(r_i, r_l)| \leqslant B_\psi\left\{h_{il} + h_{li} + \sum_j (h_{ii}h_{ij}h_{ll}h_{jj}w_j^2 w_j^{-1} w_l^{-1})^{1/2}\right\}$$

by Theorem 5.3.1(ii) and Lemma 5.3.1(g)

$$\leqslant B_\psi\left[h_{il} + h_{li} + \frac{B_W}{b_w} p \max(h_{ii})\right]$$

by Lemma 5.3.1(e) and Theorem 5.3.1(iii)

$$\leqslant B_\psi \max(h_{ii})[2B_W^{1/2}b_W^{-1/2} + B_W b_W^{-1} p],$$

for p the rank of $\mathbf{X}$. Then the left-hand side of (5.4.4) equals $A + B$, where

$$A = \frac{\Sigma h_{ii}^2 a_i^2 \operatorname{var}(r_i)/(1-h_{ii})^2}{n\Sigma a_i^2\psi_i} \leqslant \max\left(\frac{h_{ii}}{1-h_{ii}}\right)^2 n^{-1} b_\psi^{-1} \max[\operatorname{var}(r_i)] \to 0$$

and

$$\begin{aligned}B &= \frac{\Sigma_{l \neq i}\Sigma\, h_{ii}h_{ll}a_i a_l \operatorname{cov}(r_i, r_l)/\{(1-h_{ii})(1-h_{ll})\}}{n\Sigma a_i^2 \psi_i} \\ &\leqslant \frac{\Sigma_{l \neq i}\Sigma h_{ii}h_{ll}|a_i||a_l||\operatorname{cov}(r_i, r_l)|/\{(1-h_{ii})(1-h_{ll})\}}{n\Sigma a_i^2 \psi_i} \\ &\leqslant \max\left(\frac{h_{ii}}{1-h_{ii}}\right)^2 b_\psi^{-1} \max[|\operatorname{cov}(r_i, r_l)|]\left[\frac{(\Sigma|a_i|)^2 - \Sigma a_i^2}{n\Sigma a_i^2}\right].\end{aligned} \tag{5.4.5}$$

Next, we have $(\Sigma|a_i|)^2 - \Sigma a_i^2 \leqslant (\Sigma|a_i|)^2 \leqslant p^2 c B_W \max(h_{ii})$ by Lemma 5.3.1(i) and $\Sigma a_i^2 \geqslant cnb_w$ by Lemma 5.3.1(j). Continuing from (5.4.5), we have

$$B \leqslant \max\left(\frac{h_{ii}}{1-h_{ii}}\right)^2 b_\psi^{-1} \max[|\operatorname{cov}(r_i, r_l)|]\frac{p^2 B_W}{nb_w}\max(h_{ii}) \to 0$$

which completes the proof. □

Consequently, v_R, v_D, v_H, v_{J*}, and v_J are all consistent and will often be numerically similar in large samples. Note, however, that unlike v_R, each of v_D, v_H, v_{J*}, and v_J involves a division by $1 - h_{ii}$ or $(1 - h_{ii})^2$. By Lemma 5.3.1(d) all of the h_{ii}'s will be between 0 and 1, and in large samples, we expect h_{ii} to approach 0. However, in small to moderate size samples, values of h_{ii} near 1 can seriously inflate the variance estimates. Thus, checking whether $1 - h_{ii}$ is near zero is important in numerical calculations. Large values of the leverages h_{ii} may also cause v_{J*} and v_J to be more unstable than the other alternatives because both use $(1 - h_{ii})^{-2}$.

The proof of Theorem 5.4.1 also contains this practically useful formula for the jackknife:

$$v_J = \frac{n-1}{n}\left\{\sum_s \frac{a_i^2 r_i^2}{(1-h_{ii})^2} - n^{-1}\left[\sum_s \frac{a_i r_i}{(1-h_{ii})}\right]^2\right\}. \tag{5.4.6}$$

We have inserted the factor $(n - 1)/n$ in (5.4.6) to conform to the standard definition. Expression (5.4.6) allows the jackknife to be computed directly from the estimation coefficients, residuals, and leverages rather than going through the potentially laborious process of computing the $\hat{\Theta}_{(i)}$ individually. Thus, a common set of ingredients goes into calculating v_R, v_D, v_H, v_{J*}, and v_J—a feature that facilitates computer programming.

The jackknife is an example of what are known as *replication estimators*. Other examples are balanced repeated replication (or balanced half-sampling), which is studied in Chapter 10, and the bootstrap. The general idea is to form subsamples or replicates of the full sample and then to compute the point estimator, $\hat{\Theta}$ for example, from each replicate. The variance among the replicate estimates is then calculated and multiplied by an appropriate constant to get an estimator of variance for the full-sample point estimator. Be aware, however, that the jackknife may not work for all types of estimators. In particular, the delete-one jackknife is not a consistent estimator of the variance of an estimated quantile of a distribution (Efron, 1981).

5.5. VARIANCE ESTIMATION FOR TOTALS

After the general preliminaries of the preceding section, we now return to the sampling context, and to the question of robust variance estimation for the error $\hat{T} - T$ or, equivalently, of $\hat{T}_r - T_r$. As in the preceding section, our working model is $M = M(\mathbf{X}:\mathbf{V})$, and the underlying model is $\tilde{M} = M(\mathbf{X}:\boldsymbol{\Psi})$. The notation is the same as in the preceding section, except that now the vector $\mathbf{q}'$ is $\mathbf{q}' = \mathbf{1}_r'\mathbf{X}_r$, so that

$$a_i = \mathbf{1}_r'\mathbf{X}_r(\mathbf{X}_s'\mathbf{W}_s\mathbf{X}_s)^{-1}\mathbf{x}_i w_i \qquad \text{for } i \in s, \tag{5.5.1}$$

$\hat{T}_r = \Sigma_s a_i Y_i$, and $\hat{T} = \Sigma_s Y_i + \Sigma_s a_i Y_i$. We want to estimate $\text{var}_M(\hat{T}_r - T_r) = \text{var}_M(\hat{T}_r) + \text{var}_M(T_r) = \Sigma_s a_i^2 \psi_i + \Sigma_r \psi_j$. We term the first component of the variance, $\text{var}_M(\hat{T}_r)$, the *sample* (*based*) *component* and the second, $\text{var}_M(T_r)$, the *nonsample component*.

For purposes of asymptotic results, we make the following reasonable set of assumptions:

(i) $n, N \to \infty$.

(ii) As n and N grow, the assumptions of Theorem 5.3.1 hold and, in addition, the absolute values of the x_{ij} predictor variables are bounded above.

(iii) The sum of L terms in the variates and weights is of order L, for $L = n$, N, or $N - n$ (e.g. $\Sigma_s \mathbf{x}_i/n$ is bounded).

(iv) The sampling fraction $f \equiv n/N \to 0$.

Assumption (iv) is probably not entirely necessary, and in 5.5.1 we suggest methods for when it does not hold. Note that for $N_r = N - n$, condition (iv) implies that N_r and N have the same order of magnitude and $N_r/n \to \infty$.

The above assumptions imply that a_i is of order N_r/n. From this, the sample component $\text{var}_{\tilde{M}}(\hat{T}_r) = \Sigma_s a_i^2 \psi_i$ is of order N_r^2/n, and so of greater order than the nonsample component $\text{var}_M(T_r) = \Sigma_r \psi_j$, which is of order N_r. That is, under (iv), the nonsample component of the variance is relatively negligible. This is in contrast to the typical prediction problem in standard regression, where $N = n + 1$, and the prediction (nonsample) component dominates.

The estimators of Sections 5.3 and 5.4 suffice to estimate the sample component $\text{var}_{\tilde{M}}(\hat{T}_r) = \Sigma_s a_i^2 \psi_i$. We need not rewrite any of the formulas therein; thus, for example, $v_D(\hat{T}_r) = \Sigma a_i^2 \hat{\psi}_i$, with $\hat{\psi}_i = (1 - h_{ii})^{-1} r_i^2$, where residuals and leverages are defined as in the previous sections for the general case, and only the definition of the estimation coefficients a_i reflects the fact that we are estimating here the variance of an estimator of *total*.

Since we assume $f = n/N$ is small, the nonsample component, $\text{var}_M(T_r)$, is an order of magnitude lower than the sample component, and we are safe in using reasonable approximations to estimate it. The basic approximation, good for n/N small, will be called the *working model adjustment*, given by $\hat{\text{var}}_M(T_r) = \Sigma_r v_j (\Sigma_s v_i)^{-1} \Sigma_s \hat{\psi}_i$, where $\hat{\psi}_i$ is one of the estimates of ψ_i discussed above, namely,

$$\hat{\psi}_i = r_i^2,$$
$$\hat{\psi}_i = r_i^2 (1 - h_{ii})^{-1},$$
$$\hat{\psi}_i = r_i^2 \frac{\Sigma_s a_i^2 v_i}{\Sigma_s a_i^2 v_i (1 - h_{ii})}, \qquad \text{and}$$
$$\hat{\psi}_i = r_i^2 (1 - h_{ii})^{-2}$$

for v_R, v_D, v_H, and v_{J*}, respectively. The example of Section 5.1.2 for the ratio estimator typifies this approach. A robust estimator of the error variance of $\hat{T}$ is then

$$\hat{\text{var}}_M(\hat{T} - T) = \hat{\text{var}}_M(\hat{T}_r) + \frac{\Sigma_r v_j}{\Sigma_s v_j} \sum_s \hat{\psi}_i, \tag{5.5.2}$$

where $\hat{\text{var}}_M(\hat{T}_r) = v_R(\hat{T}_r)$, $v_D(\hat{T}_r)$, $v_H(\hat{T}_r)$, or $v_{J*}(\hat{T}_r)$. A "quick-and-dirty" variant is just $\hat{\text{var}}_M(\hat{T} - T) = (N/N_r)\hat{\text{var}}_M(\hat{T}_r)$.

For the jackknife, we will use $v_J(\hat{T}_R)$ as defined in (5.4.6) plus the estimator of $\text{var}_M(T_R)$ that accompanies v_{J*} so that

$$v_J(\hat{T} - T) = \frac{n-1}{n} \left\{ \sum_s \frac{a_i^2 r_i^2}{(1 - h_{ii})} - n^{-1} \left[\sum_s \frac{a_i r_i}{(1 - h_{ii})} \right]^2 \right\} + \frac{\Sigma_r v_j}{\Sigma_s v_j} \sum_s \hat{\psi}_i \tag{5.5.3}$$

with $\hat{\psi}_i = r_i^2 (1 - h_{ii})^{-2}$.

In practice, for programming purposes, we shall not need to specify details of the variance estimators under various models. A "modular" approach suffices: we use the general formulas for the robust variance estimators, which can be programmed as functions of the vectors $\mathbf{a} = (a_1, \ldots, a_n)'$, $\mathbf{h} = (h_{11}, \ldots, h_{nn})'$, and $\mathbf{r} = (r_1, \ldots, r_n)'$. (See the S-PLUS™ programs *ahr*, *vR*, *vD*, *vH*, *vJ.star*, and *vJ* in Appendix C.) General programs for calculation of the vectors, **a**, **h**, and **r**, can be written to accommodate the specific working model either directly, if formulas have been worked out for the model in question, or using their general formulas in terms of the response variable **Y**, the predictor matrix **X**, and weights **W** implied by the working model. This approach provides great clarity and flexibility of operation, and saves us from getting lost in details. The following examples and Exercises 5.16 and 5.17 illustrate this approach.

5.5.1 Some Simple Examples

This section illustrates the robust variance estimators for a particular minimal model. The exercises ask you to work out the details for several other working models. We also note at the end of this section that, although estimators of the total may reduce to simpler forms with balanced sampling, the variance estimators do not.

Example 5.5.1. Minimal Model Estimator

Consider the model $M(0, 1, 1{:}x^2)$, that is, the *minimal quadratic model*:

$$Y_i = \beta_1 x_i + \beta_2 x_i^2 + e_i, \qquad e_i \sim (0, v_i) \qquad i = 1, 2, \ldots, N \tag{5.5.4}$$

with variances proportional to *x-squared*: $v_i = x_i^2 \sigma^2$. Least squares estimators of β_1 and β_2 are

$$\hat{\beta}_1 = \frac{1}{\Sigma_s (x_i - \bar{x}_s)^2} \left\{ \bar{x}_s^{(2)} \sum_s \frac{Y_i}{x_i} - \bar{x}_s \sum_s Y_i \right\}$$

$$\hat{\beta}_2 = \frac{1}{\Sigma_s (x_i - \bar{x}_s)^2} \left\{ -\bar{x}_s \sum_s \frac{Y_i}{x_i} + \sum_s Y_i \right\}$$

It follows that the fitted values are

$$\hat{Y}_i = \hat{\beta}_1 x_i + \hat{\beta}_2 x_i^2, \qquad i = 1, 2, \ldots, N$$

and the residuals are

$$r_i = Y_i - \hat{Y}_i = Y_i - \hat{\beta}_1 x_i - \hat{\beta}_2 x_i^2, \qquad \text{for } i \in s.$$

The leverages are

$$h_{ii} = \frac{1}{\Sigma_s (x_i - \bar{x}_s)^2} \{ \bar{x}_s^{(2)} - 2 x_i \bar{x}_s + x_i^2 \}, \qquad i \in s,$$

and the estimator of total is

$$\hat{T} = \sum_s Y_i + \sum_r \hat{Y}_i = \sum_s Y_i + \sum_r (\hat{\beta}_1 x_i + \hat{\beta}_2 x_i^2).$$

Some algebra shows that another way to write the second term is

$$\sum_r \hat{Y}_i = \sum_s a_i Y_i, \qquad \text{where}$$

$$a_i = \frac{N-n}{\Sigma_s(x_i - \bar{x}_s)^2}\left\{(\bar{x}_s^{(2)}\bar{x}_r - \bar{x}_s\bar{x}_r^{(2)})\frac{1}{x_i} + (-\bar{x}_s\bar{x}_r + \bar{x}_r^{(2)})\right\}.$$

Thus the error variance is

$$\begin{aligned}\text{var}_M(\hat{T} - T) &= \text{var}_M\left(\sum_{i\in r}\hat{Y}_i - \sum_{i\in r}Y_i\right) = \text{var}_M\left(\sum_{i\in r}\hat{Y}_i\right) + \text{var}_M\left(\sum_{i\in r}Y_i\right)\\ &= \sum_{i\in s}a_i^2 v_i + \sum_r v_i,\end{aligned}$$

and to get v_R, v_D, v_{J*}, we set $\hat{v}_i = r_i^2,\ r_i^2/(1-h_{ii}),\ r_i^2/(1-h_{ii})^2$, respectively, replacing v_i in the first term by $\hat{v}_i$, and the second term as a whole by $(\Sigma_r x_i^2/\Sigma_s x_i^2)\Sigma_s\hat{v}_i$. All told, this gives an estimate of the error variance such as

$$v_D = \sum_s a_i^2\frac{r_i^2}{1-h_{ii}} + \frac{\Sigma_r x_i^2}{\Sigma_s x_i^2}\sum_s\frac{r_i^2}{1-h_{ii}}.$$

Exercise 5.15 asks to show that

$$v_H = \frac{\Sigma_s a_i^2 x_i^2}{\Sigma_s a_i^2 x_i^2(1-h_{ii})}\sum_s a_i^2 r_i^2 + \frac{\Sigma_r x_i^2}{\Sigma_s x_i^2(1-h_{ii})}\sum_s r_i^2.$$

Next, consider the case of the minimal model

$$Y_i = \beta_1 x_i^{1/2} + \beta_2 x_i + e_i, \qquad e_i \sim (0, v_i) \qquad i = 1, 2, \ldots, N \tag{5.5.5}$$

with variances proportional to x: $v_i = x_i\sigma^2$, leading to the minimal model estimator of total that serves as a rival to the ratio estimator. All of the variance expressions above hold with x_i replaced by $x_i^{1/2}$.

Example 5.5.2. Numerical Illustration with a Minimal Model Estimator

The table below lists a sample of $n = 20$ units selected from the Hospitals population with probabilities proportional to x. The value of the minimal model estimator under $M(0,1,1{:}x^2)$ is $\hat{T}(0,1,1{:}x^2) = 309{,}623.2$. The values of square roots of the robust variance estimates from this sample are $\sqrt{v_R} = 17{,}885.06$, $\sqrt{v_D} = 19{,}140.38$, $\sqrt{v_H} = 19{,}238.43$, $\sqrt{v_{J*}} = 20{,}494.77$, $\sqrt{v_J} = 20{,}494.77$. The estimates $\sqrt{v_{J*}}$ and $\sqrt{v_J}$ are identical to two decimals, indicating that the subtractive term in (5.4.6) is negligible.

The table also lists the sample leverages. A rule-of-thumb based on the leverages (Belsley, Kuh, and Welsch, 1980, p.17), is to look for points where $h_{ii} > 2p/n$; such "high leverage" points can have a disproportionate effect on estimates. In this sample all units fall well within this cutoff since $2p/n = 0.20$.

Unit No.	h_{ii}	Unit No.	h_{ii}
223	0.15	345	0.15
231	0.13	367	0.14
238	0.14	372	0.14
240	0.10	377	0.13
252	0.05	380	0.13
276	0.05	381	0.09
283	0.03	387	0.11
300	0.02	389	0.15
326	0.02	390	0.12
336	0.01	391	0.14

Cautionary Note: We learned in Chapter 3 that, in the case of balanced samples of sufficiently high order, subject to certain restrictions on the variance structure, the best linear unbiased estimators reduce to the simple expansion estimator or, in the case of weighted balance, to a weighted expansion estimator. This does *not* mean that the corresponding *variance* estimators reduce to the variance estimators that would be appropriate to the expansion estimator. What happens is that, under balance, the vector **a** simplifies to its value under the expansion estimator model, but neither **h** nor, more crucially, **r** reduces. In particular, the values of **r** can be very different under the assumed and reduced models, tending to be much larger (in absolute value) under the simpler expansion model. Thus, even under balance we are not quite off the hook from doing our best to specify the model correctly.

Example 5.5.3. Ratio Estimator

Under (standard, unweighted) balance the ratio estimator, with $a_i = N\bar{x}/(n\bar{x}_s)$, reduces to the expansion estimator, with $a_i = N/n$, the value that would be appropriate under $M(1:1)$. But, $h_{ii} = x_i/n\bar{x}_s$ does not reduce to $1/n$ nor does $r_i = y_i - x_i\bar{y}_s/\bar{x}_s$ reduce to $y_i - \bar{y}_s$, the corresponding values appropriate under $M(1:1)$. As a result, the appropriate construction of a robust variance estimator is unaffected by whether we have a balanced sample or not.

5.5.2* Effect of a Large Sampling Fraction

In unstratified samples, it is rare that the sampling fraction is large. In stratified sampling, however, that we will consider in the next chapter, it is common that a large proportion is sampled in some strata. When the sampling fraction n/N is large, it no longer is the case that $\text{var}_M(\hat{T}_r)$ dominates $\text{var}_M(T_r)$.

In this situation, the sample residuals can be expected to provide information for estimating nonsample variances, so long as some functional relation exists between the auxiliary variable(s) and the variances. If such a relation does exist, then one can apply either a parametric or nonparametric regression model fitting the squared residuals to the auxiliary. The latter might be the safest approach if each nonsample unit has several nearby (with respect to values of the auxiliary) sample units.

Probably the simplest nonparametric regression approach, familiar to samplers, is "post-stratification"; the population is divided into mutually exclusive sub-groups g, and the average of sample values of r_i^2, $i \in g \cap s$, assigned to ψ_i, $i \in g \cap r$. Having a balanced sample, giving a spread of sample points through the population, should facilitate such a procedure.

The effect of such procedures on the validity of confidence intervals has not, to our knowledge, been investigated.

5.6 MISSPECIFICATION OF THE REGRESSION COMPONENT

The robust variance estimators that we have been considering deal with the possibility that the *variance component* of the working model is incorrect. Having adopted these variance estimators, we now address the important question of the consequences for variance estimation and inference of deviation from the *regression component* of the model. We can expect the consequences to be more serious. The analysis here is fairly informal; Royall and Cumberland (1978, sec. 4) study the issues in more detail. Under reasonable conditions, when the regression component of the model is misspecified:

1. A robust variance estimator will *over*estimate the variance of the estimator of T, but will still be of the same order of magnitude.
2. The mean square error will in general be of a higher order than the variance estimator, due to the presence of bias in $\hat{T}_r$.
3. When the bias is uncontrolled, the coverage of confidence intervals can be far less than the nominal value.
4. However, as outlined in Chapters 3 and 4, under appropriate conditions of *balance*, the bias of $\hat{T}_r$ shrinks to zero.

The implications of these points are rather important, since, as we have emphasized, it may be taken as a maxim that the regression component of the working model typically deviates from truth to some extent. On the one hand, if no precautions are taken regarding the design of the sample, use of the robust variance estimators will *not* guarantee sound or conservative confidence intervals. On the other hand, if the conditions on balance *are* met, as delineated in Section 4.2.1, then coverage will be *conservative*, since then the bias shrinks to zero, and the robust variance estimators will tend to overestimate the variance. While not ideal, this situation may, in fact, be the best we can hope for in practice.

Point (4) above was established in Chapter 3. It remains to elaborate the other three points.

In what follows we ignore, for simplicity, the second-order term $\text{var}_M(T_r)$ in $\text{var}_M(\hat{T}_r - T_r) = \text{var}_M(\hat{T} - T)$. We consider the *robust* variance estimators only. We assume the conditions of Theorem 5.3.1 and, in addition, other mild conditions, as stated below. Suppose the working model is $M(\mathbf{X}:\mathbf{V})$ but the correct underlying model is $\tilde{M}(\mathbf{X}, \mathbf{Z}:\Psi)$, where $\mathbf{V}$ and Ψ are assumed diagonal. That is, $Y_i = \mathbf{x}_i'\boldsymbol{\beta} + \mathbf{z}_i'\gamma + \psi_i^{1/2}\varepsilon_i$, for $i = 1, 2, \ldots, N$, with the $\varepsilon_i \sim (0, \sigma^2)$. The matrix $\mathbf{X}$ often represents several powers of a single variate x, and $\mathbf{Z}$ might represent higher powers or an intercept term, but also conceivably an extraneous variable. Conceivably, too, the regression coefficients β corresponding to the columns of $\mathbf{X}$ might be zero, meaning that the variates in the working model are totally irrelevant to the behavior of the variable of interest.

The estimator of the nonsample total is $\hat{T}_r = \Sigma_s a_i Y_i$, where the a_i are determined by $\mathbf{X}$ and $\mathbf{V}$. Then, $\text{var}_{\tilde{M}}(\hat{T}_r) = \Sigma_s a_i^2 \psi_i$, regardless of misspecification of the regression component of the model. On the other hand, each robust variance estimator has the form $\hat{\text{var}}(\hat{T}_r) = \Sigma_s a_i^2 \hat{\psi}_i$ with $\hat{\psi}_i \cong r_i^2$, and

$$E_{\tilde{M}}[\hat{\text{var}}(\hat{T}_r)] \cong \sum_s a_i^2 E_{\tilde{M}}(r_i^2) \cong \sum_s a_i^2 \{\psi_i + [E_{\tilde{M}}(r_i)]^2\}$$
$$= \text{var}_{\tilde{M}}(\hat{T}_r) + \sum_s a_i^2 [E_{\tilde{M}}(r_i)]^2.$$

Thus, as a *variance* estimator a robust variance estimator is positively biased. Additionally, $E_{\tilde{M}}(r_i) = \mathbf{z}_i'\gamma - \mathbf{x}_i'(\mathbf{X}_s'\mathbf{W}_s\mathbf{X}_s)^{-1}\mathbf{X}_s'\mathbf{W}_s\mathbf{Z}_s\gamma$ is of the order of a constant, so long as the weights, and the (absolute value of the) variates making up $\mathbf{X}$ and $\mathbf{Z}$ are bounded above and away from zero. Thus, the bias of $\hat{\text{var}}(\hat{T}_r)$ as an estimator of $\text{var}_{\tilde{M}}(\hat{T}_r)$ is of the same order as $\text{var}_{\tilde{M}}(\hat{T}_r)$ itself, namely, $O(N_r^2/n)$. This establishes point (1).

Now,

$$E_{\tilde{M}}(\hat{T}_r - T_r) = \mathbf{a}'(\mathbf{X}_s\boldsymbol{\beta} + \mathbf{Z}_s\gamma) - \mathbf{1}_r'(\mathbf{X}_r\boldsymbol{\beta} + \mathbf{Z}_r\gamma) = (\mathbf{a}'\mathbf{Z}_s - \mathbf{1}_r'\mathbf{Z}_r)\gamma,$$

since $\mathbf{a}'\mathbf{X}_s = \mathbf{1}_r'\mathbf{X}_r(\mathbf{X}_s'\mathbf{W}_s\mathbf{X}_s)^{-1}\mathbf{X}_s'\mathbf{W}_s\mathbf{X}_s = \mathbf{1}_r'\mathbf{X}_r$. Under the same weak assumptions on $\mathbf{X}$ and $\mathbf{Z}$ and the weights, $\mathbf{1}_r'\mathbf{Z}_r = O(N_r)$, and

$$\mathbf{a}'\mathbf{Z}_s = \mathbf{1}_r'\mathbf{X}_r(\mathbf{X}_s'\mathbf{W}_s\mathbf{X}_s)^{-1}\mathbf{X}_s'\mathbf{W}_s\mathbf{Z}_s = O(N_r)O(n^{-1})O(n) = O(N_r).$$

Thus, in general, the bias of $\hat{T}_r$ is of order N_r, and the full error variance (or mean square error) is $E_{\tilde{M}}(\hat{T}_r - T_r)^2 = O(N_r^2) > O(N_r^2/n)$, the order of the variance and of the variance estimators, establishing point (2).

Next, consider whether the presence of $\mathbf{Z}$ and Ψ in the model affects the coverage of confidence intervals. The pivot used for constructing a confidence interval can be written as

$$\begin{aligned}\frac{\hat{T}-T}{[\hat{\text{var}}(\hat{T}-T)]^{1/2}} &= \frac{\hat{T}_r - T_r}{[\hat{\text{var}}(\hat{T}_r - T_r)]^{1/2}} \cong \frac{\hat{T}_r - T_r}{[\hat{\text{var}}(\hat{T}_r)]^{1/2}} \\ &= \frac{\hat{T}_r - T_r - E_{\tilde{M}}(\hat{T}_r - T_r)}{[\text{var}_{\tilde{M}}(\hat{T}_r)]^{1/2}} \frac{[\text{var}_{\tilde{M}}(\hat{T}_r)]^{1/2}}{[\hat{\text{var}}(\hat{T}_r)]^{1/2}} + \frac{E_{\tilde{M}}(\hat{T}_r - T_r)}{[\hat{\text{var}}(\hat{T}_r)]^{1/2}}. \\ &\equiv A \cdot B + C. \end{aligned} \tag{5.6.1}$$

The terms A, B, and C have the obvious definitions and will be useful for later reference.

Because the bias of $\hat{T}_r$ is, in general, of order N_r while the variance estimators have order $O(N_r^2/n)$, the C term on the right-hand side of (5.6.1) can be asymptotically unbounded. Thus, in this worst case, the coverage probability of a confidence interval would approach 0 in large samples, supporting point (3) above. Generally, the pivot in (5.6.1) can have a distribution that is far from even an approximation to the standard normal distribution that yields sound confidence intervals. Thus precautions are needed.

The principal precaution we can take is to seek that our sample be *balanced*, in the sense of Chapters 3 and 4. If, in the notation of Section 4.2.1, we have $\mathbf{V1}_N$, $\mathbf{V}^{1/2}\mathbf{1}_N \in \mathcal{M}(\mathbf{X})$ and $s \in B(\mathbf{X}, \mathbf{Z}:\mathbf{V})$, then the bias $E_{\tilde{M}}(\hat{T} - T) = [\mathbf{1}_r'\mathbf{X}_r(\mathbf{X}_s'\mathbf{W}_s\mathbf{X}_s)^{-1}\mathbf{X}_s'\mathbf{W}_s\mathbf{Z}_s - \mathbf{1}_r'\mathbf{Z}_r]\gamma$ equals *zero* (see Exercise 5.18(a)). In this case, C in (5.6.1) disappears and the standardized t-statistic $(\hat{T}_r - T_r)/[\hat{\text{var}}(\hat{T}_r)]^{1/2}$ can be written as the product

$$\frac{\hat{T}_r - T_r}{[\text{var}_{\tilde{M}}(\hat{T}_r)]^{1/2}} \frac{[\text{var}_{\tilde{M}}(\hat{T}_r)]^{1/2}}{[\hat{\text{var}}(\hat{T}_r)]^{1/2}}. \tag{5.6.2}$$

The first factor in (5.6.2) will be asymptotically standard normal, under the conditions of Theorem 5.3.1 and, by the discussion above, the second factor will tend to be less than one. This yields conservative confidence intervals.

Since it is likely that the regression component of our working model is incorrect, not to have balance (weighted or unweighted) risks undercoverage of confidence intervals of T. Under balance, our confidence intervals may be larger than necessary, yielding conservative coverage, but this is preferable to leaving ourselves vulnerable to undercoverage, and claiming a greater precision than actually exists.

An important question is whether random sampling (or *pps* sampling where weighted balance is at stake) yields sufficient balance, to justify inference using the robust variance estimators. The answer is in general, *no*. This was already suggested earlier in Section 3.2. We here restate what was indicated there, in slightly more general terms, considering whether simple random sampling, in general, provides unweighted balance sufficiently closely that the bias is asymptotically negligible. Assume that $\mathbf{V}_{ss}\mathbf{1}_s \in \mathscr{M}(\mathbf{X})$. Then, since $\mathbf{V}_{ss}\mathbf{1}_s = \mathbf{X}_s\mathbf{c}$ for some vector $\mathbf{c}$,

$$E_{\tilde{M}}(\hat{T}_r - T_r) = [\mathbf{1}_r'\mathbf{X}_r(\mathbf{X}_s'\mathbf{W}_s\mathbf{X}_s)^{-1}\mathbf{X}_s'\mathbf{W}_s\mathbf{Z}_s - \mathbf{1}_r'\mathbf{Z}_r]\gamma$$
$$= N_r[\Delta_X(\mathbf{X}_s'\mathbf{W}_s\mathbf{X}_s)^{-1}\mathbf{X}_s'\mathbf{W}_s\mathbf{Z}_s - \Delta_Z]\gamma, \tag{5.6.3}$$

where $\Delta_X = \bar{\mathbf{x}}_r - \bar{\mathbf{x}}_s$ and $\Delta_Z = \bar{\mathbf{z}}_r - \bar{\mathbf{z}}_s$ (see Exercise 5.18(b)). By Theorem A.9 in the Appendix, $n^{1/2}\Delta_X$ and $n^{1/2}\Delta_Z$ are asymptotically normal under simple random sampling, implying that both are $O_p(n^{-1/2})$ with respect to the probability structure of simple random sampling. Consequently, $E_{\tilde{M}}(\hat{T}_r - T_r) = O_p(N_r/n^{1/2})$ with respect to *srs*. This is the same order of magnitude as the root of the error variance. Asymptotically, expression (5.6.1) will then be a multiple of a standard normal random variable, $A \cdot B$, plus a finite quantity, C, that is not necessarily zero or even approaching zero as the sample size increases. As a result, the t-statistic will not, in general, behave like a standardized normal random variable.

Thus the troublesome question arises of the effect on confidence intervals of lurking independent variables, about which the investigator is ignorant, and on which, therefore, the sample cannot be deliberately balanced. We consider this issue in the next section, with no final answers proffered.

5.7 HIDDEN REGRESSION COMPONENTS

We consider a simple situation to exemplify ideas. Suppose the working model is $M(0, 1\!:\!x)$, the through the origin ratio estimator model, but that actually Y depends on an additional, unknown variable z:

$$M^*\!: \; Y_i = \beta x_i + \gamma z_i + \psi_i^{1/2}\varepsilon_i. \tag{5.7.1}$$

We take a sample balanced on x, estimate T by the ratio estimator $\hat{T}(0, 1\!:\!x)$, its variance by v_{J*}, and so get a nominal 95% confidence interval, $\hat{T} \pm 2\sqrt{v_{J*}}$, all ignoring z. What are the likely characteristics of such intervals? In particular, what is the actual coverage going to be like?

The answer is that the coverage depends on the nature of z and on its relationship with the known auxiliary x. With respect to the latter, we can distinguish two extreme cases:

(a) z is highly (positively or negatively) correlated with x (e.g., z some power of x). In this case, balance on x will give approximate balance on

z. Thus, under balance on x, the ratio estimator should be close to unbiased, and use of a robust variance estimator should give adequate coverage.

(b) z is not correlated with x. In this case, z is like an error term plus an intercept that is not necessarily zero. Again, since the ratio estimator will be unbiased under balanced samples, despite the intercept, the combination of balanced sample, and robust variance estimator should give adequate, even conservative, coverage.

Since the relation of z to x must fall somewhere in between zero and total correlation, and balance yields adequate coverage in either case, it would seem that not knowing a relevant variable will exact a price in lesser efficiency but nonetheless yield adequate confidence intervals.

However, this sanguine conclusion gets considerably tempered when we come to consider the nature of the lurking variable z itself, apart from its relation to x. In this regard, the main question becomes the degree of *skewness* of z. If z has a normal or reasonably symmetric distribution, then coverage should be adequate based on the consideration of (a) and (b) above.

However, variates in finite populations are often skewed. Skewness of errors can undermine coverage of confidence intervals, even when the first and second moments of Y are adequately modeled. Since the unknown z in effect becomes part of the error component of the model, a skewed z implies skewed errors, and this can seriously undermine coverage.

Say the skewness is positive, so that a multitude of small negative errors is compensated by a few large positive errors. Then the danger is of frequently *omitting* any of the positive errors from the sample. In this case, the estimate of T will be low, and the available residuals, reflecting the small negative errors, will tend to yield an estimate of variance that is small. The net effect can be a confidence interval lying well below T.

Skewed variables can further be divided into the continuous, such as the lognormal, and the discrete. In the case of the continuous distribution there may be adequate indication from the sample data of the existence of the skewness, and steps may be taken to account for it. The discrete case is more pernicious, since if only the below-zero errors are present in the sample, there will be no sign of skewness at all, and a confidence interval that is too short will be readily accepted.

5.7.1 Some Artificial Examples

We illustrate the above points by a simulation study in which populations are generated of the form (5.7.1), taking x to be the x-variable (number of beds) in the Hospitals population, sorted by size, and z is a variate generated by one of the methods given in Table 5.1. In all cases the errors were set equal to $\sqrt{x_i}\varepsilon_i$, where the ε_i were generated as independent and identically distributed standard normal random variables, and both β and γ were set equal to 1.

Table 5.1. Six z Variates, $N = 393$

Name of z	Method of Generation	Correlation of z with x
Norm	$Z_i \sim N(300,70)$	0.07
Norm.sorted	Z_i's from "norm" sorted in increasing order	0.95
Log	$Z_i \sim 100\exp[N(0, 1)]$	0.13
Log.sorted	Z_i's from "log" sorted in increasing order	0.90
Bin	$Z_i \sim 600\,Bern(0.04)$	0.08
Bin.trunc	$Z_i \sim 600\,Bern(0.08)I(x_i \geqslant \bar{x})$	0.15

Bern(p) denotes the Bernoulli distribution with parameter p.

For each z-vector, a corresponding Y was generated according to (5.7.1) as already described, giving six populations, which we refer to as the "norm", "norm.sorted", and so on, populations. Scatterplots of the populations are shown in Figure 5.1. The line in each panel is for the model $M(0, 1\!:\!x)$ fitted to the whole population. In each case, the straight line through the origin model clearly does not fit well for the whole population but might be less obviously wrong when only a sample is available. Note that the "skewness" in bin and bin.trunc manifests itself as a small set of outliers in each population. These types of contaminated populations have often been studied in the literature on estimation in the presence of outliers (see this volume, Chapter 11 and Tukey, 1960).

From each population, 1000 simple random samples of size $n = 32$ were taken without replacement, the ratio estimator $\hat{T}(0, 1\!:\!x)$ calculated in each, and confidence intervals constructed using the robust variance estimators. Overall coverage is given in Table 5.2.

Despite omission from analysis of the z variate, the overall confidence level of coverage appears to be sound, when z is normal. The soundness arises from a mixture of conservative coverage under balance, and weaker coverage off balance, as seen in Figure 5.2, which graphs standardized t-statistics $t_{J*} = (\hat{T} - T)/\sqrt{v_{J*}}$ based on the (near) jackknife variance estimator, v_{J*}, against the sample mean for norm and norm.sorted. The solid vertical line in each figure indicates where the population mean $\bar{x}$ is, and the vertical dashed lines contain values for samples with $0.975\bar{x} \leqslant \bar{x}_s \leqslant 1.025\bar{x}$; there were 152 such "closest to balance" samples. The bias tends downwards as $\bar{x}_s$ increases; conservative coverage around balance is quite marked in this artificial example.

Figures 5.3 and 5.4 for the populations with skewed "error" give a quite different picture. Although bias trends are discernible, the t-statistics can be markedly low throughout the range of sample x's, even in the 152 relatively balanced samples. The peculiar strata-like appearance in Figure 5.4 for the discrete z arises from the effect of the different number of non-zero z_i's that make it into the sample, with 0 such sample z_i's yielding the lowest cloud, 1 such z_i the second lowest cloud, and so forth.

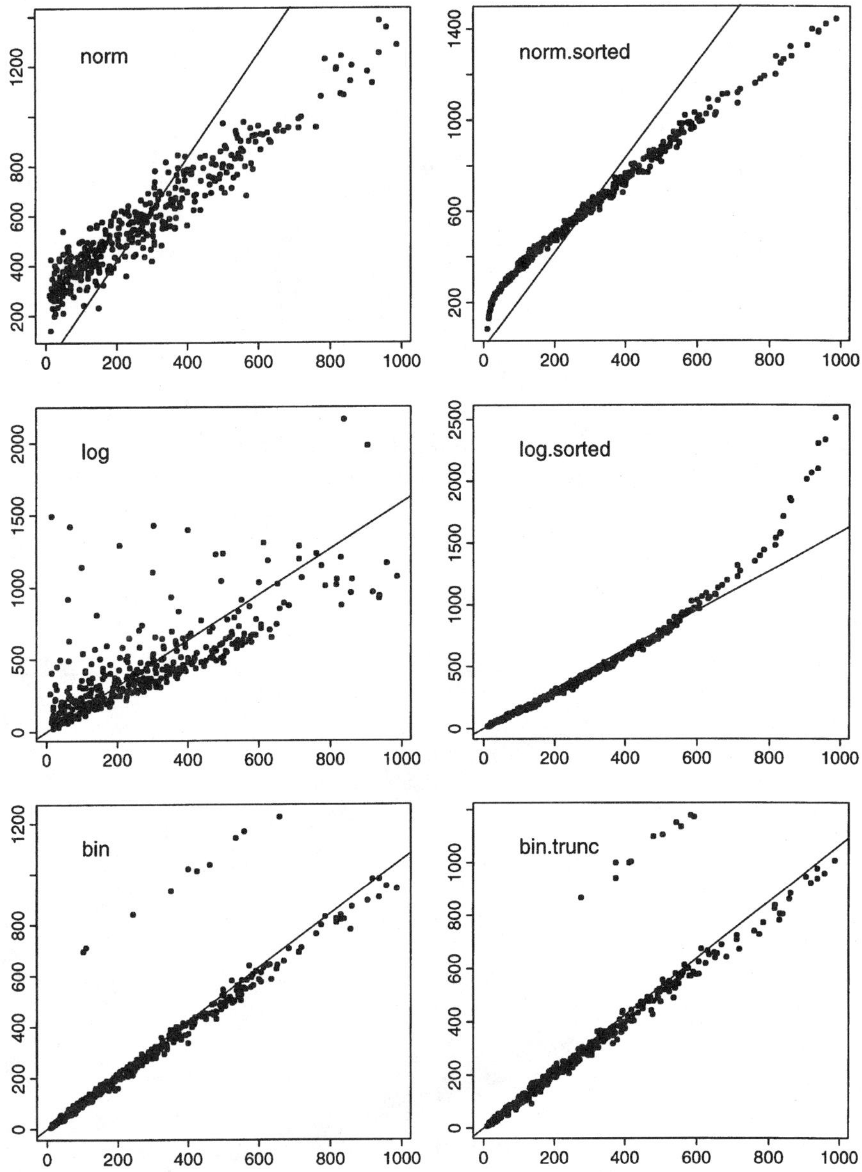

Figure 5.1 x-y scatterplots of six artificial populations used in the ratio estimator simulations. Populations are generated as shown in Table 5.1. Lines are through the origin with slope $\bar{y}/\bar{x}$.

Table 5.2. Percent Coverage of 95% Confidence Intervals, Six Artificial Populations

Population	v_R	v_D	v_H	v_{J*}
Norm	93.5	94.6	94.5	95.0
Norm.sorted	93.8	94.2	94.2	94.8
Log	90.0	90.1	90.1	90.5
Log.sorted	73.3	74.7	74.3	75.8
Bin	54.9	54.9	54.9	55.0
Bin.trunc	60.3	60.3	60.3	60.5

The main conclusion these examples lead us to is that the combination of balance (or, more generally, weighted balance) and robust variance estimation will suffice to yield adequate confidence intervals so long as any lurking variable—one with an impact on the variable of interest, but whose presence or values are unknown to the sample analyst—is not markedly skewed. In some cases there will be indications in the sample data that a skew variable is

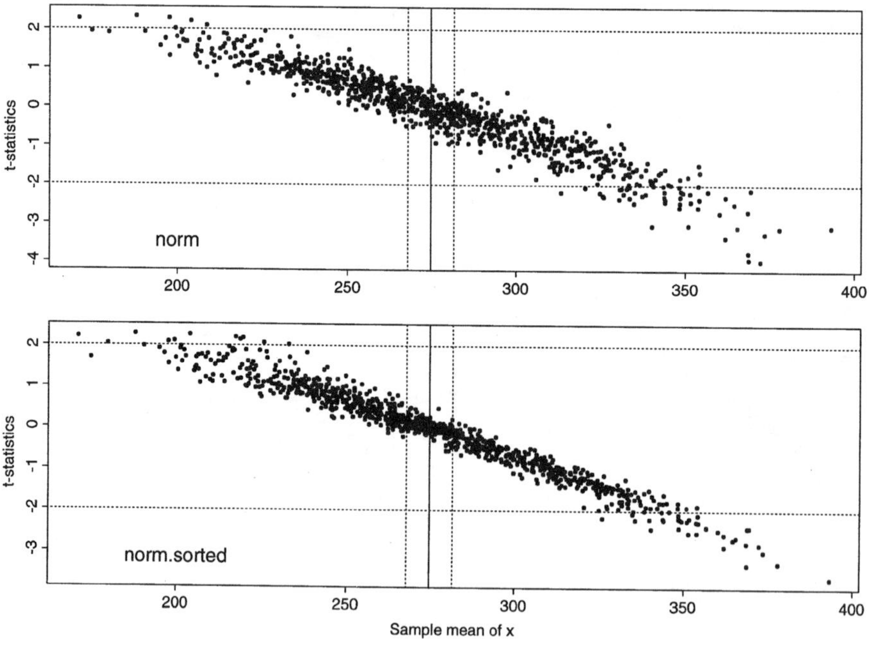

Figure 5.2 Values of $t_{J*} = (\hat{T} - T)/\sqrt{v_{J*}}$ plotted versus $\bar{x}_s$ for 1000 *srswor*'s from the norm and norm.sorted populations. The solid vertical line is drawn at $\bar{x}_s = \bar{x}$. Vertical dotted lines are drawn at $0.975\bar{x}$ and $1.025\bar{x}$. Horizontal dotted lines are shown at $t = \pm 1.96$.

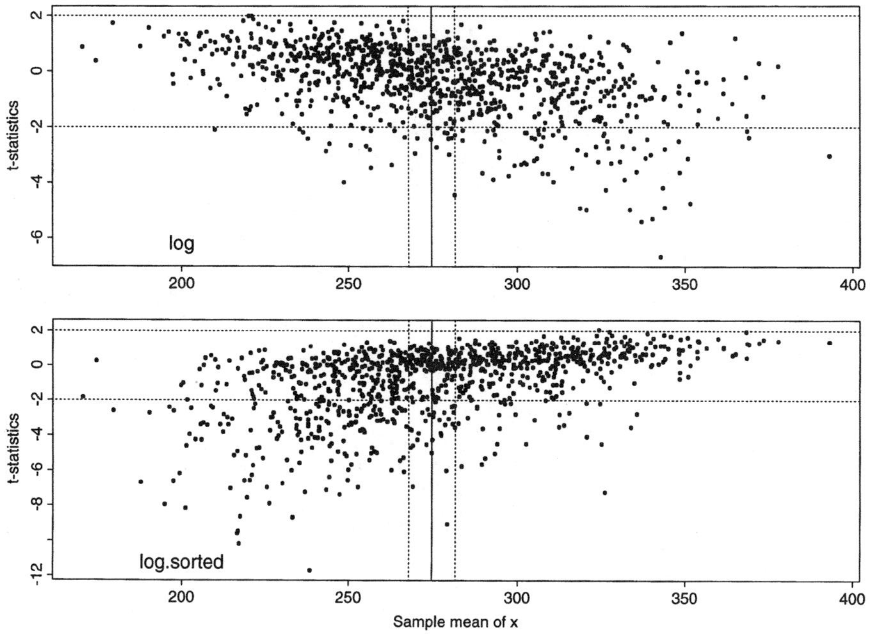

Figure 5.3 Values of $t_{J*} = (\hat{T} - T)/\sqrt{v_{J*}}$ plotted versus $\bar{x}_s$ for 1000 *srswor*'s from the log and log.sorted populations. The solid vertical line is drawn at $\bar{x}_s = \bar{x}$. Vertical dotted lines are drawn at $0.975\bar{x}$ and $1.025\bar{x}$. Horizontal dotted lines are shown at $t = \pm 1.96$.

present, for example, from an outlier, from skewness in the pattern of residuals, or because a *transformation* of the sample data gives a linear, homoscedastic look to the residuals. Where such indications exist, it will be well to take the revealed skewness into account. One method has been suggested by Chen and Chen (1996).

However, signs of lurking skewness need not manifest themselves in the sample. Especially when the hidden variable is an indicator variable, the possibly large resultant outliers in the population may not be selected for the sample. We return to this point after considering a population well known in the literature as Counties 70 (Royall and Cumberland, 1981a).

5.7.2 Counties 70 Population

Figure 5.5 plots the population which is listed in full in Appendix B.4. The variable of interest Y is the 1970 population in each of $N = 304$ counties in three southeastern states of the United States, and the auxiliary variable x is the number of households in each county in 1960. The line in the figure is the least squares fit of $M(0, 1{:}x)$. We note the apparent strong linear relationship, but we also see that certain of the (x, Y) points lie distinctly above the line.

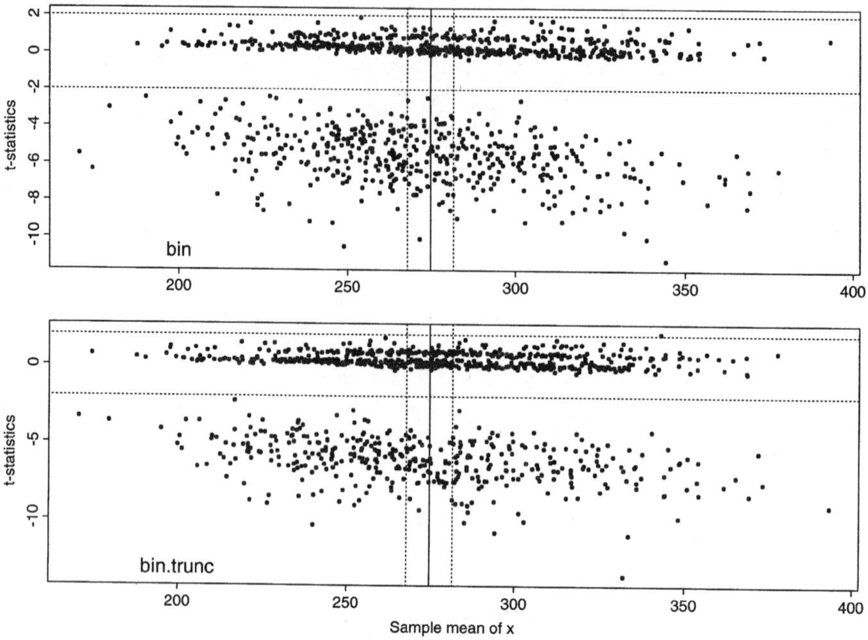

Figure 5.4 Values of $t_{J*} = (\hat{T} - T)/\sqrt{v_{J*}}$ plotted versus $\bar{x}_s$ for 1000 *srswor*'s from the bin and bin.sorted populations. The solid vertical line is drawn at $\bar{x}_s = \bar{x}$. Vertical dotted lines are drawn at $0.975\bar{x}$ and $1.025\bar{x}$. Horizontal dotted lines are shown at $t = \pm 1.96$.

Thus, if we had the full population data, we would see a suggestion of skewed errors. This would not be entirely surprising, since it is reasonable to expect that the majority of counties will have grown in an orderly way over the course of a decade, but a handful have undergone some sharp shift, due to relocation of industry, immigration, and so forth.

Under repeated simple random sampling from this population ($n = 32$), Royall and Cumberland (1985) found severe undercoverage of confidence intervals. In particular, coverage for samples having $\bar{x}_s \approx \bar{x}$ was only about 70–80%. Chen and Chen (1996) developed a somewhat complicated transformation method which seemed to take care of the problem. This suggests that the skew component of the errors is continuous.

Nevertheless, as we shall show, it is when a few high outliers are missing from the sample that the *t*-statistic is vulnerable to very low values.

The variance of Y is clearly increasing with x, so that, in the light of Chapters 3 and 4, we might want to select a weighted balanced sample. This suggests it might be of interest to see the effect of sampling with probabilities proportional to $\sqrt{x}$, combined with the minimal estimator $\hat{T}(x^{1/2}, x : x)$. This should yield an estimator more efficient than the ratio estimator under *srs* and possibly improve coverage.

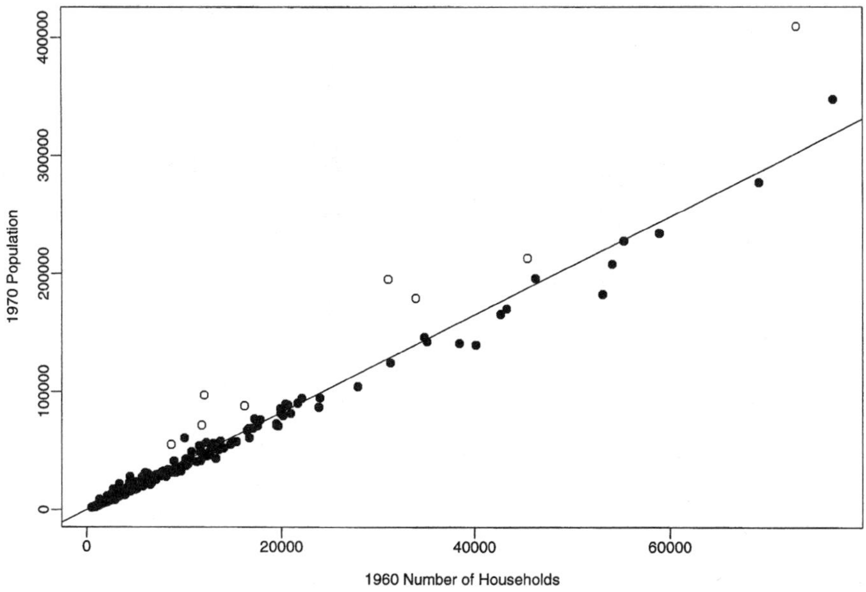

Figure 5.5 Scatterplot of the Counties70 population. Potential outliers are shown as circles.

We selected 1000 samples of size $n = 32$ from Counties 70 with probabilities proportional to $\sqrt{x}$ using the algorithm *P*1*:random*, described in Section 3.4.5. $\hat{T}(x^{1/2}, x:x)$ was calculated in each sample along with the robust variance estimators. Overall coverage for 1000 samples was 85.7%, 87%, 88.7%, and 88.9% for v_R, v_D, v_H, and v_{J*} respectively. This is somewhat better than is achieved using *srswor*, where, for example, Royall and Cumberland (1985) found about 80% coverage for the jackknife. The improvement may have to do with the outliers tending to occur at higher x values, so that sampling *pps* tends to increase their chance of being in the sample.

With units in order of increasing x, the following units, as numbered in Appendix B.4, may be regarded as high outliers: 215, 238, 241, 262, 287, 289, 296, 303, with one or two others borderline. Figure 5.6 plots the standardized t-values based on v_{J*} against the number of outliers in the sample. The pattern is clear: extreme t-values and misleading intervals are most likely to occur when there are 0 or 1 outliers in the sample. When few outliers are in the sample, the prediction of the nonsample total tends to be too small, leading to the excessive number of negative t-statistics seen in the figure. Since balance on x (in this case, requiring $\bar{x}_s = \bar{x}^{(3/2)}/\bar{x}^{(1/2)}$) need have no effect on the number of outliers, we should not be surprised that balance in itself, does not give sound coverage. This is shown in Figure 5.7, which plots the t-values against $\bar{x}_s$ across samples. Coverage for v_{J*} in the region around balance, namely, in $0.975\,(\bar{x}^{(3/2)}/\bar{x}^{(1/2)}) \leqslant \bar{x}_s \leqslant 1.025\,(\bar{x}^{(3/2)}/\bar{x}^{(1/2)})$, is $113/129 = 87.6\%$, which is slightly worse, in fact, than the overall coverage.

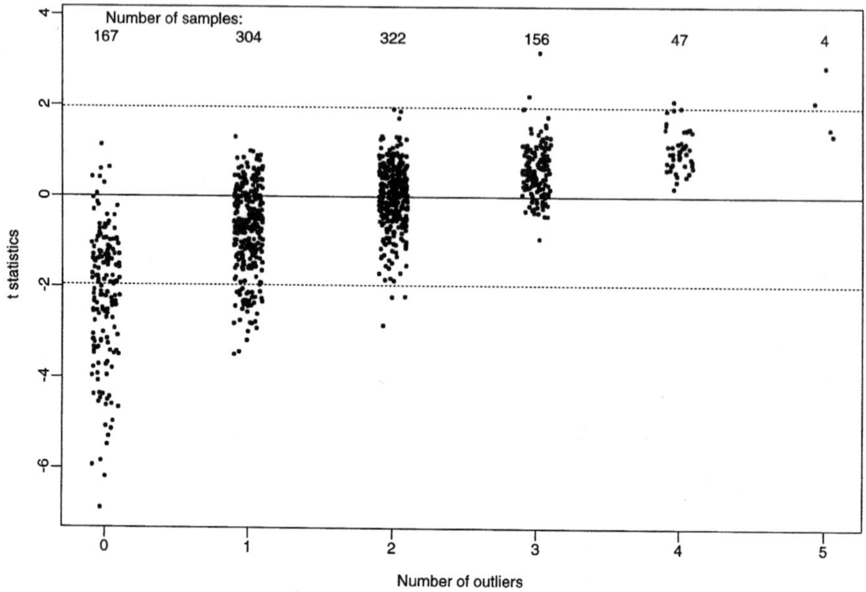

Figure 5.6 Values of $t_{J*} = (\hat{T} - T)/\sqrt{v_{J*}}$ plotted versus the number of sample outliers for 1000 $pp(x^{1/2})$ samples where $\hat{T} = \hat{T}(x^{1/2}, x{:}x)$. Horizontal dotted lines are shown at $t = \pm 1.96$.

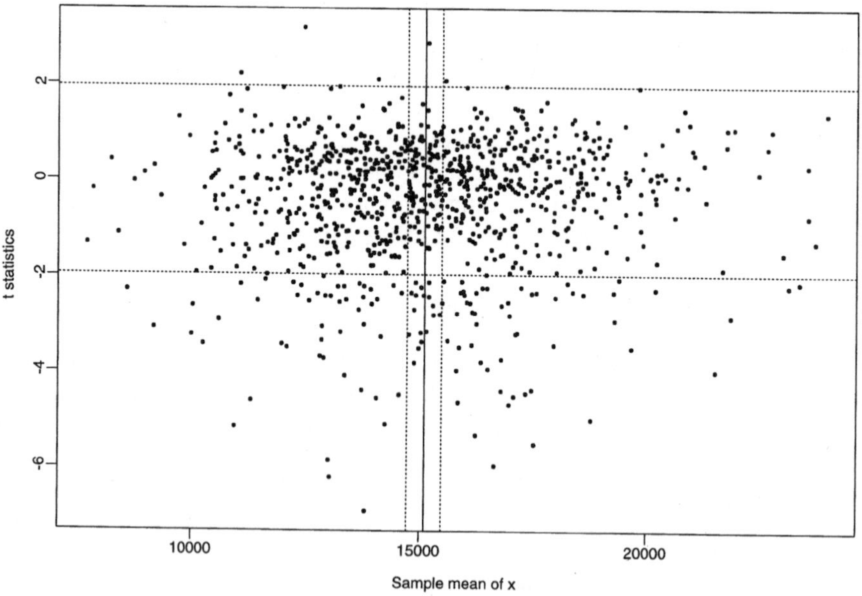

Figure 5.7 Values of $t_{J*} = (\hat{T} - T)/\sqrt{v_{J*}}$ plotted versus $\bar{x}_s$ for 1000 $pp(x^{1/2})$ samples where $\hat{T} = \hat{T}(x^{1/2}, x{:}x)$. The solid vertical line is drawn at $\bar{x}_s = \bar{x}^{(3/2)}/\bar{x}^{(1/2)}$. Vertical dotted lines are drawn at $0.975\bar{x}^{(3/2)}/\bar{x}^{(1/2)}$ and $1.025\bar{x}^{(3/2)}/\bar{x}^{(1/2)}$. Horizontal dotted lines are shown at $t = \pm 1.96$.

It is clear that there is an additional dimension to sound inference, beyond using (weighted) balance to reduce bias, and robust variance estimation to protect against a misspecified variance in the working model.

5.7.3 Lurking Discrete Skewed Variables

We must face squarely the following question. Is there *any* remedy for the lurking discrete skewed variable problem? Random sampling clearly provides no panacea, but neither does balance and robust variance estimation. If there is no guaranteed remedy, and if there could *always* be such a lurking variable, then does this not imply that *all* survey sampling data must contain the potential to be misleading in terms of the precision it claims?

In practice, common sense, knowledge, and experience are needed. In the Counties 70 Population, it would not have been hard to guess that over a ten-year period, *some* of the counties would take a leap in population quite out of proportion to the rest. Furthermore, the demographer experienced in the populations in question might well have been able to suggest which. As samplers, our duty would then be to take steps at the design stage to assure that (at least some of) the potential outliers are included in the sample.

In general, our inferences will always carry the tacit or, perhaps better, explicit tag "subject to the absence of outlying points in the population," and our procedures should always take into account whatever expertise is available that can advise on relevant variables to include in the model. In other words, inference in which we can be confident cannot be solely mathematically grounded, but requires the knowledge of the subject matter expert.

5.8 COMPARISONS WITH DESIGN-BASED VARIANCE ESTIMATION

When considering an estimator $\hat{T}$ of a population total, the variance estimation target in the design-based approach is

$$\text{var}_\pi(\hat{T}) = \sum_s p(s)\{\hat{T}(s) - E_\pi[\hat{T}(s)]\}^2,$$

where we write $\hat{T}(s)$ on the right-hand side to emphasize that the variance is an average over the possible set of samples. For example, as noted in Section 1.3, the design-variance of the π-estimator, $\hat{T}_\pi = \Sigma_s y_i/\pi_i$, is

$$\text{var}_\pi(\hat{T}_\pi) = \sum_{i=1}^{N}\sum_{j=1}^{N}(\pi_{ij} - \pi_i\pi_j)\frac{y_i}{\pi_i}\frac{y_j}{\pi_j}. \tag{5.8.1}$$

Similar formulas apply to more complicated estimators. If, for example, the GREG, $\hat{T}_{yr} = \hat{T}_{y\pi} + \hat{\mathbf{B}}'(\mathbf{T}_x - \hat{\mathbf{T}}_{x\pi})$, defined in (2.8.1), is approximated by a first-order Taylor expansion and the general design-variance computed, then

$$\operatorname{var}_\pi(\hat{T}_{yr}) \cong \sum_{i=1}^{N} \sum_{j=1}^{N} (\pi_{ij} - \pi_i \pi_j) \frac{g_i r_i}{\pi_i} \frac{g_j r_j}{\pi_j} \tag{5.8.2}$$

where the g_i coefficients are elements of the vector $\mathbf{g}_{\pi s} = \mathbf{V}_{ss}^{-1} \mathbf{X}_s \mathbf{A}_{\pi s}^{-1} \times (\mathbf{T}_x - \hat{\mathbf{T}}_{x\pi}) + \mathbf{1}_s$, $r_i = y_i - \mathbf{x}_i' \hat{\mathbf{B}}$, and the other terms are as defined in Section 2.8. This kind of variance—representing variation of estimates across samples—can differ substantially from a model variance that conditions on the set of units in a particular sample.

Given these different goals it would be no surprise to find that the two theories lead to different variance estimators. An interesting commentary on what is and is not accomplished by different variance estimators is found in Royall and Cumberland (1981a) and the five discussants of that paper. The ratio estimator provides a simple illustration. Two variance estimators from the design-based literature are

$$v_c = \frac{N^2}{n}\left(1 - \frac{n}{N}\right)\frac{\Sigma_s r_i^2}{n-1} \quad \text{and} \quad v_2 = \left(\frac{\bar{x}}{\bar{x}_s}\right)^2 v_c,$$

where $r_i = Y_i - x_i \bar{Y}_s / \bar{x}_s$. Both of these are approximately design unbiased under *srswor* for the design-variance of $\hat{T}(0, 1\!:\!x)$. A strictly design-based approach does not lead to an unambiguous choice between v_c and v_2. However, v_c has a bias under $M(0, 1\!:\!x)$ as an estimator of the model error variance (see Exercise 5.19.) The relative model bias is approximately

$$\frac{\bar{x}_s^2}{\bar{x}\bar{x}_r} - 1,$$

which increases with $\bar{x}_s$ and vanishes only in balanced samples. The alternative v_2 is virtually the same as v_D or v_H in moderate-to-large samples and thus shares their robustness under the alternative model $M(0, 1\!:\!\psi)$.

Since v_2, v_D, and v_H are all design-consistent in large simple random samples, any of them is clearly preferable to v_c when the additional information obtained through model-based analysis is considered. Note, however, that neither *srswor* nor any other type of probability sampling plan is required for model-based analysis of v_c and v_2.

The jackknife was another variance estimator that we found to be robust against departures from the working model variance specification. The jackknife is also one of the standard tools in the design-based approach (see, e.g., Krewski and Rao 1981; Rao, and Wu, 1988; Stukel, Hidiroglou, and Särndal, 1996; Rust, 1985). It is recommended in this context for its ability to estimate the design variance of nonlinear estimators. Recollect, however, that what is meant by "linear" is different in design-based theory than in prediction theory.

The GREG is nonlinear in the quantities that vary from sample to sample because $\hat{\mathbf{B}}'(\mathbf{T}_x - \hat{\mathbf{T}}_{x\pi})$ involves products and ratios of sample-dependent terms. In contrast, $\hat{T}_{yr} = \hat{T}_{y\pi} + \hat{\mathbf{B}}'(\mathbf{T}_x - \hat{\mathbf{T}}_{x\pi})$ is just a linear function of the sample Y's, and so is a linear estimator as far as prediction theory is concerned. The jackknife is an important instance of harmony between estimation methods recommended in the two theories—albeit for different reasons.

The design-based theory for the jackknife, when applied to complex estimators and sample designs, is often derived under the assumption that sample units are selected with replacement (e.g., Krewski and Rao, 1981). Designs in which units are selected without replacement, but in which the sampling fraction is small, may often be safely treated as with-replacement designs, so that this type of assumption is not too unrealistic. In situations where the sampling fraction is large, formulas like (5.8.1) and (5.8.2) need to be used that involve the π_{ij} joint selection probabilities. In practice, it is common to select samples in such a way (e.g., by systematic selection from a sorted list) that the necessary selection probabilities are unknown, so that it is not even feasible to calculate the appropriate design-based variance estimators.

This cumbersome requirement of knowing the second-order inclusion probabilities led Särndal (1996) to suggest the use of selection schemes where the π_{ij}'s are easy to compute and variance estimators like (5.8.1) and (5.8.2) simplify. Two fairly simple schemes that can be implemented in ways that retain many of the benefits of unequal probability sampling are stratified simple random sampling and Poisson sampling. The latter, which we have not covered in this text, is a list sequential procedure. Each unit i is examined and independently given some probability π_i of being selected. The joint selection probability of two units is simply $\pi_{ij} = \pi_i \pi_j$. An approximately design-unbiased variance estimator for the GREG is then

$$\begin{aligned} v_\pi(\hat{T}_{yr}) &\cong \sum_{i \in s} (1 - \pi_i) \left(\frac{g_i r_i}{\pi_i} \right)^2 \\ &= \tilde{\mathbf{g}}' \operatorname{diag}(r_i^2/\pi_i^2)\, \tilde{\mathbf{g}}, \end{aligned}$$

where $\tilde{\mathbf{g}}' = [g_1\sqrt{1-\pi_1}, \ldots, g_n\sqrt{1-\pi_n}]$. This is clearly a type of sandwich estimator. The thinking leading to the estimator is based on the design rather than a model and is unaffected by any concern for model-robustness.

Relatively little research has been directed toward deriving variance estimators that explicitly incorporate both design-based and model-based thinking. However, as the ratio estimator case and the jackknife illustrate, there are examples where estimators have properties that are desirable under both theories. For the GREG estimator, Kott (1990) studied a variance estimator of the form

$$v_K(\hat{T}_{yr}) = \frac{\operatorname{var}_M(\hat{T}_{yr} - T)}{E_M[v_{YG}(\hat{T}_{yr})]} v_{YG}(\hat{T}_{yr}),$$

where $\text{var}_M(\hat{T}_{yr} - T)$ is a working model error variance, $v_{YG}(\hat{T}_{yr})$ is a type of Yates-Grundy design-variance estimator like the one defined in (1.3.8), and $E_M[v_{YG}(\hat{T}_{yr})]$ is its expectation under the working model. The estimator v_K is clearly model unbiased under the working model. If the ratio $\text{var}_M(\hat{T}_{yr} - T)/E_M[v_{YG}(\hat{T}_{yr})]$ converges to 1, then v_K will also be design consistent. For example, when the sample design is *srswor* and $\hat{T}_{yr}$ is the ratio estimator, v_K reduces to v_H, given in Section 5.1.2.

EXERCISES

5.1 Prove Lemma 5.1.1 using mathematical induction.

5.2 Show that if $E_M(Y_i) = \beta x_i$ and $\text{var}_M(Y_i) = \psi_i$, then $E_M(\tilde{\psi}_i) \geqslant \psi_i$ where $\tilde{\psi}_i = r_i^2/(1 - h_{ii})^2$, $r_i = Y_i - \hat{\beta} x_i$, $\hat{\beta} = \Sigma_s x_i Y_i / \Sigma_s x_i^2$, and $h_{ii} = x_i^2/\Sigma_s x_i^2$.

5.3 **(a)** Derive the jacknife estimator of $\text{var}_M(\hat{\beta})$ under the model $E_M(Y_i) = \beta x_i$, $\text{var}_M(Y_i) = \sigma^2 x_i$.

(b) Show directly that as $n \to \infty$, $v_J/v_{J*} \to 1$ where v_{J*} was defined by (5.4.1).

5.4 Show that under the working model $E_M(Y_i) = \beta x_i$, $\text{var}_M(Y_i) = \sigma^2 x_i$, the variance estimators $v_H(\hat{T}_r)$ and $v_H(\hat{T}_r - T_r)$ are as given by (5.1.11) and (5.1.14).

5.5 Prove Lemma 5.3.1, parts (a)–(e), (g)–(i), (k)–(n). (*Hint:* Consult Belsley, Kuh, and Welsch, 1980, Appendix 2A; use $\mathbf{A} = \mathbf{A}^{1/2}\mathbf{A}^{1/2}$ and the Cauchy-Schwarz inequality.)

5.6 Suppose that $Y_1, Y_2, \ldots$ are random variables such that $Y_i > 0$ and that $E(Y_i) \to 0$ as $n \to \infty$, Show that $Y_n \xrightarrow{p} 0$. (*Hint:* See Serfling (1980), sec. 1.3.)

5.7 Assume that the model is $Y_i = \mathbf{x}_i'\boldsymbol{\beta} + \varepsilon_i$ where $\varepsilon_i \sim (0, \sigma^2)$ and are independent with distribution functions $F_1, F_2, \ldots$. Show that if $\varepsilon_i \leqslant B$ for some constant B, then $\sup_{1 \leqslant i \leqslant n;\, n = p+1, \ldots} (\int_{|u|>d} u^2 dF_i) \to 0$ as $d \to \infty$.

5.8 Assume that the working model $M(\mathbf{X}:\boldsymbol{\Psi})$ is correct. Show that $E_M(r_i^2) = \psi_i(1 - h_{ii})$, where h_{ii} is the ith diagonal element of the hat matrix $\mathbf{H} = \mathbf{X}(\mathbf{X}'\boldsymbol{\Psi}^{-1}\mathbf{X})^{-1}\mathbf{X}'\boldsymbol{\Psi}^{-1}$ and $r_i = Y_i - \mathbf{x}_i'\hat{\boldsymbol{\beta}}$. (*Hint:* Use expression (5.1.2) and the form of h_{ii} given in Lemma 5.3.1(a).)

5.9 Suppose that $M(\mathbf{X}:\mathbf{V})$ is the working model and that $\tilde{M}(\mathbf{X}:\boldsymbol{\Psi})$ is the true, unknown model. Show that $E_{\tilde{M}}(v_{J*}) \geqslant \text{var}_{\tilde{M}}(\hat{T} - T)$ where v_{J*} is defined by (5.4.3) and $\hat{T}$ is the *BLU* predictor under the working model.

5.10 Prove Lemma 5.4.1.

5.11 Show that, when the ith sample observation is deleted, $\mathbf{X}'_{(i)}\mathbf{W}_{(i)}\mathbf{X}_{(i)} = \mathbf{X}'\mathbf{W}\mathbf{X} - w_i\mathbf{x}_i'\mathbf{x}_i$, where $\mathbf{x}_i$ is the ith row of $\mathbf{X}$ and the other terms are as defined as in Section 5.4.2. Hence, show that $\hat{\Theta}_{(i)} - \hat{\Theta} = -\mathbf{q}'(\mathbf{X}'\mathbf{W}\mathbf{X})^{-1} \times \mathbf{x}_i w_i r_i/(1-h_{ii}) = -a_i r_i/(1-h_{ii})$.

5.12 Generate your own realizations of the populations in Table 5.1. Do all pairwise scatterplots of y, x, and z for each population.

5.13 Work out explicit formulas for **a**, **h**, **r**, and, consequently, v_R, v_D, v_H, v_{J*}, and v_J for the following working models and estimators:

(a) $M(1,1{:}1)$, $\hat{T}_{LR} = N\bar{Y}_s + b(\bar{x} - \bar{x}_s)$ with $b = \Sigma_s(x_i - \bar{x}_s)Y_i/\Sigma_s(x_i - \bar{x}_s)^2$.

(b) $M(0,1{:}x^2)$, $\hat{T}_{MR} = N\bar{x}\Sigma_s Y_i/nx_i$.

(c) $M(x^{1/2}, x{:}x)$, $\hat{T}(x^{1/2}, x{:}x)$.

5.15 Verify the formula for v_H in Example 5.5.1.

5.16 The following is a sample of $n = 20$ units selected with probability proportional to $x^{1/2}$ from the Cancer population. (a) Compute the estimator $\hat{T}(x^{1/2}, x{:}x)$ and the standard error estimates $\sqrt{v_R}$, $\sqrt{v_D}$, $\sqrt{v_H}$, $\sqrt{v_{J*}}$, and $\sqrt{v_J}$. (b) Compute the leverages, i.e., the diagonal of the hat matrix. Are there any high leverage points? Sample units: 77, 87, 91, 98, 102, 126, 133, 142, 164, 194, 225, 237, 240, 248, 258, 262, 265, 266, 280, 301.

5.17 The following is a sample of $n = 20$ units selected with probability proportional to x from the Hospitals population. (a) Compute the estimator $\hat{T}(0, 1, 1{:}x^2)$ and the standard error estimates $\sqrt{v_R}$, $\sqrt{v_D}$, $\sqrt{v_H}$, $\sqrt{v_{J*}}$, and $\sqrt{v_J}$. (b) Compute the leverages, i.e., the diagonal of the hat matrix. Are there any high leverage points? Sample units: 79, 117, 134, 175, 223, 273, 274, 311, 314, 328, 352, 354, 355, 356, 364, 373, 382, 386, 392, 393.

5.18 Suppose that the working model is $M(\mathbf{X}{:}\mathbf{V})$ but the correct underlying model is $\tilde{M}(\mathbf{X}, \mathbf{Z}{:}\Psi)$. Consider the BLU predictor under $M(\mathbf{X}{:}\mathbf{V})$, $\hat{T} = \Sigma_s Y_i + \mathbf{a}'\mathbf{Y}_s$ with $\mathbf{a}' = \mathbf{1}_r'\mathbf{X}_r(\mathbf{X}_s'\mathbf{W}_s\mathbf{X}_s)^{-1}\mathbf{X}_s'\mathbf{W}_s$, with $\mathbf{W}_s = \mathbf{V}_{ss}^{-1}$. (a) Show that if $\mathbf{V}\mathbf{1}_N$, $\mathbf{V}^{1/2}\mathbf{1}_N \in \mathscr{M}(\mathbf{X})$ and $s \in B(\mathbf{X}, \mathbf{Z}{:}\mathbf{V})$, then the bias $E_{\tilde{M}}(\hat{T} - T) = [\mathbf{1}_r'\mathbf{X}_r(\mathbf{X}_s'\mathbf{W}_s\mathbf{X}_s)^{-1}\mathbf{X}_s'\mathbf{W}_s\mathbf{Z}_s - \mathbf{1}_r'\mathbf{Z}_r]\gamma$ under $\tilde{M}(\mathbf{X}, Z{:}\Psi)$ equals zero. (b) Assuming only that $\mathbf{V}_{ss}\mathbf{1}_s \in \mathscr{M}(\mathbf{X}_s)$, show that

$$E_{\tilde{M}}(\hat{T}_r - T_r) = N_r[\Delta_X(\mathbf{X}_s'\mathbf{W}_s\mathbf{X}_s)^{-1}\mathbf{X}_s'\mathbf{W}_s\mathbf{Z}_s - \Delta_Z]\gamma$$

with the terms defined as in (5.6.3).

5.19 Two variance estimators for the ratio estimator, $\hat{T}(0, 1:x)$, found in standard texts like Cochran (1977) are

$$v_c = \frac{N^2}{n}\left(1 - \frac{n}{N}\right)\frac{\Sigma_s r_i^2}{n-1} \quad \text{and} \quad v_2 = \left(\frac{\bar{x}}{\bar{x}_s}\right)^2 v_c,$$

where $r_i = Y_i - x_i \bar{y}_s/\bar{x}_s$. (a) Are these estimators model-unbiased or approximately so under $M(0, 1:x)$? (b) Are they robust if the correct model is $M(0, 1:\psi)$?

5.20 Consider the GREG estimator, $\hat{T}_{yr} = \hat{T}_{y\pi} + \hat{\mathbf{B}}'(\mathbf{T}_x - \hat{\mathbf{T}}_{x\pi})$, defined in Section 2.8 and the variance estimator

$$v_\pi(\hat{T}_{yr}) \cong \sum_{i \in s}(1 - \pi_i)\left(\frac{g_i r_i}{\pi_i}\right)^2,$$

where $r_i = Y_i - \mathbf{x}_i \boldsymbol{\beta}$. Is this estimator approximately unbiased under the model $M(\mathbf{X}:\boldsymbol{\Psi})$ where $\boldsymbol{\Psi} = \text{diag}(\psi_i)$?

5.21 What does $v_\pi(\hat{T}_{yr})$ equal in the following cases?

(i) $\hat{T}_0$, *srs*, and $M(1:1)$;

(ii) $\hat{T}_R$, *srs*, and $M(0, 1:x)$.

How do these variance estimators compare to the model-based ones associated with $\hat{T}_0$ under $M(1:1)$ and $\hat{T}_R$, under $M(0, 1:x)$?

CHAPTER 6

Stratified Populations

Stratification is one of the most widely used techniques in finite population sampling. It has both statistical and administrative functions though the distinction is often not clearly made in textbooks. In this chapter we attempt to clarify the roles that stratification plays and build upon the mathematical results of earlier chapters. Strata are disjoint subdivisions of a population, each of which contains a portion of the sample, and the union of which composes the population. We have already met stratification in a subordinate role in Chapter 3 where it was used as a mechanism for implementing a particular sampling scheme. There the aim was to achieve an appropriately balanced sample. Stratification's primary and more traditional purposes are to:

1. Deal statistically with subpopulations;
2. Allow for efficient estimation; and
3. Contribute to efficient administration of large surveys.

We discuss each of these in more detail below.

(1) *Subpopulations.* Many populations are, in effect, collections of populations, and a target variable in a survey may follow a different model in each of the subpopulations. In a survey of households to estimate average income, for example, the income levels may vary widely among different demographic groups and regions of a country. The sample data from one subgroup may be of limited use in making estimates for other subgroups. In these populations, estimates may be required both for the full population and for some or all of the subpopulations. In either case, it is desirable to take each subpopulation as a stratum and so require a sample in each subpopulation. Our main focus will be on full population estimates that are combinations of the subpopulation estimates.

(2) *Efficient estimation.* Costs of conducting the survey may differ substantially among the strata. An optimum allocation of the sample to the strata will consider both the cost and the variability of the target variable in each stratum. When a single model for the whole population fits a target variable, strata can also be used for selecting a sample with overall optimal balance, as in Section

3.3, or separate balanced samples can be selected in each stratum, a topic we cover in this chapter.

(3) *Administration.* Practical problems related to response and measurement may differ considerably among subpopulations. Stratification allows some flexibility in the choice of data-collection procedures that are used for different subpopulations. Telephone data collection may be adequate for some groups while personal interviews may be needed for others, for instance. For operational convenience, the survey organization may also be divided into geographic districts with a field office supervising work in each district. It is natural to let each district be an administrative stratum.

Distinguishing between the administrative and the statistical roles of stratification is important to understanding its functions. The former can be essential for the efficient planning and execution of fieldwork. But administrative strata need not enter into estimation unless they coincide with subpopulations that require separate models. This last observation reflects the fundamental distinction between the prediction approach and the design-based approach to sampling. The latter *requires* that estimators take into account the stratification scheme, even if it was merely the result of administrative convenience.

Stratified sampling is also distinguished from *cluster sampling*, which will be covered in Chapter 8. In both cases the population is divided into disjoint groups but in a cluster sample only a subset of the groups is included in the sample. In a stratified sample each group must be represented in the sample, by definition.

6.1 STRATIFICATION WITH HOMOGENEOUS SUBPOPULATIONS

Let h denote a stratum and i a unit within the stratum. The target variable for unit hi is Y_{hi}. Assume that the population contains H strata with the number of units in each stratum being N_h $(h = 1, \ldots, H)$ and the population size being $N = \Sigma_{h=1}^{H} N_h$. A sample of n_h units is selected from stratum h with the total sample size being $n = \Sigma_h n_h$. Denote the set of sample units in stratum h as s_h and the set of nonsample units as r_h. A simple model that reflects homogeneity within strata but heterogeneity between strata is

$$E_M(Y_{hi}) = \mu_h \qquad \mathrm{var}_M(Y_{hi}) = \sigma_h^2 \tag{6.1.1}$$

with different units being uncorrelated. The *BLU* predictor of the population total—the stratified expansion estimator—is easily calculated (see Exercise 6.1) as

$$\hat{T}_{0S} = \sum_{h=1}^{H} N_h \bar{Y}_{hs},$$

where $\bar{Y}_{hs} = \Sigma_{i \in s_h} Y_{hi}/n_h$, with error variance

$$\begin{aligned}\operatorname{var}_M(\hat{T}_{0S} - T) &= \sum_{h=1}^{H} \frac{N_h^2}{n_h}(1 - f_h)\sigma_h^2 \\ &= \sum_{h=1}^{H} \frac{N_h^2}{n_h}\sigma_h^2 - \sum_{h=1}^{H} N_h \sigma_h^2, \qquad (6.1.2)\end{aligned}$$

where $f_h = n_h/N_h$ is the within-stratum sampling fraction. An unbiased estimator of this error variance under (6.1.1) is

$$v(\hat{T}_{0S}) = \sum_{h=1}^{H} \frac{N_h^2}{n_h}(1 - f_h) \sum_{i \in s_h} \frac{(Y_{hi} - \bar{Y}_{hs})^2}{n_h - 1}.$$

Suppose that the overall sample size is fixed at n. How should we allocate the n units among the strata, that is, assign the values n_h with $n = \Sigma_h n_h$? One simple method is *proportional allocation*, which takes $n_h \propto N_h$. Specifically,

$$n_h = n\frac{N_h}{N}.$$

In that case $\hat{T}_{0S} = \hat{T}_0 = N\bar{Y}_s$, the simple expansion estimator from earlier chapters. Another simple possibility is *equal allocation*, which divides the sample equally among strata with each getting $n_h = n/H$ sample units.

A more efficient distribution of the sample for estimating the population total is *optimum allocation*, which minimizes the error variance subject to cost constraints. In previous chapters we did not consider costs—implicitly assuming that the cost of conducting the survey was the same for every unit. Now, we make the formulation a bit more realistic by supposing that the cost of conducting the survey is $C = C_0 + \Sigma_h c_h n_h$, where C_0 is a fixed cost and c_h is the cost per unit sampled in stratum h. The fixed cost C_0 might cover administrative costs for a project manager and central office staff plus any other costs that *do not* vary with the size of sample. The unit cost c_h could reflect interviewing cost, data processing expenses, and other expenditures that *do* depend on the sample size. A good reference for the types of costs involved in conducting surveys is Groves (1989).

Optimal allocation for a given cost function is determined by Theorem 6.1.1. The theorem shows how to parcel out a sample of size n to the H strata. How large n should be is governed by whether the budget is fixed or a target variance is specified.

Theorem 6.1.1. The allocation of the sample to the strata that minimizes the error variance of $\hat{T}_{0S}$ under model (6.1.1), subject to the cost function $C = C_0 + \Sigma_h c_h n_h$, is

$$\frac{n_h}{n} = \frac{N_h \sigma_h / \sqrt{c_h}}{\Sigma_{h'} N_{h'} \sigma_{h'} / \sqrt{c_h}} \quad \text{for } h = 1, \ldots, H.$$

Proof: To find the optimum set of n_h's, we need only consider the first term in (6.1.2). The Lagrange function to be minimized is then

$$\Phi = \sum_h \frac{N_h^2}{n_h} \sigma_h^2 + \lambda \left(C_o + \sum_h c_h n_h - C \right).$$

The derivative with respect to a particular n_h is

$$\frac{\partial \Phi}{\partial n_h} = -\frac{N_h^2}{n_h^2} \sigma_h^2 + \lambda c_h.$$

Setting this derivative equal to 0, implies that $n_h = N_h \sigma_h / \sqrt{\lambda c_h}$ or that $n_h \propto N_h \sigma_h / \sqrt{c_h}$. Since $\Sigma_h n_h = n$, the result follows. □

The total sample size n will depend on whether the total cost is fixed or the error variance of $\hat{T}_{0S}$ is fixed at some desired level. The result of Theorem 6.1.1 can be applied in either case.

Corollary 6.1.1. Under the conditions of Theorem 6.1.1, if the total cost $C = C_0 + \Sigma_h c_h n_h$ is fixed at a particular value C^*, then the optimum total sample size is

$$n = (C^* - C_0) \frac{\Sigma_h N_h \sigma_h / \sqrt{c_h}}{\Sigma_{h'} N_{h'} \sigma_{h'} \sqrt{c_{h'}}}.$$

Proof: The result follows by substituting the optimal values of n_h into the cost function. □

When the total sample size is given by Corollary 6.1.1, the variance is

$$\text{var}_M(\hat{T}_{0S} - T) = \frac{(\Sigma_h N_h \sigma_h \sqrt{c_h})^2}{C^* - C_0} - \sum_h N_h \sigma_h^2. \quad (6.1.3)$$

Corollary 6.1.2. Under the conditions of Theorem 6.1.1, if the error variance $\text{var}_M(\hat{T}_{0S} - T) = V_0$ is fixed, then the optimum total sample size is

$$n = \frac{(\Sigma_h N_h \sigma_h / \sqrt{c_h})(\Sigma_h N_h \sigma_h \sqrt{c_h})}{V_0 + \Sigma_h N_h \sigma_h^2}.$$

Proof: The result follows by substituting the optimal values of n_h into the variance (6.1.2). □

The expression for the optimum given in Corollary 6.1.1 will be more often used in practice than that in Corollary 6.1.2, because the determining factor in most surveys is the budget.

When the unit costs are constant ($c_h = \bar{c}$) and the variance is minimized for a fixed total sample size, Theorem 6.1.1 yields

$$n_h = n \frac{N_h \sigma_h}{\Sigma_{h'} N_{h'} \sigma_{h'}}, \tag{6.1.4}$$

which is known as the *Neyman allocation*, after Neyman (1934), whose proof was done in a design-based context and gave the result prominence. For the Neyman allocation the error variance is

$$\text{var}_M(\hat{T}_{OS} - T) = \frac{1}{n}\left(\sum_h N_h \sigma_h\right)^2 - \sum_h N_h \sigma_h^2, \tag{6.1.5}$$

which follows directly from (6.1.2) and (6.1.4). Expression (6.1.5) is also a special case of (6.1.3) because $C^* - C_0 = \bar{c}n$ when costs are constant.

In the equal cost case, proportional allocation is optimal when the stratum standard deviations are equal. Otherwise, proportional allocation puts too small a sample in strata with large values of σ_h. Equal allocation, on the other hand, is optimal if $N_h \sigma_h$ is constant.

An example will be helpful in comparing different allocations. Suppose a business population has the five strata and population characteristics given in Table 6.1. The data are artificial but the industry subpopulations are standard classifications used in establishment surveys. Costs per unit vary from \$25 to \$75 and, as the last three columns of the table show, allocations proportional to N_h, $N_h \sigma_h$, and $N_h \sigma_h / \sqrt{c_h}$ will be somewhat different. Assume that a total variable cost of $C^* - C_0 = \$5{,}250$ is available for the survey.

Table 6.2 lists four different allocations. From Corollary 6.1.1 the optimal total sample size is $n = 141$, whose optimal allocation is in column (1) of the table. The other three columns are for Neyman, proportional and equal

Table 6.1. Population Characteristics Used for an Example of Optimal Allocation Subject to a Cost Constraint for a Case with Homogeneous Subpopulations

Stratum	N_h	σ_h	c_h	$\frac{N_h}{N}$	$\frac{N_h \sigma_h}{\Sigma_{h'} N_{h'} \sigma_{h'}}$	$\frac{N_h \sigma_h / \sqrt{c_h}}{\Sigma_{h'} N_{h'} \sigma_{h'} / \sqrt{c_h}}$
1. Manufacturing	680	432	30	0.30	0.36	0.40
2. Retail trade	552	229	75	0.24	0.16	0.11
3. Wholesale trade	429	344	25	0.19	0.18	0.22
4. Services	363	196	50	0.16	0.09	0.07
5. Government	251	686	40	0.11	0.21	0.20

Table 6.2. Sample Allocations for a Business Population with a Cost Constraint of $C^* - C_0 = \$5{,}250$

Stratum	Stratum Samples n_h			
	(1) Optimal with Cost Constraint	(2) Optimal Ignoring Cost Constraint	(3) Proportional Allocation	(4) Equal Allocation
1. Manufacturing	56	48	35	24
2. Retail trade	15	20	29	24
3. Wholesale trade	31	24	22	24
4. Services	11	12	19	24
5. Government	28	28	13	24
Total sample	141	132	118	120
Total cost	5,250	5,260	5,245	5,280
$\sqrt{\text{var}_M(\hat{T} - T)}/10^3$	67.0	68.2	78.8	79.2

allocations—none of which explicitly accounts for costs. To compute a total sample size for these three allocations that adheres to the budget constraint, note that the cost function implies that $C^* - C_0 = \Sigma_h n_h c_h$. For the allocations in columns (2)–(4) the stratum sample size can be written as $n_h = nw_h$, where $w_h = N_h\sigma_h/\Sigma_h N_h\sigma_h$ for Neyman allocation, $w_h = N_h/N$ for proportional, and $w_h = 1/H$ for equal allocation, respectively. As a result, we have $\Sigma_h n_h c_h = n\bar{c}$, where $\bar{c} = \Sigma_h w_h c_h$ is a weighted average of the unit costs, and the implied total sample size is

$$n = \frac{C^* - C_0}{\bar{c}}.$$

For the example, the resulting values of n are 132 for Neyman, 118 for proportional, and 120 for equal. The total costs in the table for these three allocations are slightly different from \$5,250 because stratum sample sizes are rounded to integers.

The cost-constrained optimal allocation in column (1) produces the smallest standard error which is 67 thousand. The Neyman allocation gives only a mildly higher standard error at 68.2 thousand, illustrating the common phenomenon that cost differentials among strata must be fairly large to be worth accounting for. Proportional and equal allocations do, in contrast, give standard errors that are about 18% higher than the optimal allocation with the cost constraint, their distributions among the strata being far from the optimal.

6.2 STRATIFIED LINEAR MODEL AND WEIGHTED BALANCED SAMPLES

We can generalize the results of the last section, to the circumstance where a separate linear regression model holds within each stratum:

$$E_M(\mathbf{Y}_h) = \mathbf{X}_h\boldsymbol{\beta}_h, \quad \text{var}_M(\mathbf{Y}_h) = \mathbf{V}_h\sigma_h^2, \tag{6.2.1}$$

where $\mathbf{Y}_h$ is $N_h \times 1$, $\mathbf{X}_h$ is $N_h \times p_h$, $\mathbf{V}_h = \text{diag}(v_{hi})$ is $N_h \times N_h$, and $\boldsymbol{\beta}_h$ is a $p_h \times 1$ parameter vector. Using the same notation as introduced in Chapter 3, the model in stratum h is $M(\mathbf{X}_h{:}\mathbf{V}_h)$ and the *BLU* predictor is then the sum of the *BLU* predictors in each stratum:

$$\hat{T} = \sum_{h=1}^{H} \hat{T}(\mathbf{X}_h{:}\mathbf{V}_h).$$

The predictor in stratum h is $\hat{T}(\mathbf{X}_h{:}\mathbf{V}_h) = \Sigma_{i\in s_h} Y_{hi} + \mathbf{1}'_{rh}\mathbf{X}_{rh}\hat{\boldsymbol{\beta}}_h$, where $\mathbf{1}_{rh}$ is a vector of $N_h - n_h$ 1's and $\mathbf{X}_{rh}$ is the $(N_h - n_h) \times p_h$ matrix of auxiliaries for the nonsample units. The estimator of the slope in stratum h is $\hat{\boldsymbol{\beta}}_h = (\mathbf{X}'_{sh}\mathbf{W}_{sh}\mathbf{X}_{sh})^{-1}\mathbf{X}'_{sh}\mathbf{W}_{sh}\mathbf{Y}_{sh}$, where $\mathbf{X}_{sh}$ is the $n_h \times p_h$ matrix of auxiliaries for the sample units, $\mathbf{W}_{sh}$ is the inverse of $\mathbf{V}_{sh}$, the $n_h \times n_h$ diagonal covariance matrix for the sample units, and the vector of sample Y's from stratum h is $\mathbf{Y}_{sh}$. We can also write the *BLU* predictor as

$$\hat{T} = \sum_h \sum_{i\in s_h} Y_{hi} + \sum_h \sum_{i\in s_h} a_{hi} Y_{hi}$$

with $a_{hi} = \mathbf{1}'_{rh}\mathbf{X}_{rh}(\mathbf{X}'_{sh}\mathbf{W}_{sh}\mathbf{X}_{sh})^{-1}\mathbf{x}_{hi}w_{hi}$ and $w_{hi} = v_{hi}^{-1}$. This form will be useful for variance estimation in Section 6.7.

In stratum h define a *root*(v)-balanced sample to be one that satisfies

$$\frac{1}{n_h}\mathbf{1}'_{sh}\mathbf{V}_{sh}^{-1/2}\mathbf{X}_{sh} = \frac{\mathbf{1}'_{Nh}\mathbf{X}_h}{\mathbf{1}'_{Nh}\mathbf{V}_h^{1/2}\mathbf{1}_{Nh}}, \tag{6.2.2}$$

where $\mathbf{1}_{sh}$ is a vector of n_h 1's and $\mathbf{1}_{Nh}$ is a vector of N_h 1's. Any sample satisfying (6.2.2) will be denoted by $B(\mathbf{X}_h{:}\mathbf{V}_h)$ and, when (6.2.2) is satisfied in each stratum, the entire sample is a *stratified weighted balanced sample*.

As in the unstratified case introduced in Chapter 3, there are compelling reasons here for using weighted balanced samples within each stratum. If the model has the structure given in Theorem 6.2.1 below, then a weighted balanced sample is the best that can be selected in the sense of making the error variance of the *BLU* predictor smallest. To guard against bias under more elaborate models, we can balance on additional x factors that are not in the working model, without introducing any bias or losing any precision under the working model. A straightforward application of Theorem 4.2.1 yields the following result.

Theorem 6.2.1. Suppose that model (6.2.1) holds in stratum h for $h = 1, \ldots, H$. If both $\mathbf{V}_h \mathbf{1}_{Nh}$ and $\mathbf{V}_h^{1/2} \mathbf{1}_{Nh} \in \mathscr{M}(\mathbf{X}_h)$, then the *BLU* predictor achieves its minimum variance when each stratum sample is $B(\mathbf{X}_h : \mathbf{V}_h)$. In that case, the *BLU* predictor reduces to

$$\hat{T} = \sum_{h=1}^{H} N_h \bar{v}_h^{(1/2)} \frac{1}{n_h} \sum_{i \in s_h} \frac{Y_{hi}}{v_{hi}^{1/2}}. \tag{6.2.3}$$

and the error variance is

$$\text{var}_M(\hat{T} - T) = \sum_h \left[\frac{1}{n_h} (N_h \bar{v}_h^{(1/2)})^2 - N_h \bar{v}_h \right] \sigma_h^2, \tag{6.2.4}$$

where $\bar{v}_h^{(1/2)} = \Sigma_{i=1}^{N_h} v_{hi}^{1/2} / N_h$ and $\bar{v}_h = \Sigma_{i=1}^{N_h} v_{hi} / N_h$.

In a stratified weighted balanced sample, the optimal estimator thus reduces to a sum of mean-of-ratios estimators, which we can write as $\hat{T}_{MRS}(v^{1/2}) = \Sigma_h \hat{T}_{MRh}(v_h^{1/2})$, using the notation introduced in Chapter 3.

6.2.1. Optimal Allocation for Stratified Balanced Sampling

The optimum allocation to the strata of a weighted balanced sample is calculated in much the same way as in Section 6.1. We again assume that the cost of sampling is $C = C_0 + \Sigma_h c_h n_h$, where C_0 is a fixed cost and c_h is the cost per unit sampled in stratum h.

Theorem 6.2.2. Assume that model (6.2.1) holds, that $\mathbf{V}_h \mathbf{1}_{Nh}$ and $\mathbf{V}_h^{1/2} \mathbf{1}_{Nh} \in \mathscr{M}(\mathbf{X}_h)$, and that a weighted balanced sample $B(\mathbf{X}_h : \mathbf{V}_h)$ is selected in each stratum. The allocation of the sample to the strata that minimizes the error variance of the *BLU* predictor, subject to the cost constraint $C = C_0 + \Sigma_h c_h n_h$, is

$$\frac{n_h}{n} = \frac{N_h \bar{v}_h^{(1/2)} \sigma_h / \sqrt{c_h}}{\Sigma_{h'} N_{h'} \bar{v}_{h'}^{(1/2)} \sigma_{h'} / \sqrt{c_{h'}}} \quad \text{for } h = 1, \ldots, H.$$

Proof: The proof is similar to that of Theorem 6.1.1 (see Exercise 6.7). □

As in Section 6.1, we have two corollaries giving the optimal total sample size n, depending on whether the total cost or the desired error variance is fixed. The proofs are also left to the exercises.

Corollary 6.2.1. Under the conditions of Theorem 6.2.1, if the total cost $C = C_0 + \Sigma_h c_h n_h$ is fixed at C^*, then the optimum total sample size is

$$n = (C^* - C_0) \frac{\Sigma_h N_h \bar{v}_h^{(1/2)} \sigma_h / \sqrt{c_h}}{\Sigma_h N_h \bar{v}_h^{(1/2)} \sigma_h \sqrt{c_h}}.$$

When the total sample size is given by Corollary 6.2.1, the variance is

$$\mathrm{var}_M(\hat{T} - T) = \frac{(\Sigma_h N_h v_h^{(1/2)} \sigma_h \sqrt{c_h})^2}{C^* - C_0} - \sum_h N_h \bar{v}_h \sigma_h^2. \qquad (6.2.5)$$

Corollary 6.2.2. Under the conditions of Theorem 6.2.1, if the error variance $\mathrm{var}_M(\hat{T} - T) = V_0$ is fixed, then the optimum total sample size is

$$n = \frac{(\Sigma_h N_h \bar{v}_h^{(1/2)} \sigma_h / \sqrt{c_h})(\Sigma_h N_h \bar{v}_h^{(1/2)} \sigma_h \sqrt{c_h})}{V_0 + \Sigma_h N_h \bar{v}_h \sigma_h^2}.$$

When optimal allocation is used and all costs are equal, the *BLU* predictor (6.2.3) becomes

$$\hat{T} = \frac{1}{n}\left(\sum_h N_h \bar{v}_h^{(1/2)} \sigma_h\right) \sum_h \sum_{s_h} \frac{Y_{hi}}{v_{hi}^{1/2} \sigma_h} \qquad (6.2.6)$$

and its error variance (6.2.4) can be rewritten as

$$\mathrm{var}_M(\hat{T} - T) = \frac{1}{n}\left(\sum_h N_h \bar{v}_h^{(1/2)} \sigma_h\right)^2 - \sum_h N_h \bar{v}_h \sigma_h^2. \qquad (6.2.7)$$

Expression (6.2.7) comes from substituting the expression for n_h from Theorem 6.2.2 for the equal cost case into (6.2.4) or from the reduction of (6.2.5) when costs are equal.

6.2.2. Case of a Single Model for the Population

An important special case is having a single model that fits the whole population. Assume the model in each stratum is

$$E_M(\mathbf{Y}_h) = \mathbf{X}_h \boldsymbol{\beta}, \quad \mathrm{var}_M(\mathbf{Y}_h) = \mathbf{V}_h \sigma^2 \qquad (6.2.8)$$

with $\mathbf{X}_h$ and $\mathbf{V}_h$ defined as in (6.2.1). Expression (6.2.8) is just another way of writing the model $M(\mathbf{X}:\mathbf{V})$. Thus one does not have to consider strata explicitly, and then the *BLU* predictor and its error variance are given by Theorem 2.1. If $\mathbf{V1}_N$ and $\mathbf{V}^{1/2}\mathbf{1}_N \in \mathscr{M}(\mathbf{X})$, then Theorem 4.2.1 showed that a weighted balanced sample $s \in B(\mathbf{B}:\mathbf{V})$ was optimal for the *BLU* predictor, in which case, the *BLU* reduces to

$$\hat{T} = \frac{1}{n} N \bar{v}^{(1/2)} \sum_s Y_i / v_i^{1/2} \qquad (6.2.9)$$

with error variance $\mathrm{var}_M(\hat{T} - T) = \sigma^2[n^{-1}(N\bar{v}^{(1/2)})^2 - N\bar{v}]$.

On the other hand, suppose we do select a stratified weighted balanced sample and use the optimal allocation given in Theorem 6.2.2 for the equal cost case. Using (6.2.6) with $\sigma_h = \sigma$, the *BLU* predictor with the optimal allocation is

$$\hat{T} = \frac{1}{n}\left(\sum_h N_h \bar{v}_h^{(1/2)}\right)\sum_h \sum_{s_h} Y_{hi}/v_{hi}^{1/2}$$

$$= \frac{1}{n}(N\bar{v}^{(1/2)}) \sum_h \sum_{s_h} Y_{hi}/v_{hi}^{1/2}.$$

This is exactly equal to (6.2.9). In other words, stratification with optimal allocation of a stratified weighted balanced sample gains nothing at all compared to the strategy of selecting an unstratified sample with overall weighted balance when a single model fits the whole population.

The ineffectiveness of stratification, in this case, can also be deduced by noting that an optimally allocated stratified weighted balanced sample has overall weighted balance for the equal cost, equal σ_h case. Let $\mathbf{x}_{hi}$ be the ith row of $\mathbf{X}_h$, and define the vectors of population means $\bar{\mathbf{x}}_h = \Sigma_{i=1}^{N_h} \mathbf{x}_{hi}/N_h$ and $\bar{\mathbf{x}} = \Sigma_{h=1}^{H} \Sigma_{i=1}^{N_h} \mathbf{x}_{hi}/N$. Beginning with the left-hand side of the within-stratum balance conditions in (6.2.2), multiplying by n_h/n, summing over h, and using the right-hand side of (6.2.2) together with the value of n_h/n from Theorem 6.2.2, we get

$$\frac{1}{n}\sum_h \sum_{s_h} \frac{\mathbf{x}_{hi}}{v_{hi}^{1/2}} = \frac{1}{n}\sum_h n_h \frac{\bar{\mathbf{x}}_h}{\bar{v}_h^{(1/2)}}$$

$$= \frac{\bar{\mathbf{x}}}{\bar{v}^{(1/2)}},$$

that is, overall balance is achieved.

This result is rather important. Within the context of a single hypothetical model, a great deal of effort in the sampling literature has been devoted to questions such as:

(a) How many strata should there be, that is, how big should H be?

(b) Given H and a size variable x, where along the x-axis should the strata boundaries be drawn? In particular, what should the relative sizes of the N_h be?

See, for example, Cochran (1977, sections 5A.7, 5A.8) and references in Section 6.5 below. For completeness we will examine these questions in later sections. However, we emphasize the above result that, with optimal allocation and balanced sampling, *it does not matter how we choose our strata, or even whether we stratify at all*, as long as the single model (6.2.8) adequately characterizes the population and has the given variance structure.

There still may be reasons for stratification. For example, with differential costs an overall balanced sample for a given value of n may violate the cost constraint even though it is optimal in the sense of Theorem 4.2.1. A stratified allocation is then needed to account for costs even when a single model fits the entire population. Also, if domains are to be compared, then stratification may be useful to control the sample size in each domain.

6.2.3. Case of a Single Auxiliary Variable

Consider the situation studied in earlier chapters, where a common model holds for the whole population and a single auxiliary variable x is available. The auxiliary can be used for stratification as well as estimation. Strata are formed by ordering the units from low to high based on x so that the first stratum contains the N_1 units with the smallest x values, the second stratum contains the next N_2 smallest units, and so on. This method is also known as *stratification by size* and x is referred to as the *size variable*. Take the special case of model (6.2.8) given by

$$E_M(\mathbf{Y}_h) = \mathbf{X}_h\boldsymbol{\beta}, \quad \text{var}_M(Y_{hi}) = \sigma^2 x_{hi}^{\gamma}, \tag{6.2.10}$$

where, in many populations, $0 \leqslant \gamma \leqslant 2$ (see, e.g., Brewer, 1963b; Scott, Brewer, and Ho, 1978). An instance of (6.2.10) would be the minimal model with $E_M(Y_{hi}) = \beta_{\gamma/2}x_{hi}^{\gamma/2} + \beta_\gamma x_{hi}^{\gamma}$. Assume that the cost per unit is c_h, which may differ across the strata. (Indeed, with the common model (6.2.10) only differing costs supply a reason to stratify.) With the variance specification in (6.2.10), the optimum allocation in Theorem 6.2.2 becomes

$$\frac{n_h}{n} = \frac{N_h \bar{x}_h^{(\gamma/2)}/\sqrt{c_h}}{\Sigma_{h'} N_{h'}\bar{x}_{h'}^{(\gamma/2)}/\sqrt{c_{h'}}}, \tag{6.2.11}$$

and the error variance (6.2.4) of the *BLU* predictor in a stratified weighted balanced sample is

$$\text{var}_M(\hat{T} - T) = \sigma^2 \sum_h \left[\frac{1}{n_h}(N_h\bar{x}_h^{(\gamma/2)})^2 - N_h\bar{x}_h^{(\gamma)}\right]. \tag{6.2.12}$$

When the optimal allocation (6.2.10) is used and costs are all equal, this reduces to the variance for an unstratified, weighted balanced sample:

$$\text{var}_M(\hat{T} - T) = \sigma^2\left[\frac{1}{n}(N\bar{x}^{(\gamma/2)})^2 - N\bar{x}^{(\gamma)}\right]. \tag{6.2.13}$$

The expressions (6.2.12) or (6.2.13) can be useful for anticipating, at the design stage of a sample, the variance that will be obtained for a particular set of strata definitions and a given allocation. When a frame is available with individual values of x for each unit, the terms in brackets in (6.2.12) and (6.2.13) can be computed and only σ^2 must be estimated.

For illustration, suppose that a population of business establishments or institutions, like hospitals or colleges, is divided into five strata based on a size measure x. The auxiliary x might be the number of employees per establishment, beds per hospital, or students in each school, with the target variable Y the amount of annual exports, the charges for surgical procedures, or the amount expended on faculty salaries, respectively. Table 6.3 gives the population stratum sizes and averages for x and x^2. In this example, there is a gradation between many small (with respect to x) units having low cost, and a relatively few large units with high cost—a pattern met often in practice.

Suppose that Y follows the model $E_M(Y_{hi}) = \beta_0 + \beta_1 x_{hi} + \beta_2 x_{hi}^2$, $\text{var}_M(Y_{hi}) = \sigma^2 x_{hi}^2$, with $\sigma^2 = 1$. A stratified *root*(x^2)-balanced sample satisfying (6.2.2) will have $\bar{x}_{hs}^{(-1)} = 1/\bar{x}_h$ and $\bar{x}_{hs} = \bar{x}_h^{(2)}/\bar{x}_h$. Theorem 6.2.2 gives the optimal allocation of a stratified balanced sample to strata as being proportional to $N_h\bar{x}_h$, ignoring costs, or proportional to $N_h\bar{x}_h/\sqrt{c_h}$, accounting for them. Assume that $C^* - C_0 = \$5{,}250$ is the budget. As in Section 6.1, the implied total sample size for a given budget is $n = (C^* - C_0)/\bar{c}$, where $\bar{c} = \Sigma_k w_h c_h$ is a weighted average of the unit costs. For Neyman allocation, $w_h = N_h\bar{x}_h/\Sigma_h N_h\bar{x}_h$, while $w_h = N_h/N$ for proportional, and $w_h = 1/H$ for equal allocation.

Table 6.4 shows the results for four allocations. The total costs for the four differ somewhat from \$5,250 because stratum sample sizes are rounded to integers. The optimal total sample from Corollary 6.2.1 is $n = 98$. Unlike the earlier example, the cost-constrained optimal directs much more effort to particular strata—almost half the sample of $n = 98$ is assigned to stratum 5, which contains the largest units. On the other hand, at least in this example, the proportional and equal allocations have standard errors that are much higher than that of the column (1) allocation. For proportional allocation, which puts most of its sample in the strata of small size units, the total sample size is 181—85% more than the optimal of 98—but the standard error of 36.0

Table 6.3. Stratum Population Characteristics for an Example of Stratification by Size

h	N_h	$\bar{x}_h$	$\bar{x}_h^{(2)}$	c_h
1	680	8.7	81.1	10
2	552	15.5	263.3	20
3	429	76.1	6,727.9	30
4	363	190.8	37,641.6	50
5	251	621.6	406,747.2	70

Table 6.4. Sample Allocations for a Population Stratified by Size with a Cost Constraint of $C^* - c_0 = \$5{,}250$

	Stratum Samples n_h			
Stratum	(1) Optimal with Cost Constraint	(2) Neyman Allocation	(3) Proportional Allocation	(4) Equal Allocation
1	5	2	54	29
2	5	3	44	29
3	15	11	34	29
4	25	23	29	29
5	48	53	20	29
Total sample	98	94	181	145
Total cost	5,210	5,270	5,290	5,220
$\sqrt{\text{var}_M(\hat{T} - T)}/10^3$	25.9	26.2	36.0	30.4

thousand is about 39% larger than that of the column (1) optimal. The total sample size for equal allocation in column (4) is $n = 145$, also much larger than the 98 in column (1), but, because the sample is poorly allocated, the standard error is 30.4 thousand, 17% larger than the column (1) optimal.

The Neyman allocation in column (2) loses little in terms of standard error compared to the column (1) allocation (25.9 vs. 26.2). But the unconstrained, column (2) allocation is equivalent to a sample with overall root(x^2) weighted balance. In other words, with no stratification at all, but merely using root(x^2) balance on the whole, one achieves an *rmse* close to the optimum under the cost constraints.

The S-Plus™ function *strat.opt.alloc.x* will allow more exploration of the effects of different N_h, $\bar{x}_h$, and c_h vectors on allocations. There are combinations of costs where the cost-constrained optimal (column 1) and the unconstrained optimal (column 2) will differ more sharply—a point explored in Exercise 6.12.

6.3. SAMPLING FRACTIONS GREATER THAN 1

In some populations, it may be possible that the sampling fraction implied by the optimal allocation in Theorem 6.1.1 or 6.2.2 is greater than or equal to 1 in some strata. For example, in the case of homogeneous subpopulations, it could happen that for some h

$$\frac{n_h}{N_h} = \frac{n\sigma_h/\sqrt{c_h}}{\Sigma_{h'} N_{h'}\sigma_{h'}/\sqrt{c_{h'}}} \geqslant 1 \tag{6.3.1}$$

in Theorem 6.1.1. This is similar to the situation described in Section 3.3.7 and occurs when the number of units, N_h, in a stratum is relatively small and the sample size n is very large. In such a case, the units in that stratum are said to be *take-alls* or to be selected with certainty: all units are selected from the stratum. The stratum then contributes nothing to the error variance of $\hat{T}$ in equation (6.1.2). Given that the units in one stratum are completely sampled, Theorem 6.1.1 or 6.2.2 can be applied to the remaining strata to determine their allocation and the corollaries to those theorems can be used to find the optimal total sample size.

As an example, consider the population in Table 6.1, and suppose that we set the desired error variance of $\hat{T}$ to be $V_0 = (12{,}000)^2$. In practice, it is most natural to set a target value for the coefficient of variation, $\sqrt{\text{var}_M(\hat{T} - T)}/E_M(T)$, as in Section 3.5, but for the sake of illustration we assign a value to the error variance here. The value V_0 is considerably smaller than the error variance of $(67{,}000)^2$ in Table 6.2 for the optimal allocation with a budget of $C^* - C_0 = \$5{,}250$ and will lead to a much larger sample size. Corollary 6.1.2 gives the total sample size as $n = 1408$. Applying Theorem 6.1.1, we compute the sample sizes for strata 1–5 as 559, 152, 308, 105, and 284, respectively. But, $n_5 = 284$ exceeds the population size of 251. Designating stratum 5 as a certainty and applying Corollary 6.1.2 to strata 1–4, we still have a target of $V_0 = (12{,}000)^2$ and obtain $n' = 1{,}166$ as the optimal number to distribute among those strata. Again invoking Theorem 6.1.1, we get the sample sizes shown in Table 6.5. The cost of this sample is \$52,715 and achieves a standard error of 12,002.

Another useful formulation of the allocation problem is as a constrained, nonlinear programming problem. The problem of finding the allocation to minimize cost while achieving the fixed variance of $V_0 = (12{,}000)^2$ might be stated as:

$$\text{Find } \{n_1, \ldots, n_H\} \text{ to minimize } C = C_0 + \sum_h c_h n_h$$

Subject to the constraints:

$$\text{(i) } n_0 \leqslant n_h \leqslant N_h \text{ for all } h$$

$$\text{(ii) } \sum_h \left(\frac{N_h^2}{n_h} - N_h \right) \sigma_h^2 \leqslant V_0.$$

Constraint (i) requires the sample size in each stratum to be greater than or equal to n_0, some desired minimum sample size, and less than or equal to the population size N_h. The second constraint requires the variance of the estimator of the total to meet the target value V_0.

Spreadsheet software packages like Microsoft Excel™ and Lotus 1-2-3™ often include algorithms for solving this type of problem. Rather than applying

Table 6.5. Optimal Allocation in an Example Where One Stratum is a Certainty

Stratum	N_h	σ_h	c_h	n_h	n_h/N_h
1. Manufacturing	680	432	30	580	0.85
2. Retail trade	552	229	75	158	0.29
3. Wholesale trade	429	344	25	319	0.74
4. Services	363	196	50	109	0.30
5. Government	251	686	40	251	1.00
Total sizes	2,275			1,417	
Total cost				$52,715	

Theorem 6.1.1 and Corollary 6.1.2 in two steps to get an allocation, these algorithms iteratively solve the constrained problem given above. In the example above, this approach will yield about the same solution as in Table 6.5 since the problem is simple. But in more complicated problems having a number of constraints or involving multiple Y variables, the nonlinear programming approach has important advantages. Section 6.4.2 gives some further examples where it can be applied.

When a common model holds across strata but costs vary, and stratification by size is used, then (6.2.10) implies that a stratum h will be a certainty if

$$\frac{\bar{x}_h^{(\gamma/2)}}{\sqrt{c_h}} \geqslant \frac{1}{n}\sum_{h'} \frac{N_{h'}\bar{x}_{h'}^{(\gamma/2)}}{\sqrt{c_{h'}}}. \tag{6.3.2}$$

Note that, if costs are all equal, then (6.3.2) reduces to a comparison depending on the means of functions of the auxiliaries. However, in the equal-cost/common-model case, stratification with optimal allocation of a weighted balanced achieves exactly the same error variance as an unstratified, weighted balanced sample, as we noted in Section 6.2.2. If a weighted balanced sample is impossible because of large n, the method of partial balancing described in Section 3.3.7 can be used to designate individual units as certainties.

6.4. ALLOCATION TO STRATA IN MORE COMPLICATED CASES

Rare is the survey whose goals are so limited that only a single Y variable is collected and only the population total is estimated. Comparisons between domains are usually wanted for several different variables. We might, for example, be interested in recording the blood cholesterol level, weight, blood pressure level, and other physical measurements on a sample of persons and then comparing the averages per person for different racial/ethnic groups. If domains are strata or groups of strata, the optimal allocations in Theorems 6.1.1 and 6.2.2 are likely to put too few sample cases in some strata to permit reliable domain estimation.

6.4.1. Contrasts Between Strata

As a simple example of the need to depart from the optimum, consider estimating the difference between the means in two subpopulations, which are strata. As Cochran (1977, sec. 5A.13) notes, the real goal in making such a comparison is usually to estimate the difference in the superpopulation means, not in the finite population means. For example, if we want to estimate the difference in the incidence of a disease between two racial groups, the question of scientific interest is not how large the difference is in a finite population as it exists at a specific point in time but in a more general phenomenon likely to be present at other times also. For that reason, we omit finite population correction factors in the variances computed below. Suppose that the model is

$$\begin{aligned} E_M(Y_{hi}) &= \mu_h \\ \text{var}_M(Y_{hi}) &= \sigma_h^2 \end{aligned} \tag{6.4.1}$$

and different Y_{hi}'s are independent. We want to determine how to allocate a sample of $n = n_1 + n_2$ to strata 1 and 2, assuming that costs are equal in the two strata. The estimator of the difference in the strata superpopulation means, $\mu_1 - \mu_2$, using the optimal expansion estimator $\hat{T}_{0h}$ in each stratum, is

$$\hat{\bar{Y}}_1 - \hat{\bar{Y}}_2 = \frac{\hat{T}_{01}}{N_1} - \frac{\hat{T}_{02}}{N_2},$$

where $\hat{T}_{0h} = N_h \Sigma_{s_h} Y_{hi}/n_h$. The Lagrange function to be minimized is

$$\begin{aligned} \Phi &= \text{var}_M(\hat{\bar{Y}}_1 - \hat{\bar{Y}}_2) + \lambda(n_1 + n_2 - n) \\ &= \sum_{h=1}^{2} \frac{\sigma_h^2}{n_h} + \lambda(n_1 + n_2 - n). \end{aligned}$$

Differentiating Φ with respect to n_h leads to the equations

$$\frac{\sigma_h^2}{n_h^2} - \lambda = 0.$$

The optimum proportion of the sample to allocate to stratum h is then

$$\frac{n_h}{n} = \frac{\sigma_h}{\sigma_1 + \sigma_2}.$$

If the model parameters σ_1 and σ_2 are about equal, then allocating half the sample to each stratum will be close to optimal for minimizing the variance of $\hat{\bar{Y}}_1 - \hat{\bar{Y}}_2$. Notice that the population sizes N_h do not enter the calculation. The optimal allocation for estimating the total across the two strata, in contrast, has $n_h \propto N_h \sigma_h$, which is affected by the stratum size N_h.

If there are more than two strata, several alternatives might be used for allocation. Cochran (1977, sec. 5A.13) suggests two. One is to minimize the average variance of the estimator of the difference in means between all $H(H-1)/2$ pairs of strata:

$$\bar{V} = \frac{2}{H(H-1)} \sum_{h<h'} \left(\frac{\sigma_h^2}{n_h} + \frac{\sigma_{h'}^2}{n_{h'}} \right)$$
$$= \frac{2}{H} \sum_{h=1}^{H} \frac{\sigma_h^2}{n_h}.$$

Again assuming equal costs, the stratum sample sizes that minimize $\bar{V}$ are easily shown to be

$$n_h = n \frac{\sigma_h}{\Sigma_h \sigma_h}. \tag{6.4.2}$$

The minimum value of the average variance is then $\bar{V}_{\min} = 2(\Sigma_h \sigma_h)^2 / Hn$.

Another route to allocation would be to require that the variance of the estimator of the difference in means between any two strata be the same, that is, $\sigma_h^2/n_h + \sigma_{h'}^2/n_{h'} = V$. This implies that

$$\frac{H(H-1)}{2} V = \sum_{h<h'} \left(\frac{\sigma_h^2}{n_h} + \frac{\sigma_{h'}^2}{n_{h'}} \right) \quad \text{or} \quad \frac{1}{H} \sum_{h=1}^{H} \frac{\sigma_h^2}{n_h} = \frac{V}{2}.$$

One solution is to set $n_h = 2\sigma_h^2/V$, which implies that $V = 2\,\Sigma_h \sigma_h^2/n$ and

$$n_h = n \frac{\sigma_h^2}{\Sigma_h \sigma_h^2}. \tag{6.4.3}$$

It follows from the Cauchy-Schwartz inequality that $\bar{V}_{\min} \leqslant V$ with equality only when σ_h is a constant.

We illustrate these allocations using the population data in Table 6.1 and assuming equal costs for all strata. The desired total sample size is set at $n = 150$. The allocations in columns (2) and (3) in Table 6.6 are for $\bar{V}_{\min}$ and V as defined above. In accordance with the theory above, the value of $\bar{V}/10^3$ for the $\bar{V}_{\min}$ allocation is 9.5—less than $\bar{V}/10^3 = 11.6$ for the allocation in column (3). The reader can verify that the pairwise differences in means are not equally precise with the $\bar{V}_{\min}$ allocation, however. Thus, if specific comparisons are of paramount interest, the $\bar{V}_{\min}$ allocation may need adjustment.

Table 6.6. Sample Allocations in a Business Population for Estimating Differences in Stratum Means

			Stratum Samples n_h		
Stratum	N_h	σ_h	(1) Neyman Allocation	(2) $\bar{V}_{\text{min}}$ Allocation	(3) V Allocation
1. Manufacturing	680	432	54	34	32
2. Retail trade	552	229	23	18	9
3. Wholesale trade	429	344	27	27	20
4. Services	363	196	13	16	7
5. Government	251	686	32	55	81
Total sample			149	150	149
$\sqrt{\text{var}_M(\hat{T}_{0S} - T)}/10^3$			63.8	68.9	79.4
$\bar{V}/10^3$			11.1	9.5	11.6

Total Sample Size is $\boldsymbol{n} = 150$. The Allocation in Columns (1) and (3) Differ from 150 Due to Rounding

The Neyman allocation, defined in (6.1.4), in column (1) of Table 6.6 is shown for comparison. Note that its value of $\bar{V}/10^3$ is larger at 11.1 than the 9.5 for the $\bar{V}_{\text{min}}$ allocation. On the other hand, the standard error of the expansion estimator of the population total is smaller for the Neyman allocation (63.8 thousand) than for either the $\bar{V}_{\text{min}}$ or V allocation. Hence, we cannot be "best" for both a full population estimator and for subpopulation comparisons.

6.4.2. More Than One Target Variable

Multivariate allocation problems are also common and require deviation from the optimum for a single variable. When multiple Y variables are collected, the optimum for estimating the total for one may be suboptimal for others. Specific precision levels, defined in terms of the coefficient of variation or relvariance, may also be desired for a set of domain and total population estimates.

One approach to formulating the problem with multiple survey goals is *multicriteria mathematical programming*. This is a fairly difficult and specialized area that we will describe here only generally. Arthanari and Dodge (1981), Narula and Weistroffer (1989), Steuer (1986), or Weistroffer and Narula (1991) survey the area more thoroughly.

For the multivariate sample allocation problem, a weighted combination of the relvariances of different estimators is formed with each weight being the "importance" of each statistic in the overall survey design. Using a mathematical programming algorithm, the weighted combination is minimized subject to a variety of constraints, including, for example, one on total cost or sample size, minimum and maximum sample size constraints in each stratum, and coefficient of variation constraints on individual estimators. Let $l = 1, \ldots, L$ index different estimators. In a business survey we might, for example, be interested in separately estimating the costs to employers of providing paid vacations, health insurance, and retirement plans. We might also want these costs estimated separately for businesses in manufacturing, retail trade, and construction.

Letting w_l be the importance weight assigned to estimator l and ϑ_l the relvariance for estimator l, an example of an optimization problem can be stated formally as

$$\text{minimize } \phi = \sum_{l=1}^{L} w_l \vartheta_l \tag{6.4.4}$$

subject to

1. $n_{h,\min} \leqslant n_h \leqslant N_h$, where $n_{h,\min}$ is a specified minimum sample size,
2. $C_0 + \Sigma_h c_h n_h \leqslant C^*$, a bound on the total cost,
3. $\vartheta_l^{1/2} \leqslant K_{l0}$ for $l \in S_E$, i.e., the coefficient of variation of an estimator l is bounded by some target level K_{l0} for each estimator in some set S_E.

The use of the relvariance, a unitless measure, allows estimates that have different units (e.g., dollars, milligrams per deciliter, inches of mercury, and proportion of persons) to be weighted together in a single objective function. Use of the set S_E in (3) simply means that the coefficients of variation are not necessarily constrained for all estimators in the survey.

Because the sample sizes enter the objective function and the constraint (3) in the denominators, both the objective and the constraints are nonlinear in the target variables n_h. Exact analytic solutions as in Theorems 6.2.1 and 6.2.2 and in the stratum comparison example above are usually not available. A variety of algorithms has been developed for solving this and similar kinds of optimization problems (Fiacco and McCormick, 1968; Lasdon and Waren, 1978; Moré and Wright, 1993; Schittkowski, 1985). Some of these techniques are also readily available for solving problems of limited size in spreadsheet software like Microsoft Excel™ and Lotus 1-2-3™. The algorithms iteratively search for feasible solutions, that is, ones that meet all the constraints, and then, among the feasible solutions, attempt to find one that minimizes the objective function. This type of optimization problem can be quite difficult to solve because there may be many local minima and the global minimum, if it exists, may be little different from other feasible solutions.

A number of applications of mathematical programming to sample allocation problems have been reported in the literature. Leaver et al. (1987, 1996) used the technique in the sample design for the U.S. Consumer Price Index. Valliant and Gentle (1997) applied the multicriteria approach in a case where two surveys use the same sample design to make estimates of a variety of characteristics in a business establishment universe. Bethel (1985, 1989) and Chromy (1987) discussed algorithms that are designed to solve problems somewhat simpler than (6.4.2). Some of the earliest work on this type of problem was done by Kokan (1963) and Kokan and Khan (1967).

6.5. TWO TRADITIONAL TOPICS

In this section we discuss two topics—the separate ratio estimator and the formation of strata based on a size variable—which traditionally receive attention in sampling texts. In large measure what is said about them is rendered moot by the results of the preceding sections. Being cognizant of these topics will, however, be helpful in understanding techniques that are often used in practice.

6.5.1. Efficiency of the Separate Ratio Estimator

When strata are formed on the basis of a size measure x, the separate ratio estimator is frequently recommended and is defined as

$$\hat{T}_{RS} = \sum_{h=1}^{H} N_h \bar{Y}_{hs} \frac{\bar{x}_h}{\bar{x}_{hs}},$$

where $\bar{x}_h = \Sigma_{i=1}^{N_h} x_{hi}/N_h$, $\bar{Y}_{hs} = \Sigma_{i \in s_h} Y_{hi}/n_h$, and $\bar{x}_{hs} = \Sigma_{i \in s_h} x_{hi}/n_h$. Under the working model $M(0, 1{:}x)$, $\hat{T}_{RS}$ is unbiased with variance equal to

$$\text{var}_M(\hat{T}_{RS} - T) = \sigma^2 \sum_h \frac{N_h^2}{n_h}(1 - f_h)\frac{\bar{x}_{hr}\bar{x}_h}{\bar{x}_{hs}},$$

where $f_h = n_h/N_h$ and $\bar{x}_{hr} = \Sigma_{i \notin s_h} x_{hi}/(N_h - n_h)$. If one is completely confident that $M(0, 1{:}x)$ is correct, then the optimal sample for $\hat{T}_{RS}$ would be to pick the n_h units with the largest x's in each stratum. Even more extreme is the globally optimal strategy of the simple ratio estimator and the n largest units in the population.

Confidence in any single model is seldom this high and having protection against model failure is usually prudent. Using notation from Chapter 3, if the true model is $M(\delta_0, \ldots, \delta_J{:}x)$, then the estimator has a bias:

$$E_M(\hat{T}_{RS} - T) = \sum_h N_h \bar{x}_h \sum_{j=0}^{J} \delta_j \beta_j \left[\frac{\bar{x}_{hs}^{(j)}}{\bar{x}_{hs}} - \frac{\bar{x}_h^{(j)}}{\bar{x}_h}\right],$$

where $\bar{x}_h^{(j)} = \Sigma_{i=1}^{N_h} x_{hi}^j/N_h$ and $\bar{x}_{hs}^{(j)} = \Sigma_{i \in s_h} x_{hi}^j/n_h$. If a *stratified (unweighted) bal-*

anced sample, that is, one that is balanced in each stratum ($\bar{x}_{hs}^{(j)} = \bar{x}_h^{(j)}$ for $j = 1, \ldots, J$), is selected, then $\hat{T}_{RS}$ will be unbiased and reduces to the stratified expansion estimator

$$\hat{T}_{0S} = \sum_{h=1}^{H} N_h \bar{Y}_{hs}.$$

A stratified balanced sample will be denoted by $s^*(J)$. As in Chapter 3, $s(J)$ denotes a simple balanced sample of order J. When the working model $M(0, 1{:}x)$ holds, the prediction variance and the optimal allocation to strata depend on the stratum means of the auxiliary, as noted in Lemmas 6.5.1 and 6.5.2.

Lemma 6.5.1. If $M(\delta_0, \ldots, \delta_J{:}x)$ is the correct model, then the error variance of $\hat{T}_{RS}$, in a stratified balanced sample, is

$$E_M(\hat{T}_{RS} - T)^2 = \sigma^2 \left[\sum_h \frac{N_h^2}{n_h} \bar{x}_h - N\bar{x} \right].$$

Lemma 6.5.2. Under $M(\delta_0, \ldots, \delta_J{:}x)$ the allocation of a stratified balanced sample that minimizes the error variance of $\hat{T}_{RS}$, subject to the cost constraint $C = C_0 + \Sigma_h c_h n_h$, is

$$\frac{n_h}{n} = \frac{N_h\sqrt{\bar{x}_h/c_h}}{\Sigma_{h'} N_{h'}\sqrt{\bar{x}_{h'}/c_{h'}}} \quad \text{for } h = 1, \ldots, H.$$

When costs are equal, the error variance with the optimal allocation is

$$E_M(\hat{T}_{RS} - T)^2 = \sigma^2 \left[\frac{1}{n} \left(\sum_h N_h \sqrt{\bar{x}_h} \right)^2 - N\bar{x} \right]. \tag{6.5.1}$$

The proofs of both the Lemmas are left to Exercise 6.16.

Protection against bias under the polynomial model $M(\delta_0, \ldots, \delta_J{:}x)$ can be achieved either by simple balanced sampling with the ratio estimator or stratified balanced sampling with the separate ratio estimator. Which is more efficient? Does it pay to stratify on the size variable if we confine ourselves to these two types of ratio estimator? The answer to the latter question is yes when optimal allocation is used, as shown in Theorem 6.5.1.

Theorem 6.5.1 (Royall and Herson, 1973b). If $n_h \propto N_h\sqrt{\bar{x}_h}$, then under $M(\delta_0, \ldots, \delta_J{:}x)$ the strategy $[s^*(J), \hat{T}_{RS}]$ is more efficient than $[s(J), \hat{T}(0, 1{:}x)]$ in the sense that

$$E_M[\hat{T}(0, 1{:}x) - T]^2 \geqslant E_M(\hat{T}_{RS} - T)^2. \tag{6.5.2}$$

Proof: With a simple balanced sample $s(J)$, the left-hand side of (6.5.2) is $\sigma^2\bar{x}(N^2/n)(1-f)$ while the right is given by (6.5.1). Their difference is $\sigma^2/n[N^2\bar{x} - (\Sigma_h N_h\sqrt{\bar{x}_h})^2]$, which is nonnegative since

$$\left(\sum_h N_h\sqrt{\bar{x}_h}\right)^2 \leqslant \left(\sum_h N_h\right)\left(\sum_h N_h\bar{x}_h\right) = N^2\bar{x}$$

by the Cauchy-Schwarz inequality (Appendix A). Strict inequality holds except in the case where all the x's are equal. □

Proportional allocation is another simple allocation that might be used, that is, $n_h \propto N_h$. But, proportional allocation gains nothing. For, one readily shows that, with proportional allocation, a stratified balanced sample is also a simple balanced sample, in which case $\hat{T}_{RS} = \hat{T}(0, 1{:}x)$. Thus nothing is gained beyond the strategy $[s(J), \hat{T}(0, 1{:}x)]$. Proportional allocation thus nullifies the possible gains in efficiency that stratification can yield.

However, because the separate ratio estimator does not flow from a model satisfying the conditions of Theorem 6.2.2, the strategy $[s^*(J), \hat{T}_{RS}]$ is not the best that we can do. When $\text{var}_M(Y_i) = \sigma^2 x_i$, as in $M(0, 1{:}x)$, the minimal model is $M(x^{1/2}, x{:}x)$. Let $M(\delta_0, \delta_{1/2}, \ldots, \delta_J{:}x)$ denote the model with $E_M(Y_i) = \delta_0 + \delta_{1/2}x_i^{1/2} + \delta_1 x_i + \cdots + \delta_J x_i^J$ and $\text{var}_M(Y_i) = \sigma^2 x_i$. $\hat{T}_{RS}$ will be unbiased under $M(\delta_0, \delta_{1/2}, \ldots, \delta_J{:}x)$ if the sample is $s^*(J)$ and, in addition, $\bar{x}_{hs}^{(1/2)} = \bar{x}_h^{(1/2)}$ for all h. Call these combined balance conditions $s^{**}(J)$.

If the sample has weighted balance in the sense that

$$\frac{1}{n}\sum_s \frac{x_i^j}{x_i^{1/2}} = \frac{\bar{x}^{(j)}}{\bar{x}^{(1/2)}} \quad \text{for } j = 0, \tfrac{1}{2}, 1, \ldots, J, \tag{6.5.3}$$

then the *BLU* predictor $\hat{T}(x^{1/2}, x{:}x)$ under $M(x^{1/2}, x{:}x)$ is protected against bias if the model is really $M(\delta_0, \delta_{1/2}, \ldots, \delta_J{:}x)$. By Theorem 4.2.1, when (6.5.3) is satisfied, $\hat{T}(x^{1/2}, x{:}x)$ reduces to the mean-of-ratios estimator $\hat{T}_{MR}(x^{1/2}) = N\bar{x}^{(1/2)}n^{-1}\Sigma_s Y_i/x_i^{1/2}$ and has error variance

$$\text{var}_M[\hat{T}_{MR}(x^{1/2}) - T] = \sigma^2\left[\frac{1}{n}(N\bar{x}^{(1/2)})^2 - N\bar{x}\right]. \tag{6.5.4}$$

Denote the strategy of using a sample satisfying (6.5.3) and the estimator $\hat{T}(x^{1/2}, x{:}x)$ as $[s(\mathbf{X}{:}\mathbf{V}), \hat{T}(x^{1/2}, x{:}x)]$, where $\mathbf{V} = \text{diag}(x_i)$. Theorem 4.2.1 implies that $[s(\mathbf{X}{:}\mathbf{V}), \hat{T}(x^{1/2}, x{:}x)]$ is optimal and, in particular, is better than $[s^{**}(J), \hat{T}_{RS}]$. This result can be shown directly:

Theorem 6.5.2. Suppose that $n_h \propto N_h\sqrt{\bar{x}_h}$ for the separate ratio estimator $\hat{T}_{RS}$. Under $M(\delta_0, \delta_{1/2}, \ldots, \delta_J{:}x)$ the strategy $[s(\mathbf{X}{:}\mathbf{V}), \hat{T}(x^{1/2}, x{:}x)]$ is more efficient than $[s^{**}(J), \hat{T}_{RS}]$ in the sense that

$$E_M(\hat{T}_{RS} - T)^2 \geqslant E_M[\hat{T}(x^{1/2}, x{:}x) - T]^2. \tag{6.5.5}$$

Proof: Both strategies provide unbiased estimators under

$$M(\delta_0, \delta_{1/2}, \ldots, \delta_J{:}x).$$

Using (6.5.1) and (6.5.4), the difference in the left-hand and right-hand sides of (6.5.5) is $\sigma^2/n[(\Sigma_h N_h\sqrt{\bar{x}_h})^2 - (N\bar{x}^{(1/2)})^2]$. This difference is nonnegative because

$$\begin{aligned}\sum_h N_h\sqrt{\bar{x}_h} &\geqslant \sum_h N_h \bar{x}_h^{(1/2)}\\ &= N\bar{x}^{(1/2)}\end{aligned}$$

by the concavity of the square root function. □

A question that traditionally arises when considering stratification is how many strata to form when stratifying by size. The foregoing results imply that, under the conditions of Theorem 6.5.2, this question is not pertinent. We achieve optimality with no stratification. However, note that when using a suboptimal strategy like $[s^*(J), \hat{T}_{RS}]$, the result in Theorem 6.5.1 can be applied within any substratum to which more than one observation is allocated. Unless the size of x is constant within a stratum, substratification reduces the variance of $\hat{T}_{RS}$. Pursuit of this argument leads to the conclusion that the strata be as small as possible while still compatible with within-stratum balance.

6.5.2. Formation of Strata

Another question traditionally posed when stratifying by size is how to form the strata. When a common model holds for the entire population as in Section 6.2.2 and $\mathbf{V1}_N$, $\mathbf{V}^{1/2}\mathbf{1}_N \in \mathscr{M}(\mathbf{X})$, we know that the *BLU* predictor with a weighted balanced sample is the best strategy. That is, stratification in this common circumstance is unnecessary. Nonetheless, for completeness, we mention various methods of strata formation that are likely to be met in current practice.

One set of methods are known as *equal aggregate size* rules. Units are sorted from low to high based on x. Strata are then formed in such a way that each contains about the same total of the size variable or a monotone transformation of it. Equalizing $N_h\bar{x}_h^{(\gamma/2)}$ leads to several stratification rules that have been in the literature for many years, for example, Cochran (1977, p.172), Hansen, Hurwitz, and Madow (1953), and Mahalanobis (1952). When $\gamma = 0$, equal values of $N_h\bar{x}_h^{(\gamma/2)}$ correspond to equal numbers of units N_h in each stratum. When $\gamma = 1$, we have equal aggregate square root of size, and $\gamma = 2$ gives equal aggregate x.

These rules are justified in the more recent design-based literature as ways of approximating optimum selection probabilities when a type of regression estimator is used. Godfrey, Roshwalb, and Wright (1984), Särndal, Swensson, and Wretman (1992, sec. 12.4) and Wright (1983) discuss these matters in detail. Equal aggregate size rules can also be derived using only model-based arguments (see Exercise 6.17). In particular, when the model is $E_M(\mathbf{Y}_h) = \mathbf{X}_h\boldsymbol{\beta}$, $\text{var}_M(Y_{hi}) = \sigma^2 x_{hi}^{\gamma}$ with $\mathbf{V1}_N$, $\mathbf{V}^{1/2}\mathbf{1}_N \in \mathscr{M}(\mathbf{X})$, the sample is $s_h \in B(\mathbf{X}_h:\mathbf{V}_h)$, and an equal number of units is allocated to each stratum, then the error variance of the *BLU* predictor is minimized if strata are constructed to have equal $N_h\bar{x}_h^{(\gamma/2)}$ in each. Moreover, if strata are constructed in this manner, an equal allocation is the optimal allocation in the equal cost case. However, an optimally allocated stratified, weighted balanced sample yields exactly the same variance as an unstratified sample with weighted balance.

If we again confine ourselves to using the separate ratio estimator, we obtain the specialized, model-based Theorem 6.5.3 below on strata formation. Assume that the estimator is $\hat{T}_{RS}$, the model is $M(0, 1:x)$ and that a stratified balanced sample is selected. We assume that the same number of sample units is allocated per stratum. An equal allocation, sufficiently large in each stratum, will permit a within-stratum balanced sample to be selected and will allow some model checking also.

When a stratified balanced sample $s^*(J)$ is selected with $n_h = n_0$ for $h = 1, \ldots, H$, the prediction variance under any model $M(\delta_0, \delta_1, \ldots, \delta_J:x)$ is, from Lemma 6.5.1,

$$E_M(\hat{T}_{RS} - T)^2 = \sigma^2\left\{\sum_h N_h^2\bar{x}_h/n_0 - N\bar{x}\right\}. \tag{6.5.6}$$

Optimal stratification for equal allocation occurs when stratum boundaries are chosen to minimize $\Sigma_h N_h^2\bar{x}_h$.

Theorem 6.5.3 (Royall and Herson 1973b). When the size measures $x_1, \ldots, x_N$ all have different values and the stratified balanced strategy $[s^*(J), \hat{T}_{RS}]$ is employed: (i) A necessary condition for optimal stratification under any model $M(\delta_0, \delta_1, \ldots, \delta_J:x)$ is

$$N_1 \geqslant N_2 \geqslant \cdots \geqslant N_H. \tag{6.5.7}$$

(ii) The condition

$$N_1\bar{x}_1 \leqslant N_2\bar{x}_2 \leqslant \cdots \leqslant N_H\bar{x}_H \tag{6.5.8}$$

is also necessary when it is compatible with (6.5.7). The proof of the theorem can be found in Royall and Herson (1973b).

Inequalities (6.5.7) and (6.5.8) give an indication of the essential features of a good stratification scheme for use with equal allocation, stratified balanced sampling, if we were to use the separate ratio estimator. The strata should be constructed so that there are more units in stratum h than in stratum $h+1$ for $h=1,\ldots,H$. But there should not be so many more units in h that the sum of the sizes in h exceeds those in $h+1$. Three special cases in which both (6.5.7) and (6.5.8) are satisfied are

1. $N_1 = N_2 = \cdots = N_H$;
2. $N_1\bar{x}_1 = N_2\bar{x}_2 = \cdots = N_H\bar{x}_H$; and
3. $N_1^2\bar{x}_1 = N_2^2\bar{x}_2 = \cdots = N_H^2\bar{x}_H$.

Case (1) obviously satisfies (6.5.7) and, since $\bar{x}_1 < \bar{x}_2 < \cdots < \bar{x}_H$, (6.5.8) holds. The second case implies $N_h = N_{h+1}\bar{x}_{h+1}/\bar{x}_h > N_{h+1}$ so that both (6.5.7) and (6.5.8) hold. Case (3) implies that $N_h^2 = N_{h+1}^2\bar{x}_{h+1}/\bar{x}_h > N_{h+1}^2$ and that $N_h\bar{x}_h = (N_{h+1}/N_h)N_{h+1}\bar{x}_{h+1}$ so that again both conditions in Theorem 6.5.3 hold. In case (1) equal allocation is proportional allocation, while in (3) equal allocation is optimal allocation. The second case corresponds to having equal total sizes in the strata.

Another method of stratification is known as the cum $\sqrt{f}$ rule after Dalenius and Hodges (1959), which we will sketch here. Cochran (1977, Section 5A.6) gives more details. The technique is derived from a hybrid of design-based and model-based thinking. Suppose that all Y's in the population follow a common distribution with density function $f_Y(Y)$. Units are assigned to strata, an *stsrs* is selected without replacement, and the estimator $\hat{\bar{Y}} = \sum_{h=1}^{H} W_h\bar{Y}_{hs}$ is used to estimate the mean, where $W_h = N_h/N$. The design-variance with Neyman allocation is given by

$$\mathrm{var}_\pi(\hat{\bar{Y}}) = \frac{1}{n}\left(\sum_{h=1}^{H} W_h S_h\right)^2 - \frac{1}{N}\sum_{h=1}^{H} W_h S_h^2$$

with the parameter S_h^2 being the finite population variance among units assigned to stratum h. For a given density $f(\cdot)$ and a fixed number of strata H, the idea is to compute boundaries that minimize $\mathrm{var}_\pi(\hat{\bar{Y}})$. When the sampling fractions, n_h/N_h, are negligible, it is sufficient to minimize $\sum_{h=1}^{H} W_h S_h$. The calculation of the minimum is unmanageable for most densities and the cum $\sqrt{f}$ rule, to be described, is an approximation.

Suppose that X is a random auxiliary variable whose realization x is known for all units in the population and closely related to Y. Since Y is unknown, we form strata using the distribution of X. Let $f_X(x)$ be the density of X with range $b_L \leqslant X \leqslant b_U$. Define the lower boundary of the first stratum to be b_L and the upper boundary of the last stratum to be b_U. The intermediate stratum boundaries, which are to be determined, are $b_1, \ldots, b_{H-1}$ in order from low to high. The cum $\sqrt{f}$ rule creates strata in such a way that

$$\int_{b_{h-1}}^{b_h} [f_X(x)]^{1/2}\,dx = \frac{1}{H}\int_{b_L}^{b_U} [f_X(x)]^{1/2}\,dx. \tag{6.5.9}$$

In words, the total integral of $[f_X(x)]^{1/2}$ is divided equally among the strata. An approximation to (6.5.9) is based on the frequency distribution of $x_1, x_2, \ldots, x_N$. First, we create a histogram with a number J of equally wide intervals, where J is considerably larger than H, the number of strata. Let f_j be the number of units in the jth interval. Compute $f_j^{1/2}$ for each interval and the sum $F = \Sigma_j f_j^{1/2}$ over all the intervals. Then, starting at the left end of the histogram, combine the intervals to form H strata in such a way that each stratum contains about the same amount, F/H, of the sum of the $f_j^{1/2}$. Then, either an equal allocation is used in all strata or the Neyman allocation based on the within-stratum variances of x is used.

6.6. SOME EMPIRICAL RESULTS ON STRATA FORMATION

In this section we will illustrate the different methods of strata formation and their effects on estimation. Recall that, in theory, when the single model (6.2.10) holds for an entire population, stratification by size and optimal allocation of weighted balanced within-stratum samples gain nothing compared to overall weighted balanced sampling. Whether this theory holds in real populations will be examined here. Figure 6.1 shows stratum boundaries in the Hospitals and Cancer populations that result from using four methods of formation with $H = 5$:

1. Equal numbers of units N_h in each stratum;
2. Equal cum $\sqrt{f}$ in each stratum;
3. Equal aggregate total of $\sqrt{x}$ (cum $\sqrt{x}$) in each stratum; and
4. Equal aggregate total of x (cum x) in each stratum.

Aggregation for all methods was done after sorting each population in ascending order on the auxiliary x. For the cum $\sqrt{f}$ rule, 100 equal length x-intervals were formed, the running totals of $f_j^{1/2}$ were computed, and the intervals then grouped into five strata. The numbers of units in each stratum that result from the four methods are listed in Table 6.7. For these populations, the cum $\sqrt{f}$ and cum $\sqrt{x}$ rules give similar strata definitions. The cum x rule puts fewer units in the stratum with the largest x's than the other two methods. With a larger number of strata this effect would be even more pronounced. Creating a stratum with relatively few units may cause difficulties in selecting a balanced sample.

To further illustrate the differences among methods of strata formation, we conducted a simulation study using the Hospitals and Cancer populations. The four methods of stratification listed above were used along with unstratified sampling. The particular combinations of stratification, estimators, and selection methods that were used are listed in the rows of Figure 6.2, which will be described in detail later in this section. Five combinations of estimators and

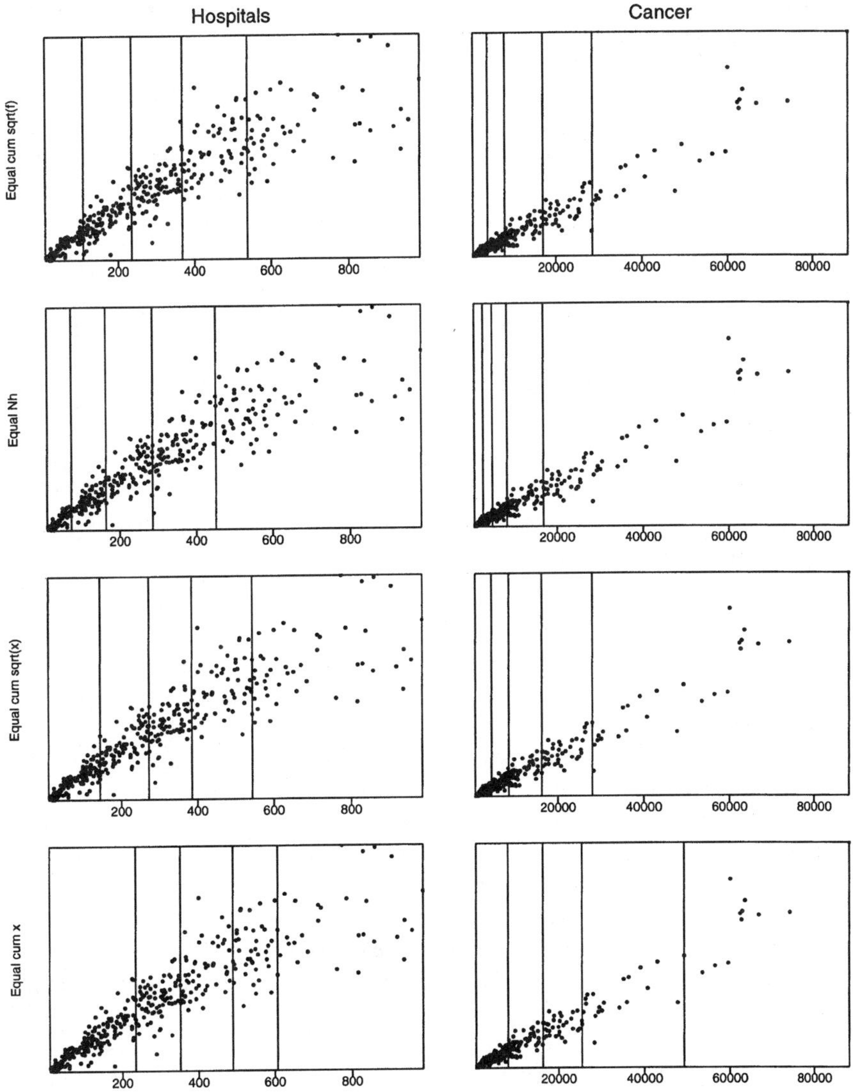

Figure 6.1 Stratum boundaries in the Hospitals and Cancer populations with five strata using four methods of strata formation.

sample selection methods were used:

1. $\hat{T}(x^{\gamma/2}, x^{\gamma}:x^{\gamma})$, which is minimal when $\text{var}_M(Y_i) = \sigma^2 x_i^{\gamma}$, and $pp(x^{\gamma/2})$ sampling with $\gamma = 1, 2$;
2. $\hat{T}_{MRS}(x^{\gamma/2})$, the stratified mean-of-ratios estimator defined in Section 6.2, and $pp(x^{\gamma/2})$ sampling with $\gamma = 1, 2$;

Table 6.7. Numbers of Units per Stratum with Three Methods of Strata Formation in the Hospitals and Cancer Populations

	Hospitals				Cancer			
Stratum	cum $\sqrt{f}$	Equal N_h	cum $\sqrt{x}$	cum x	cum $\sqrt{f}$	Equal N_h	cum $\sqrt{x}$	cum x
1	107	78	143	199	107	60	114	177
2	92	79	82	74	71	60	70	60
3	85	78	67	52	63	60	52	33
4	59	79	55	40	36	60	39	20
5	49	79	46	28	24	61	26	11

3. $\hat{T}_{0S}$, the stratified expansion estimator, and *stsrs* without replacement;
4. $\hat{T}_{RS}$, the separate ratio estimator, and *stsrs* without replacement; and
5. $\hat{T}_{LS}$, the separate regression estimator, defined below, and *stsrs* without replacement.

The separate regression estimator is defined as $\hat{T}_{LS} = \hat{T}_{0S} + \Sigma_h N_h b_{hs}(\bar{x}_h - x_{hs})$ with $b_{hs} = \Sigma_{s_h}(x_{hi}-\bar{x}_{hs})(y_{hi}-\bar{y}_{hs})/\Sigma_{s_h}(x_{hi}-\bar{x}_{hs})^2$. This estimator may be nearly model-unbiased if a model with some curvature can be reasonably approximated by piecewise straight lines within the size strata. Recall that $\hat{T}_{MRS}(x^{\gamma/2})$ is also the Horvitz-Thompson estimator in $pp(x^{\gamma/2})$ sampling.

For each method of stratification a sample of $n = 30$ was divided equally among the five strata in Table 6.6 giving $n_h = 6$ in each stratum. When strata are formed to equalize $N_h\bar{x}_h^{(\gamma/2)}$, costs are all equal, and $\text{var}_M(Y_{hi}) = \sigma^2 x_{hi}^{\gamma}$, then an equal allocation is optimal (see Exercise 6.18). Thus there is a logical consistency to using an equal allocation in each of methods (1), (3), and (4) of strata formation noted above. In addition, equal allocation is one method traditionally used with the cum $\sqrt{f}$ method.

Both unrestricted and restricted sampling techniques were used in the simulation. Unrestricted $pp(x^{\gamma/2})$ was implemented using the random order, systematic method described in Section 3.3.5. Restricted $pp(x^{\gamma/2})$ sampling was done by selecting a sample with the random-order method and then checking its closeness to weighted balance on four moments within each stratum. The balance measures

$$e_j(s_h) = \left| \frac{\sqrt{n}(\bar{x}_{sh}^{(j-\gamma/2)} - \bar{x}_h^{(j)}/\bar{x}_h^{(\gamma/2)})}{s_{jxh}} \right|, \qquad j = 0, \tfrac{1}{2}, 1, 2$$

were calculated in each stratum, where $s_{jxh} = [\Sigma_{i=1}^{N_h} \pi_{hi}(x_{hi}^{j-\gamma/2} - \bar{x}_h^{(j)}/\bar{x}_h^{(\gamma/2)})^2]^{1/2}$ and $\pi_{hi} = x_{hi}^{\gamma/2}/(N_h\bar{x}_h^{(\gamma/2)})$. For the pairs $(j = \frac{1}{2}, \gamma = 1)$ and $(j = 1, \gamma = 2)$,

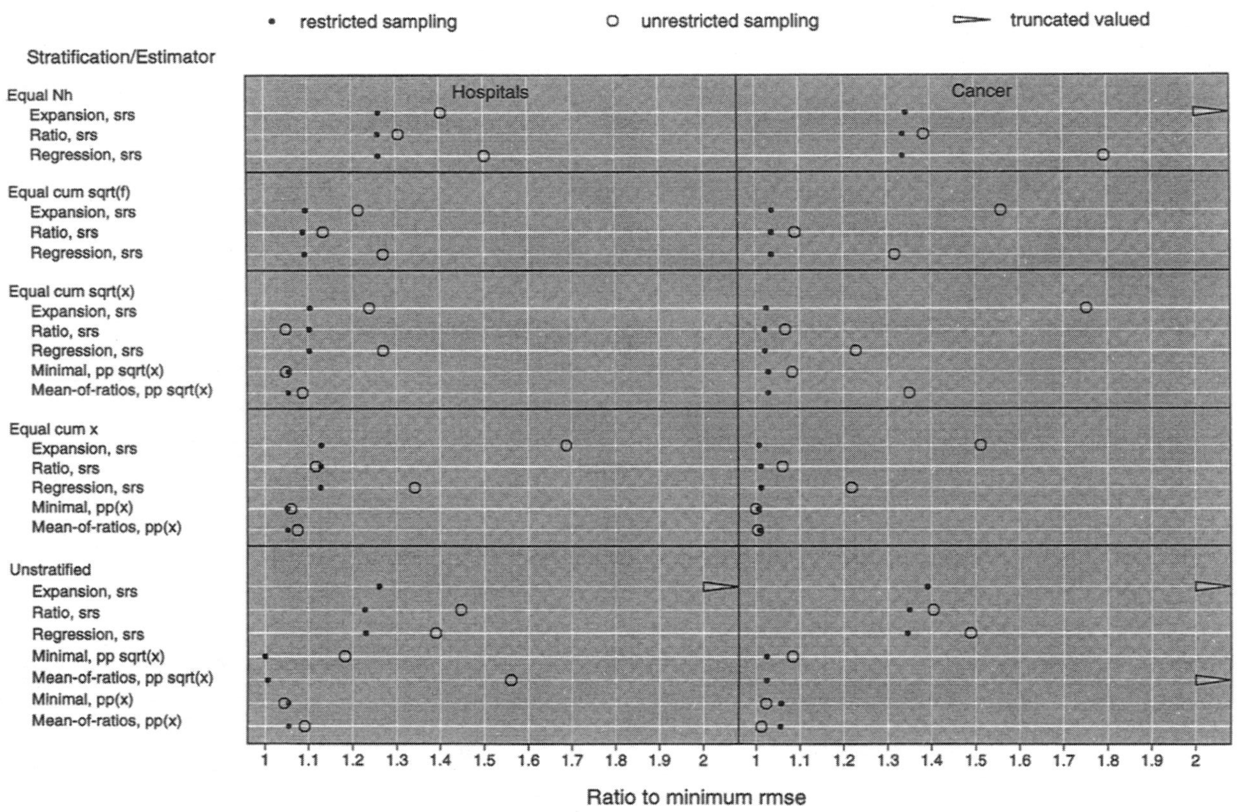

Figure 6.2 Ratios of *rmse*'s for estimators in stratified and unstratified sampling to the minimum *rmse* for the Hospitals and Cancer populations; $S = 1000$, $n = 30$.

$e_j(s_h) = 0$, and balance on those moments is trivially satisfied. For the non-trivial cases, if $e_j(s_h) \leqslant 0.126$ for all measures in every stratum, then the sample was retained; otherwise, it was discarded and another drawn. With unstratified sampling, $H = 1$, so that only overall sample balance was checked. The 55th quantile of the standard normal distribution is $q_{.55} = 0.126$. Thus this technique retains only about 10% of the best-balanced samples. Under a model where $\text{var}_M(Y_i) = \sigma^2 x_i^\gamma$ and the **X** matrix consists of a column of $x_i^{\gamma/2}$'s and a column of x_i^γ's, the minimal estimator will be *BLU* and the best sample will have weighted-balance on the $j = \gamma$ sample moment. When $\gamma = 1$, for example, the best sample is one with $\bar{x}_s^{(1/2)} = \bar{x}/\bar{x}^{(1/2)}$, as in Example 4.2.2. When $\gamma = 2$, any sample with $\bar{x}_s = \bar{x}^{(2)}/\bar{x}$ is optimal, as in Example 4.2.1. Recall that an optimally allocated, stratified, weighted-balanced sample is also one with overall weighted balance.

Balancing on the other moments above, in addition to $j = \gamma$, protects the minimal estimator against different polynomial terms not in a minimal working model. For instance, when $\gamma = 1$, balancing on the $j = 0$ and 2 terms protects the minimal estimator against the possibility that the correct model contains an intercept and a quadratic term. With the weighted balance conditions above, the mean-of-ratios estimator $\hat{T}_{MRS}(x^{\gamma/2})$ is equal to the minimal estimator $\hat{T}(x^{\gamma/2}, x^\gamma : x^\gamma)$, but in unbalanced samples there may be important differences—a point that the simulation results illustrate.

Unrestricted and restricted *stsrs* samples were used for estimators (c)–(e) above. In the unrestricted samples, a simple random sample was selected without replacement in each stratum and retained regardless of its configuration. For restricted samples, a without-replacement *srs* was selected in each stratum and checked for simple balance on four moments using

$$\tilde{e}_j(s_h) = \left| \frac{\sqrt{n}(\bar{x}_{sh}^{(j)} - \bar{x}_h^{(j)})}{\tilde{s}_{jxh}} \right|, \qquad j = 0, \tfrac{1}{2}, 1, 2,$$

where $\tilde{s}_{jxh} = [\Sigma_{i=1}^{N_h} (x_{hi}^j - x_h^{(j)})^2/(N_h - 1)]^{1/2}$. If each stratum sample had $e_j(s_h) \leqslant 0.126$ for $j = 0, \frac{1}{2}, 1, 2$, the sample was retained, and rejected otherwise. As above, this technique retains only about 10% of the best-balanced samples.

For each combination of stratification, sampling method, and estimator, 1000 samples were selected. For restricted samples this means that samples were selected until 1000 were retained. The root mean square error for each estimator was computed as $rmse(\hat{T}) = [\Sigma_{s=1}^{1000} (\hat{T} - T)^2/1000]^{1/2}$. Figure 6.2 presents results, using a rowplot of the type devised by Carr (1994). In each column, the ratio of each *rmse* to the minimum *rmse* among the estimators for the population is plotted. Black dots represent restricted samples while open circles are for unrestricted samples. The narrow triangles are cases where the ratio was truncated at 2 to avoid scaling problems. Some observations are:

- In both populations, the minimal estimator with unstratified, restricted $pp(x^{1/2})$ sampling has the smallest *rmse* or very near it.
- Unrestricted sampling is generally inferior to restricted, balanced sampling.
- The minimal and mean-of-ratios estimators have about the same *rmse*'s in weighted balanced samples as expected. In contrast, $\hat{T}_{MRS}(x^{1/2})$ can have much higher *rmse*'s than $\hat{T}(x^{1/2}, x:x)$ in unrestricted $pp(x^{1/2})$ sampling in both populations especially when no strata are used.
- The estimators used when sampling is *stsrs*—expansion, ratio, and regression—are improved by balanced sampling, but are generally inferior to the minimal estimator with weighted balance.
- For a given selection method ($pp(x^{1/2})$ or $pp(x)$), stratification with weighted balance within strata yields *rmse*'s very near those of unstratified sampling and weighted balance for the minimal or mean-of-ratios estimator. This is expected since the minimal and mean-of-ratios estimators are equal in weighted balanced samples, and an optimally allocated, stratified, weighted balance sample also has overall balance.
- In contrast, stratification with balanced sampling does substantially improve the expansion, ratio, and regression estimators. In the Cancer population, these three estimators are competitive with the minimal estimator when the cum $\sqrt{f}$ or cum $\sqrt{x}$ rule is used for strata formation.

6.7. VARIANCE ESTIMATION IN STRATIFIED POPULATIONS

Variance estimation for estimators in stratified populations requires only minor modifications to the methods given earlier in Chapter 5. Under model (6.2.1) the estimator of the population total is

$$\begin{aligned}\hat{T} &= \sum_{h=1}^{H} \hat{T}(\mathbf{X}_h:\mathbf{V}_h) \\ &= \sum_h \left[\sum_{s_h} Y_{hi} + \sum_{s_h} a_{hi} Y_{hi} \right],\end{aligned}$$

where a_{hi} is a coefficient appropriate for estimating the nonsample total in stratum h. Under model (6.1.1), for example, with a common mean and variance within each stratum, $a_{hi} = N_h/n_h - 1$. Analogous to results in Section 5.5, for the more general model $M(\mathbf{X}_h:\mathbf{V}_h)$, we have $a_{hi} = \mathbf{1}'_{rh}\mathbf{X}_{rh}\mathbf{A}_h^{-1}\mathbf{x}_{hi}w_{hi}$ with $\mathbf{A}_h = \mathbf{X}'_{sh}\mathbf{W}_{sh}\mathbf{X}_{sh}$ and $\mathbf{W}_{sh} = \mathbf{V}_{sh}^{-1}$, as in Section 6.2. The error variance of $\hat{T}$, when the working model is $M(\mathbf{X}_h:\mathbf{V}_h)$ in each stratum h, is

$$\begin{aligned}\text{var}_M(\hat{T} - T) &= \text{var}(\hat{T}_r - T_r) \\ &= \sum_h \sum_{s_h} a_{hi}^2 v_{hi} + \sum_h \sum_{r_h} v_{hi}.\end{aligned}$$

A variance estimator is just the sum of variance estimators across the strata,

$$v(\hat{T} - T) = \sum_{h=1}^{H} v[\hat{T}(\mathbf{X}_h:\mathbf{V}_h)]. \tag{6.7.1}$$

If, within each stratum, we use one of the robust variance estimators from Chapter 5, the result will be consistent even if the working model variance specification, $\mathbf{V}_h$, is wrong in one or more strata. Simplifying the notation a bit, so that $\hat{T}_h = \hat{T}(\mathbf{X}_h:\mathbf{V}_h)$ and $v_h(\hat{T}_h) = v[\hat{T}(\mathbf{X}_h:\mathbf{V}_h)]$, the stratum h variance estimator is

$$v_h(\hat{T}_h) = \sum_{s_h} a_{hi}^2 \hat{\psi}_{hi} + \frac{\Sigma_{r_h} v_{hj}}{\Sigma_{s_h} v_{hj}} \sum_{s_h} \hat{\psi}_{hi}, \tag{6.7.2}$$

where $\hat{\psi}_{hi}$ is one of the adjusted squared residual estimators from Chapter 5. First, define $r_{hi} = Y_{hi} - \mathbf{x}_{hi}'\hat{\boldsymbol{\beta}}_h$ with $\boldsymbol{\beta}_h = \mathbf{A}_h^{-1}\mathbf{X}_{sh}'\mathbf{W}_{sh}\mathbf{Y}_{sh}$. The sample hat matrix in stratum h is $\mathbf{H}_{sh} = \mathbf{X}_{sh}\mathbf{A}_h^{-1}\mathbf{X}_{sh}'\mathbf{W}_{sh}$. Denote the leverages, or diagonal elements of $\mathbf{H}_{sh}$ by h_{hii}. The double use of h may, at first, appear somewhat confusing, but the convention here is that h, when used as a subscript, will always stand for a stratum. Then, our alternatives for $\hat{\psi}_{hi}$ are

$$\hat{\psi}_{hi} = r_{hi}^2$$

$$\hat{\psi}_{hi} = r_{hi}^2(1 - h_{hii})^{-1}$$

$$\hat{\psi}_{hi} = r_{hi}^2 \frac{\Sigma_{s_h} a_{hi}^2 v_{hi}}{\Sigma_{s_h} a_{hi}^2 v_{hi}(1 - h_{hii})}$$

and

$$\hat{\psi}_{hi} = r_{hi}^2(1 - h_{hii})^{-2},$$

corresponding to v_R, v_D, v_H, and v_{J*} for the unstratified case. The estimators given by (6.7.2) all are unbiased or approximately unbiased estimators of the error variance under the working model (6.1.1), but are also consistent estimators under the more general model $M(\mathbf{X}_h:\boldsymbol{\Psi}_h)$ in each stratum.

There is also a version of the jackknife variance estimator for $\hat{T}_r$ whose stratified form is

$$v_J(\hat{T}_r) = \sum_h \frac{n_h - 1}{n_h} \sum_{s_h} [\hat{T}_{r(hi)} - \hat{T}_{r(h)}]^2. \tag{6.7.3}$$

The estimator of the nonsample total based on deleting the single unit hi is

$$\hat{T}_{r(hi)} = \sum_{h' \neq h} \mathbf{1}_{rh'}'\mathbf{X}_{rh'}\hat{\boldsymbol{\beta}}_{h'} + \mathbf{1}_{rh}'\mathbf{X}_{rh}\hat{\boldsymbol{\beta}}_{h(hi)},$$

where

$$\hat{\boldsymbol{\beta}}_{h(hi)} = (\mathbf{X}'_{sh(hi)}\mathbf{W}_{sh(hi)}\mathbf{X}_{sh(hi)})^{-1}\mathbf{X}'_{sh(hi)}\mathbf{W}_{sh(hi)}\mathbf{Y}_{sh(hi)} \quad \text{and} \quad \hat{T}_{r(h)} = \Sigma_{s_h}\hat{T}_{r(hi)}/n_h.$$

The subscript (hi) means that the single, sample unit hi is omitted from the various sample matrices and vectors in computing $\hat{\boldsymbol{\beta}}_{h(hi)}$. Analogous to what was done in Chapter 5, we will pair with the jackknife, the estimator of $\text{var}_M(T_R)$ that goes with v_{J*}, leading to

$$v_J(\hat{T} - T) = v_J(\hat{T}_r) + \sum_h \frac{\Sigma_{r_h} v_{hj}}{\Sigma_{s_h} v_{hj}} \sum_{s_h} \hat{\psi}_{hi}, \tag{6.7.4}$$

where $\hat{\psi}_{hi} = r_{hi}^2(1 - h_{hii})^{-2}$. As in Chapter 5, the jackknife can also be rewritten in terms of a_{hi}, r_{hi}, and h_{hii}.

Lemma 6.7.1. The stratified jackknife variance estimator for $\hat{T}_r$ is equal to

$$v_J(\hat{T}_r) = \sum_h \frac{n_h - 1}{n_h}\left\{\sum_{s_h}\frac{a_{hi}^2 r_{hi}^2}{(1 - h_{hii})^2} - \frac{1}{n_h}\left[\sum_{s_h}\frac{a_{hi} r_{hi}}{(1 - h_{hii})}\right]^2\right\}. \tag{6.7.5}$$

Proof: The derivation involves some of the same steps as Theorem 5.4.1. We provide a sketch of some of the details here. Using the fact that

$$\mathbf{X}'_{sh(hi)}\mathbf{W}_{sh(hi)}\mathbf{X}_{sh(hi)} = \mathbf{X}'_{sh}\mathbf{W}_{sh}\mathbf{X}_{sh} - w_{hi}\mathbf{x}_{hi}\mathbf{x}'_{hi}$$

and applying Lemma 5.4.1, the slope estimator in stratum h after omitting unit hi is $\hat{\boldsymbol{\beta}}_{h(hi)} = \hat{\boldsymbol{\beta}}_h - \mathbf{A}_h^{-1}\mathbf{x}_{hi}w_{hi}\mathbf{r}_{hi}/(1 - h_{hii})$. Substituting this into the expression for $\hat{T}_{r(hi)}$, we obtain $\hat{T}_{r(hi)} = \hat{T}_r - a_{hi}r_{hi}/(1 - h_{hii})$, which leads directly to (6.7.5). □

Since each of the robust variance estimators is the sum over strata, the S-Plus™ functions *vR*, *vD*, *vH*, *vJ.star*, and *vJ* can be used in a loop to calculate the stratified versions.

Much of the previous material in this chapter concerned allocating the sample to the strata. Note that regardless of the allocation, the five variance estimators above apply. The allocation only serves to determine the values of n_h in the variance formulae. In the case of a single model for the entire population, as in Section 6.2.2, strata do not have to be considered in variance estimation, and we are back to the case covered in Chapter 5.

Example 6.7.1. Stratified Expansion Estimator. For $\hat{T}_{0S}$, $a_{hi} = N_h/n_h - 1$, and under model (6.1.1) $h_{hii} = n_h^{-1}$. The sandwich estimator is then $v_R(\hat{T}_{0S}) = \Sigma_h(N_h^2/n_h)(1 - f_h)\Sigma_{s_h}(Y_{hi} - \bar{Y}_{hs})^2/n_h$. The estimators v_D, v_H, and v_J each reduce to $\Sigma_h (N_h^2/n_h)(1 - f_h)\Sigma_{s_h}(Y_{hi} - \bar{Y}_{hs})^2/(n_h - 1)$. The other choice, v_{J*}, does not reduce to anything that is particularly informative, but retains its general property of positive bias.

Example 6.7.2. Separate Ratio Estimator. If the working model in each stratum is $M(0, 1{:}x)$ with a different slope β_h and variance parameter σ_h^2 in each stratum, then the robust estimators are just the summation over the strata of the estimators defined in Section 5.1.2. For example, the sandwich estimator is

$$v_R(\hat{T}_{RS} - T) = \sum_n \frac{N_h^2}{n_h}(1 - f_h)\frac{\bar{x}_h\bar{x}_{hr}}{\bar{x}_{hs}^2}\left(\frac{1}{n_h}\sum_{s_h} r_{hi}^2\right),$$

where $\bar{x}_{hr}$ is the x-mean for the nonsample units in stratum h and $r_{hi} = y_{hi} - x_{hi}\bar{y}_{hs}/\bar{x}_{hs}$.

6.8. STRATIFICATION IN DESIGN-BASED THEORY

The practical uses of stratification in design-based sampling are much the same as noted in the introduction to this chapter—administrative convenience and control of allocation to subpopulations or groups that differ in their data-collection costs. Another function of stratification, recognized by practitioners, is in restricting the sets of samples that can be obtained by random selection methods. One common practice is to create many strata based on an auxiliary x and then select only two sample units per stratum, thereby substantially reducing the number of possible samples. Based on the results in Chapter 3, we can recognize this so-called "deep stratification" as a way of approximately aiming at certain types of balanced samples.

Stratification by size has also been studied in the design-based literature as a means of approximating optimum selection probabilities, as we noted in Section 6.5.2. Suppose we estimate the total by the GREG, $\hat{T}_{yr} = \hat{T}_{y\pi} + \hat{\mathbf{B}}'(\mathbf{T}_x - \hat{\mathbf{T}}_{x\pi})$, defined by (2.8.1). Särndal, Swensson, and Wretman (1992, sec. 12.2) show that if the model is $M(\mathbf{X}{:}\mathbf{V})$, then in large samples the anticipated variance, defined as $E_M E_\pi(\hat{T}_{yr} - T)^2$, is minimized when the selection probabilities are

$$\pi_i = \frac{n v_i^{1/2}}{\Sigma_{i=1}^N v_i^{1/2}},$$

that is, proportional to the model standard deviations. A method of stratification and sample selection that approximates these optimum probabilities was described in Section 3.3.5. Sort the population in random order, divide the population into H strata so that each has about the same total of $v^{1/2}$, and then select an equal-sized *srswor* in each stratum. This technique is thus a way of both obtaining approximately optimum selection probabilities and of selecting weighted balanced samples.

Though stratified random sampling will restrict the configuration of samples that can be selected, conditional biases can still occur. The recognition of this possibility of bias is again a feature that separates the prediction approach from the design-based approach. We illustrate this with three estimators that are often studied in the design-based literature (e.g., Hansen, Madow, and Tepping 1983). The three are the stratified expansion estimator, $\hat{T}_{0S} = \Sigma_h N_h \bar{Y}_{hs}$, introduced in Section 6.1, the combined ratio estimator, and the combined regression estimator. The combined ratio estimator and the combined regression estimator are defined as

$$\hat{T}_{RC} = \hat{T}_{0S}\bar{x}/\bar{x}_{0S} \qquad \text{and} \qquad \hat{T}_{LC} = \hat{T}_{0S} + Nb(\bar{x} - \bar{x}_{0S}),$$

where

$$b = \sum_h (N_h/n_h) \sum_{s_h} (Y_{hi} - \bar{Y}_{0S})(x_{hi} - \bar{x}_{0S}) \bigg/ \sum_h (N_h/n_h) \sum_{s_h} (x_{hi} - \bar{x}_{0S})^2,$$

$\bar{x}_{0S} = \Sigma_h N_h \bar{x}_{hs}/N$, and $\bar{Y}_{0S} = \Sigma_h N_h \bar{Y}_{hs}/N$. $\hat{T}_{RC}$ is the GREG if the model is $M(0, 1{:}x)$ and the sampling plan is *stsrs* (or any stratified plan with equal probability sampling within strata). $\hat{T}_{LC}$ is the GREG under $M(1, 1{:}1)$ and *stsrs*. This is the version of $\hat{T}_{LC}$ defined in Särndal, Swensson, and Wretman (1992, p. 275). Another version in Cochran (1977, Chapter 5) uses

$$b = \sum_h K_h \sum_{s_h} (Y_{hi} - \bar{Y}_{hs})(x_{hi} - \bar{x}_{hs}) \bigg/ \sum_h K_h \sum_{s_h} (x_{hi} - \bar{x}_{hs})^2$$

with $K_h = N_h^2(1 - f_h)/[n_h(n_h - 1)]$. This value of b is the one that minimizes the *stsrs* design variance of $\hat{T}_{0S} + Nb(\bar{x} - \bar{x}_{0S})$ (Cochran 1977, sec. 7.7). The combined ratio or regression estimators are sometimes recommended when the stratum sample sizes are small. In a stratified balanced sample, the combined ratio and combined regression estimators both reduce to the stratified expansion estimator $\hat{T}_{0S}$. All three estimators are approximately normally distributed in large *stsrs*'s and are design-consistent in the sense that $(\hat{T} - T)/N \xrightarrow{p} 0$ as $n \to \infty$ under some reasonable conditions on the distribution created by *stsrs*.

From the prediction point-of-view, when the model is $M(0, 1{:}x)$ or $M(1, 1{:}1)$, the strata are strictly a mechanism for selecting the sample. The strata do not have to be considered when constructing an estimator of the total. The optimal estimators under $M(0, 1{:}x)$ and $M(1, 1{:}1)$, as we know from earlier work, are the ratio estimator $\hat{T}(0, 1{:}x)$ and the simple linear regression estimator $\hat{T}(1, 1{:}1)$. The fact that the GREG is constructed to be approximately design-unbiased and, thus, explicitly uses the strata, leads to weights that can be considerably different from those in a purely model-based estimator. For example, the weight for unit hi in $\hat{T}_{RC}$ is $N_h\bar{x}/(n_h\bar{x}_{OS})$ compared to $N\bar{x}/(n\bar{x}_s)$ in $\hat{T}(0,1{:}x)$. If a proportional allocation ($n_h/n = N_h/N$) is made to strata, then the weights are the same and $\hat{T}_{RC} = \hat{T}(0,1{:}x)$. But for other allocations the estimators may be quite different.

The stratified expansion estimator, $\hat{T}_{OS}$, and the two combined estimators also have a potential for conditional biases in the types of populations where they are often recommended. Suppose that size strata are formed in one of the ways described in Section 6.5.2. Assume that the relationship between Y_{hi} and x_{hi} can be approximated by piecewise straight lines within the strata:

$$E_M(Y_{hi}) = \beta_{0h} + \beta_{1h}x_{hi}$$
$$\text{var}_M(Y_{hi}) = v_{hi} \qquad (6.8.1)$$

and that the Y's are independent under the model.

Now, suppose the number of strata H is fixed and the following three conditions hold as $N_h, n_h \to \infty$ in each stratum:

(i) $f_h = n_h/N_h \to 0$;
(ii) $n_h/n \to$ a constant; and
(iii) $W_h = N_h/N \to$ a constant.

Note that each of the three estimators can be written as

$$\hat{T} = \sum_h \sum_{i \in s_h} g_{hi} Y_{hi}$$

as we did for the analysis in Chapter 2, or as

$$\hat{T} = \sum_h \sum_{i \in s_h} Y_{hi} + \sum_h \sum_{i \in s_h} a_{hi} Y_{hi}, \qquad (6.8.2)$$

where $a_{hi} = g_{hi} - 1$, as we did in Chapter 5 and Section 6.7 for variance estimation. For $\hat{T}_{OS}$, $a_{hi} = N_h/n_h - 1$; for $\hat{T}_{RC}$, $a_{hi} = N_h\bar{x}/(n_h\bar{x}_{OS}) - 1$; and for $\hat{T}_{LC}$,

$$a_{hi} = N_h/n_h \left[1 + (\bar{x} - \bar{x}_{OS})(x_{hi} - x_{OS}) \Big/ \sum_h (N_h/n_h) \sum_{s_h} (x_{hi} - \bar{x}_{OS})^2 \right] - 1.$$

Note that, in each case, a_{hi} is of order N_h/n_h under reasonable conditions.

The general form of the prediction variance of $\hat{T}$ under model (6.8.1) is

$$\text{var}_M(\hat{T} - T) = \sum_h \sum_{i \in s_h} a_{hi}^2 v_{hi} + \sum_h \sum_{i \in r_h} v_{hi}.$$

If $\bar{v}_r = \Sigma_h \Sigma_{i \in r_h} v_{hi}/(N - n)$ is bounded as $N_h, n_h \to \infty$, then under (i)–(iii) the two terms above are $O(N^2/n)$ and $O(N - n)$, and an asymptotically equivalent form of the prediction variance is

$$\text{var}_M(\hat{T} - T) \approx \sum_h \sum_{i \in s_h} a_{hi}^2 v_{hi}. \qquad (6.8.3)$$

Under model (6.8.1) it is easy to show that, for the stratified expansion estimator,

$$E_M(\hat{T}_{0S} - T)/N = \sum_h W_h \beta_{1h}(\bar{x}_{hs} - \bar{x}_h) \qquad (6.8.4)$$

and

$$\text{var}_M(\hat{T}_{0S} - T)/N^2 = \sum_h W_h^2(1 - f_h)^2 \frac{\bar{v}_{hs}}{n_h} + \frac{1 - f}{N} \bar{v}_r, \qquad (6.8.5)$$

where $\bar{v}_{hs} = \Sigma_{s_h} v_{hi}/n_h$, and $\bar{v}_r = \Sigma_h \Sigma_{r_h} v_{hi}/(N - n)$. Asymptotically,

$$\text{var}_M(\hat{T}_{0S} - T)/N^2 \approx \sum_h W_h^2 \bar{v}_{hs}/n_h = O(n^{-1})$$

as long as $\bar{v}_{hs}$ and $\bar{v}_r$ are bounded. Since $\sqrt{n_h}(\bar{x}_{hs} - \bar{x}_h)$ is approximately normal in large simple random samples and (ii) holds, the square of the bias in (6.8.4) is also $O(n^{-1})$ if the selection method is *stsrs*. In other words, the squared bias of $\hat{T}_{0S}$ and its error variance have the same order with an *stsrs* plan, regardless of how large the sample is. Valliant (1987a) reports on several simulation studies, illustrating that this can occur in practice in real populations for $\hat{T}_{0S}$ and other stratified estimators. Stratified balanced sampling, on the other hand, would eliminate the bias (6.8.4).

Similar calculations show that the model bias and variance of $\hat{T}_{RC}$ under (6.8.1) and conditions (i)–(iii) are

$$E_M(\hat{T}_{RC} - T)/N = \left(\frac{\bar{x}}{\bar{x}_{0S}} - 1\right) \sum_h W_h \beta_{0h} + \sum_h W_h \beta_{1h} \bar{x}_{hs} \left(\frac{\bar{x}}{\bar{x}_{0S}} - \frac{\bar{x}_h}{\bar{x}_{hs}}\right) \qquad (6.8.6)$$

$$\begin{aligned} \text{var}_M(\hat{T}_{RC} - RT)/N^2 &= N^{-2} \sum_h \left(\frac{N_h \bar{x}}{n_h \bar{x}_{0S}} - 1\right)^2 n_h \bar{v}_{hs} + \frac{1 - f}{N} \bar{v}_r \\ &\approx \left(\frac{\bar{x}}{\bar{x}_{0S}}\right)^2 \sum_h W_h^2 \frac{\bar{v}_{hs}}{n_h}. \end{aligned} \qquad (6.8.7)$$

Both the error variance and the square of the model bias of the combined ratio estimator are $O(n^{-1})$ under *stsrs*. Consequently, the same problem exists as for the stratified expansion estimator—even in large randomly selected *stsrs*'s, the squared bias never becomes an unimportant part of the *mse*.

The stratified expansion estimator and the combined ratio estimator with *stsrs* meet the design-based criteria for robustness that we summarized in Section 3.7: random selection, consistency, and asymptotic normality. Nevertheless, $\hat{T}_{0S}$ and $\hat{T}_{RC}$ have such obvious conditional bias problems that we cannot call them robust. Restricted random sampling or systematic sampling that produces balance on x within each stratum, on the other hand, would protect both against model-bias. Unrestricted *stsrs* is an improvement over unstratified *srs* because it guarantees the sample is spread over the size strata. But *stsrs* does not lead to balance fast enough to make the squared model-bias a negligible part of the mean squared error.

The same conclusions apply to the combined regression estimator under model (6.8.1) and *stsrs* (see Exercise 6.25).

Design-based theory has also produced its own variance estimators for $\hat{T}_{0S}$, $\hat{T}_{RC}$, and $\hat{T}_{LC}$. The variance estimators are, naturally, derived based on the form of the estimator of the total and the type of sampling plan rather than based on the estimator and the population structure as approximated by a model.

For $\hat{T}_{0S}$ and *stsrs* the standard variance estimator is

$$v_0 = \sum_h K_h \Sigma_{i \in s_h} (y_{hi} - \bar{y}_{hs})^2 \qquad \text{with} \qquad K_h = N_h^2(1 - f_h)/[n_h(n_h - 1)],$$

which is the same as v_D, v_H, and v_J in Example 6.7.1. This estimator is prediction-unbiased if $M(1{:}1)$ holds in each stratum and design-unbiased under *stsrs*. However, if model (6.8.1) holds, then

$$E_M(v_0)/N^2 = \sum_h \frac{W_h^2}{n_h}(1 - f_h)(\bar{v}_{hs} + \beta_{1h}^2 s_{xhs}^2),$$

where $s_{xhs}^2 = \Sigma_{s_h}(x_{hi} - \bar{x}_{hs})^2/(n_h - 1)$. As an estimator of the error variance, v_0 has a positive bias that persists even in stratified balanced samples in which $\hat{T}_{0S}$ itself is unbiased. Although v_0 is on the average an overestimate of the model variance, it is also a biased estimator of the model *mse*, which is apparent from (6.8.4) and (6.8.5). This will be true even in balanced samples, reinforcing a key point made in Chapter 5. Protecting an estimator of a total against bias by balancing does not necessarily protect the variance estimator against bias.

Recall that if the squared bias does not converge to zero faster than the variance of an estimator, confidence intervals will not cover at the desired rate. This is because the distribution of the estimation error, $\hat{T}_{0S} - T$ in this case, will not be centered at zero. In extreme samples where $\hat{T}_{0S}$ was negatively biased, the simulation study in Valliant (1987a) showed cases in which confidence intervals of the form $\hat{T}_{0S} \pm 1.96 v_0^{1/2}$ covered the population total much less than 50% of the time. In samples where the bias was small, v_0 overestimated the error variance so badly that the 95% confidence interval coverage was virtually 100% in the simulation.

Even though the combined estimators, $\hat{T}_{RC}$ and $\hat{T}_{LC}$, do not flow directly from a working model (except as being GREGs), the fact that the estimators can be written in the form (6.8.2) means that the general approach to variance estimation described in Section 6.7 can be used. Each estimator has an associated set of a_{hi}'s and a model robust variance estimator is

$$v(\hat{T}) = \sum_h \left(\sum_{s_h} a_{hi}^2 \hat{\psi}_{hi} + \frac{\Sigma_{r_h} v_{hj}}{\Sigma_{s_h} v_{hj}} \sum_{s_h} \hat{\psi}_{hi} \right) \tag{6.8.8}$$

as in (6.7.2). For $\hat{T}_{RC}$ we have $a_{hi} = N_h \bar{x}/(n_h \bar{x}_{0S}) - 1$ and under the $M(0, 1{:}x)$ working model, $r_{hi} = Y_{hi} - \bar{Y}_s \bar{x}/\bar{x}_s$, where $\bar{Y}_s$ and $\bar{x}_s$ are the unweighted sample means across all stratum samples. The residual r_{hi} would be used in constructing the various choices of $\hat{\psi}_{hi}$ defined following (6.7.2).

Other design-based variance estimators are, naturally, available, but they do not necessarily have any model-robustness properties. For the combined ratio estimator, for example, a standard choice is

$$v_{RC0} = \sum_h K_h \sum_{s_h} r_{Chi}^2, \tag{6.8.9}$$

where $r_{Chi} = (Y_{hi} - \bar{Y}_{hs}) - (\bar{Y}_{0S}/\bar{x}_{0S})(x_{hi} - \bar{x}_{hs})$. Since $(N_h^2/n_h)(1 - f_h)/(n_h - 1) \approx N_h^2/n_h^2$ in v_{RC0} above and $a_{hi} \approx (N_h^2/n_h^2)(\bar{x}/\bar{x}_{0S})^2$, the general robust variance estimator in (6.8.8) contains a key factor, $(\bar{x}/\bar{x}_{0S})^2$, needed for estimating the large sample variance in (6.8.7). This factor is omitted from the conventional estimator v_{RC0} in (6.8.9). Although the residuals in the two estimators differ, both can be shown to be approximately unbiased for v_{hi}.

In a stratified balanced sample, the variance estimators (6.8.8) and (6.8.9) for $\hat{T}_{RC}$ will be very similar. However, in unbalanced samples, the term $(\bar{x}/\bar{x}_{0S})^2$ in the model-derived estimator will lead to better conditional performance. This again illustrates that the design-based goal of estimating properties across all samples can lead to estimates that do not account for properties of the particular sample that has been selected. Valliant (1990) gives a variety of empirical results on the performance of different conventional and model-based variance estimators in different types of populations. He also compares their performance in *stsrs*, a plan that can produce badly unbalanced samples, and in stratified systematic sampling (*stsys*) from a list sorted on the auxiliary x—a plan that leads to better balance.

EXERCISES

6.1 Show that the *BLU* predictor of the population total under model (6.1.1) is $\hat{T}_{0S} = \Sigma_{h=1}^{H} N_h \bar{Y}_{hs}$ with error variance $\text{var}_M(\hat{T}_{0S} - T) = \Sigma_{h=1}^{H} N_h^2/n_h(1 - f_h)\sigma_h^2$.

6.2 Show that an unbiased estimator of the error variance of the stratified expansion estimator under model (6.1.1) is

$$v(\hat{T}_{0S}) = \sum_{h=1}^{H} \frac{N_h^2}{n_h}(1 - f_h) \sum_{i \in s_h} \frac{(Y_{hi} - \bar{Y}_{hs})^2}{n_h - 1}.$$

6.3 Prove Corollaries 6.1.1 and 6.1.2.

6.4 Suppose that a population contains five strata, which are subpopulations, having the following characteristics:

Stratum	N_h	σ_h	c_h
1	77	45	30
2	53	22	75
3	48	34	25
4	32	19	50
5	29	60	10

Assume that the stratified expansion estimator, $\hat{T}_{0S} = \Sigma_{h=1}^{H} N_h \bar{Y}_{hs}$, will be used. Use the result in Section 6.1 that the total sample size for various allocations, that adheres to a constraint on total costs, can be expressed as $n = (C^* - C_0)/\bar{c}$, where $\bar{c} = \Sigma_h w_h c_h$ with w_h being the fraction of the sample allocated to stratum h. Compute:

(a) The total sample sizes for optimal allocation accounting for costs, Neyman allocation, proportional allocation, and equal allocation, assuming that the budget for variable costs is $C^* - C_0 = \$2{,}100$. Round n to the nearest integer.

(b) The allocations to each of the five strata for the four types of allocations in part (a). Round each stratum sample size to the nearest integer.

(c) The value of $\sqrt{\text{var}_M(\hat{T}_{0S} - T)}$ for each of the four types of allocation in part (a).

(d) Comment on the differences among the allocations.

6.5 In Exercise 6.4, suppose that you have the following prior estimates of the stratum means of Y: 40, 30, 40, 20, 60. Assume that the goals of the survey are to achieve coefficients of variation of 0.10 for the estimated means for each of strata 1 and 2, and to minimize the error variance of the overall estimator of the total $\hat{T}$ for a budget of $C - C_0 = \$5{,}000$. Describe the method you would use to allocate the sample, and then carry out the computations.

6.6 Assume that the model is (6.2.1), that is, $E_M(\mathbf{Y}_h) = \mathbf{X}_h \boldsymbol{\beta}_h$, $\text{var}_M(\mathbf{Y}_h) = \mathbf{V}_h \sigma_h^2$, where $h = 1, \ldots, H$, $\mathbf{Y}_h$ is $N_h \times 1$, $\mathbf{X}_h$ is $N_h \times p_h$, $\mathbf{V}_h = \text{diag}(v_{hi})$ is $N_h \times N_h$, and $\boldsymbol{\beta}_h$ is a $p_h \times 1$ parameter vector. Express as a single combined model and apply Theorem 4.2.1 to show that the *BLU* predictor is $\hat{T} = \Sigma_{h=1}^{H} \hat{T}(\mathbf{X}_h:\mathbf{V}_h)$.

6.7 Prove Theorem 6.2.2, that is, when a weighted balanced sample $B(\mathbf{X}_h:\mathbf{V}_h)$ is selected in each stratum, the allocation of the sample to the strata that minimizes the error variance of the *BLU* predictor under model (6.2.1) is $n_h/n = \{N_h \bar{v}_h^{(1/2)} \sigma_h/\sqrt{c_h}\}/\{\Sigma_{h'} N_{h'} \bar{v}_{h'}^{(1/2)} \sigma_{h'}/\sqrt{c_{h'}}\}$ for $h = 1, \ldots, H$.

6.8 Using Theorem 6.2.2, prove Corollaries 6.2.1 and 6.2.2.

6.9 Suppose that a stratified sample, balanced in the sense that $\bar{x}_s = \bar{x}$ and $\bar{x}_s^{(\gamma)} = \bar{x}^{(\gamma)}$ (for some $\gamma \neq 1$), is selected. Derive the optimum allocation to strata for the estimator $\hat{T}_{RS}$ under the model $M(0, 1:x^\gamma)$. What is the allocation when $\gamma = 0$?

6.10 One measure of the gain due to the strategy $[s^*(J), \hat{T}_{RS}]$ compared to the strategy $[s(J), \hat{T}(0, 1:x)]$ is the ratio of the sample size n^* necessary for a given *mse* with the former strategy divided by the sample n needed for the same *mse* with the latter strategy. Compute the theoretical ratio n^*/n when the model is $M(0, 1:x)$.

6.11 Suppose that model (6.1.1) holds and that the desired error variance for $\hat{T}_{0S}$ is fixed at V_0. Derive a formula for the total sample size $n = \Sigma_h n_h$ that applies to Neyman, proportional, and equal allocation and that assures that the target variance V_0 is achieved.

6.12 In the example in Table 6.3 suppose that the cost vector is $\mathbf{c} = (200, 150, 30, 20, 10)$. Recompute Table 6.4 assuming a budget of $C^* - C_0 =$ \$5,250. Comment on the differences in the allocations and why they occur.

6.13 Suppose that a population has two strata with the model within stratum h being $E_M(Y_{hi}) = \beta_1 x_{hi} + \beta_2 x_{hi}^2$, $\text{var}_M(Y_{hi}) = \sigma_h^2 x_{hi}^2$ with the Y_{hi}'s uncorrelated. Determine how to allocate a sample of $n = n_1 + n_2$ units to strata 1 and 2 in order to minimize the error variance of the estimator of the difference in strata means, $\hat{\bar{Y}}_1 - \hat{\bar{Y}}_2 = \hat{T}_{MR1}(x)/N_1 - \hat{T}_{MR2}(x)/N_2$.

6.14 In a stratified population with H strata, suppose that all pairwise differences in stratum means are to be estimated and that model 6.1.1 holds. Assume that the cost function is $C = C_0 + \Sigma_h c_h n_h$ and that, for reasons given in Section 6.4, finite population correction factors can be ignored when calculating error variances. (a) Find the allocation that minimizes the average error variance,

$$\bar{V} = \frac{2}{H(H-1)} \sum_{h<h'} \left(\frac{\sigma_h^2}{n_h} + \frac{\sigma_{h'}^2}{n_{h'}} \right)$$

of the estimators of the differences in stratum means subject to a fixed total cost C^*. Compute the total sample size and minimum value $\bar{V}_{\min}$ for that allocation. (b) Suppose we fix the desired variance of the estimator of the difference at V for each pair of strata. Find the allocation that achieves this and the implied values of n and V for a fixed total cost C^*.

6.15 Consider the population in Table 6.1. Use the answers to (a) and (b) of the previous problem to compute the two allocations for estimating differences between strata means subject to a cost constraint of $C^* - C_0 = \$5{,}250$. Compute the standard error that each of these allocations produces for the stratified expansion estimator, $\hat{T}_{0S} = \Sigma_{h=1}^{H} N_h \bar{Y}_{hs}$. Compare those standard errors to that of the cost-constrained optimal allocation given in Table 6.2.

6.16 Prove Lemmas 6.5.1 and 6.5.2.

6.17 Assume that model (6.2.10) holds, that is, $E_M(\mathbf{Y}_h) = \mathbf{X}_h \boldsymbol{\beta}$, $\mathrm{var}_M(Y_{hi}) = \sigma^2 x_{hi}^{\gamma}$. Suppose that $\mathbf{V1}_N$, $\mathbf{V}^{1/2}\mathbf{1}_N \in \mathscr{M}(\mathbf{X})$, and the sample is $s_h \in B(\mathbf{X}_h : \mathbf{V}_h)$. Assume that there are H strata, and an equal number of sample units, $n_h = n_0$, is allocated to each stratum. (a) What is the error variance of the *BLU* predictor of the total? (b) If optimal stratification occurs when this error variance is minimized, which term in the answer to (a) should be minimized? (c) Show that the term in (b) is minimized by making $Z_h = N_h \bar{x}_h^{(\gamma/2)}$ the same for each stratum. (d) Assuming that strata are constructed as in (c), a weighted balanced sample is selected in each stratum, and all stratum costs are equal, show that the optimal allocation of a total sample of n to the strata is an equal allocation $n_h = n_0$. (e) Show that with strata constructed as in part (c) and with the optimal, equal allocation as in part (d), the error variance of the *BLU* is exactly the same as for an unstratified sample with weighted balance.

6.18 Show that (a) the separate ratio estimator, $\hat{T}_{RS}$, is the generalized regression (GREG) estimator under model (6.8.1) if $\beta_{0h} = 0$, $v_{hi} = \sigma_h^2 x_{hi}$, and the sampling plan is *stsrs*, and (b) the combined ratio estimator is the GREG when $\beta_{0h} = 0$, $\beta_{1h} = \beta_1$, $v_{hi} = \sigma^2 x_{hi}$, and the plan is *stsrs*.

6.19 Show that (a) the separate linear regression estimator,

$$\hat{T}_{LS} = \sum_h N_h[\bar{Y}_{hs} + b_{hs}(\bar{x} - \bar{x}_{hs})] \quad \text{with} \quad b_{hs} = \sum_{s_h} (x_{hi} - \bar{x}_{hs}) Y_{hi} \Big/ \sum_{s_h} (x_{hi} - \bar{x}_{hs})^2,$$

is the generalized regression (GREG) estimator under model (6.8.1) if $v_{hi} = \sigma_h^2$, and the sampling plan is *stsrs*, and (b) the combined regression estimator is the GREG when $\beta_{0h} = \beta_0$, $\beta_{1h} = \beta_1$, $v_{hi} = \sigma^2$, and the plan is *stsrs*.

6.20 Show that these relationships to the stratified expansion estimator hold for $\hat{T}_{RC}$, $\hat{T}_{LS}$, and $\hat{T}_{LC}$: $\hat{T}/N = \hat{T}_{0S}/N + O_\pi(n^{-1/2})$ under *stsrs* and $\hat{T}/N = \hat{T}_{0S}/N + O(n^{-1})$ under *stsys* where $\hat{T}$ is one of the three estimators, and the sampling frame is sorted by the auxiliary x used in $\hat{T}_{RC}$, $\hat{T}_{LS}$, and $\hat{T}_{LC}$. (O_π denotes probabilistic order with respect to *stsrs*. Use Lemma 3.4.1 for systematic sampling.)

6.21 Verify expressions (6.8.4) and (6.8.5) for the model bias and error variance of the stratified expansion estimator under model (6.8.1).

6.22 Verify expressions (6.8.6) and (6.8.7) for the model bias and error variance of the combined ratio estimator under model (6.8.1).

6.23 Show that, under model (6.8.1) and conditions (i)–(iii) of Section 6.8, the model bias and error variance of $\hat{T}_{RS}$ are:

$$E_M(\hat{T}_{RS} - T)/N = \sum_h W_h \beta_{0h}(\bar{x}_{hs} - \bar{x}_h)/\bar{x}_{hs}$$

and

$$\begin{aligned}\text{var}_M(\hat{T}_{RS} - T)/N^2 &= N^{-2} \sum_h \left(\frac{N_h \bar{x}_h}{n_h \bar{x}_{hs}} - 1 \right)^2 n_h \bar{v}_{hs} + \frac{1-f}{N} \bar{v}_r \\ &\approx \sum_h W_h^2 \left(\frac{\bar{x}_h}{\bar{x}_{hs}} \right)^2 \frac{\bar{v}_{hs}}{n_h}.\end{aligned}$$

What is the value of a_{hi} for $\hat{T}_{RS}$? What conditions are sufficient for $1/a_{hi} \to 0$ as N_h, $n_h \to \infty$? What are the orders of magnitude of the bias squared and the error variance? As the sample size increases in each stratum, does the squared bias diminish in importance compared to the error variance?

6.24 Show that, under model (6.8.1) and conditions (i)–(iii) of Section 6.8, the model bias and error variance of $\hat{T}_{RC}$ are:

$$\begin{aligned}E_M(\hat{T}_{RC} - T)/N &= \left(\frac{\bar{x}}{\bar{x}_0} - 1 \right) \sum_h W_h \beta_{0h} + \sum_h W_h \beta_{1h} \bar{x}_{hs} \left(\frac{\bar{x}}{\bar{x}_0} - \frac{\bar{x}_h}{\bar{x}_{hs}} \right) \\ \text{var}_M(\hat{T}_{RC} - T)/N^2 &= N^{-2} \sum_h \left(\frac{N_h \bar{x}}{n_h \bar{x}_0} - 1 \right)^2 n_h \bar{v}_{hs} + \frac{1-f}{N} \bar{v}_r \\ &\approx \left(\frac{\bar{x}}{\bar{x}_0} \right)^2 \sum_h W_h^2 \frac{\bar{v}_{hs}}{n_h}.\end{aligned}$$

6.25 Compute the bias and error variance of $\hat{T}_{LC}$ under model (6.8.1). What are the orders of magnitude of the squared bias and error variance?

6.26 Consider the separate regression estimator,

$$\hat{T}_{LS} = \sum_h N_h[\bar{Y}_{hs} + b_{hs}(\bar{x}_h - \bar{x}_{hs})]$$

Derive v_R and v_H when $E_M(Y_{hi}) = \beta_{0h} + \beta_{1h}x_{hi}$ and $\text{var}_M(Y_{hi}) = \sigma_h^2$. Show that both are approximately unbiased if $\text{var}_M(Y_{hi}) = v_{hi}$, $n_h \rightarrow \infty$, and $f_h \rightarrow 0$. What sample moments must be bounded in order for approximate unbiasedness to hold?

6.27 The following variance estimator for the combined regression estimator is usually given in design-based texts:

$$v_{LCO} = \sum_h K_h \sum_{i \in s_h} d_{2hi}^2,$$

where

$$d_{2hi} = (y_{hi} - \bar{y}_{hs}) - b(x_{hi} - \bar{x}_{hs}),\ K_h = N_h^2(1 - f_h)/[n_h(n_h - 1)],$$

and

$$b = \left[\sum_{h,s_h} K_h(x_{hi} - \bar{x}_{hs})Y_{hi}\right] \Big/ \left[\sum_{h,s_h} K_h(x_{hi} - \bar{x}_{hs})^2\right].$$

In design-based analysis, the estimator v_{LCO} is used when the sampling is *stsrs* without replacement. (a) Compute the model expectation of v_{LCO} when $E_M(y_{hi}) = \beta_0 + \beta_1 x_{hi}$ and $\text{var}_M(Y_{hi}) = v_{hi}$ assuming conditions (i)–(iii) of Section 6.8. (b) Compare your answer to the variance to $\hat{T}_{LC}$ in Exercise 6.25 to determine the bias of v_{LCO}. What are the orders of magnitude of the bias of v_{LCO} and the variance of $\hat{T}_{LC}$?

CHAPTER 7

Models with Qualitative Auxiliaries

In surveys of many populations, some of the most useful auxiliary variables are qualitative. For example, in a survey of a human population, we may know the age, race, and sex of each sample person and the total numbers of persons in the population in each of those groups. Quantitative auxiliaries may or may not be available. When x's are qualitative, situations can occur where the mathematical conditions required for Theorem 2.2.1 may not hold; in particular, $\mathbf{X}_s'\mathbf{V}_{ss}^{-1}\mathbf{X}_s$ may not be invertible. The special methods for handling this situation are covered in this chapter.

7.1. SIMPLE EXAMPLE

Regression on qualitative variables is standard practice in the study of linear models and experimental designs. Searle (1971, 1987) gives especially thorough coverage of the mathematics involved. As in the experimental design literature, we will refer to qualitative, categorical variables as *factors* and the categories into which each factor is divided as *levels*. Terminology is discussed in more detail in Section 7.2. To fix ideas, we use an example from Searle (1971, sec. 4.4). Suppose we have a single factor which is a person's highest educational attainment, and the levels are (a) high school degree or less, (b) college degree, or (c) graduate degree. To set up a regression of a target variable Y, which might be the person's income, on the three levels of educational attainment, we create three "dummy" variables which take on the values 0 or 1:

$$x_{1i} = \begin{cases} 1 & \text{if person } i \text{ has a high school degree or less} \\ 0 & \text{if not} \end{cases}$$

$$x_{2i} = \begin{cases} 1 & \text{if person } i \text{ has a college degree} \\ 0 & \text{if not} \end{cases}$$

$$x_{3i} = \begin{cases} 1 & \text{if person } i \text{ has a graduate degree} \\ 0 & \text{if not.} \end{cases}$$

Each person falls into one and only one of the educational attainment categories. As a result, one of x_{1i}, x_{2i}, and x_{3i} will equal 1 and the other two will be 0. Even though the dummy variables are considerably different from quantitative ones like hospital bed size, we can still fit a regression model of the form

$$Y_i = \beta_0 + \beta_1 x_{1i} + \beta_2 x_{2i} + \beta_3 x_{3i} + \varepsilon_i. \tag{7.1.1}$$

For simplicity in this illustration, assume that the errors are independent with mean 0 and variance σ^2. As a practical matter, in the applications where only qualitative auxiliaries are available, we often have little evidence of a heterogeneous variance structure, so that assuming a common variance is reasonable. The more general case is covered later in this chapter.

Suppose that there are $n = 6$ persons in the sample and that three have high school degrees, two have college degrees, and one has a graduate degree. It will be notationally convenient to index persons within each of the factor levels. To that end, Y_{ij} will be the income of the jth person in the ith educational level. In the form $\mathbf{Y}_s = \mathbf{X}_s\boldsymbol{\beta} + \boldsymbol{\varepsilon}$, the model for the sample units is then

$$\begin{bmatrix} Y_{11} \\ Y_{12} \\ Y_{13} \\ Y_{21} \\ Y_{22} \\ Y_{31} \end{bmatrix} = \begin{bmatrix} 1 & 1 & 0 & 0 \\ 1 & 1 & 0 & 0 \\ 1 & 1 & 0 & 0 \\ 1 & 0 & 1 & 0 \\ 1 & 0 & 1 & 0 \\ 1 & 0 & 0 & 1 \end{bmatrix} \begin{bmatrix} \beta_0 \\ \beta_1 \\ \beta_2 \\ \beta_3 \end{bmatrix} + \begin{bmatrix} \varepsilon_{11} \\ \varepsilon_{12} \\ \varepsilon_{13} \\ \varepsilon_{21} \\ \varepsilon_{22} \\ \varepsilon_{31} \end{bmatrix}. \tag{7.1.2}$$

Note that the expected value for any person in the ith educational level is $E_M(Y_{ij}) = \beta_0 + \beta_i$, which we could denote alternatively as $E_M(Y_{ij}) = \mu_i$. Model (7.1.2) is then just another way of writing the stratified model in Example 2.3.3. Later in this chapter we will consider some models that do not reduce to simpler full rank models.

Least squares estimation can be applied to (7.1.2) to obtain the estimating equations $\mathbf{X}_s'\mathbf{X}_s\hat{\boldsymbol{\beta}} = \mathbf{X}_s'\mathbf{Y}_s$. However, in this case $\mathbf{X}_s$ does not have full column rank since the sum of the last three columns equals the first. Because of this, the normal equations will have infinitely many solutions for $\hat{\boldsymbol{\beta}}$. For the case of $\mathbf{Y}_s' = (16, 10, 19, 11, 13, 27)$, the normal equations are

$$\begin{bmatrix} 6 & 3 & 2 & 1 \\ 3 & 3 & 0 & 0 \\ 2 & 0 & 2 & 0 \\ 1 & 0 & 0 & 1 \end{bmatrix} \begin{bmatrix} \beta_0 \\ \beta_1 \\ \beta_2 \\ \beta_3 \end{bmatrix} = \begin{bmatrix} 96 \\ 45 \\ 24 \\ 27 \end{bmatrix}. \tag{7.1.3}$$

$\mathbf{X}_s'\mathbf{X}_s$ is singular since the sum of the last three columns equals the first column.

One solution is $\boldsymbol{\beta}^0 = (16, -1, -4, 11)'$, but $\boldsymbol{\beta}^o = \boldsymbol{\beta}^o + (d, -d, -d, -d)'$, for example, is also a solution for any choice of d. Because of the non-uniqueness of $\boldsymbol{\beta}^o$, it is referred to as a solution to the normal equations and not as an estimator of $\boldsymbol{\beta}$.

7.2. FACTORS, LEVELS, AND EFFECTS

The terms factor, level, and effect that are common in the experimental design literature are also useful when qualitative auxiliaries are used in finite population sampling. For example, suppose that a household survey collects the age, race, and sex of every person in the sample. The counts of sample persons might be summarized in an array like Table 7.1. The categorical variables age, race, and sex are *factors* and may be related to quantitative Y variables like personal income and expenditures on food, clothing, and housing. The categories of age, race, and sex are called *levels.*

In methods like the analysis of variance, the interest is in whether the levels of a factor affect the response variable equally. This is referred to as the *effect* of the level of a factor on the variable of interest. In finite populations the goal is to select factors and levels that are useful predictors of most, if not all, of the Y variables that are measured. In the example in the introduction to this chapter, there is one factor—educational attainment—and three levels—high school degree or less, college degree, and graduate degree. Model (7.1.1) can be written as

$$Y_{ij} = \beta_0 + \beta_i + \varepsilon_{ij}, \quad \varepsilon_{ij} \sim (0, \sigma^2), \qquad i = 1, 2, 3$$

The parameters β_1, β_2, and β_3 are the effects of the three levels of the educational attainment factor.

Table 7.1. Example of Cross-Classified Categorical, Survey Variables

	Race			
	White		Non-white	
	Sex		Sex	
Age	Male	Female	Male	Female
<10				
10–19				
20–34				
35–54				
55–64				
65+				

Two types of effects are known as *fixed* and *random*. *Fixed effects* are associated with factors having a finite number of levels. In experimental designs, the levels are there because of an interest in comparing the size of the effects on a response variable. For instance, three dosage levels of a drug might be compared in a clinical trial. The trial is conducted to discover whether the levels have different effects, or any effect at all, on some health condition. In a finite population, on the other hand, the levels of a factor—like male and female—are often known in advance to have different sizes of fixed effects. Thus there may be no interest in testing whether the effects are different, but there is a need to account for the known differences when estimating a population mean or total.

The other kind of effects are *random effects* associated with factors having a very large (possibly infinite) number of levels. Only a finite number of levels will occur in any particular survey. For example, because of differential skill levels, interviewers are known to affect the data values collected. The set of interviewers in a survey are only a subset of the ones that might have been used and are themselves usually of no intrinsic interest. An interviewer effect may then be modeled as a random effect with mean 0 and some unknown variance (see, e.g., Biemer and Stokes, 1985, 1989, 1991). The model variance is then the sum of the variance of the error term plus the variance of the interviewer random effect.

A model that contains only fixed effects is called a *fixed effects model*. One that has only random effects is a *random effects model*, and a model that has both is termed a *mixed model*. We will deal mainly with fixed effects models in this chapter, though there are important applications, like the ones studied by Biemer and Stokes, where random effects models or mixed models are appropriate. In Chapter 8 on clustered populations, a random effects model will also be a flexible way of studying those populations.

In experimental design a distinction is also made between cases where all cells in a table like 7.1 have the same sample size and cases where they do not. The former case is usually described as *balanced data* and the latter as *unbalanced data*. However, since we have reserved the terms *balanced* and *unbalanced* for the purposes described in earlier chapters, we will refer here to equal cell-size and unequal cell-size cases.

In finite population samples, unequal cell sizes are the norm because sample allocations usually are not controlled in detailed cells like those in Table 7.1. Estimation of model parameters is often considerably simpler in the equal cell-size case, but the general methods described later in this chapter are needed for the unequal cell-size case typical in samples from finite populations.

7.3. GENERALIZED INVERSES

As in Chapter 2, consider the general linear model

$$E_M(\mathbf{Y}) = \mathbf{X}\boldsymbol{\beta}$$
$$\text{var}_M(\mathbf{Y}) = \mathbf{V}, \qquad (7.3.1)$$

where $\mathbf{X}$ is an $N \times p$ matrix of auxiliaries, $\boldsymbol{\beta}$ is a $p \times 1$ vector of unknown parameters, and $\mathbf{V}$ is a positive definite covariance matrix. Given a sample s, we also have the various sample and nonsample components of $\mathbf{X}$ and $\mathbf{V}$ as defined in Chapter 2: $\mathbf{X}_s$, $\mathbf{X}_r$, $\mathbf{V}_{ss}$, $\mathbf{V}_{rr}$, $\mathbf{V}_{sr}$, and $\mathbf{V}_{rs} = \mathbf{V}'_{sr}$. We assume that $\mathbf{V}_{ss}$ is positive definite.

When $\mathbf{X}'_s\mathbf{V}_{ss}^{-1}\mathbf{X}_s$ does not have full rank, the estimating equations $\mathbf{X}'_s\mathbf{V}_{ss}^{-1}\mathbf{X}_s\hat{\boldsymbol{\beta}} = \mathbf{X}'_s\mathbf{V}_{ss}^{-1}\mathbf{Y}_s$ can be solved using what are known as *generalized inverses*. Searle (1971) is again a good source for the details which we will only sketch here. A *generalized inverse* (or g-inverse) of a matrix $\mathbf{A}$ is defined to be any matrix $\mathbf{G}$ that satisfies the equation

$$\mathbf{AGA} = \mathbf{A}. \tag{7.3.2}$$

The matrix $\mathbf{A}$ does not have to be either symmetric or square but for the applications here will be both since we are interested in $\mathbf{A}_s = \mathbf{X}'_s\mathbf{V}_{ss}^{-1}\mathbf{X}_s$. If $\mathbf{G}$ is a g-inverse of $\mathbf{X}'_s\mathbf{V}_{ss}^{-1}\mathbf{X}_s$, then a solution to the normal equations for estimating $\boldsymbol{\beta}$ is given by the following lemma.

Lemma 7.3.1. If $\mathbf{G}$ is a g-inverse of $\mathbf{A}_s = \mathbf{X}'_s\mathbf{V}_{ss}^{-1}\mathbf{X}_s$, then a solution to the normal equations $\mathbf{X}'_s\mathbf{V}_{ss}^{-1}\mathbf{X}_s\hat{\boldsymbol{\beta}} = \mathbf{X}'_s\mathbf{V}_{ss}^{-1}\mathbf{Y}_s$ is $\boldsymbol{\beta}^o = \mathbf{G}\mathbf{X}'_s\mathbf{V}_{ss}^{-1}\mathbf{Y}_s$.

Proof: First, note that since $\mathbf{A}_s$ is symmetric, $\mathbf{G}'$ is also a g-inverse. Next, write $\mathbf{A}_s = \mathbf{Z}'_s\mathbf{Z}_s$ where $\mathbf{Z}_s = \mathbf{V}_{ss}^{-1/2}\mathbf{X}_s$ and $\mathbf{V}_{ss}^{-1/2}$ is the square root of $\mathbf{V}_{ss}^{-1}$. By Lemma A.10.2(ii), $\mathbf{Z}_s\mathbf{G}'\mathbf{A}_s = \mathbf{Z}_s$. Multiplying both sides of this equation on the left by $\mathbf{Y}'_s\mathbf{V}_{ss}^{-1/2}$ and transposing gives the result. □

There are various ways of computing a $\mathbf{G}$. The S-PLUS™ function *ginv* will return the g-inverse of a general, real-valued rectangular matrix. The function *ginv* uses the singular value decomposition (SVD) of a matrix defined in Appendix A.10, where the construction of a g-inverse using the SVD is also described.

A more specialized algorithm for square matrices is the following: Suppose that the rank of a $p \times p$ matrix $\mathbf{A}$ is $q < p$, and the rows and columns of $\mathbf{A}$ can be permuted in such a way that $\mathbf{A}$ can be partitioned so that its leading $q \times q$ minor, $\mathbf{A}_{11}$, is nonsingular, that is,

$$\mathbf{A} = \begin{bmatrix} \mathbf{A}_{11} & \mathbf{A}_{12} \\ \mathbf{A}_{21} & \mathbf{A}_{22} \end{bmatrix}$$

Because the rank of $\mathbf{A}$ is q, $[\mathbf{A}_{21} \;\; \mathbf{A}_{22}] = \mathbf{K}[\mathbf{A}_{11} \;\; \mathbf{A}_{12}]$ for some matrix $\mathbf{K}$; that is, the rows in the second tier of submatrices in $\mathbf{A}$ are some linear combination of the rows in the first tier. Then a g-inverse of $\mathbf{A}$ is

$$\mathbf{G} = \begin{bmatrix} \mathbf{A}_{11}^{-1} & \mathbf{0} \\ \mathbf{0} & \mathbf{0} \end{bmatrix}, \tag{7.3.3}$$

where the null matrices are of appropriate order to make $\mathbf{G}$ be $p \times p$. Note that if $\mathbf{A}_{11}$ is symmetric, then so is $\mathbf{G}$. To see that $\mathbf{G}$ is a g-inverse, first observe that

$$\mathbf{AGA} = \begin{bmatrix} \mathbf{A}_{11} & \mathbf{A}_{12} \\ \mathbf{A}_{21} & \mathbf{A}_{21}\mathbf{A}_{11}^{-1}\mathbf{A}_{12} \end{bmatrix}.$$

Using $[\mathbf{A}_{21} \ \ \mathbf{A}_{22}] = \mathbf{K}[\mathbf{A}_{11} \ \ \mathbf{A}_{12}]$, we have $\mathbf{K} = \mathbf{A}_{21}\mathbf{A}_{11}^{-1}$ and, thus, $\mathbf{A}_{21}\mathbf{A}_{11}^{-1}\mathbf{A}_{12} = \mathbf{KA}_{12} = \mathbf{A}_{22}$, verifying that $\mathbf{AGA} = \mathbf{A}$.

There is no need for the nonsingular minor to be in the leading position as it is above. The rows and columns of $\mathbf{A}$ can be permuted to put the nonsingular minor in that position as in Searle (1971, sec 1.b), and a g-inverse constructed for the permuted matrix. Or, if an invertible submatrix exists in a non-leading position, like $\mathbf{A}_{22}$, then it can be used. In (7.1.3), for example, the lower 3×3 minor of $\mathbf{X}_s'\mathbf{X}_s$ is full rank, and a g-inverse is

$$\mathbf{G} = \begin{bmatrix} 0 & 0 & 0 & 0 \\ 0 & \frac{1}{3} & 0 & 0 \\ 0 & 0 & \frac{1}{2} & 0 \\ 0 & 0 & 0 & 1 \end{bmatrix}. \tag{7.3.4}$$

The reader can verify by direct multiplication that $\mathbf{X}_s'\mathbf{X}_s\mathbf{G}\mathbf{X}_s'\mathbf{X}_s = \mathbf{X}_s'\mathbf{X}_s$ for this example.

Note that the variance of $\boldsymbol{\beta}^o$ under model (7.3.1) is $\text{var}_M(\boldsymbol{\beta}^o) = \mathbf{G}\mathbf{A}_s\mathbf{G}'$. This is not an analog of its counterpart, $\text{var}_M(\hat{\boldsymbol{\beta}}) = (\mathbf{X}_s'\mathbf{V}_{ss}^{-1}\mathbf{X}_s)^{-1}$, in the full rank model, as $\text{var}_M(\boldsymbol{\beta}^o) = \mathbf{G}$ would be. However, it is possible to construct a g-inverse of $\mathbf{A}_s$ with the properties

$$\mathbf{G}\mathbf{A}_s\mathbf{G}' = \mathbf{G}, \tag{7.3.5}$$

$$\mathbf{G}\mathbf{G}' \text{ is symmetric,} \tag{7.3.6}$$

and

$$\mathbf{G}'\mathbf{G} \text{ is symmetric.} \tag{7.3.7}$$

A g-inverse that satisfies (7.3.2), (7.3.5), (7.3.6), and (7.3.7) is called a *Moore-Penrose* inverse and is unique (Gentle, 1998, sec. 2.1.9). The g-inverses for symmetric matrices produced by the SVD method in Appendix A.10 are Moore-Penrose inverses and are themselves symmetric. In the remainder of this chapter, it will be convenient for some results to assume symmetry, $\mathbf{G}' = \mathbf{G}$, and that (7.3.5) holds. This will not result in any loss of generality but will simplify certain formulas.

Although the g-inverse of a matrix is not unique, certain functions of any solution $\boldsymbol{\beta}^o = \mathbf{G}\mathbf{X}_s'\mathbf{V}_{ss}^{-1}\mathbf{Y}_s$ will be unique. The ones for which unique, unbiased estimators can be constructed are known as *estimable functions* and are discussed in Searle (1971, sec. 5.4). In the finite population case, we will be able to uniquely estimate $\mathbf{1}_r'\mathbf{X}_r\boldsymbol{\beta}$ with $\mathbf{1}_r'\mathbf{X}_r\boldsymbol{\beta}^o$, as shown in the next section.

7.4. ESTIMATING LINEAR COMBINATIONS OF THE Y's

In this section, we consider the general model (7.3.1). Assume that $\mathbf{V}$ has an inverse but that the column rank of $\mathbf{X}$ is $q < p$. As in Chapter 2, we consider predicting a linear combination $\theta = \boldsymbol{\gamma}'\mathbf{Y}$ of all Y's in the population. The analog to the predictor in Theorem 2.2.1, using a g-inverse, is $\hat{\theta}_{\text{opt}} = \boldsymbol{\gamma}_s'\mathbf{Y}_s + \boldsymbol{\gamma}_r'[\mathbf{X}_r\boldsymbol{\beta}^o + \mathbf{V}_{rs}\mathbf{V}_{ss}^{-1}(\mathbf{Y}_s - \mathbf{X}_s\boldsymbol{\beta}^o)]$. We will show that the predictor $\hat{\theta}_{\text{opt}}$ is the best linear unbiased predictor of θ regardless of the choice of g-inverse used in computing the solution to the normal equations. First, we need the additional results in Lemma 7.4.1.

Lemma 7.4.1. When $\mathbf{G}$ is a g-inverse of $\mathbf{A}_s = \mathbf{X}_s'\mathbf{V}_{ss}^{-1}\mathbf{X}_s$, then

(i) $\mathbf{X}_r\mathbf{G}\mathbf{X}_s'$ is invariant to the choice of $\mathbf{G}$; and

(ii) $\mathbf{X}_r\mathbf{G}\mathbf{A}_s = \mathbf{X}_r$.

Proof: To prove (i), suppose that $\mathbf{F}$ is some other g-inverse of $\mathbf{A}_s$. Let q be the column rank of $\mathbf{X}$ and write $\mathbf{X}$ as $\mathbf{X} = [\mathbf{X}_1 \ \ \mathbf{X}_2]$, where $\mathbf{X}_1$ is $N \times q$, $\mathbf{X}_2$ is $N \times (p-q)$ and $\mathbf{X}_1$ has full column rank. Because the columns of $\mathbf{X}_2$ are linearly dependent on those of $\mathbf{X}_1$, $\mathbf{X}_2 = \mathbf{X}_1\mathbf{K}$ for some $q \times (p-q)$ matrix $\mathbf{K}$. We can write $\mathbf{X} = \mathbf{X}_1\mathbf{K}^*$, where $\mathbf{K}^* = [\mathbf{I}_q \ \ \mathbf{K}]$ and $\mathbf{I}_q$ is the $q \times q$ identity matrix. Splitting $\mathbf{X}$ into sample and nonsample pieces gives $\mathbf{X} = [\mathbf{X}_s' \ \ \mathbf{X}_r']'$ and $\mathbf{X}_s = [\mathbf{X}_{1s} \ \ \mathbf{X}_{2s}] = \mathbf{X}_{1s}\mathbf{K}^*$, where $\mathbf{X}_{1s}$ is $n \times q$. Similarly, $\mathbf{X}_r = \mathbf{X}_{1r}\mathbf{K}^*$. Since $\mathbf{F}$ and $\mathbf{G}$ are g-inverses of $\mathbf{A}_s$, Lemma A.10.2(iii) implies that $\mathbf{V}_{ss}^{-1/2}\mathbf{X}_s\mathbf{G}\mathbf{X}_s'\mathbf{V}_{ss}^{-1/2} = \mathbf{V}_{ss}^{-1/2}\mathbf{X}_s\mathbf{F}\mathbf{X}_s'\mathbf{V}_{ss}^{-1/2}$. It follows that

$$\mathbf{V}_{ss}^{-1/2}\mathbf{X}_{1s}\mathbf{K}^*\mathbf{G}\mathbf{X}_s'\mathbf{V}_{ss}^{-1/2} = \mathbf{V}_{ss}^{-1/2}\mathbf{X}_{1s}\mathbf{K}^*\mathbf{F}\mathbf{X}_s'\mathbf{V}_{ss}^{-1/2}. \tag{7.4.1}$$

By construction, $\mathbf{X}_{1s}$ has full column rank q and on multiplying both sides of (7.4.1) on the left by $\mathbf{X}_{1r}\mathbf{A}_{1s}^{-1}\mathbf{X}_{1s}'\mathbf{V}_{ss}^{-1/2}$ with $\mathbf{A}_{1s} = \mathbf{X}_{1s}'\mathbf{V}_{ss}^{-1}\mathbf{X}_{1s}$ and on the right by $\mathbf{V}_{ss}^{1/2}$, we obtain $\mathbf{X}_r\mathbf{G}\mathbf{X}_s' = \mathbf{X}_r\mathbf{F}\mathbf{X}_s'$.

To prove (ii), note that by Lemma A.10.2(ii),

$$\mathbf{V}_{ss}^{-1/2}\mathbf{X}_s\mathbf{G}\mathbf{A}_s = \mathbf{V}_{ss}^{-1/2}\mathbf{X}_s. \tag{7.4.2}$$

Using the same decomposition of $\mathbf{X}$ as in part (i),

$$\mathbf{V}_{ss}^{-1/2}\mathbf{X}_{1s}\mathbf{K}^*\mathbf{G}\mathbf{A}_s = \mathbf{V}_{ss}^{-1/2}\mathbf{X}_{1s}\mathbf{K}^*. \tag{7.4.3}$$

Multiplying each side on the left by $\mathbf{X}_{1r}\mathbf{A}_{1s}^{-1}\mathbf{X}'_{1s}\mathbf{V}_{ss}^{-1/2}$ implies that $\mathbf{X}_r\mathbf{G}\mathbf{A}_s = \mathbf{X}_r$ as required. □

Given Lemma 7.4.1, we show in the next theorem that $\hat{\theta}_{\text{opt}}$ will be the same regardless of the choice of g-inverse used to generate the solution vector $\boldsymbol{\beta}^o$.

Theorem 7.4.1. The estimator

$$\hat{\theta}_{\text{opt}} = \boldsymbol{\gamma}'_s\mathbf{Y}_s + \boldsymbol{\gamma}'_r[\mathbf{X}_r\boldsymbol{\beta}^o + \mathbf{V}_{rs}\mathbf{V}_{ss}^{-1}(\mathbf{Y}_s - \mathbf{X}_s\boldsymbol{\beta}^o)],$$

where $\boldsymbol{\beta}^o = \mathbf{G}\mathbf{X}'_s\mathbf{V}_{ss}^{-1}\mathbf{Y}_s$, is invariant to the choice of $\mathbf{G}$, the g-inverse of $\mathbf{X}'_s\mathbf{V}_{ss}^{-1}\mathbf{X}_s$.

Proof: Note that $\hat{\theta}_{\text{opt}}$ can be written as

$$\hat{\theta}_{\text{opt}} = \boldsymbol{\gamma}'_s\mathbf{Y}_s + \boldsymbol{\gamma}'_r[\mathbf{X}_r\mathbf{G}\mathbf{X}'_s + \mathbf{V}_{rs} - \mathbf{V}_{rs}\mathbf{V}_{ss}^{-1}\mathbf{X}_s\mathbf{G}\mathbf{X}'_s]\mathbf{V}_{ss}^{-1}\mathbf{Y}_s.$$

$\mathbf{X}_r\mathbf{G}\mathbf{X}'_s$ is invariant by Lemma 7.4.1(i). Lemma A.10.2(iii) implies that $\mathbf{V}_{ss}^{-1}\mathbf{X}_s\mathbf{G}\mathbf{X}'_s\mathbf{V}_{ss}^{-1}$ is also invariant to the choice of $\mathbf{G}$, and the theorem is proved. □

Not only is $\hat{\theta}_{\text{opt}}$ invariant to the choice of generalized inverse, but it is also the *BLU* predictor of θ as the next theorem demonstrates. Notice that if $\mathbf{X}'_s\mathbf{V}_{ss}^{-1}\mathbf{X}_s$ has an inverse, then Theorem 7.4.2 reduces to Theorem 2.2.1.

Theorem 7.4.2. The best linear unbiased predictor of $\theta = \boldsymbol{\gamma}'\mathbf{Y}$ under model (7.3.1) is

$$\hat{\theta}_{\text{opt}} = \boldsymbol{\gamma}'_s\mathbf{Y}_s + \boldsymbol{\gamma}'_r[\mathbf{X}_r\boldsymbol{\beta}^o + \mathbf{V}_{rs}\mathbf{V}_{ss}^{-1}(\mathbf{Y}_s - \mathbf{X}_s\boldsymbol{\beta}^o)] \tag{7.4.4}$$

and has error variance

$$\begin{aligned}\text{var}_M(\hat{\theta}_{\text{opt}} - \theta) &= \boldsymbol{\gamma}'_r(\mathbf{V}_{rr} - \mathbf{V}_{rs}\mathbf{V}_{ss}^{-1}\mathbf{V}_{sr})\boldsymbol{\gamma}_r \\ &\quad + \boldsymbol{\gamma}'_r(\mathbf{X}_r - \mathbf{V}_{rs}\mathbf{V}_{ss}^{-1}\mathbf{X}_s)\mathbf{G}(\mathbf{X}_r - \mathbf{V}_{rs}\mathbf{V}_{ss}^{-1}\mathbf{X}_s)'\boldsymbol{\gamma}_r,\end{aligned} \tag{7.4.5}$$

where $\mathbf{G}$ is a g-inverse of $\mathbf{A}_s = \mathbf{X}'_s\mathbf{V}_{ss}^{-1}\mathbf{X}_s$.

Proof: To show unbiasedness, note that the estimation error is

$$\hat{\theta}_{\text{opt}} - \theta = \boldsymbol{\gamma}'_r[(\mathbf{X}_r - \mathbf{V}_{rs}\mathbf{V}_{ss}^{-1}\mathbf{X}_s)\mathbf{G}\mathbf{X}'_s + \mathbf{V}_{rs}]\mathbf{V}_{ss}^{-1}\mathbf{Y}_s - \boldsymbol{\gamma}'_r\mathbf{Y}_r$$

with expectation $\boldsymbol{\gamma}'_r[\mathbf{X}_r\mathbf{G}\mathbf{X}'_s - \mathbf{V}_{rs}\mathbf{V}_{ss}^{-1}\mathbf{X}_s\mathbf{G}\mathbf{X}'_s + \mathbf{V}_{rs}]\mathbf{V}_{ss}^{-1}\mathbf{X}_s\boldsymbol{\beta} - \boldsymbol{\gamma}'_r\mathbf{X}_r\boldsymbol{\beta}$. Unbiasedness follows by using $\mathbf{X}_r\mathbf{G}\mathbf{X}'_s\mathbf{V}_{ss}^{-1}\mathbf{X}_s = \mathbf{X}_r$ from Lemma 7.4.1(ii) and (7.4.2).

As in the proof of Theorem 2.2.1, the error variance of a general estimator with the form $\mathbf{g}_s'\mathbf{Y}_s$ is

$$E_M(\mathbf{g}_s'\mathbf{Y}_s - \gamma'\mathbf{Y})^2 = \mathbf{a}'\mathbf{V}_{ss}\mathbf{a} - 2\mathbf{a}'\mathbf{V}_{sr}\gamma_r + \gamma_r'\mathbf{V}_{rr}\gamma_r + [(\mathbf{a}'\mathbf{X}_s - \gamma_r'\mathbf{X}_r)\boldsymbol{\beta}]^2, \tag{7.4.6}$$

where $\mathbf{a} = \mathbf{g}_s - \gamma_s$. The requirement of unbiasedness means that the last term in brackets in (7.4.6) is zero. The Lagrange function to be minimized is

$$\Phi = \mathbf{a}'\mathbf{V}_{ss}\mathbf{a} - 2\mathbf{a}'\mathbf{V}_{sr}\gamma_r + 2(\mathbf{a}'\mathbf{X}_s - \gamma_r'\mathbf{X}_r)\boldsymbol{\lambda}.$$

Differentiating with respect to $\mathbf{a}$ and equating to zero leads to the two equations

$$\mathbf{X}_s\boldsymbol{\lambda} = \mathbf{V}_{sr}\gamma_r - \mathbf{V}_{ss}\mathbf{a} \tag{7.4.7}$$

$$\mathbf{a} = \mathbf{V}_{ss}^{-1}(\mathbf{V}_{sr}\gamma_r - \mathbf{X}_s\boldsymbol{\lambda}). \tag{7.4.8}$$

Note that these are the same as (2.2.4) and (2.2.5) in the proof of Theorem 2.2.1. But, because $\mathbf{A}_s$ is not invertible, we solve for $\boldsymbol{\lambda}$ as $\boldsymbol{\lambda} = \mathbf{G}(\mathbf{X}_s'\mathbf{V}_{ss}^{-1}\mathbf{V}_{sr} - \mathbf{X}_r')\gamma_r$ and obtain the solution for $\mathbf{a}$ as $\mathbf{a}_{\text{opt}} = \mathbf{V}_{ss}^{-1}[\mathbf{V}_{sr} + \mathbf{X}_s\mathbf{G}(\mathbf{X}_r' - \mathbf{X}_s'\mathbf{V}_{ss}^{-1}\mathbf{V}_{sr})]\gamma_r$. The optimal estimator is then $\gamma_s'\mathbf{Y}_s + \mathbf{a}_{\text{opt}}'\mathbf{Y}_s$, which, after some rearrangement, is seen to be equal to $\hat{\theta}_{\text{opt}}$.

The error variance can be derived by substituting $\mathbf{a}_{\text{opt}}$ into (7.4.6) to yield

$$\begin{aligned} \text{var}_M(\hat{\theta}_{\text{opt}} - \theta) = {} & \gamma_r'(\mathbf{V}_{rr} - \mathbf{V}_{rs}\mathbf{V}_{ss}^{-1}\mathbf{V}_{sr})\gamma_r \\ & + \gamma_r'(\mathbf{B}\mathbf{G}'\mathbf{X}_s'\mathbf{V}_{ss}^{-1}\mathbf{V}_{sr} + \mathbf{B}\mathbf{G}'\mathbf{X}_s'\mathbf{V}_{ss}^{-1}\mathbf{X}_s\mathbf{G}\mathbf{B}' \\ & + \mathbf{V}_{rs}\mathbf{V}_{ss}^{-1}\mathbf{X}_s\mathbf{G}\mathbf{B}' - 2\mathbf{B}\mathbf{G}'\mathbf{X}_s'\mathbf{V}_{ss}^{-1}\mathbf{V}_{sr})\gamma_r, \end{aligned} \tag{7.4.9}$$

where $\mathbf{B} = \mathbf{X}_r - \mathbf{V}_{rs}\mathbf{V}_{ss}^{-1}\mathbf{X}_s$. Next, note that the invariance results in Lemmas A.10.2(iii) and 7.4.1(i) imply that $\mathbf{X}_s\mathbf{G}'\mathbf{X}_s' = \mathbf{X}_s\mathbf{G}\mathbf{X}_s'$ and $\mathbf{X}_r\mathbf{G}'\mathbf{X}_s' = \mathbf{X}_r\mathbf{G}\mathbf{X}_s'$. By using (7.4.2) and Lemma 7.4.1(ii), we have $\mathbf{B}\mathbf{G}'\mathbf{X}_s'\mathbf{V}_{ss}^{-1}\mathbf{X}_s = \mathbf{B}$. Substitution in (7.4.7) and some rearrangement produces the error variance in the statement of the theorem. □

When the model is formulated in such a way that some of the columns of $\mathbf{X}$ are linear combinations of others, Theorem 7.4.2 says that the optimal estimator and its error variance from the full-rank case of Theorem 2.2.1 can be modified to fit the new situation. The modification consists simply of replacing $\mathbf{A}_s^{-1}$ by a generalized inverse. As in the full rank case, when all units have a common variance the *BLU* predictor and its error variance simplify as noted in the following corollary.

Corollary 7.4.1. When $\mathbf{V} = \sigma^2\mathbf{I}_N$, the normal equations are $\mathbf{X}_s'\mathbf{X}_s\boldsymbol{\beta} = \mathbf{X}_s'\mathbf{Y}_s$. The *BLU* predictor in Theorem 7.4.2 and its error variance reduce to

$$\hat{\theta}_{\text{opt}} = \gamma_s'\mathbf{Y}_s + \gamma_r'\mathbf{X}_r\boldsymbol{\beta}^o$$

$$\text{var}_M(\hat{\theta}_{\text{opt}} - \theta) = \sigma^2\gamma_r'\gamma_r + \sigma^2\gamma_r'\mathbf{X}_r\mathbf{G}\mathbf{X}_r'\gamma_r$$

with $\mathbf{G}$ being the generalized inverse of $\mathbf{X}_s'\mathbf{X}_s$.

When $\theta = T$, $\mathbf{X}_s'\mathbf{V}_{ss}^{-1}\mathbf{X}_s$ is singular, and a g-inverse is used, the estimator of the total in Theorem 7.4.2 can be written as $\hat{T} = \mathbf{g}_s'\mathbf{Y}_s$ with $\mathbf{g}_s = \mathbf{V}_{ss}^{-1}[\mathbf{V}_{sr} + \mathbf{X}_s\mathbf{G}(\mathbf{X}_r' - \mathbf{X}_s'\mathbf{V}_{ss}^{-1}\mathbf{V}_{sr})]\mathbf{1}_r + \mathbf{1}_s$. From Lemmas A.10.2(ii) and 7.4.1(i), $\mathbf{X}_s\mathbf{G}\mathbf{X}_s'$ and $\mathbf{X}_s\mathbf{G}\mathbf{X}_r'$ are invariant to the choice of g-inverse and thus so is the weight vector $\mathbf{g}_s$. If we use $\text{var}_M(\mathbf{Y}) = \sigma^2\mathbf{I}_N$ to generate $\hat{T}$, the weight vector simplifies to

$$\dot{\mathbf{g}}_s = \mathbf{X}_s\mathbf{G}\mathbf{X}_r'\mathbf{1}_r + \mathbf{1}_s. \tag{7.4.10}$$

The $\hat{T}$ using these weights is not optimal under model (7.3.1) with a general covariance matrix but will still be unbiased.

We now cover estimation of the total T under some common models having a homogeneous variance structure, that are usually associated with experimental designs.

7.5. ONE-WAY CLASSIFICATION

The example using educational attainment in the beginning of this chapter involved what is known as a one-way classification, that is, a single factor with several levels. Since that case was equivalent to a stratified model, it could be analyzed without recourse to g-inverses. However, addressing the one-way model in more detail will allow some general points to be made on the solution of the normal equations that will be useful for more elaborate models. Suppose that there are I levels of the single factor, that there are N_i units in the population in level i, and that the model is

$$Y_{ij} = \beta_0 + \beta_i + \varepsilon_{ij},\ \varepsilon_{ij} \sim (0, \sigma^2), \quad i = 1, \ldots, I; j = 1, \ldots, N_i. \tag{7.5.1}$$

Assume that the sample contains n_i units in level i, that all levels are represented in the sample, and that the total sample size is $n = \Sigma_{i=1}^I n_i$. Having a sample from each level is critical to estimating the population total because the model provides no inferential link between units in different levels.

To write the general form of the normal equations, first, let s_i be the set of sample units in level i and define $Y_{si\cdot} = \Sigma_{j\in s_i} Y_{ij}$ and $Y_{s\cdot} = \Sigma_{i=1}^I Y_{si\cdot}$. The normal equations are then

$$\mathbf{X}_s'\mathbf{X}_s\boldsymbol{\beta}^o = \left[\begin{array}{c|cccc} n & n_1 & n_2 & \cdots & n_I \\ \hline n_1 & n_1 & 0 & \cdots & 0 \\ n_2 & 0 & n_2 & \cdots & 0 \\ \vdots & & & \ddots & \\ n_I & 0 & 0 & \cdots & n_I \end{array}\right]\begin{bmatrix} \beta_0^o \\ \beta_1^o \\ \beta_2^o \\ \vdots \\ \beta_I^o \end{bmatrix} = \begin{bmatrix} Y_{s\cdot} \\ Y_{s1\cdot} \\ Y_{s2\cdot} \\ \vdots \\ Y_{sI\cdot} \end{bmatrix} = \mathbf{X}_s'\mathbf{Y}_s. \tag{7.5.2}$$

$\mathbf{X}_s'\mathbf{X}_s$ is $(I+1)\times(I+1)$ and has rank I. Since the lower $I \times I$ submatrix of $\mathbf{X}_s'\mathbf{X}_s$ is invertible, a g-inverse is

$$\mathbf{G} = \begin{bmatrix} 0 & \mathbf{0}' \\ \mathbf{0} & \text{diag}\{1/n_i\} \end{bmatrix}, \tag{7.5.3}$$

where $\mathbf{0}$ is an I-vector of zeroes and $\text{diag}\{1/n_i\}$ is an $I \times I$ diagonal matrix with the ith diagonal element equal to $1/n_i$. A solution to the normal equations is then

$$\boldsymbol{\beta}^o = \mathbf{G}\mathbf{X}_s'\mathbf{Y}_s = \begin{bmatrix} 0 \\ \bar{Y}_{s1} \\ \bar{Y}_{s2} \\ \vdots \\ \bar{Y}_{sI} \end{bmatrix}. \tag{7.5.4}$$

Notice that using this $\mathbf{G}$ is equivalent to crossing out the equation for β_0 in (7.5.2) and solving for the remainder of the parameter vector. This is an example of a more general prescription for parameter estimation that is followed by many software packages:

- Determine the number of estimable parameters in the model, that is, the rank of $\mathbf{X}_s'\mathbf{X}_s$ denoted as $r(\mathbf{X}_s'\mathbf{X}_s)$.
- Compute the difference between the order of $\mathbf{X}_s'\mathbf{X}_s$ and its rank, $p - r(\mathbf{X}_s'\mathbf{X}_s)$.
- Set $p - r(\mathbf{X}_s'\mathbf{X}_s)$ parameters to 0 and reduce $\mathbf{X}_s$ by removing the columns corresponding to the parameters that were set to 0. This must be done in such a way that the reduced matrix $\mathbf{X}_{s(R)}$ has full column rank.
- Solve for the remaining parameters as $\hat{\boldsymbol{\beta}}_{(R)} = (\mathbf{X}_{s(R)}'\mathbf{X}_{s(R)})^{-1}\mathbf{X}_{s(R)}'\mathbf{Y}_s$.

This is equivalent to using a g-inverse of the type given in (7.3.3).

The estimator of the finite population total, corresponding to parameter estimator (7.5.4), is then

$$\begin{aligned}\hat{T} &= \mathbf{1}_s'\mathbf{Y}_s + \mathbf{1}_r'\mathbf{X}_r\boldsymbol{\beta}^o \\ &= \sum_{i=1}^{I} n_i\bar{Y}_{si} + \sum_{i=1}^{I}(N_i - n_i)\bar{Y}_{si} \\ &= \sum_{i=1}^{I} N_i\bar{Y}_{si},\end{aligned}$$

which is, of course, the same as the stratified estimator in Example 2.3.3 and Section 6.1.

There are other ways in which a solution $\boldsymbol{\beta}^o$ can be found. A standard method is to add restrictions to parameters in the model. For example, if we require that $\Sigma_{i=1}^{I} n_i\beta_i = 0$, then the solution to (7.5.2) is

$$\boldsymbol{\beta}^o = \begin{bmatrix} \bar{Y}_s \\ \bar{Y}_{s1} - \bar{Y}_s \\ \bar{Y}_{s2} - \bar{Y}_s \\ \vdots \\ \bar{Y}_{sI} - \bar{Y}_s \end{bmatrix}.$$

The accompanying estimator of the total is then

$$\begin{aligned}\hat{T} &= \mathbf{1}_s'\mathbf{Y}_s + \mathbf{1}_r'\mathbf{X}_r\boldsymbol{\beta}^o \\ &= \sum_{i=1}^{I} n_i\bar{Y}_{si} + (N - n)\bar{Y}_s + \sum_{i=1}^{I}(N_i - n_i)(\bar{Y}_{si} - \bar{Y}_s) \\ &= \sum_{i=1}^{I} N_i\bar{Y}_{si},\end{aligned}$$

illustrating that the estimator of the total is invariant to the particular solution used for $\boldsymbol{\beta}^o$. One of the exercises is to show that the stratified expansion estimator is also obtained if the restriction is $\Sigma_{i=1}^{I}\beta_i = 0$.

Using the expression in Theorem 7.4.2 and the g-inverse in (7.5.3), the error variance of $\hat{T}$ is

$$\begin{aligned}\mathrm{var}_M(\hat{T} - T) &= \mathbf{1}_r'\mathbf{X}_r\mathbf{G}\mathbf{X}_r'\mathbf{1}_r\sigma^2 + (N - n)\sigma^2 \\ &= \sum_{i=1}^{I}\frac{(N_i - n_i)^2}{n_i}\sigma^2 + \sum_{i=1}^{I}(N_i - n_i)\sigma^2 \\ &= \sum_{i=1}^{I}\frac{N_i^2}{n_i}\left(1 - \frac{n_i}{N_i}\right)\sigma^2,\end{aligned}$$

which is again equal to the result in Example 2.3.3 for the special case of a common value of the model variance in every stratum.

7.6. TWO-WAY NESTED CLASSIFICATION

When there are two factors, several models are conceivable depending on the structure of the population. Beginning in this section, we cover three of them—(1) a model in which one factor is nested within the other, (2) a model where there is no interaction between the factors, and (3) a model with interaction.

A two-factor model is said to have a *two-way nested classification* if every level of the first factor does not occur with every level of the second. For example, suppose that the population consists of students in the tenth grade of high school and that the interest is in estimating the average test score per student on a standardized mathematics test. Assume there are a number of schools and that within each school there are several different teachers of mathematics. The two factors are school and teacher with teachers being nested within schools.

In a two-factor model it will be convenient to use α for one factor, β for the other, and μ for the overall mean, as is customary in experimental design. Assume that the model is

$$Y_{ijk} = \mu + \alpha_i + \beta_{ij} + \varepsilon_{ijk},\ \varepsilon_{ijk} \sim (0, \sigma^2) \tag{7.6.1}$$

with $i = 1, \ldots, I$; $j = 1, \ldots, B_i$; and $k = 1, \ldots, N_{ij}$. The number of sample units in cell (ij) is n_{ij}. In the student example, Y_{ijk} is the test score of the kth student of the jth teacher in the ith school, α_i is the effect of school i, and β_{ij} is the effect of teacher j in school i.

To illustrate, suppose that $I = 2$, $B_1 = 2$, $B_2 = 3$, $n_{11} = 1$, $n_{12} = 2$, $n_{21} = 3$, $n_{22} = 2$, and $n_{23} = 1$. These sizes will be sufficient to show patterns in $\mathbf{X}_s$ and $\mathbf{X}_s'\mathbf{X}_s$ for more general sample sizes. The model equations are

$$\begin{bmatrix} Y_{111} \\ Y_{121} \\ Y_{122} \\ Y_{211} \\ Y_{212} \\ Y_{213} \\ Y_{221} \\ Y_{222} \\ Y_{231} \end{bmatrix} = \left[\begin{array}{c|cc|ccccc} 1 & 1 & 0 & 1 & 0 & 0 & 0 & 0 \\ 1 & 1 & 0 & 0 & 1 & 0 & 0 & 0 \\ 1 & 1 & 0 & 0 & 1 & 0 & 0 & 0 \\ 1 & 0 & 1 & 0 & 0 & 1 & 0 & 0 \\ 1 & 0 & 1 & 0 & 0 & 1 & 0 & 0 \\ 1 & 0 & 1 & 0 & 0 & 1 & 0 & 0 \\ 1 & 0 & 1 & 0 & 0 & 0 & 1 & 0 \\ 1 & 0 & 1 & 0 & 0 & 0 & 1 & 0 \\ 1 & 0 & 1 & 0 & 0 & 0 & 0 & 1 \end{array}\right] \begin{bmatrix} \mu \\ \alpha_1 \\ \alpha_2 \\ \beta_{11} \\ \beta_{12} \\ \beta_{21} \\ \beta_{22} \\ \beta_{23} \end{bmatrix} + \begin{bmatrix} \varepsilon_{111} \\ \varepsilon_{121} \\ \varepsilon_{122} \\ \varepsilon_{211} \\ \varepsilon_{212} \\ \varepsilon_{213} \\ \varepsilon_{221} \\ \varepsilon_{222} \\ \varepsilon_{231} \end{bmatrix}.$$

Next, define $n = \Sigma_{i=1}^{I} \Sigma_{j=1}^{B_i} n_{ij}$, $n_{i\cdot} = \Sigma_{j=1}^{B_i} n_{ij}$, $Y_{s\cdots} = \Sigma_{i=1}^{I} \Sigma_{j=1}^{B_i} \Sigma_{k \in s_{ij}} Y_{ijk}$, $Y_{si\cdot\cdot} = \Sigma_{j=1}^{B_i} \Sigma_{k \in s_{ij}} Y_{ijk}$, and $Y_{sij\cdot} = \Sigma_{k \in s_{ij}} Y_{ijk}$. The normal equations are then

$$\left[\begin{array}{c|cc|ccccc} n & n_{1\cdot} & n_{2\cdot} & n_{11} & n_{12} & n_{21} & n_{22} & n_{23} \\ \hline n_1 & n_{1\cdot} & 0 & n_{11} & n_{12} & 0 & 0 & 0 \\ \hline n_2 & 0 & n_{2\cdot} & 0 & 0 & n_{21} & n_{22} & n_{23} \\ \hline n_{11} & n_{11} & 0 & n_{11} & 0 & 0 & 0 & 0 \\ n_{12} & n_{12} & 0 & 0 & n_{12} & 0 & 0 & 0 \\ n_{21} & 0 & n_{21} & 0 & 0 & n_{21} & 0 & 0 \\ n_{22} & 0 & n_{22} & 0 & 0 & 0 & n_{22} & 0 \\ n_{23} & 0 & n_{23} & 0 & 0 & 0 & 0 & n_{23} \end{array}\right] \begin{bmatrix} \mu \\ \alpha_1 \\ \alpha_2 \\ \beta_{11} \\ \beta_{12} \\ \beta_{21} \\ \beta_{22} \\ \beta_{23} \end{bmatrix} = \begin{bmatrix} Y_{s\cdots} \\ Y_{s1\cdot\cdot} \\ Y_{s2\cdot\cdot} \\ Y_{s11\cdot} \\ Y_{s12\cdot} \\ Y_{s21\cdot} \\ Y_{s22\cdot} \\ Y_{s23\cdot} \end{bmatrix}.$$

As in the one-way case, the model is overparameterized. $\mathbf{X}_s'\mathbf{X}_s$ has order 8 but rank 5. Rows 2 and 3 sum to row 1; rows 4 and 5 sum to row 2; and rows 6, 7, and 8 sum to row 3. In general, the rank of $\mathbf{X}_s'\mathbf{X}_s$ will be $B_{\cdot} = \Sigma_{i=1}^{I} B_i$, the number of subclasses across the first-level factor. Notice that the lower-right 5×5 submatrix of $\mathbf{X}_s'\mathbf{X}_s$ is $\text{diag}\{n_{ij}\}$ which is invertible. Thus a g-inverse is

$$\mathbf{G} = \begin{bmatrix} \mathbf{0} & \mathbf{0} \\ \mathbf{0} & \text{diag}\{1/n_{ij}\} \end{bmatrix}$$

and a solution to the normal equations is

$$\boldsymbol{\beta}^o = [\mathbf{0}' \quad \bar{\mathbf{Y}}_s']',$$

where $\bar{\mathbf{Y}}_s = [\bar{Y}_{s11} \ \bar{Y}_{s12} \ \bar{Y}_{s21} \ \bar{Y}_{s22} \ \bar{Y}_{s23}]'$ and $\bar{Y}_{sij} = Y_{sij\cdot}/n_{ij}$. The *BLU* predictor of T and its error variance are then

$$\begin{aligned} \hat{T} &= \sum_{i=1}^{I} \sum_{j=1}^{B_i} n_{ij} \bar{Y}_{sij} + \sum_{i=1}^{I} \sum_{j=1}^{B_i} (N_{ij} - n_{ij}) \bar{Y}_{sij} \\ &= \sum_{i=1}^{I} \sum_{j=1}^{B_i} N_{ij} \bar{Y}_{sij} \end{aligned}$$

and

$$\text{var}_M(\hat{T} - T) = \sum_{i=1}^{I} \sum_{j=1}^{B_i} \frac{N_{ij}^2}{n_{ij}} \left(1 - \frac{n_{ij}}{N_{ij}}\right) \sigma^2.$$

As is clear from these formulas, the nested classification is simply a type of stratified model in which major strata are formed by the levels of the α-factor and substrata are the β-levels within each level of α.

Note, that to estimate T in the fixed effects model, we must have a sample in every one of the (ij) cells in the population because each cell has its own mean $\mu + \alpha_i + \beta_{ij}$ that cannot be inferred from data in other cells. This can have serious practical implications in some populations. If, for example, the population in our student example consists of all tenth graders in the United States and there is an important fixed, nested, teacher effect, the sample would have to contain students from every mathematics teacher in the country—something that is clearly not feasible in any survey with a limited budget. In such a situation, it may be reasonable to treat both the school and the teacher effect as random effects with mean zero, in which case estimation is possible.

7.7. TWO-WAY CLASSIFICATION WITHOUT INTERACTION

We now consider two factors where the levels of each factor are completely crossed with each other. For example, one factor might be age, broken into several categories, and the other factor race, categorized as white and non-white. The factors are completely crossed in a population if all age × race cells contain at least one unit. A model with no interaction between the factors is

$$Y_{ijk} = \mu + \alpha_i + \beta_j + \varepsilon_{ijk},\ \varepsilon_{ijk} \sim (0, \sigma^2) \qquad (7.7.1)$$

with $i = 1, \ldots, I$; $j = 1, \ldots, J$; and $k = 1, \ldots, N_{ij}$. The number of sample units in cell (ij) is again n_{ij} with n and $n_{i\cdot}$ defined as in the last section, and $n_{\cdot j} = \Sigma_{i=1}^{I} n_{ij}$. The sample sums of Y for each level of the α-factor are $Y_{si\cdot\cdot}$ $(i = 1, \ldots, I)$ and for each level of the β-factor are $Y_{s\cdot j\cdot}$ $(j = 1, \ldots, J)$. Defining $\mathbf{X}_s$ in the obvious way, the normal equations are

$$\left[\begin{array}{c|ccc|ccc} n & n_{1\cdot} & \cdots & n_{I\cdot} & n_{\cdot 1} & \cdots & n_{\cdot J} \\ \hline n_{1\cdot} & & & & & & \\ \vdots & & \mathrm{diag}\{n_{i\cdot}\} & & & \{n_{ij}\} & \\ n_{I\cdot} & & & & & & \\ \hline n_{1} & & & & & & \\ \vdots & & \{n_{ji}\} & & & \mathrm{diag}\{n_{\cdot j}\} & \\ n_{\cdot J} & & & & & & \end{array}\right] \begin{bmatrix} \mu \\ \alpha_1 \\ \vdots \\ \alpha_I \\ \beta_1 \\ \vdots \\ \beta_J \end{bmatrix} = \begin{bmatrix} Y_{s\cdots} \\ Y_{s1\cdot\cdot} \\ \vdots \\ Y_{sI\cdot\cdot} \\ Y_{s\cdot 1\cdot} \\ \vdots \\ Y_{s\cdot J\cdot} \end{bmatrix},$$

where $\mathrm{diag}\{n_{i\cdot}\}$ is the $I \times I$ diagonal matrix of the $n_{i\cdot}$, $\mathrm{diag}\{n_{\cdot j}\}$ is the $J \times J$ diagonal matrix of the $n_{\cdot j}$, $\{n_{ij}\}$ is the $I \times J$ matrix of the cell sample sizes n_{ij}, and $\{n_{ji}\}$ is its transpose.

In this case $\mathbf{X}_s'\mathbf{X}_s$ has order $I + J + 1$ but rank $I + J - 1$ since there are two dependencies. Rows 2 through $I + 1$ sum to the first row as do rows $I + 2$ through $I + J + 1$. An $(I + J - 1) \times (I + J - 1)$ invertible submatrix is not easily available here, and simple analytic solutions for the estimator of the total and its error variance cannot be obtained—unlike the cases of the one-way model and the two-way nested model. Searle (1971, sec. 7.1.d) describes how the normal equations can be solved by a so-called "absorption process," in which the equations for μ and one of the β's are eliminated, the α's are solved for in terms of the β's, and then an analytic solution for the β's is found in terms of the $Y_{si\cdot\cdot}$'s and $Y_{s\cdot j\cdot}$'s. As Searle notes, the absorption method may not work for all data sets.

However, given a sample, we can always compute a g-inverse from the singular value decomposition, as described in Appendix A.10, and then numerically evaluate the formulas for $\hat{T}$ and its error variance in Theorem 7.4.2. Another alternative is to set two of the parameters to 0 (say, $\alpha_I = 0$ and $\beta_J = 0$) and then solve for the remainder as described in Section 7.6. This is an example of using a g-inverse of the form (7.3.3).

In the model with no interaction, the vector $\mathbf{1}_r'\mathbf{X}_r$, a key as always in getting $\hat{T}$, consists of the number of nonsample units, $N - n$, the nonsample x-total at each level of the α-factor, $N_{i\cdot} - n_{i\cdot}$, and the similar, nonsample sums at each level of the β-factor, $N_{\cdot j} - n_{\cdot j}$. Thus only the marginal totals of the x's are needed, not the cell totals, $N_{ij} - n_{ij}$, in the table formed by crossing the levels of α and β.

7.8. TWO-WAY CLASSIFICATION WITH INTERACTION

The two-way classification with interaction is actually much easier to handle computationally than the model with no interaction, although it involves more parameters. Suppose that the mean is affected by the particular cell a unit falls into and that the model is

$$Y_{ijk} = \mu + \alpha_i + \beta_j + \gamma_{ij} + \varepsilon_{ijk},\ \varepsilon_{ijk} \sim (0, \sigma^2). \tag{7.8.1}$$

The problem will be easier to understand with an example. Table 7.2 gives sample cell sizes that can be used to construct $\mathbf{X}_s$ in

$$\begin{bmatrix} Y_{111} \\ Y_{112} \\ Y_{121} \\ Y_{122} \\ Y_{211} \\ Y_{212} \\ Y_{221} \\ Y_{311} \\ Y_{321} \\ Y_{322} \\ Y_{323} \end{bmatrix} = \left[\begin{array}{c|ccc|cc|cccccc} 1 & 1 & 0 & 0 & 1 & 0 & 1 & 0 & 0 & 0 & 0 & 0 \\ 1 & 1 & 0 & 0 & 1 & 0 & 1 & 0 & 0 & 0 & 0 & 0 \\ 1 & 1 & 0 & 0 & 0 & 1 & 0 & 1 & 0 & 0 & 0 & 0 \\ 1 & 1 & 0 & 0 & 0 & 1 & 0 & 1 & 0 & 0 & 0 & 0 \\ 1 & 0 & 1 & 0 & 1 & 0 & 0 & 0 & 1 & 0 & 0 & 0 \\ 1 & 0 & 1 & 0 & 1 & 0 & 0 & 0 & 1 & 0 & 0 & 0 \\ 1 & 0 & 1 & 0 & 0 & 1 & 0 & 0 & 0 & 1 & 0 & 0 \\ 1 & 0 & 0 & 1 & 1 & 0 & 0 & 0 & 0 & 0 & 1 & 0 \\ 1 & 0 & 0 & 1 & 0 & 1 & 0 & 0 & 0 & 0 & 0 & 1 \\ 1 & 0 & 0 & 1 & 0 & 1 & 0 & 0 & 0 & 0 & 0 & 1 \\ 1 & 0 & 0 & 1 & 0 & 1 & 0 & 0 & 0 & 0 & 0 & 1 \end{array}\right] \begin{bmatrix} \mu \\ \alpha_1 \\ \alpha_2 \\ \alpha_3 \\ \beta_1 \\ \beta_2 \\ \gamma_{11} \\ \gamma_{12} \\ \gamma_{21} \\ \gamma_{22} \\ \gamma_{31} \\ \gamma_{32} \end{bmatrix} + \begin{bmatrix} \varepsilon_{111} \\ \varepsilon_{112} \\ \varepsilon_{121} \\ \varepsilon_{122} \\ \varepsilon_{211} \\ \varepsilon_{212} \\ \varepsilon_{221} \\ \varepsilon_{311} \\ \varepsilon_{321} \\ \varepsilon_{322} \\ \varepsilon_{323} \end{bmatrix}.$$

Table 7.2. Illustrative Cell Sample Sizes for the Two-Way Model with Interaction

i	$j=1$	$j=2$	$n_{i\cdot}$
1	2	2	4
2	2	1	3
3	1	3	4
$n_{\cdot j}$	5	6	11

Of course, in practice, we would not want to fit a 12 parameter model with 11 observations, but creating a large, more realistic example is unnecessary to illustrate how $\mathbf{X}_s$ and the normal equations are constructed. The estimating equations for the parameters are

$$\left[\begin{array}{c:ccc:cc:cccccc}
n & n_{1\cdot} & n_{2\cdot} & n_{3\cdot} & n_{\cdot 1} & n_{\cdot 2} & n_{11} & n_{12} & n_{21} & n_{22} & n_{31} & n_{32} \\ \hdashline
n_{1\cdot} & n_{1\cdot} & 0 & 0 & n_{11} & n_{12} & n_{11} & n_{12} & 0 & 0 & 0 & 0 \\
n_{2\cdot} & 0 & n_{2\cdot} & 0 & n_{21} & n_{22} & 0 & 0 & n_{21} & n_{22} & 0 & 0 \\
n_{3\cdot} & 0 & 0 & n_{3\cdot} & n_{31} & n_{32} & 0 & 0 & 0 & 0 & n_{31} & n_{32} \\ \hdashline
n_{\cdot 1} & n_{11} & n_{21} & n_{31} & n_{\cdot 1} & 0 & n_{11} & 0 & n_{21} & 0 & n_{31} & 0 \\
n_{\cdot 2} & n_{12} & n_{22} & n_{32} & 0 & n_{\cdot 2} & 0 & n_{12} & 0 & n_{22} & 0 & n_{32} \\ \hdashline
n_{11} & n_{11} & 0 & 0 & n_{11} & 0 & n_{11} & 0 & 0 & 0 & 0 & 0 \\
n_{12} & n_{12} & 0 & 0 & 0 & n_{12} & 0 & n_{12} & 0 & 0 & 0 & 0 \\
n_{21} & 0 & n_{21} & 0 & n_{21} & 0 & 0 & 0 & n_{21} & 0 & 0 & 0 \\
n_{22} & 0 & n_{22} & 0 & 0 & n_{22} & 0 & 0 & 0 & n_{22} & 0 & 0 \\
n_{31} & 0 & 0 & n_{31} & n_{31} & 0 & 0 & 0 & 0 & 0 & n_{31} & 0 \\
n_{32} & 0 & 0 & n_{32} & 0 & n_{32} & 0 & 0 & 0 & 0 & 0 & n_{32}
\end{array}\right]$$

$$\times \begin{bmatrix} \mu \\ \alpha_1 \\ \alpha_2 \\ \alpha_3 \\ \beta_1 \\ \beta_2 \\ \gamma_{11} \\ \gamma_{12} \\ \gamma_{21} \\ \gamma_{22} \\ \gamma_{31} \\ \gamma_{32} \end{bmatrix} = \begin{bmatrix} Y_{s\cdots} \\ Y_{s1\cdot\cdot} \\ Y_{s2\cdot\cdot} \\ Y_{s\cdot 1\cdot} \\ Y_{s\cdot 2\cdot} \\ Y_{s\cdot 3\cdot} \\ Y_{s11\cdot} \\ Y_{s12\cdot} \\ Y_{s21\cdot} \\ Y_{s22\cdot} \\ Y_{s31\cdot} \\ Y_{s32\cdot} \end{bmatrix}. \tag{7.8.2}$$

In general, we have $1 + I + J + IJ$ equations but the rank of $\mathbf{X}_s'\mathbf{X}_s$ is less because there are a variety of dependencies among the equations. The last IJ equations are linearly independent while the other $1 + I + J$ will depend upon them. In (7.8.2), for example, the lower right 6×6 submatrix of $\mathbf{X}_s'\mathbf{X}_s$ is invertible. The rank of $\mathbf{X}_s'\mathbf{X}_s$ is $IJ = 6$. Thus, a g-inverse of $\mathbf{X}_s'\mathbf{X}_s$ is

$$\mathbf{G} = \begin{bmatrix} \mathbf{0} & \mathbf{0} \\ \mathbf{0} & \text{diag}\{1/n_{ij}\} \end{bmatrix},$$

where the $\mathbf{0}$ matrices are of appropriate size to make $\mathbf{G}$ have order $(1 + I + J + IJ) \times (1 + I + J + IJ)$. A particularly simple solution to the normal equations is then

$$\boldsymbol{\beta}^o = (\mathbf{0}' \quad \bar{\mathbf{Y}}_s')' \tag{7.8.3}$$

with $\mathbf{0}$ a vector of $1 + I + J$ zeros and $\bar{\mathbf{Y}}_s$ the $IJ \times 1$ vector of sample cell means $\bar{Y}_{sij}$. The BLU predictor and its error variance are then

$$\begin{aligned} \hat{T} &= \sum_{i=1}^{I} \sum_{j=1}^{J} n_{ij}\bar{Y}_{sij} + \sum_{i=1}^{I} \sum_{j=1}^{J} (N_{ij} - n_{ij})\bar{Y}_{sij} \\ &= \sum_{i=1}^{I} \sum_{j=1}^{J} N_{ij}\bar{Y}_{sij} \end{aligned}$$

and

$$\text{var}_M(\hat{T} - T) = \sum_{i=1}^{I} \sum_{j=1}^{J} \frac{N_{ij}^2}{n_{ij}} \left(1 - \frac{n_{ij}}{N_{ij}}\right) \sigma^2$$

These results are again much like the ones for stratified populations. All ij cells must be represented in the sample in order to estimate T, and the number of population units N_{ij} must be known for each cell. If, because the allocation of the sample could not be controlled, there is no sample for one or more cells and model (7.8.1) is correct, then, strictly speaking, we have no way of inferring what the total is for the missing cells.

Suppose, as an example, that we have no sample in cell (IJ). Then, it is easy to show that the bias of $\hat{T}$ is $E_M(\hat{T} - T) = -N_{IJ}\mu_{IJ}$, where $\mu_{IJ} = \mu + \alpha_I + \beta_J + \gamma_{IJ}$. This is a case where we may be overtaken by practical exigencies and have to impute an estimate of μ_{IJ} based on other cells or to fall back to a model with no interaction or to a random effects model just in order to calculate a $\hat{T}$.

The next example uses the Labor Force population in Appendix B to illustrate the two-way model. This population actually involves clustering that

Table 7.3. Average Usual Weekly Wages (Rounded to the Nearest Dollar) for Persons in the Labor Force Population Categorized by Age and Sex

Age	Male	Female
19 and under	135	111
20–24	225	182
25–34	370	253
35–64	443	253
65 and over	307	91

will be addressed in Chapters 8 and 9, but will provide some useful illustrations in this chapter. Table 7.3 gives the population average of the variable "usual weekly wages" categorized by age and sex. There are clear differences in males and females and in age groups which should be accounted for in estimation.

Example 7.8.1. Two-Way Model. Suppose we want to estimate the total of weekly wages in the Labor Force population and select a systematic sample that consists of units 1, 11, 21, 31, and so on through unit 471, giving a total of $n = 48$. The sample numbers of persons by age category and sex are given in Table 7.3. We have collapsed together the 35–64 and 65+ age groups, that are separated in Table 7.3 because of the small sample size. (In practice, we would want to keep these age groups separate, if possible, since their means in Table 7.3 were quite different.)

We can consider fitting a model either with or without interaction since we have sample units from each of the age × sex cells. A rough check on whether there is interaction or not can be had by graphing the cell means as is done in Figure 7.1. There the average weekly wages for males and females across the entire population are plotted versus age category. If there were no interaction between age and sex, the curves would be spaced an equal distance apart.

Table 7.4. Distribution of the Sample by Age and Sex for Example 7.8.1

	Age Category				
Sex	19 and Under	20–24	25–34	35+	Total
Male	2	5	6	11	24
Female	2	6	5	11	24
Total	4	11	11	22	48

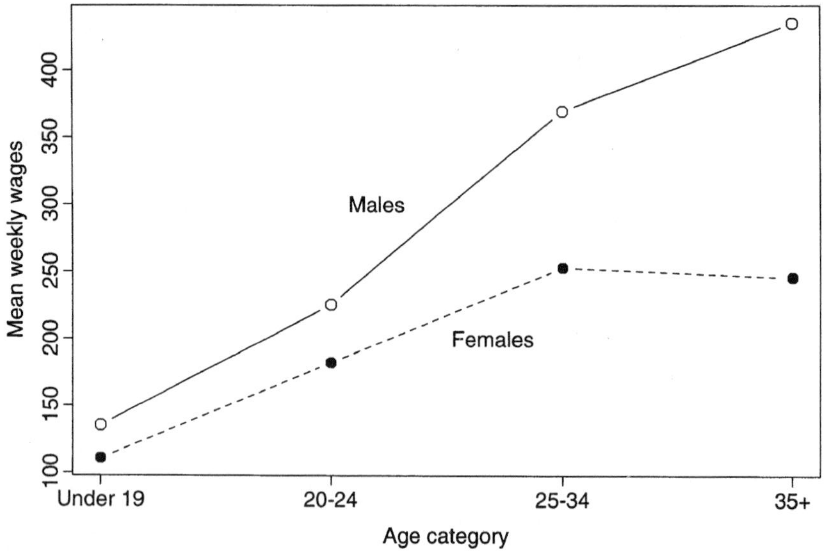

Figure 7.1 Average weekly wages plotted versus age category for males and females in the Labor Force population.

There is some evidence of interaction, but, of course, we never have the luxury of doing a full population plot in a real survey.

Table 7.5 shows solutions to the normal equations for models without and with interaction. The factor for sex is denoted by α and that for age by β. Solution 1 for the with-interaction model is computed using the S-PLUS™ function *ginv*; solution 2 uses the cells means as in (7.8.3). Note that, even though solutions 1 and 2 are different, cell means are estimable and the estimates from solutions 1 and 2 are the same. For example, the estimated (1, 3) cell mean is $\mu^o + \alpha_1^o + \beta_3^o + \gamma_{13}^o = 133.07 + 91.05 + 92.05 + 115.99 = 432.16$ for solution 1 and is $\gamma_{13}^o = 432.17$ for solution 2—with the difference due only to rounding. For this sample, we have $\mathbf{1}_s'\mathbf{Y}_s = 13{,}635$ and for the model with no interaction

$$\begin{aligned}\mathbf{1}_r'\mathbf{X}_r &= (N - n, N_{1\cdot} - n_{1\cdot}, N_{2\cdot} - n_{2\cdot}, N_{\cdot 1} - n_{\cdot 1}, N_{\cdot 2} - n_{\cdot 2}, N_{\cdot 3} - n_{\cdot 3}, N_{\cdot 4} - n_{\cdot 4}) \\ &= (430, 208, 222, 43, 43, 118, 226).\end{aligned}$$

For the model with interaction,

$$\begin{aligned}\mathbf{1}_r'\mathbf{X}_r = (&N - n, N_{1\cdot} - n_{1\cdot}, N_{2\cdot} - n_{2\cdot}, N_{\cdot 1} - n_{\cdot 1}, N_{\cdot 2} - n_{\cdot 2}, N_{\cdot 3} - n_{\cdot 3}, N_{\cdot 4} - n_{\cdot 4}, \\ &N_{11} - n_{11}, N_{12} - n_{12}, N_{13} - n_{13}, N_{14} - n_{14}, \\ &N_{21} - n_{21}, N_{22} - n_{22}, N_{23} - n_{23}, N_{24} - n_{24}) \\ = (&430, 208, 222, 43, 43, 118, 226, 13, 23, 66, 106, 30, 20, 52, 120).\end{aligned}$$

Table 7.5. Solutions to the Normal Equations for Two-Way Models Without and With Interaction

Solution	Model Without Interaction	Model With Interaction	
		Solution 1	Solution 2
μ^o	143.24	133.07	0
α_1^o	128.41	91.05	0
α_2^o	14.83	42.02	0
β_1^o	−68.86	−35.74	0
β_2^o	−29.06	−10.02	0
β_3^o	126.25	92.05	0
β_4^o	114.91	86.78	0
γ_{11}^o		−128.39	60.00
γ_{12}^o		13.89	228.00
γ_{13}^o		115.99	432.17
γ_{14}^o		89.55	400.45
γ_{21}^o		92.65	232.00
γ_{22}^o		−23.91	141.17
γ_{23}^o		−23.94	243.20
γ_{24}^o		2.78	259.09

The estimates of T are then $\hat{T} = 141{,}887.00$ for the no-interaction model and $\hat{T} = 144{,}151$ for the model with interaction. These compare to the actual total of $T = 140{,}818$.

In the preceding material, we have not discussed the usual questions that an analysis of variance is normally designed to answer. For example, are the α-effects all equal to each other? The sums of squares and F-statistics used for testing effects assume that Y is normally distributed. When cell sample sizes are unequal the "order" of parameter fitting must be considered—the F for testing the significance of β after fitting μ and α is different from the F for testing β after fitting only μ, for example. Given a model, these issues need not concern us here, but at the model-fitting stage, we should be aware of the complications associated with unequal cell sizes. Searle (1971) covers these issues in depth.

7.9. COMBINING QUALITATIVE AND QUANTITATIVE AUXILIARIES

Qualitative and quantitative auxiliaries can both be used in a model. There is no difference in parameter estimation for this model than for the others we have considered in earlier sections. Using a generalized inverse to solve for the parameter vector and for estimating the total works in this case, also. It will, however, be instructive to write the model in a way that distinguishes between the two types of parameters.

7.9.1. General Covariance Model

Let $\mathbf{X}$ be the $N \times p$ matrix of qualitative auxiliaries and $\mathbf{Z}$ be the $N \times q$ matrix of quantitative auxiliaries. Suppose that the model for $\mathbf{Y}$ is

$$\mathbf{Y} = \mathbf{X}\boldsymbol{\beta} + \mathbf{Z}\boldsymbol{\gamma} + \boldsymbol{\varepsilon} \tag{7.9.1}$$

where $\boldsymbol{\beta}$ is $p \times 1$, $\boldsymbol{\gamma}$ is $q \times 1$, and $\boldsymbol{\varepsilon}$ is $N \times 1$ with mean $\mathbf{0}$ and variance $\text{var}_M(\boldsymbol{\varepsilon}) = \sigma^2\mathbf{I}$. We assume that $\mathbf{X}$ does not necessarily have full column rank but that $\mathbf{Z}$ does. We also assume that the columns of $\mathbf{Z}$ are linearly independent of those of $\mathbf{X}$, that is, the columns of $\mathbf{Z}$ cannot be expressed as a linear combination of the columns of $\mathbf{X}$ and that the matrix of qualitative auxiliaries for the sample units, $\mathbf{X}_s$, does not have full column rank. Thus $\mathbf{X}_s'\mathbf{X}_s$ has no inverse but $\mathbf{Z}_s'\mathbf{Z}_s$ is invertible. The normal equations can be written as

$$\begin{bmatrix} \mathbf{X}_s'\mathbf{X}_s & \mathbf{X}_s'\mathbf{Z}_s \\ \mathbf{Z}_s'\mathbf{X}_s & \mathbf{Z}_s'\mathbf{Z}_s \end{bmatrix} \begin{bmatrix} \boldsymbol{\beta} \\ \boldsymbol{\gamma} \end{bmatrix} = \begin{bmatrix} \mathbf{X}_s'\mathbf{Y}_s \\ \mathbf{Z}_s'\mathbf{Y}_s \end{bmatrix}. \tag{7.9.2}$$

We can get a solution vector directly as

$$\begin{bmatrix} \boldsymbol{\beta}^o \\ \boldsymbol{\gamma}^o \end{bmatrix} = \mathbf{G}^* \begin{bmatrix} \mathbf{X}_s'\mathbf{Y}_s \\ \mathbf{Z}_s'\mathbf{Y}_s \end{bmatrix},$$

where $\mathbf{G}^*$ is a g-inverse of

$$\begin{bmatrix} \mathbf{X}_s'\mathbf{X}_s & \mathbf{X}_s'\mathbf{Z}_s \\ \mathbf{Z}_s'\mathbf{X}_s & \mathbf{Z}_s'\mathbf{Z}_s \end{bmatrix}.$$

Alternatively, we can treat (7.9.2) as a set of two equations in two unknowns as follows: Suppose that solution vectors are $\boldsymbol{\beta}^o$ and $\boldsymbol{\gamma}^o$ and that $\mathbf{G}_x$ is a g-inverse of $\mathbf{X}_s'\mathbf{X}_s$. Then the first equation in (7.9.2) gives

$$\begin{aligned} \boldsymbol{\beta}^o &= \mathbf{G}_x(\mathbf{X}_s'\mathbf{Y}_s - \mathbf{X}_s'\mathbf{Z}_s\boldsymbol{\gamma}^o) \\ &= \boldsymbol{\beta}^* - \mathbf{G}_x\mathbf{X}_s'\mathbf{Z}_s\boldsymbol{\gamma}^o \end{aligned} \tag{7.9.3}$$

where $\boldsymbol{\beta}^* = \mathbf{G}_x\mathbf{X}_s'\mathbf{Y}_s$ is a solution for the model having no $\mathbf{Z}$ covariates. Substituting $\boldsymbol{\beta}^o$ into the second equation in (7.9.2) and doing some rearranging gives

$$\mathbf{Z}_s'\mathbf{P}_s\mathbf{Z}_s\boldsymbol{\gamma}^o = \mathbf{Z}_s'\mathbf{P}_s\mathbf{Y}_s,$$

where $\mathbf{P}_s = \mathbf{I}_n - \mathbf{X}_s\mathbf{G}_x\mathbf{X}_s'$. Thus a solution for $\boldsymbol{\gamma}^o$ is

$$\boldsymbol{\gamma}^o = \mathbf{G}_1\mathbf{Z}_s'\mathbf{P}_s\mathbf{Y}_s, \tag{7.9.4}$$

where $\mathbf{G}_1$ is a g-inverse of $\mathbf{Z}_s'\mathbf{P}_s\mathbf{Z}_s$. Solutions (7.9.3) and (7.9.4) are exactly the same as would be obtained if we used the formula for the generalized inverse of a partitioned matrix given in Lemma A.10.3 in the Appendix (see Exercise 7.7).

The solution γ^o is unique. We will only sketch why this is the case here. Notice that although $\mathbf{G}_x$ is not unique, it only enters into the equation for γ^o in the form $\mathbf{X}_s\mathbf{G}_x\mathbf{X}_s'$, which is invariant to whatever g-inverse is used for $\mathbf{X}_s'\mathbf{X}_s$. Thus, the fact that $\mathbf{X}_s'\mathbf{X}_s$ does not have full rank does not lead to multiple solutions for γ^o. The fact that $\mathbf{P}_s$ is symmetric and idempotent (i.e., $\mathbf{P}_s^2 = \mathbf{P}_s$) and that $(\mathbf{Z}_s'\mathbf{Z}_s)^{-1}$ exists can be used to show that $\mathbf{Z}_s'\mathbf{P}_s\mathbf{Z}_s$ is nonsingular (Searle, 1971, sec. 8.2). As a result, the unique solution for the estimator of γ is

$$\hat{\gamma} = (\mathbf{Z}_s'\mathbf{P}_s\mathbf{Z}_s)^{-1}\mathbf{Z}_s'\mathbf{P}_s\mathbf{Y}_s. \tag{7.9.5}$$

The estimator of the total is thus

$$\hat{T} = \mathbf{1}_s'\mathbf{Y}_s + \mathbf{1}_s'\mathbf{X}_r\boldsymbol{\beta}^o + \mathbf{1}_r'\mathbf{Z}_r\hat{\gamma}. \tag{7.9.6}$$

The error variance requires some computation and is given in the lemma below.

Lemma 7.9.1. The error variance of $\hat{T} = \mathbf{1}_r'\mathbf{Y}_s + \mathbf{1}_r'\mathbf{X}_r\boldsymbol{\beta}^o + \mathbf{1}_r'\mathbf{Z}_r\hat{\gamma}$ under model (7.9.1) is

$$\text{var}_M(\hat{T} - T) = \mathbf{1}_r'\mathbf{V}_{rr}\mathbf{1}_r + \mathbf{1}_r'[\mathbf{X}_r \;\; \mathbf{Z}_r]\,\text{var}_M\begin{pmatrix}\boldsymbol{\beta}^o\\ \hat{\gamma}\end{pmatrix}\begin{bmatrix}\mathbf{X}_r'\\ \mathbf{Z}_r'\end{bmatrix}\mathbf{1}_r,$$

where

$$\text{var}_M\begin{pmatrix}\boldsymbol{\beta}^o\\ \hat{\gamma}\end{pmatrix} = \begin{bmatrix}\mathbf{G}_x + \mathbf{G}_x\mathbf{X}_s'\mathbf{Z}_s\mathbf{A}_z^{-1}\mathbf{Z}_s'\mathbf{X}_s\mathbf{G}_x' & -\mathbf{G}_x\mathbf{X}_s'\mathbf{Z}_s\mathbf{A}_z^{-1}\\ -\mathbf{A}_z^{-1}\mathbf{Z}_s'\mathbf{X}_s\mathbf{G}_x' & \mathbf{A}_z^{-1}\end{bmatrix}\sigma^2,$$

where $\mathbf{A}_z = \mathbf{Z}_s'\mathbf{P}_s\mathbf{Z}_s$.

Proof: Using the solution vectors in (7.9.3) and (7.9.5), we have

$$\text{var}_M(\boldsymbol{\beta}^o) = \mathbf{G}_x\mathbf{X}_s'\mathbf{Q}\mathbf{Q}'\mathbf{X}_s\mathbf{G}_x\sigma^2,$$

where $\mathbf{Q} = \mathbf{I}_n - \mathbf{Z}_s\mathbf{A}_z^{-1}\mathbf{Z}_s'\mathbf{P}_s$. By Lemma A.10.2(ii), we have $\mathbf{X}_s'\mathbf{P}_s = \mathbf{0}$. From this fact and (7.3.5), we get

$$\mathbf{G}_x\mathbf{X}_s'\mathbf{Q}\mathbf{Q}'\mathbf{X}_s\mathbf{G}_x = \mathbf{G}_x + \mathbf{G}_x\mathbf{X}_s'\mathbf{Z}_s\mathbf{A}_z^{-1}\mathbf{Z}_s'\mathbf{X}_s\mathbf{G}_x'$$

and thus the form required for $\text{var}_M(\boldsymbol{\beta}^o)$. The covariance of the two parameter solutions is

$$\text{cov}_M(\boldsymbol{\beta}^o, \hat{\gamma}) = \mathbf{G}_x\mathbf{X}_s'\mathbf{Q}\mathbf{P}_s\mathbf{Z}_s\mathbf{A}_z^{-1}\sigma^2.$$

Expanding $\mathbf{QP}_s$ and using (7.3.5) yields $\text{cov}_M(\boldsymbol{\beta}^o, \hat{\gamma}) = -\mathbf{G}_x\mathbf{X}_s'\mathbf{Z}_s\mathbf{A}_z^{-1}$. Finally, $\text{var}_M(\hat{\gamma}) = \mathbf{A}_z^{-1}\sigma^2$, which follows directly from the formula for $\hat{\gamma}$. Combining results gives the desired error variance. □

7.9.2. One-Way Classification with a Single Covariate

To illustrate the general results in the preceding section, we consider a model with a single qualitative factor with several levels, together with one quantitative variable. Suppose that there are $i = 1, \ldots, I$ levels of the factor and that the quantitative variable for unit j in level i is denoted z_{ij}. Let there be $j = 1, \ldots, n_i$ sample units having level i of the factor and denote the set of these units by s_i. Assume that the model is

$$Y_{ij} = \mu + \beta_i + \gamma z_{ij} + \varepsilon_{ij}, \varepsilon_{ij} \sim (0, \sigma^2), \tag{7.9.8}$$

where, as before, the errors are assumed to be independent. Note that we could write this model as $Y_{ij} = \tilde{\beta}_i + \gamma z_{ij} + \varepsilon_{ij}$ with $\tilde{\beta}_i = \mu + \beta_i$ and have a full rank model that would avoid the need for a generalized inverse. Formulating the model as in (7.9.8) allows the general formulas in Section 7.9.1 to be illustrated fairly simply.

The $\mathbf{X}$ matrix has the same form as in the one-way classification of Section 7.5, and the g-inverse $\mathbf{G}_x$ is given by (7.5.3). The solution for the one-way model is $\boldsymbol{\beta}^* = \mathbf{G}_x\mathbf{X}_s'\mathbf{Y}_s = [0, \bar{Y}_{s1}, \ldots, \bar{Y}_{sI}]'$. Using the definitions of $\mathbf{X}_s$ and $\mathbf{G}_x$, the matrix $\mathbf{P}_s$ is

$$\mathbf{P}_s = \mathbf{I} - \text{blkdiag}(n_i^{-1}\mathbf{1}_{n_i}\mathbf{1}_{n_i}'),$$

where $\mathbf{1}_{n_i}$ is an n_i-vector of all 1's. The sample vector of quantitative auxiliaries is $\mathbf{Z}_s = [z_{11}, \ldots, z_{1n_1}, \ldots, z_{I1}, \ldots, z_{In_I}]'$. Making the appropriate substitutions, the estimator of γ is then

$$\hat{\gamma} = \frac{\sum_{i=1}^{I} \sum_{j \in s_i} (z_{ij} - \bar{z}_{si}) Y_{ij}}{\sum_{i=1}^{I} \sum_{j \in s_i} (z_{ij} - \bar{z}_{si})^2}. \tag{7.9.9}$$

which is a slope estimator pooled across the levels of the qualitative factor. We also have $\mathbf{G}_x\mathbf{X}_s'\mathbf{Z}_s = [0, \bar{z}_{s1}, \ldots, \bar{z}_{sI}]'$, where $\bar{z}_{si} = \sum_{j \in s_i} z_{ij}/n_i$. Substituting the preceding expressions in (7.9.3), a solution for $\boldsymbol{\beta}$ is

$$\boldsymbol{\beta}^o = \boldsymbol{\beta}^* - \mathbf{G}_x \mathbf{X}_s' \mathbf{Z}_s \gamma^o = \begin{bmatrix} 0 \\ \bar{Y}_{s1} - \hat{\gamma}\bar{z}_{s1} \\ \vdots \\ \bar{Y}_{sI} - \hat{\gamma}\bar{z}_{sI} \end{bmatrix}. \tag{7.9.10}$$

After some rearrangement, it follows that the predictor of the total is

$$\hat{T} = \mathbf{1}_s'\mathbf{Y}_s + \mathbf{1}_r'\mathbf{X}_r\boldsymbol{\beta}^o + \mathbf{1}_r'\mathbf{Z}_r\hat{\gamma} = \sum_{i=1}^{I} N_i[\bar{Y}_{si} + \hat{\gamma}(\bar{z}_i - \bar{z}_{si})].$$

This is a stratified regression estimator that uses a pooled slope estimator rather than a separate slope for each stratum. The predictor can also be written as

$$\hat{T} = \sum_{i=1}^{I} \sum_{j \in s_i} N_i \left[\frac{1}{n_i} + A_z^{-1}(\bar{z}_i - \bar{z}_{si})(z_{ij} - \bar{z}_{si}) \right] Y_{ij}, \tag{7.9.11}$$

where $A_z = \Sigma_{i=1}^{I} \Sigma_{j \in s_i} (z_{ij} - \bar{z}_{si})^2$. Expression (7.9.11) is a useful form for variance calculation.

To compute the error variance, we could use the general formula in Lemma 7.9.1, but in this case it is simpler to work directly from (7.9.11) to get an analytic form. Note that the estimation error is

$$\hat{T} - T = \sum_{i=1}^{I} \sum_{j \in s_i} (N_i d_{ij} - 1) Y_{ij} - \sum_{i=1}^{I} \sum_{j \in r_i} Y_{ij}$$

with $d_{ij} = n_i^{-1} + A_z^{-1}(\bar{z}_i - \bar{z}_{si})(z_{ij} - \bar{z}_{si})$ and r_i the nonsample units that have level i of the factor. The error variance is then $\text{var}_M(\hat{T} - T) = \Sigma_{i=1}^{I} \Sigma_{j \in s_i} (N_i d_{ij} - 1)^2 \sigma^2 + \Sigma_{i=1}^{I} (N_i - n_i)\sigma^2$. Using the facts that $\Sigma_{j \in s_i} d_{ij}^2 = n_i^{-1} + A_z^{-1}(\bar{z}_i - \bar{z}_{si})^2$ and $\Sigma_{j \in s_i} d_{ij} = 1$, we have

$$\text{var}_M(\hat{T} - T) = \sigma^2 \sum_{i=1}^{I} \frac{N_i^2}{n_i}(1 - f_i) \left[1 + \frac{n_i}{(1 - f_i)A_z}(\bar{z}_i - \bar{z}_{si})^2 \right], \tag{7.9.12}$$

which is quite similar to the result for the simple linear regression estimator in Example 2.3.2 in Chapter 2. Note that the optimal sample is one that is balanced on the first moment of z for each level of the factor, that is, $\bar{z}_{si} = \bar{z}_i$ for $i = 1, \ldots, I$.

7.9.3. Examples

The Labor Force Population will be used to illustrate combining qualitative and quantitative variables in estimation. Using the systematic sample of $n = 48$

persons from the Labor Force population in Example 7.8.1, consider three models that use the following variables: (1) a person's sex and the number of hours the person works per week, (2) age, sex, and hours worked, and (3) age, sex, age-by-sex interaction, and hours worked. In each case, assume that model errors are uncorrelated with a common variance.

Based on the previous development, we have two choices for calculating estimators of the total. One is to construct the full matrix of qualitative and quantitative auxiliaries, as in Section 7.4 and to compute $\hat{T}$ based on a g-inverse of the full auxiliary matrix, as in Corollary 7.4.1. The second is to partition the matrix of auxiliaries as in Section 7.9.1 and compute the estimator as in expression (7.9.6). With either computational approach, the same estimate of T is obtained, as we will illustrate. As in Example 7.8.1, let α denote the 2-level factor for sex and β the 4-level factor for age. Denote the interaction between level i of sex and level j of age as $(\alpha\beta)_{ij}$. The number of hours worked by a person will be denoted by z, and the coefficient for hours worked per week, the quantitative variable, will be γ.

For the case using sex and hours worked to predict weekly wages, the model is

$$Y_{ij} = \mu + \alpha_i + \gamma z_{ij} + \varepsilon_{ij},$$

where z_{ij} is the hours worked for the jth person in the ith level of sex. If we let the sample matrix of auxiliaries be $\mathbf{X}_s^* = [\mathbf{X}_s \;\; \mathbf{Z}_s]$, then

$$\mathbf{X}_s^{*\prime}\mathbf{X}_s^* = [\mathbf{X}_s' \;\; \mathbf{Z}_s'] \begin{bmatrix} \mathbf{X}_s \\ \mathbf{Z}_s \end{bmatrix} = \begin{bmatrix} n & n_1 & n_2 & n\bar{z}_s \\ n_1 & n_1 & 0 & n_1\bar{z}_{s1} \\ n_2 & 0 & n_2 & n_2\bar{z}_{s2} \\ n\bar{z}_s & n_1\bar{z}_{s1} & n_2\bar{z}_{s2} & \Sigma_{i,s_i} z_{ij}^2 \end{bmatrix}$$

with $\bar{z}_s = \Sigma_{i,s_i} z_{ij}/n$. We can either estimate the total using a g-inverse of the full matrix $\mathbf{X}_s^{*\prime}\mathbf{X}_s^*$, as in Theorem 7.4.2 and its corollary, or we can use a g-inverse of $\mathbf{X}_s'\mathbf{X}_s$ and solve separately for $\boldsymbol{\beta}^o = [\mu^o \;\; \alpha_1^o \;\; \alpha_2^o]'$ and $\hat{\gamma}$. Using the data in Example 7.8.1, $n_1 = n_2 = 24$, $\bar{z}_{s1} = 38.29$, $\bar{z}_{s2} = 35.38$, and $\Sigma_{i,s_i} z_{ij}^2 = 70{,}296$. Using the function *ginv* in Appendix C, a g-inverse of $\mathbf{X}_s^{*\prime}\mathbf{X}_s^*$ is

$$\mathbf{G}_x^* = \begin{bmatrix} 0.1281 & 0.0711 & 0.0570 & -0.0048 \\ 0.0711 & 0.0603 & 0.0108 & -0.0027 \\ 0.0570 & 0.0108 & 0.0462 & -0.0021 \\ -0.0048 & -0.0027 & -0.0021 & 0.0002 \end{bmatrix}.$$

The parameter solutions derived from this g-inverse are listed in Table 7.6 as Solution 1. Alternate parameter solutions are given in expressions (7.9.9) and (7.9.10) and are listed under Solution 2. The estimate $\hat{\gamma}$, being unique, is the same for both solutions. Note that for either solution $\mu^o + \alpha_i^o$ ($i = 1$ or 2) is the

Table 7.6. Solutions to the Normal Equations for the Model with Weekly Wages Predicted from Sex and Hours Worked

Parameter	Solution 1	Solution 2
μ^o	−34.84	0
α_1^o	29.29	−5.55
α_2^o	−64.12	−98.96
$\hat{\gamma}$	9.13	9.13

same. For example, $\mu^o + \alpha_1^o = -34.84 + 29.99 = -5.55$ for solution 1 and equals $0 - 5.55 = -5.55$ for solution 2. For this sample, $\Sigma_{i,j\in s_i} y_{ij} = 13{,}635$ and $\mathbf{T}_{xr} = (430,\ 208,\ 222,\ 16526)$. The estimated total for both solutions is $\hat{T} = 141{,}407$ (see Exercise 7.9).

The second model considered here predicts wages based on age, sex, and hours worked, that is, for person k in level i of sex and level j of age the model is

$$Y_{ijk} = \mu + \alpha_i + \beta_j + \gamma z_{ijk} + \varepsilon_{ij}.$$

The third model adds the age-by-sex interaction:

$$Y_{ijk} = \mu + \alpha_i + \beta_j + (\alpha\beta)_{ij} + \gamma z_{ijk} + \varepsilon_{ij}.$$

The solution vectors for these two models are listed in Table 7.7. Both solutions were obtained by applying the *ginv* function to evaluate the estimator of the total in (7.4.4). Exactly the same solution vectors would be obtained if the age, sex, and age-by-sex interaction parameter solutions are computed from (7.9.3), using *ginv* to get $\mathbf{G}_s$ (see Exercise 7.10).

For model 2 the nonsample total for the full auxiliary vector is $\mathbf{T}_{sr} =$ (430, 208, 222, 43, 43, 118, 226, 16,526). For model 3, we have $\mathbf{T}_{sr} =$ (430, 208, 222, 43, 43, 118, 226, 13, 23, 66, 106, 30, 20, 52, 120, 16,526). The estimated totals using models 2 and 3 are then, respectively, $\hat{T} = 147{,}441$ and 147,507.

7.10. VARIANCE ESTIMATION

Variance estimators from Chapter 5 can also be used when some of the explanatory variables in a model are qualitative. We will consider estimators of the total under the working model $M(\mathbf{X}:\mathbf{I})$:

$$E_M(\mathbf{Y}) = \mathbf{X}\boldsymbol{\beta} \qquad \text{var}_M(\mathbf{Y}) = \sigma^2\mathbf{I}. \tag{7.10.1}$$

The predictor of the total is

$$\hat{T} = \mathbf{1}_s'\mathbf{Y}_s + \hat{T}_r,$$

Table 7.7. Solutions to the Normal Equations for Models that Include Sex, Age, and Hours Worked as Auxiliaries for Predicting the Total of Weekly Wages

Parameter Solution	Model 2	Model 3
μ^o	−28.32	−23.44
α_1^o	31.19	17.49
α_2^o	−59.51	−40.94
β_1^o	−50.09	−34.55
β_2^o	−77.55	−54.19
β_3^o	35.31	22.68
β_4^o	64.01	42.61
$(\alpha\beta)_{11}^o$		−35.58
$(\alpha\beta)_{12}^o$		−50.00
$(\alpha\beta)_{13}^o$		51.18
$(\alpha\beta)_{14}^o$		51.90
$(\alpha\beta)_{21}^o$		1.03
$(\alpha\beta)_{22}^o$		−4.18
$(\alpha\beta)_{23}^o$		−28.50
$(\alpha\beta)_{24}^o$		−9.29
$\hat{\gamma}$	8.45	8.25

where $\hat{T}_r = \mathbf{1}_r'\mathbf{X}_r\boldsymbol{\beta}^o$ and, as in earlier sections, $\boldsymbol{\beta}^o = \mathbf{G}\mathbf{X}_s'\mathbf{Y}_s$ with $\mathbf{G}$ being a g-inverse of $\mathbf{X}_s'\mathbf{X}_s$. The matrix of auxiliaries can contain both qualitative and quantitative variables. We seek variance estimators that are unbiased under this working model, but are robust in the sense of being consistent when a more general variance structure, $\text{var}_M(\mathbf{Y}) = \text{diag}(\psi_i)$, holds. As in Chapter 5, we can construct a sandwich variance estimator based on squared residuals and modify it using the diagonal elements of the hat matrix to create alternative variance estimators.

The estimation error for the predictor of a total is $\hat{T} - T = \mathbf{a}'\mathbf{Y}_s - \mathbf{1}_r'\mathbf{Y}_r$ where $\mathbf{a} = \mathbf{X}_s\mathbf{G}\mathbf{X}_r'\mathbf{1}_r$. The error variance under model (7.10.1) is

$$\text{var}_M(\hat{T} - T) = \sigma^2\mathbf{a}'\mathbf{a} + \sigma^2(N - n). \tag{7.10.2}$$

Under the more general model $\tilde{M} = M(\mathbf{X}:\boldsymbol{\Psi})$ with variance specification, $\text{var}_{\tilde{M}}(\mathbf{Y}) = \text{diag}(\psi_i)$, the error variance is

$$\text{var}_{\tilde{M}}(\hat{T} - T) = \sum_s a_i^2\psi_i + \sum_r \psi_i. \tag{7.10.3}$$

Under the same conditions as in Section 5.5, for example, $a_i = O(N_r/n)$, ψ_i bounded, and $n/N \rightarrow 0$, the orders of the terms in (7.10.3) are N_r^2/n and N_r with

$N_r = N - n$. Thus the first term dominates the error variance and is the more important one to estimate.

7.10.1. Basic Robust Alternatives

Even though $\mathbf{X}$ may not have full column rank, the robust variance estimators introduced in Sections 5.3 and 5.4 can still be constructed from squared residuals and the diagonal elements of the hat matrix, as we describe in this section. The residual associated with the working model is $r_i = Y_i - \mathbf{x}_i'\boldsymbol{\beta}^o$ where $\mathbf{x}_i' = (x_{i1}, \ldots, x_{ip})$ is the ith row of $\mathbf{X}_s$, $\boldsymbol{\beta}^o = \mathbf{G}\mathbf{X}_s'\mathbf{Y}_s$, and $\mathbf{G}$ is a g-inverse of $\mathbf{X}_s'\mathbf{X}_s$. Since $\mathbf{X}_s\mathbf{G}\mathbf{X}_s'\mathbf{X}_s = \mathbf{X}_s$ by Lemma A.10.2(ii), we have $E_M(r_i) = 0$. We can also write the vector of residuals as

$$\mathbf{r} = \mathbf{Y}_s - \mathbf{H}\mathbf{Y}_s,$$

where $\mathbf{H} = \mathbf{X}_s\mathbf{G}\mathbf{X}_s'$. Because $\mathbf{H}$ is invariant to the choice of $\mathbf{G}$ by Lemma A.10.2(iii), the vector of residuals is unique. The variance estimator

$$v_R(\hat{T} - T) = \sum_s a_i^2 r_i^2 + \frac{N-n}{n}\sum_s r_i^2,$$

is approximately unbiased under either $M(\mathbf{X}:\mathbf{I})$ or $M(\mathbf{X}:\boldsymbol{\Psi})$. Under conditions like those in Theorem 5.3.1, $v_R(\hat{T} - T)$ is also consistent under $M(\mathbf{X}:\boldsymbol{\Psi})$. We sketch the arguments here, which are similar to those in Chapter 5.

As in Chapter 5, $r_i = Y_i(1 - h_{ii}) + \Sigma_{j \neq i} h_{ij} Y_j$ and under the working model $M(\mathbf{X}:\mathbf{I})$, the expectation of the squared residual is

$$E_M(r_i^2) = \sigma^2\left[1 - 2h_{ii} + \sum_{j \in s} h_{ij}^2\right]. \tag{7.10.4}$$

Next, again using $\mathbf{X}_s\mathbf{G}\mathbf{X}_s'\mathbf{X}_s = \mathbf{X}_s$, we have $\mathbf{x}_i'\mathbf{G}\mathbf{X}_s'\mathbf{X}_s = \mathbf{x}_i'$ and

$$\begin{aligned}\sum_{j \in s} h_{ij}^2 &= \mathbf{x}_i'\mathbf{G}\mathbf{X}_s'\mathbf{X}_s\mathbf{G}\mathbf{x}_i \\ &= \mathbf{x}_i'\mathbf{G}\mathbf{x}_i \\ &= h_{ii},\end{aligned}$$

which is a special case of Lemma 5.3.1(h). Substituting this result in (7.10.4), we have $E_M(r_i^2) = \sigma^2(1 - h_{ii})$ under the working model.

If the elements of $\mathbf{G}$ are $O(n^{-1})$, analogous to the assumption that the elements of $(\mathbf{X}_s'\mathbf{X}_s)^{-1}$ are $O(n^{-1})$ in the full rank case, then $h_{ij} = O(n^{-1})$. In that circumstance, $E_M(r_i^2) \cong \sigma^2$ in large samples. Thus $\Sigma_s a_i^2 r_i^2$ is approximately unbiased for the first term in (7.10.2) and $(N - n)\,\Sigma_s r_i^2/n$ is for the second. The sandwich estimator $v_R(\hat{T} - T)$ is then approximately unbiased for the error variance under $M(\mathbf{X}:\mathbf{I})$.

When the model is $\tilde{M}$, the expectation of the squared residual is

$$E_{\tilde{M}}(r_i^2) = \psi_i(1 - h_{ii})^2 + \sum_{j \neq i} h_{ij}^2 \psi_j. \tag{7.10.5}$$

When $h_{ij} = O(n^{-1})$, $E_{\tilde{M}}(r_i^2) \cong \psi_i$ in large samples and $\Sigma_s a_i^2 r_i^2$ is also approximately unbiased for the first term in (7.10.3). Thus $\Sigma_s a_i^2 r_i^2$, with residuals constructed under the working model $M(\mathbf{X}:\mathbf{I})$, is a robust estimator of the dominant term of the error variance. Since $\Sigma_r \psi_i$ is asymptotically negligible compared to the first term in (7.10.3), $v_R(\hat{T} - T)$ is asymptotically unbiased.

Alternative robust variance estimators can be constructed using the leverage-adjustment methods in Section 5.5. The general form of the estimators, when $M(\mathbf{X}:\mathbf{I})$ is the working model, is

$$\hat{\text{var}}_M(\hat{T} - T) = \sum_s a_i^2 \hat{\psi}_i + \frac{N - n}{n} \sum_s \hat{\psi}_i, \tag{7.10.6}$$

that is, an estimator of the variance of the estimated nonsample total plus an estimator of the variance of the nonsample total itself. Our choices for $\hat{\psi}_i$ are

$$\hat{\psi}_i = r_i^2,$$
$$\hat{\psi}_i = r_i^2(1 - h_{ii})^{-1},$$
$$\hat{\psi}_i = r_i^2 \frac{\Sigma_s a_i^2}{\Sigma_s a_i^2(1 - h_{ii})},$$

and

$$\hat{\psi}_i = r_i^2(1 - h_{ii})^{-2},$$

corresponding to for v_R, v_D, v_H, and v_{J*}, respectively. Since $h_{ii} = O(n^{-1})$, each of v_D, v_H, and v_{J*} is asymptotically equivalent to v_R. These estimators can also be derived when $M(\mathbf{X}:\mathbf{V})$ is the working model rather than $M(\mathbf{X}:\mathbf{I})$; see Exercise 7.12.

7.10.2. Jackknife Variance Estimator

The jackknife variance estimator derived in Section 5.4.2 also applies here, but we need to verify that some of the results from that section can be extended to the non-full rank case. In the discussion below $\mathbf{A}^-$ denotes a g-inverse of a matrix $\mathbf{A}$. The predictor of the nonsample total after deleting the ith sample unit is

$$\hat{T}_{r(i)} = \mathbf{1}_r'\mathbf{X}_r\boldsymbol{\beta}_{(i)}^o,$$

where $\boldsymbol{\beta}_{(i)}^o = (\mathbf{X}_{s(i)}'\mathbf{X}_{s(i)})^-\mathbf{X}_{s(i)}'\mathbf{Y}_{s(i)}$ is a solution to the estimating equations for $\boldsymbol{\beta}$

after omitting the ith sample unit. A jackknife estimator of the variance of $\hat{T}_r = \mathbf{1}_r'\mathbf{X}_r\boldsymbol{\beta}^o$ is

$$v_J(\hat{T}_r) = \frac{n-1}{n}\sum_{i\in s}(\hat{T}_{r(i)} - \hat{T}_{r(*)})^2, \tag{7.10.7}$$

where $\hat{T}_{r(*)} = n^{-1}\Sigma_s \hat{T}_{r(i)}$. The estimator of the error variance of $\hat{T}$ that we will consider is

$$v_J(\hat{T} - T) = v_J(\hat{T}_r) + \frac{N-n}{n}\sum_s \hat{\psi}_i, \tag{7.10.8}$$

where $\hat{\psi}_i = r_i^2(1 - h_{ii})^{-2}$ is the choice that accompanies v_{J*}.

As in expression (5.4.5), we can also write $v_J(\hat{T}_r)$ in terms of residuals, leverages, and estimation coefficients. To rewrite (7.10.7), we need an expression for the g-inverse of

$$\begin{aligned}\mathbf{A}_{(i)} &= \mathbf{X}_{s(i)}'\mathbf{X}_{s(i)}\\ &= \mathbf{X}_s'\mathbf{X}_s - \mathbf{x}_i\mathbf{x}_i'.\end{aligned}$$

Lemma 7.10.1 below is an analog of Lemma 5.4.1, adapted to the particular purpose at hand.

Lemma 7.10.1. Let $\mathbf{A} = \mathbf{X}_s'\mathbf{X}_s$ and $\mathbf{G}$ be a g-inverse of $\mathbf{A}$. A generalized inverse of $\mathbf{A}_{(i)} = \mathbf{A} - \mathbf{x}_i\mathbf{x}_i'$ is

$$\mathbf{A}_{(i)}^- = \mathbf{G} + \frac{\mathbf{G}\mathbf{x}_i\mathbf{x}_i'\mathbf{G}}{1 - \mathbf{x}_i'\mathbf{G}\mathbf{x}_i}.$$

Proof: We will verify that $\mathbf{A}_{(i)}\mathbf{A}_{(i)}^-\mathbf{A}_{(i)} = \mathbf{A}_{(i)}$. Multiplying out the left-hand side, and using $\mathbf{AGA} = \mathbf{A}$ and $h_{ii} = \mathbf{x}_i'\mathbf{G}\mathbf{x}_i$ gives

$$\begin{aligned}\mathbf{A}_{(i)}\mathbf{A}_{(i)}^-\mathbf{A}_{(i)} = \mathbf{A} - \mathbf{x}_i\mathbf{x}_i'\mathbf{GA} &+ \frac{\mathbf{AG}\mathbf{x}_i\mathbf{x}_i'\mathbf{GA}}{1-h_{ii}} - \frac{h_{ii}\mathbf{x}_i\mathbf{x}_i'\mathbf{GA}}{1-h_{ii}}\\ &- \mathbf{AG}\mathbf{x}_i\mathbf{x}_i' + h_{ii}\mathbf{x}_i\mathbf{x}_i' - \frac{h_{ii}\mathbf{AG}\mathbf{x}_i\mathbf{x}_i'}{1-h_{ii}} + \frac{h_{ii}^2\mathbf{x}_i\mathbf{x}_i'}{1-h_{ii}}.\end{aligned} \tag{7.10.9}$$

From Lemma A.10.2, $\mathbf{X}_s\mathbf{G}\mathbf{X}_s'\mathbf{X}_s = \mathbf{X}_s$, which implies that $\mathbf{x}_i'\mathbf{GA} = \mathbf{x}_i'$. Using this fact, the last seven terms on the right-hand side of (7.10.9) simplify to $\mathbf{x}_i\mathbf{x}_i'$, giving the desired result. □

Using this lemma, the solution vector, when sample unit i is omitted, is then

$$\begin{aligned}\boldsymbol{\beta}^o_{(i)} &= \mathbf{A}^-_{(i)}\mathbf{X}'_{s(i)}\mathbf{Y}_{s(i)} \\ &= \left(\mathbf{G} + \frac{\mathbf{G}\mathbf{x}_i\mathbf{x}'_i\mathbf{G}}{1-h_{ii}}\right)(\mathbf{X}'_s\mathbf{Y}_s - \mathbf{x}_i Y_i) \\ &= \boldsymbol{\beta}^o - \frac{\mathbf{G}\mathbf{x}_i r_i}{1-h_{ii}}.\end{aligned}$$

Consequently, $\hat{T}_{r(i)} - \hat{T}_{r(*)} = -[a_i r_i(1-h_{ii})^{-1} - n^{-1}\Sigma_s a_i r_i(1-h_{ii})^{-1}]$, where a_i is the ith element of $\mathbf{1}'_r\mathbf{X}_r\mathbf{G}\mathbf{X}'_s$. Thus the jackknife estimator of the variance of $\hat{T}_r$ can be written as

$$v_J(\hat{T}_r) = \frac{n-1}{n}\left\{\sum_s \frac{a_i^2 r_i^2}{(1-h_{ii})^2} - n^{-1}\left[\sum_s \frac{a_i r_i}{(1-h_{ii})}\right]^2\right\}, \tag{7.10.10}$$

which is the same as expression (5.4.6) in Chapter 5. The jackknife estimator of the error variance of $\hat{T}$ in (7.10.8) can be evaluated using the computationally convenient form in (7.10.10). Consistency of $v_J(\hat{T} - T)$ follows from Theorems 5.3.1 and 5.4.1.

Example 7.10.1. Using the same sample of 48 persons as in Example 7.8.1 and the examples in Section 7.9.3, we evaluated the five alternative variance estimators for the three models described in Section 7.9.3. The results are shown in Table 7.8. As in the Chapter 5 examples, we have $\sqrt{v_R}$ less than $\sqrt{v_D}$, $\sqrt{v_H}$, $\sqrt{v_{J*}}$, and $\sqrt{v_J}$ because of the adjustments using the leverages included in each of the last four alternatives. The largest of the standard error estimates is $\sqrt{v_{J*}}$ since it involves $(1-h_{ii})^{-2}$. For this small sample $\sqrt{v_{J*}}$ is about 13% larger than the sandwich estimate $\sqrt{v_R}$ for Model 3, which is the largest difference in Table 7.8.

EXERCISES

7.1 Show that the formula for the error variance in Theorem 7.4.2 holds.

7.2 In the one-way classification model (7.5.1), impose the restriction on the parameters that $\Sigma_{i=1}^I \beta_i = 0$. Find the resulting solution of (7.5.2) for $\boldsymbol{\beta}^o$, and show that the corresponding estimator of T is still $\hat{T} = \Sigma_{i=1}^I N_i \bar{Y}_{si}$, the stratified expansion estimator.

Note: In Exercises 7.3–7.5 it will be efficient for the reader to write a single computer program that, with some modification, will serve for each exercise.

7.3 Suppose that the following units are in a sample of $n = 30$ from the Labor Force population in the Appendix: 4, 20, 36, 52, and so on

Table 7.8. Estimates of Standard Error from Three Models That Use Sex, Age, and Hours Worked as Auxiliaries for Predicting the Total of Weekly Wages

Standard Error Estimator	Model 1 Sex, Hours Worked	Model 2 Sex, Hours Worked	Model 3 Sex, Age-by-Sex Hours Worked
$\hat{T}$	141,407	147,441	147,507
$\sqrt{v_R}$	10,310	10,346	10,087
$\sqrt{v_D}$	10,569	10,789	10,704
$\sqrt{v_H}$	10,615	11,033	11,187
$\sqrt{v_{J*}}$	10,836	11,259	11,387
$\sqrt{v_J}$	10,722	11,151	11,278

through unit 468. Use sex and the age categories 19 and under, 20–24, 25–34, and 35+, as the auxiliaries. (a) Tabulate the number of sample units by age category and sex. (b) Fit model (7.7.1). The function *ginv* can be used to calculate a generalized inverse to get a solution to the normal equations. Report your g-inverse, solution vector $\boldsymbol{\beta}^o$, and estimate of the total $\hat{T}$. (c) Repeat part (b) for (7.8.1), the model with interaction. (d) For parts (b) and (c) calculate the individual sample unit weights described in Section 7.4. To see any differences between the weight vectors, plot one versus the other. You may have to jitter the points to eliminate overplotting.

7.4 Suppose that the following units are in a sample of $n = 30$ from the Labor Force population in the appendix: 7, 23, 39, 55, and so on through unit 471. Use sex and the age categories 19 and under, 20–24, 25–34, and 35+, as the auxiliaries. (a) Tabulate the number of sample units by age category and sex. Notice that the (19 and under)/male age × sex cell has no sample. To complete parts (b) and (c) collapse the 19 and under and the 20–24 ages together. (b) Fit model (7.7.1). The function *ginv* can be used to calculate a generalized inverse to get a solution to the normal equations. Report your g-inverse, solution vector $\boldsymbol{\beta}^o$, and estimate of the total $\hat{T}$. (c) Repeat part (a) for (7.8.1), the model with interaction. (d) For parts (b) and (c) calculate the individual sample unit weights described in Section 7.4. (e) Comment on the age group collapsing procedure. What should be considered when collapsing categories? If you collapse categories together, will $\hat{T}$ still be model-unbiased?

7.5 Quantitative auxiliaries can be combined with qualitative auxiliaries in estimating a total as demonstrated in this exercise. Suppose that the same units as in Exercise 7.3 are in a sample of $n = 30$ from the Labor Force population in the appendix: 4, 20, 36, 52, and so on through unit 468. Use as auxiliaries the quantitative variable, Hours Worked Per Week, and the same qualitative variables as in Exercise 7.3: sex and the age categories 19 and under, 20–24, 25–34, and 35+. (a) Fit model (7.7.1). The function *ginv* can be used to calculate a generalized inverse to get a solution to the normal equations. Report your g-inverse, solution vector $\boldsymbol{\beta}^o$, and estimate of the total $\hat{T}$. (b) Repeat part (b) for (7.8.1), the model with interaction. (c) For parts (a) and (b) calculate the individual sample unit weights described in Section 7.4.

7.6 Prove Lemma A.10.3, that is, a generalized inverse of

$$\mathbf{M} = \begin{bmatrix} \mathbf{X}' \\ \mathbf{Z}' \end{bmatrix} [\mathbf{X} \quad \mathbf{Z}] \equiv \begin{bmatrix} \mathbf{A} & \mathbf{B} \\ \mathbf{B}' & \mathbf{D} \end{bmatrix}$$

is

$$\mathbf{M}^- = \begin{bmatrix} \mathbf{A}^- + \mathbf{A}^-\mathbf{B}\mathbf{Q}^-\mathbf{B}'\mathbf{A}^- & -\mathbf{A}^-\mathbf{B}\mathbf{Q}^- \\ -\mathbf{Q}^-\mathbf{B}'\mathbf{A}^- & \mathbf{Q}^- \end{bmatrix}$$

$$= \begin{bmatrix} \mathbf{A}^- & \mathbf{0} \\ \mathbf{0} & \mathbf{0} \end{bmatrix} + \begin{bmatrix} -\mathbf{A}^-\mathbf{B} \\ \mathbf{I} \end{bmatrix} \mathbf{Q}^-[-\mathbf{B}'\mathbf{A}^- \quad \mathbf{I}],$$

where $\mathbf{A}^-$ is a g-inverse of $\mathbf{A}$, $\mathbf{Q} = \mathbf{D} - \mathbf{B}'\mathbf{A}^-\mathbf{B}$, and $\mathbf{Q}^-$ is a g-inverse of $\mathbf{Q}$. (*Hint*: Use Lemma A.10.2.)

7.7 Show if the formula for a g-inverse of a partitioned matrix, given in Lemma A.10.3, is applied to solve the normal equations (7.9.2), then expressions (7.9.3) and (7.9.4) are obtained as solutions for $\boldsymbol{\beta}^o$ and γ^o.

7.8 Verify that the *BLU* predictor under model (7.9.8), having one factor and one quantitative auxiliary, is given by (7.9.11) and that its error variance is expression (7.9.12).

7.9 Use the sample of $n = 48$ units in Example 7.8.1 from the Labor Force population, that is, units 1, 11, 21, 31, and so on through unit 471. Estimate the population total under the model $Y_{ij} = \mu + \alpha_i + \gamma z_{ij} + \varepsilon_{ij}$, where sex is qualitative factor and hours worked is the quantitative variable. Verify the parameter solutions given in Table 7.6 and show that the estimate of the total is the same for both solutions.

7.10 Use the same systematic sample as in Exercise 7.9 from the Labor Force population. (a) Estimate the population total of wages using the models considered in Section 7.9.3, that is,

$$Y_{ijk} = \mu + \alpha_i + \beta_j + \gamma z_{ijk} + \varepsilon_{ij}$$

and

$$Y_{ijk} = \mu + \alpha_i + \beta_j + (\alpha\beta)_{ij} + \gamma z_{ijk} + \varepsilon_{ij},$$

where the α-factor is sex, the β-factor is age, and z is hours worked per week. (b) Using the function *ginv* in Appendix C, verify that for each model the same solution vector for the model parameters is obtained whether one uses *ginv* to invert $\mathbf{X}_s^{*\prime}\mathbf{X}_s^*$ as in Theorem 7.4.2 or uses *ginv* to invert $\mathbf{X}_s'\mathbf{X}_s$ to get the partitioned form of solutions in Section 7.9.1.

7.11 Using the sample consisting of persons 1, 20, 39, 58, ..., 476 from the Labor Force population estimate the population total under the models (a) $Y_{ij} = \mu + \alpha_i + \gamma z_{ij} + \varepsilon_{ij}$, (b) $Y_{ijk} = \mu + \alpha_i + \beta_j + \gamma z_{ijk} + \varepsilon_{ij}$, and (c) $Y_{ijk} = \mu + \alpha_i + \beta_j + (\alpha\beta)_{ij} + \gamma z_{ijk} + \varepsilon_{ij}$, where $\varepsilon \sim (0, \sigma^2)$ for all errors. Let α denote the 2-level factor for sex and β the 4-level factor for age as in Example 7.8.1. Show that the same estimated total is obtained for each model whether one uses an overall g-inverse as in Theorem 7.4.2 or the partitioned solution described in Section 7.9.1.

7.12 Assume that the working model is $M(\mathbf{X}:\mathbf{V})$, that is, $E_M(\mathbf{Y}) = \mathbf{X}\boldsymbol{\beta}$ and $\mathrm{var}_M(\mathbf{Y}) = \mathbf{V}$ where $\mathbf{V} = \mathrm{diag}(v_i)$. (a) When $\mathbf{X}$ does not have full column rank, what is the *BLU* predictor of T? (b) Derive the robust variance estimators, v_R, v_D, v_H, v_{J*} and v_J, described in Section 7.10 when $M(\mathbf{X}:\mathbf{V})$ is the working model.

7.13 (a) Compute the variance estimates v_R, v_D, v_H, v_{J*}, and v_J for the estimated total $\hat{T}$ in Exercise 7.4. (b) Compute a 95% confidence interval for the total using each of the variance estimates found in (a).

7.14 (a) Compute the variance estimates v_R, v_D, v_H, v_{J*}, and v_J for the estimated total $\hat{T}$ in Exercise 7.5. (b) Compute a 95% confidence interval for the total using each of the variance estimates found in (a).

CHAPTER 8

Clustered Populations

Many naturally occurring populations exhibit clustering in which units that are near to each other (geographically or in some other respect) have similar characteristics. Households in the same neighborhood tend to have similar incomes, education levels of the heads of household, and amounts of expenditures on food and clothing. Business establishments in the same industry and geographic area will pay similar wages to a given occupation because of competition. This similarity among "nearby" units will express itself statistically as a correlation between the target variables for different units. In this chapter we introduce some models appropriate to such clustered populations and study some alternative estimators of totals.

In clustered populations, the methods of data collection may also differ from the methods used in other populations. In a household survey, for example, a complete list of households to use for sampling is usually not available, especially if the population is large. In the United States, for instance, there were nearly 100 million households in 1995. The households of interest may be geographically dispersed; field work can be more economically done when sample units are clustered together to limit travel costs. A practical and widely used technique is to select the sample in stages, using, at each stage, sampling units for which a complete list is available. In the household example, geographic areas (counties, shires, townships, etc.) may be selected at the first stage. At the second stage, each first stage sample unit may be further subdivided and a sample of the subdivisions selected. A list of the households in each sample subdivision is then compiled and data collected from each. In a business population, establishments may be selected at the first stage, a list of occupations compiled in each sample establishment, and a sample of occupations then drawn from each list. Although occupations are the units ultimately sampled, a complete list of occupations for each establishment in the universe is unlikely to be available, whereas a list of establishments often is. Selecting occupations in two stages can also allow better control over survey costs. Travel and recruitment expenses may be associated with establishments while clerical costs of extracting data from personnel records may be related to the number of sample occupations. Two-stage sampling can allow fine-tuning of the survey budget.

There is an important distinction between clusters that are used for economy of data collection and ones that must be considered for data analysis. Just because clusters are used in the sample selection does not necessarily mean that they should be accounted for when estimating means, totals, or more complicated functions. If there is no statistical correlation between units in a given sample cluster, then the estimation techniques covered in earlier chapters may be perfectly appropriate.

In this chapter, we assume that the structure of the population is such that the clusters must be accounted for when constructing estimators. Reflecting the clustered structure appropriately will be critical in deriving the variances of estimators.

Another critical difference between estimation for clustered populations and for the unclustered ones studied in earlier chapters is in the role of balanced sampling and robustness. In Chapters 3 and 4 we emphasized the importance of balanced sampling in achieving bias-robustness and efficiency. We assumed that a complete frame of units was available in advance of sampling and that control could be exercised over the sample distribution of auxiliary variables. Although certain kinds of balance are relevant in clustered populations, selecting samples with appropriate balance may be difficult or impossible in practice. In a two-stage sample, for instance, the values of auxiliaries for individual units are usually known only for units in sample clusters and only then after the clusters are selected. Careful model building becomes more important when constructing an estimator.

We will begin the study of clustered populations with a model in which all units have a common mean, but which is complicated by its covariance structure. This model is good for gaining insight into the impact of clustering on estimation and, in combination with stratification and other procedures, can be useful for describing many populations. We treat it in thorough fashion, covering relative efficiency of a variety of estimators, sample design, the question of bias-robustness, and so forth. Despite its relative simplicity, there are several important open questions which we point out. Models involving auxiliary variables are discussed in later sections. Variance estimation will be taken up in the next chapter.

8.1. INTRACLUSTER CORRELATION MODEL FOR A CLUSTERED POPULATION

The population of units is divided into N clusters. Cluster i contains M_i units with the total number of units in the population being $M = \Sigma_{i=1}^{N} M_i$. We assume that the cluster sizes M_i, and hence the population size M, are all known. Associated with unit j in cluster i is a random variable Y_{ij} whose finite population total is $T = \Sigma_{i=1}^{N} \Sigma_{j=1}^{M_i} Y_{ij}$. The working model is

$$E_M(Y_{ij}) = \mu$$

$$\text{cov}_M(Y_{ij}, Y_{i'j'}) = \begin{cases} \sigma_i^2 & i = i', j = j' \\ \sigma_i^2 \rho_i & i = i', j \neq j' \\ 0 & \text{otherwise.} \end{cases} \tag{8.1.1}$$

The model posits that units all have a common mean μ. Within cluster i, units have a common variance σ_i^2 which can be different from one cluster to another. Units in the same cluster also have a common correlation ρ_i. This type of model can be combined with stratification (to allow μ to be different for different portions of the population), but for this initial discussion of clustering the additional complication of strata is unnecessary.

Elements are selected by a two-stage sampling scheme. First a sample s of n clusters is chosen from the N. Denote the set of nonsample clusters by r. Then from the M_i elements in sample cluster i, a sample s_i of size m_i is selected. The set of nonsample units within sample cluster i will be r_i. The total number of units in the sample is $m = \Sigma_s m_i$. In some applications, we may enumerate all units in each sample cluster, in which case $m_i = M_i$; this is referred to as *one-stage sampling* (considered as a special case in Section 8.2.1, below).

The population total can be naturally represented in three parts—the total for the observed elements, the total for unobserved elements in sample clusters, and the total for nonsample clusters:

$$T = \sum_{i\in s} \sum_{j\in s_i} Y_{ij} + \sum_{i\in s} \sum_{j\in r_i} Y_{ij} + \sum_{i\in r} \sum_{j=1}^{M_i} Y_{ij} \tag{8.1.2}$$

In the case of single-stage sampling, where each sample cluster is enumerated fully, the second term in (8.1.2) is null and the total is $T = \Sigma_{i\in s} \Sigma_{j\in s_i} Y_{ij} + \Sigma_{i\in r} \Sigma_{j=1}^{M_i} Y_{ij}$.

8.1.1. Discussion of the Common Mean Model

Suppose all the parameters, except μ, of model (8.1.1) were known, and suppose the full population were available. How would the values of the cluster correlations and variances manifest themselves in the data? If all the ρ_i were nonnegative, then clusters with larger values of ρ_i would consist of units tending to be relatively close to each other, to the point that if $\rho_i = 1$, the units would coincide. Clusters with greater σ_i would have a greater opportunity to stray from the common mean μ: if large σ_i coincided with large ρ_i, the units could be all some distance from μ, but together; units of a cluster with large σ_i and small ρ_i could be scattered at a variety of distances and both above and below μ.

What would the implications be for sampling to estimate μ and T? Clearly, the higher ρ_i, the less need, having selected cluster i, to sample heavily *within* the cluster, since other units will tend to be similar. The impact of the σ_i on our strategy would depend on whether the focus was on μ or T. Selecting units from clusters with small σ_i would enable us to zero in on μ. To predict T, we would lean toward getting good estimates of totals for large clusters (large M_i), especially those with large σ_i, to capture far out points that heavily impact the sum T.

What might be the interplay between M_i, ρ_i, and σ_i? There are no hard-and-fast rules, but it is reasonable to posit as a "default guess" that clusters with larger M_i would tend to have smaller ρ_i, as allowing greater scatter among their units, and also bigger σ_i, since they could contain a greater variety. But it is easy to think of exceptions. For example, if M_i is number of families in neighborhoods of fixed size, the larger neighborhoods might be more homogeneous on a variate Y measuring the amount spent on public transportation, and deviate less from the common population mean; that is, have larger ρ_i and smaller σ_i.

If we did not *a priori* have the ρ_i and σ_i, how would we estimate their values, or at least diagnose sets of clusters with relatively large ρ_i, and so on? This is an open question, beyond the scope of this text.

It will often make sense, especially after suitable stratification, to adopt a simplified version of (8.1.1), the *constant parameter common mean model* (or just *common parameter model*),

$$E_M(Y_{ij}) = \mu$$

$$\text{cov}_M(Y_{ij}, Y_{i'j'}) = \begin{cases} \sigma^2 & i = i', j = j' \\ \sigma^2\rho & i = i', j \neq j' \\ 0 & \text{otherwise,} \end{cases} \tag{8.1.3}$$

in which all units have a common variance and the intracluster correlation is the same for all clusters. In this case, procedures for estimating the model parameters are well known (see Section 8.3).

8.1.2. Simple Sample Designs

In the presence of uncertainty about the covariance structure, the manner of cluster sampling is often dictated by convenience. Two frequent approaches to selecting clusters are by (i) simple random sampling or (ii) by *pps* sampling where M_i is the size measure. Clusters having been selected, two approaches to choosing units within clusters will be (a) *equal allocation* with an equal number of units within all sampled clusters (so that $m_i = \bar{m} \equiv \bar{m}_s \equiv m/n$) or (b) *proportional allocation* with m_i proportional to M_i.

Allocating an equal number of sample units to each sample cluster has operational advantages in many surveys. When clusters are geographic areas and data are collected by personal visits to the sample units, having the same workload in every area means that the same number of field staff can be hired in each area. This facilitates recruiting staff and maintaining the same time schedule in all areas. Thus (a) is probably the most common method of allocation in practice. See Section 8.8 for further discussion.

Allocation (a) is often used in common with (ii) $pp(M_i)$ sampling. Intuitively, it makes sense to more heavily sample those clusters which are most likely to impact the total. From a design-based point of view, there is another advantage: one readily determines that the combination (iia) (with *srs* within clusters) leads to equal selection probabilities for all units (Exercise 8.33), and hence to the estimator

$$\hat{T}_0 = \frac{M}{m} \sum_s m_i \bar{Y}_{si},$$

where $\bar{Y}_{si} = \Sigma_{j \in s_i} Y_j / m_i$. We recognize $\hat{T}_0$ as the simple expansion estimator of earlier chapters. A sample design such as this is referred to as an *epsem design* for "*e*qual *p*robability *se*lection *m*ethod" (see discussion in Section 8.10). Designs leading to mathematically simple estimators are often considered desirable because simple estimates having high "face validity" or intuitive appeal are more readily used and accepted than ones derived from apparently esoteric mathematical manipulations.

The combination (ib), in which clusters are selected with equal probabilities and the number of sampled units (selected *srs*) within clusters is proportional to cluster size, is also readily determined to be *epsem*, and thus also to lead to $\hat{T}_0$. It is not usually as convenient as (iia) from the point of view of field logistics, unless, as often happens in sampling within establishments, the data on the units within the cluster can be collected *en masse*. Note that one-stage sampling ($m_i = M_i$) is a special case of (b).

8.2. CLASS OF UNBIASED ESTIMATORS UNDER THE COMMON MEAN MODEL

The expansion estimator $\hat{T}_0$ and the best linear unbiased estimator $\hat{T}_{BLU}$, to be developed later in Theorem 8.2.2, form the simple and complex extremes of a general class of estimators, all of which are unbiased under the common mean model (8.1.1). (In anticipation, it may be noted that $\hat{T}_{BLU}$ reduces to $\hat{T}_0$ when $\rho_i = 0$ and $\sigma_i = \sigma$, a constant, for $i = 1, \ldots, N$.) Members of this class have the form

$$\begin{aligned} \hat{T} &= \sum_s \sum_{s_i} Y_{ij} + \sum_s a_i \sum_{s_i} Y_{ij} \\ &= \sum_s (1 + a_i) m_i \bar{Y}_{si}, \end{aligned} \qquad (8.2.1)$$

where a_i is a term that does not depend on the Y's. In order for estimators in class (8.2.1) to be unbiased under model (8.1.1), we require

$$\sum_s (1 + a_i)m_i = M. \tag{8.2.2}$$

We can write the error-variance of the general estimator (8.2.1) as

$$\text{var}_M(\hat{T} - T) = v - \sum_s \sigma_i^2 \rho_i (M_i - m_i)^2 + \sum_s \sigma_i^2 \rho_i [m_i a_i - (M_i - m_i)]^2 + \sum_s (1 - \rho_i)\sigma_i^2 m_i a_i^2, \tag{8.2.3}$$

where

$$v = \sum_s (M_i - m_i)\sigma_i^2[1 + (M_i - m_i - 1)\rho_i] + \sum_r M_i \sigma_i^2 [1 + (M_i - 1)\rho_i] \tag{8.2.4a}$$

is the sum of the variance for nonsample units in sample clusters and the variance for units in nonsample clusters (see Exercise 8.9). Expression (8.2.3) will be convenient for calculating and comparing variances.

Expressions of the form $v_{Ai} \equiv \sigma_i^2[1 + (A_i - 1)\rho_i]$ occur sufficiently often to warrant special notation. The most commonly used occurrence is for $A_i = m_i$, and so we write $v_i \equiv v_{mi} \equiv \sigma_i^2[1 + (m_i - 1)\rho_i]$. It can be checked that $\text{var}(\bar{Y}_{si}) = v_i/m_i$ and $\text{var}(\bar{Y}_i) = v_{Mi}/M_i$, where $\bar{Y}_i = \Sigma_{j=1}^{M_i} Y_{ij}/M_i$. (In some cases the context may dictate that the ρ_i and/or the σ_i are being assumed constant.) Thus we have

$$v = \sum_s (M_i - m_i)v_{\Delta i} + \sum_r M_i v_{Mi} \tag{8.2.4b}$$

with $\Delta = M_i - m_i$.

In addition to the expansion and *BLU* estimators, the following estimators, which all have standing in the sampling literature (see, e.g., Thompson 1992, pp. 130–131; Royall 1976b), belong to this class, which indeed comprises an infinity of possibilities (so we shall be giving only a small fraction of these):

Ratio estimator,

$$\hat{T}_R = \frac{M}{M_s} \sum_s M_i \bar{Y}_{si}$$

pps estimator

$$\hat{T}_p = \frac{M}{n} \sum_s \bar{Y}_{si}$$

Intermediate estimator

$$\hat{T}_H = \sum_{s,s_i} Y_{ij} + \sum_s (M_i - m_i)\bar{Y}_{si} + M_r\left(\frac{\Sigma_s m_i \bar{Y}_{si}}{m}\right)$$

Perfect correlation estimator

$$\hat{T}_1 = \sum_s \sum_{s_i} Y_{ij} + \sum_s (M_i - m_i)\bar{Y}_{si} + M_r \sum_s \bar{Y}_{si}/n$$

where $M_s = \Sigma_s M_i$ and $M_r = \Sigma_r M_i$. Note that these four estimators all require the same information: sample cluster means and various counts of units. None of the estimators requires detailed values of nonsample M_i.

The ratio estimator $\hat{T}_R$, patterned after the simple ratio estimator introduced in Section 1.2 with M playing the role of x and $\hat{T}_i = M_i\bar{Y}_{si}$ the role of Y, is approximately design unbiased under two-stage sampling schemes which employ *srs* at both stages. The *pps* estimator $\hat{T}_p$ is design unbiased if the first stage of sampling is $pp(M_i)$ and the second *srs*. Alone among the estimators here considered, $\hat{T}_p$ is inconsistent in the sense of not equaling T when the population is fully sampled. The perfect correlation estimator $\hat{T}_1$ will be seen to be the *BLU estimator* under the constant parameter model (8.1.3) when $\rho = 1$, as $\hat{T}_0$ is for $\rho = 0$ (see Corollary 8.2.2, below). The intermediate estimator $\hat{T}_H$, developed as a model based improvement on $\hat{T}_R$ and $\hat{T}_p$ (Royall, 1976b), has its second and third terms of the form of the *BLU* for $\rho = 0$ and $\rho = 1$, respectively (Corollary 8.2.2).

Table 8.1 gives the a-values associated with each of the estimators, and is convenient for deducing relations among the estimators under different conditions or for calculating variances using equation (8.2.3).

The simple results collected in the following theorem are immediate from the table.

Theorem 8.2.1. (a) If $m_i = \bar{m}_s$ for $i \in s$ (i.e., within-cluster sample sizes are all the same), then $\hat{T}_P = \hat{T}_0$ and $\hat{T}_H = \hat{T}_1$. (b) If $m_i \propto M_i$ for $i \in s$, then $\hat{T}_R = \hat{T}_H = \hat{T}_0$.

Corollary 8.2.1. In one-stage sampling, $\hat{T}_R = \hat{T}_H = \hat{T}_0$.

8.2.1. One-Stage Cluster Sampling

All units in selected clusters are sampled. The goal is to estimate $T = \Sigma_{i \in s}\Sigma_{j \in s_i} Y_{ij} + \Sigma_{i \in r}\Sigma_{j=1}^{M_i} Y_{ij}$. The best linear unbiased predictor under model (8.1.1) is equal to the sum for the sample clusters plus an estimator of the total

Table 8.1. Coefficients for Major Estimators that are Unbiased under Constant Mean Model

Name	Symbol	*a*-Coefficient
Expansion	$\hat{T}_0$	$a_i = (M - m)/m$
Ratio	$\hat{T}_R$	$a_i = \dfrac{M_i M}{m_i M_s} - 1$
$pp(M_i)$	$\hat{T}_P$	$a_i = \dfrac{M}{nm_i} - 1$
Intermediate	$\hat{T}_H$	$a_i = \dfrac{M_r}{m} + \dfrac{M_i}{m_i} - 1$
Perfect correlation	$\hat{T}_1$	$a_i = \dfrac{M_r}{nm_i} + \dfrac{M_i}{m_i} - 1$
BLU	$\hat{T}_{BLU}$	See Equation (8.2.9)

for the nonsample clusters,

$$\hat{T}_{BLU} = \sum_{i \in s} \sum_{j=1}^{M_i} Y_{ij} + \sum_{i \in r} M_i \hat{\mu} \tag{8.2.5}$$

where $\hat{\mu}$ is a weighted average of the sample means, $\hat{\mu} = \Sigma_s u_i \bar{Y}_{si}$, with weights

$$u_i = \frac{M_i/v_{Mi}}{\Sigma_s M_i/v_{Mi}}.$$

The error variance of the *BLU* predictor is

$$\text{var}_M(\hat{T}_{BLU} - T) = \sum_r M_i v_{Mi} + (M - m)^2 \Big/ \left[\sum_s M_i/v_{Mi}\right]. \tag{8.2.6}$$

Expressions (8.2.5) and (8.2.6) follow as special cases of Theorem 8.2.2 and Theorem 8.2.3 in the next section.

The *BLU* predictor can also be written in the form

$$\hat{T} = \sum_s (1 + a_i) M_i \bar{Y}_i, \tag{8.2.7}$$

where $a_i = u_i M_r/M_i$. Note that $\Sigma_s(1 + a_i)M_i = M$.

The expansion estimator is given by $\hat{T}_0 = (M/M_s)\Sigma_s M_i \bar{Y}_i$, and by Corollary 8.2.1, $\hat{T}_R = \hat{T}_H = \hat{T}_0$. Also, $\hat{T}_p = (M/n)\Sigma_s \bar{Y}_i$ with $a_i = M/(nM_i) - 1$, and $\hat{T}_1 = \Sigma_s \Sigma_{s_i} Y_{ij} + M_r \Sigma_s \bar{Y}_i/n$, with $a_i = M_r/(nM_i)$.

From (8.2.3) the error variance of members of the class of unbiased predictors is, in the one-stage case,

$$\operatorname{var}_M(\hat{T} - T) = \sum_r M_i v_{Mi} + \sum_s \sigma_i^2 \rho_i M_i^2 a_i^2 + \sum_s (1 - \rho_i)\sigma_i^2 M_i a_i^2. \quad (8.2.8)$$

8.2.2. *BLU* Predictor

Theorem 8.2.2 gives the optimal estimator under model (8.1.1), with the estimator for each component of (8.1.2) shown explicitly and also in the form (8.2.1). We return to the case where a two-stage sample is selected with $m_i < M_i$.

Theorem 8.2.2 (Scott and Smith, 1969; Royall, 1976b). Under model (8.1.1) the best linear unbiased predictor of the total T is

$$\begin{aligned}\hat{T}_{BLU} &= \sum_s \sum_{s_i} Y_{ij} + \sum_s (M_i - m_i)[w_i \bar{Y}_{si} + (1 - w_i)\hat{\mu}] + \sum_r M_i \hat{\mu} \\ &= \sum (1 + a_i) m_i \bar{Y}_{si} \end{aligned} \quad (8.2.9)$$

with

$$a_i = \left[M - m - \sum_{i' \in s} (M_{i'} - m_{i'}) w_{i'} \right] u_i / m_i + (M_i - m_i) w_i / m_i,$$

where $w_i = m_i \rho_i / [1 + (m_i - 1)\rho_i]$, and $\hat{\mu}$, is a weighted average of the sample means, $\hat{\mu} = \Sigma_s u_i \bar{Y}_{si}$, with weights $u_i = (m_i/v_i)/(\Sigma_s m_i/v_i)$. (Recall the definition $v_i \equiv \sigma_i^2[1 + (m_i - 1)\rho_i]$ above and note $u_i \propto \operatorname{var}(\bar{Y}_{si})^{-1}$.)

Proof: The *BLU* predictor can be derived directly from Theorem 2.2.1. Because of the form of model (8.1.1), the vector $\mathbf{X}_s$ is a vector of $\Sigma_s m_i$ 1's. Similarly, $\mathbf{X}_r$ is all 1's and has length $\Sigma_s (M_i - m_i) + \Sigma_r M_i$. Sample units in the same cluster are correlated but units in different clusters are not. Thus the covariance matrix $\mathbf{V}_{ss}$ for the sample units is block diagonal with the $m_i \times m_i$ block for sample cluster i having the form

$$\mathbf{V}_{ss(i)} = \sigma_i^2 \begin{bmatrix} 1 & \rho_i & \cdots & \rho_i \\ \rho_i & 1 & & \rho_i \\ \vdots & & \ddots & \\ \rho_i & \rho_i & \cdots & 1 \end{bmatrix}.$$

This type of patterned matrix has a simple inverse which is

$$\mathbf{V}_{ss(i)}^{-1} = \frac{1}{v_i(1-\rho_i)} \begin{bmatrix} 1+(m_i-2)\rho_i & -\rho_i & \cdots & -\rho_i \\ -\rho_i & 1+(m_i-2)\rho_i & & -\rho_i \\ \vdots & & \ddots & \\ -\rho_i & -\rho_i & \cdots & 1+(m_i-2)\rho_i \end{bmatrix}.$$

The estimator of the common mean is then

$$\hat{\mu} = [\mathbf{X}_s' \,\text{blkdiag}(\mathbf{V}_{ss(i)}^{-1})\mathbf{X}_s]^{-1}\mathbf{X}_s' \,\text{blkdiag}(\mathbf{V}_{ss(i)}^{-1})\mathbf{Y}_s = \frac{\Sigma_s m_i \bar{Y}_{si}/v_i}{\Sigma_s m_i/v_i},$$

where blkdiag is short for "block diagonal." The

$$\left[\sum_s (M_i - m_i) + \sum_r M_i\right] \times \sum_s m_i$$

matrix $\mathbf{V}_{rs}$ is composed of two types of submatrices—one giving the covariances between the nonsample and sample units in sample clusters and one giving the covariances between units in nonsample clusters with those in the sample clusters. The latter type is zero under model (8.1.1). For sample cluster i the covariance submatrix is $\mathbf{V}_{rs(i)} = \sigma_i^2\rho_i\mathbf{1}_{M_i-m_i}\mathbf{1}_{m_i}'$, where $\mathbf{1}_k$ is a k-vector of all 1's. Thus we derive

$$\mathbf{1}_r'\mathbf{X}_r\hat{\boldsymbol{\beta}} = \sum_s (M_i - m_i)\hat{\mu} + \sum_r M_i\hat{\mu}$$

and

$$\mathbf{1}_r'\mathbf{V}_{rs}\mathbf{V}_{ss}^{-1}(\mathbf{Y}_s - \mathbf{X}_s\hat{\boldsymbol{\beta}}) = \sum_s (M_i - m_i)w_i(\bar{Y}_{si} - \hat{\mu}).$$

Summing the last two results along with $\Sigma_{i\in s}\Sigma_{j\in s_i} Y_{ij}$ gives $\hat{T}_{BLU}$. □

Corollary 8.2.2. Suppose $\sigma_i = \sigma$, $i \in s$. If $\rho_i = \rho = 0$, for $i \in s$, then $\hat{T}_{BLU} = \hat{T}_0$. If $\rho_i = \rho = 1$ for $i \in s$, then $\hat{T}_{BLU} = \hat{T}_1$. The second term of $\hat{T}_H$ coincides with the second term of $\hat{T}_{BLU}$ at $\rho = 1$, the third term, with the third term of $\hat{T}_{BLU}$ at $\rho = 0$.

Corollary 8.2.3. Suppose $\sigma_i = \sigma$, and $\rho_i = \rho$, $i \in s$. If $m_i = \bar{m}$, $i \in s$, then $\hat{T}_0 \leqslant \hat{T}_{BLU} \leqslant \hat{T}_1$ or $\hat{T}_1 \leqslant \hat{T}_{BLU} \leqslant \hat{T}_0$. In other words, the *BLU* estimator lies between $\hat{T}_0$ and $\hat{T}_1$.The proof is left as Exercise 8.4.

The *BLU* need not, in general, be between $\hat{T}_0$ and $\hat{T}_1$. The most general conditions guaranteeing it will be are unknown.

Theorem 8.2.3. The error-variance of $\hat{T}_{BLU}$ under model (8.1.1) is

$$\text{var}_M(\hat{T}_{BLU} - T) = v - \sum_s (M_i - m_i)^2 w_i \sigma_i^2 \rho_i + \left[M - m - \sum_s (M_i - m_i) w_i\right]^2 \Big/ \left[\sum_s m_i/v_i\right], \quad (8.2.10)$$

where v is given by equation (8.2.4).

Proof: The proof is left to Exercise 8.2.

8.2.3. Variance Component Model

An alternative model to (8.1.1) for a clustered population is a variance components model as described in Ross and Chambers (2000). Using such a model may seem more natural to readers familiar with experimental design and the standard literature for linear models. Variance component models have been given intense study in the *analysis* of clustered surveys, wherein the focus is on estimation of superpopulation parameters (see, e.g., Goldstein, 1995; Longford, 1993; Skinner et. al., 1989). Whether elaborate variance component models can be equally useful in the estimation of totals must at this point be regarded as an open question. We indicate here the simplest case.

Suppose that the random variable Y_{ij} follows

$$Y_{ij} = \mu + u_i + w_{ij}, \quad (8.2.11)$$

where u_i and w_{ij} are independent random errors with $u_i \sim (0, \sigma_u^2)$ and $w_{ij} \sim (0, \sigma_w^2)$. Under this model we might conceive of the population as being generated by a two-stage process. At the first stage all units within cluster i have the same random error u_i. At the second-stage a different random error w_{ij} is generated for each unit. This is an example of a random effects model that was introduced more generally in Chapter 2. Model (8.2.11) leads to this structure:

$$E_M(Y_{ij}) = \mu$$

$$\text{cov}_M(Y_{ij}, Y_{i'j'}) = \begin{cases} \sigma_u^2 + \sigma_w^2 & i = i', j = j' \\ \sigma_u^2 & i = i', j \neq j' \\ 0 & \text{otherwise.} \end{cases} \quad (8.2.12)$$

The correlation between units in the same cluster—the intracluster correlation—is the same for each cluster, as it was in (8.1.3), and is

$$\rho = \frac{\text{cov}_M(Y_{ij}, Y_{ik})}{\text{var}_M(Y_{ij})} = \frac{\sigma_u^2}{\sigma_u^2 + \sigma_w^2}. \quad (8.2.13)$$

With this formulation, ρ is the proportion of the variance of Y_{ij} attributable to the variance among the cluster errors u_i. If $\sigma_w^2 = 0$, then all units in cluster i

have the same value of Y_{ij} and the intracluster correlation is 1. At the other extreme, if $\sigma_u^2 = 0$, there is no cluster component, and $\rho = 0$. In that case, the Y's are simply uncorrelated random variables.

Model (8.2.12) is more restrictive than (8.1.3) because the intracluster correlation implied by (8.2.12) must be positive. Thus in a sense, (8.2.12) is a special case of (8.1.3). A positive ρ indicates that units in a cluster tend to be alike in a statistical sense. In practice, the correlation is small and positive in many populations. There are exceptions, of course, and negative correlations are possible. Ross and Chambers (2000), for example, note that area clusters in some countries may contain both landowners and tenant farmers. If one family in a pair is rich, the other may be poor. In such cases, model (8.2.12) is too restrictive, unless one is able to stratify on types of ownership, separating out the "opposing" types. It is worth noting, too, that although model (8.1.3) allows ρ to be negative, even in that model ρ is bounded well above -1. A lower bound on the correlation in the more general model (8.1.3) is derived in Section 8.3.3.

8.3. ESTIMATION OF PARAMETERS IN THE CONSTANT PARAMETER MODEL

Almost always, the covariance parameters necessary to construct $\hat{T}_{BLU}$ are unknown. An approximation to $\hat{T}_{BLU}$ is possible by estimating these parameters, at least in the case of the constant parameter model (8.1.3), with $\sigma_1^2 = \sigma^2$ and $\rho_i = \rho$. In addition, estimation of model parameters of the common mean model is useful to get estimates of variance for planning purposes. We may, for example, be interested in designing a sample that meets specified precision goals and estimates of σ^2 and ρ are needed to predict the variance of an estimator for different sample sizes. In general, the area of variance component estimation presents special problems discussed, for example, in Harville (1977) and Searle, Casella, and McCullogh (1992). The analysis of variance and maximum likelihood methods of estimation, as applied to model (8.2.12), are covered in this section. There are other more recently developed, and possibly better, methods of model parameter estimation involving the technique of Gibbs sampling (Gelfand, Hills, Racine-Poon, and Smith, 1990; Gelfand and Smith, 1990). Use of those methods may prove to be a useful addition to the sampler's armamentarium. We here treat relatively simple, well-known methodologies applied to the constant parameter model or variance component model.

8.3.1. ANOVA Estimators

One set of parameter estimators is based on the simple analysis variance (ANOVA) estimators given in Table 8.2. In the table the overall sample mean is defined as $\bar{Y}_s = \Sigma_s m_i \bar{Y}_{si}/m$.

Table 8.2. Analysis of Variance for a Sample from a Clustered Population Following the Constant Parameter Model (8.1.3)

Source	Mean Squares	Expected Mean Square
Between clusters	$B = \dfrac{\Sigma_{i\in s} m_i(\bar{Y}_{si} - \bar{Y}_s)^2}{n-1}$	$\sigma^2(1-\rho) + \dfrac{\rho\sigma^2}{n-1}\left(m - \sum_{i\in s}\dfrac{m_i^2}{m}\right)$
Within clusters	$W = \dfrac{1}{n}\sum_{i\in s}\sum_{j\in s_i}\dfrac{(Y_{ij} - \bar{Y}_{si})^2}{m_i - 1}$	$\sigma^2(1-\rho)$

Routine calculations are enough to verify that the expected values under model (8.1.3) of the mean squares in the second column are equal to the values in the last column (see Exercise 8.6). The ANOVA table provides two equations in the two unknowns, σ^2 and ρ, that can be easily solved. In the special case of equal subsample sizes from each cluster, $m_i = \bar{m}$, the expected value of B, the mean square for clusters, reduces to $\sigma^2[1 + (\bar{m} - 1)\rho]$.

Note that for the variance component model, the between cluster and within cluster expectations are $\sigma_w^2 + [\sigma_u^2/(n-1)](m - \Sigma_s m_i^2/m)$ and σ_w^2, respectively.

Example 8.3.1. To illustrate the ANOVA procedures, consider the sample from the Labor Force population in Appendix B consisting of the 20 clusters numbered 1, 5, 7, 17, 23, 27, 28, 30, 33, 38, 43, 44, 46, 58, 61, 63, 71, 88, 101, and 114. Suppose each sample cluster is fully enumerated yielding 70 units in all. The sample ANOVA estimates for the variable labeled y are $B = 76.322$, $W = 64.103$, and $D = (m - \Sigma_s m_i^2/m)/(n-1) = 3.670$, leading to $\hat{\sigma} = 67.644$ and $\hat{\rho} = 0.052$. The S-Plus™ function *clus.parm.est.anova* can be used to generate ANOVA estimates for other samples.

It should be noted that the ANOVA estimator does not always respect the bounds which necessarily hold for ρ. If that occurs, the estimate is typically brought to or within the nearer boundary.

8.3.2.* Maximum Likelihood Estimators

We can also estimate ρ and σ^2 through maximum likelihood estimation. As in other statistical problems, if the working model is correct, the maximum likelihood estimators (MLEs) can be asymptotically optimal (see, e.g., Rao, 1973). This approach may apply even when the ρ_i and σ_i^2 are not the same across clusters (Bates and Pinheiro, 1998), although this has not been explored in the sampling context. We here limit ourselves to the constant parameter model (8.1.3). We suppose that $\mathbf{Y}_s = (Y_{11}, \ldots, Y_{1m_1}, \ldots, Y_{n1}, \ldots, Y_{nm_n})'$ is multivariate normal. The likelihood for the sample of n clusters with m_i sample units from sample cluster i is

$$L(\mu, \sigma^2, \rho; \mathbf{Y}_s) = \frac{1}{\sqrt{(2\pi)^m \det|\mathbf{V}_{ss}|}} \exp\{-\tfrac{1}{2}(\mathbf{Y}_s - \mu\mathbf{1}_s)'\mathbf{V}_{ss}^{-1}(\mathbf{Y}_s - \mu\mathbf{1}_s)\}.$$

The determinant of $\mathbf{V}_{ss}$ is

$$\det(\mathbf{V}_{ss}) = \prod_{i\in s} \det(\mathbf{V}_{ss(i)}) = \prod_{i\in s} \sigma^{2m_i}(1-\rho)^{m_i-1} b_i,$$

where $b_i = 1 + (m_i - 1)\rho$ (see Exercise 8.11). The log likelihood is then

$$l(\mu, \sigma^2, \rho; \mathbf{Y}) = -\frac{m}{2}\log(2\pi) - \frac{m}{2}\log(\sigma^2) - \frac{1}{2}\log(1-\rho)\sum_s (m_i - 1)$$
$$-\frac{1}{2}\sum_s \log(b_i) - \frac{1}{2}(\mathbf{Y}_s - \mu\mathbf{1}_s)'\mathbf{V}_{ss}^{-1}(\mathbf{Y}_s - \mu\mathbf{1}_s).$$

To facilitate calculations, the quadratic form in the last term of the log likelihood can be expanded as

$$(\mathbf{Y}_s - \mu\mathbf{1}_s)'\mathbf{V}_{ss}^{-1}(\mathbf{Y}_s - \mu\mathbf{1}_s) = \frac{1}{\sigma^2(1-\rho)}\left[\sum_s\sum_{s_i}(Y_{ij} - \mu)^2 - \rho\sum_s \frac{m_i^2(\bar{Y}_{si} - \mu)^2}{b_i}\right].$$

Differentiating $l(\mu, \sigma^2, \rho; \mathbf{Y})$ with respect to each of the parameters and equating to 0 produces, after a substantial amount of algebra, the following estimating equations:

$$\hat{\mu} = \frac{\sum_{i\in s} m_i \bar{Y}_{si}/\hat{b}_i}{\sum_{i\in s} m_i/\hat{b}_i} \tag{8.3.1}$$

$$\hat{\sigma}^2 = \frac{1}{m(1-\hat{\rho})}\left[\sum_{i\in s}\sum_{j\in s_i}(Y_{ij} - \hat{\mu})^2 - \hat{\rho}\sum_{i\in s}\frac{m_i^2(\bar{Y}_{si} - \hat{\mu})^2}{\hat{b}_i}\right] \tag{8.3.2}$$

and

$$\frac{\hat{\rho}}{1-\hat{\rho}}\sum_{i\in s}\frac{m_i(m_i - 1)}{\hat{b}_i} - \frac{1}{\hat{\sigma}^2(1-\hat{\rho})}\sum_{i\in s}(Y_{ij} - \hat{\mu})^2$$
$$+\frac{1}{\hat{\sigma}^2(1-\hat{\rho})^2}\sum_{i\in s}\frac{m_i^2(\bar{Y}_{si} - \hat{\mu})^2}{\hat{b}_i}\left\{1 - \hat{\rho}(1-\hat{\rho})\frac{m_i - 1}{\hat{b}_i}\right\} = 0, \tag{8.3.3}$$

where $\hat{b}_i = 1 + (m_i - 1)\hat{\rho}$. Notice that the form of $\hat{\mu}$ is the same as that given in Theorem 8.2.2 for the case of $\sigma_i^2 = \sigma^2$ and $\rho_i = \rho$, except that in (8.3.1) an estimator of ρ is used. Each of these equations involves all three of the parameters μ, σ^2, and ρ and an iterative process can be used to obtain estimates. The last equation is nonlinear in $\hat{\rho}$ and the Newton-Raphson method (see, e.g., Kennedy and Gentle, 1980, sec. 10.2.2) can be used to solve for $\hat{\rho}$. The method, in this case, consists of expanding (8.3.3) at the $(t+1)$th iteration step around the estimate of ρ at the tth step. Writing (8.3.3) as $F(\hat{\rho}) = 0$ and letting $\boldsymbol{\theta} = (\mu, \sigma^2, \rho)$, we expand $F(\hat{\rho}^{(t+1)})$ to obtain

$$F(\hat{\rho}^{(t+1)}) \cong F(\hat{\rho}^{(t)}) + \left.\frac{\partial F}{\partial \rho}\right|_{\boldsymbol{\theta} = \hat{\boldsymbol{\theta}}^{(t)}} (\hat{\rho}^{(t+1)} - \hat{\rho}^{(t)}) \cong 0,$$

from which it follows that

$$\hat{\rho}^{(t+1)} \cong \hat{\rho}^{(t)} - F(\hat{\rho}^{(t)}) \Big/ \left.\frac{\partial F}{\partial \rho}\right|_{\boldsymbol{\theta} = \hat{\boldsymbol{\theta}}^{(t)}} \tag{8.3.4}$$

with

$$\frac{\partial F}{\partial \rho} = \frac{1}{(1-\rho)^2} \sum_s m_i d_{1i}[1 - \rho(1-\rho)\, d_{2i}] - \frac{1}{\sigma^2(1-\rho)^2} \sum_s \sum_{s_i} (Y_{ij} - \mu)^2$$
$$+ \frac{1}{\sigma^2(1-\rho)} \left\{ \frac{1}{(1-\rho)^2} \sum_s d_{1i}[1 - \rho(1-\rho)\, d_{2i}] - \sum_s \frac{d_{1i}}{u_i^2}(m_i - 1) \right\},$$

where $d_{1i} = m_i^2(\bar{Y}_{si} - \hat{\mu})/\hat{b}_i$ and $d_{2i} = (m_i - 1)/\hat{b}_i$. Given a set of starting values $\mu^{(0)}$ $\sigma^{(0)2}$, and $\rho^{(0)}$, we can cycle through (8.3.1), (8.3.2), and (8.3.4) until the parameter estimates have changed minimally between two successive steps.

In large samples, this procedure leads to consistent parameter estimates, but in some samples the estimates may, as with ANOVA, produce unreasonable values. Constraints on the solution for $\hat{\rho}$ and $\hat{\sigma}$ can be added to the iterative procedure in such a case. Whether the iterative method converges and the values it converges to may be quite sensitive to the starting values. Performance may be improved by using a quadratic approximation instead of (8.3.4), but this complicates the formulas that must be derived. The data may also be poorly modeled by a multivariate normal distribution, in which case a different likelihood may be more appropriate.

Example 8.3.2. To illustrate the procedure, we will use the same sample from the Labor Force population as in Example 8.3.1. Each cluster was enumerated completely and the parameters μ, σ^2, and ρ for the variable labeled y were estimated using the above algorithm obtaining $\hat{\mu} = 49.80$, $\hat{\sigma} = 69.77$, and $\hat{\rho} = 0.050$. The estimates are very close to those obtained in the earlier example, but this will not necessarily be true for all samples. The S-PLUS™ function *lme*, for linear mixed model estimation, can also be used to compute ML estimates for model (8.1.3) and for more elaborate models.

The methods in this section assume that the data are multivariate normal, which often will not be the case. Some alternative ways of estimating of correlation structures for nonnormal, dependent data are given by Hall and Severini (1998), Liang, Zeger, and Qaqish (1992), and Prentice and Zhao (1991).

8.3.3.* Lower Bound on the Intracluster Correlation

Under model (8.1.3), the intracluster correlation can be negative, but there is a lower bound on its value which we derive here. Define the population between-cluster mean square as

$$B_N = \frac{\Sigma_{i=1}^N M_i(\bar{Y}_i - \bar{Y})^2}{N-1},$$

where $\bar{Y}_i = \Sigma_{j=1}^M Y_{ij}/M_i$ and $\bar{Y} = \Sigma_{i=1}^N M_i \bar{Y}_i/M$. After some algebra, we have

$$E_M(B_N) = \sigma^2(1-\rho) + \sigma^2 \rho D_N, \qquad (8.3.5)$$

where $D_N = (M - \Sigma_{i=1}^N M_i^2/M)/(N-1)$. Because B_N is a sum of squares, its expectation is positive. Setting (8.3.5) to be greater than or equal to 0 and solving for ρ gives

$$\rho \geqslant \frac{-1}{D_N - 1}, \qquad (8.3.6)$$

for $D_N > 1$ (see Exercise 8.8). If the clusters all have the same size $\bar{M}$, then the bound is $-1/(\bar{M}-1)$.

For the Labor Force population in Appendix B, $M = 478$, $N = 115$, $\Sigma_{i=1}^N M_i^2 = 2{,}412$, and the lower bound on ρ is -0.318, which, of course, is well below the small positive estimates we found in Examples 8.3.1 and 8.3.2. Though the clusters in the Labor Force population do vary in size, $-1/(\bar{M}-1) = -0.317$, very close to bound (8.3.6).

8.4. SIMULATION STUDY FOR THE COMMON MEAN MODEL

To get a better idea of the comparative merits of the several estimators we have considered in the class of unbiased estimators, we carried out a small simulation study. In all cases, the population was generated according to model (8.1.1), with $\mu = 20$, and with 100 clusters, divided into four groups. Table 8.3 gives the size breakdown. Clusters within a group were given a common variance σ_i^2 and a common correlation ρ_i. Six population types were considered, as in Table 8.4.

Table 8.3. Sizes for Different Cluster Groups, for Six Population-Types used in Simulation Study

Group	i	M_i
I	1–40	10
II	41–70	20
III	71–90	40
IV	91–100	70

Elements were selected by two *epsem* two-stage sampling schemes. In the first, "*pps*" scheme, a sample of 10 clusters was chosen with probability proportional to M_i, followed by *srs* selection of $m_i = 5$ units from each cluster, giving $m = 50$. In the second, "*srs*" scheme, 10 clusters were chosen by *srswor*, followed by a *srswor* sample of $m_i = M_i/5$ units. Note that, under this latter scheme, $E_\pi(m_i) = 5$. For each of the six population-types, 500 populations of Y values were generated (making use of the random components model within each group to generate clusters with the given values of σ_i^2 and ρ_i) and the total $T = \Sigma_{i=1}^{100} \Sigma_{j=1}^{M_i} Y_{ij}$ recorded. For each population, two samples were taken, one according to each of the above schemes, and values for several estimators calculated: $\hat{T}_{BLU(\rho)}$, $\hat{T}_{BLU(\hat{\rho})}$, $\hat{T}_0$, $\hat{T}_1$, $\hat{T}_R$, $\hat{T}_p$, $\hat{T}_H$. The *BLU* estimators were calculated assuming constant correlation and variance across clusters. The first, $\hat{T}_{BLU(\rho)}$, uses ρ or, where the ρ_i are not all equal, their population mean value $\bar{\rho}$. This assumes a priori knowledge rarely available to the survey analyst. The second, $\hat{T}_{BLU(\hat{\rho})}$, assumes the constant parameter model, and uses the ANOVA development of Section 8.3.1 to estimate ρ (a constant σ cancels out in this estimator). In the case where $\hat{\rho}$ fell below zero, it was set to 0.01. Note that $\hat{T}_{BLU(\hat{\rho})}$ is not strictly a linear estimator. Table 8.5 gives mean square error for each estimator, for each sampling scheme, for each population type.

Table 8.4. Values of Variances and Correlations for Different Cluster Groups, for Six Population-Types used in Simulation Study

	σ_i^2				ρ_i			
Population-Type	I	II	III	IV	I	II	III	IV
σ_i^2, ρ_i constant	20	20	20	20	0.2	0.2	0.2	0.2
σ_i^2 constant, $\rho_i\downarrow$	20	20	20	20	0.4	0.2	0.1	4/70
$\sigma_i^2\uparrow$, ρ_i constant	1	2	3	7	0.2	0.2	0.2	0.2
$\sigma_i^2\uparrow$, $\rho_i\downarrow$	1	2	3	7	0.4	0.2	0.1	4/70
$\sigma_i^2\downarrow$, ρ_i constant	7	3	2	1	0.2	0.2	0.2	0.2
σ_i^2 constant, $\rho_i\uparrow$	20	20	20	20	4/70	0.1	0.2	0.4

Table 8.5. Mean Squared Error/1000 of Estimators of Population Total for the Artificial, Clustered Populations of Tables 8.3 and 8.4

(a) *pps*

Population-type	$\hat{T}_{BLU(\rho)}$	$\hat{T}_{BLU(\hat{\rho})}$	$\hat{T}_0$	$\hat{T}_1$	$\hat{T}_R$	$\hat{T}_p$	$\hat{T}_H$
σ_i^2, ρ_i constant	2073	2068	**2064**	2087	*2888*	**2064**	2087
σ_i^2 constant, $\rho_i\downarrow$	1950	1947	1986	**1929**	*2169*	1986	**1929**
$\sigma_i^2\uparrow$, ρ_i constant	8761	8782	**8386**	9093	*16081*	**8386**	9093
$\sigma_i^2\uparrow$, $\rho_i\downarrow$	7393	7299	**7117**	7638	*12603*	**7117**	7638
$\sigma_i^2\downarrow$, ρ_i constant	5940	5926	*6264*	5702	**4307**	*6264*	5702
σ_i^2 constant, $\rho_i\uparrow$	1979	1988	**1933**	2025	*3324*	**1933**	2025

(b) *srs*

Population-type	$\hat{T}_{BLU(\rho)}$	$\hat{T}_{BLU(\hat{\rho})}$	$\hat{T}_0$	$\hat{T}_1$	$\hat{T}_R$	$\hat{T}_p$	$\hat{T}_H$
σ_i^2, ρ_i constant	**2347**	2409	*2810*	2479	*2810*	2561	*2810*
σ_i^2 constant, $\rho_i\downarrow$	2015	2062	**1927**	2436	**1927**	*2580*	**1927**
$\sigma_i^2\uparrow$, ρ_i constant	7333	7110	*11536*	5761	*11536*	**5510**	*11536*
$\sigma_i^2\uparrow$, $\rho_i\downarrow$	5298	5432	*6963*	4802	*6963*	**4792**	*6963*
$\sigma_i^2\downarrow$, ρ_i constant	8600	8740	**6584**	12233	**6584**	*13323*	**6584**
σ_i^2 constant, $\rho_i\uparrow$	**2441**	2462	*3155*	2484	*3155*	2523	*3155*

Comparisons are meaningful within a row, and between the same row of the two sub-tables. The best estimator in a given row is placed in bold; the worst, in italics. Because of the equalities noted in Theorem 8.2.1, there are various ties. Several observations are in order:

(i) *The BLU* is winner only once. This is not surprising when the ρ_i or σ_i are unequal, since the estimator then rests on a too simple version of the correct model. On the other hand, it is *never* the loser, nor typically close to the loser. $\hat{T}_{BLU(\hat{\rho})}$ seems like a reasonably robust, all purpose tool.

(ii) *The* π-estimator $\hat{T}_0$ can be both best and worst, depending on the variance structure, and the particular method of sampling. It does somewhat better under the *pps* sampling scheme than under the *srs* scheme. This in all likelihood is due to its closeness to $\hat{T}_{BLU}$ under equal allocation (see Exercise 8.12).

(iii) The remaining estimators seem less predictable. The ratio estimator seems especially vulnerable to being quite bad, suggesting that it would be wise *not* to use it. On the other hand, in at least one instance—*pps* schema where the variances decrease as the cluster sizes increase—it is best by a considerable margin. Exercise 8.13 suggests that the likely explanation of this lies in an accidental dovetailing of $\hat{T}_R$ with the particular heteroscedastic structure. Thus, taking into account the heteroscedastic structure might yield big gains. This suggests that it would be worthwhile to adapt the *BLU* estimator to the unequal variance case. Methods need to be developed using diagnostics on residuals, or perhaps through an extension of the maximum likelihood method of Section 8.3.2.

8.5. BIASES OF COMMON MEAN ESTIMATORS UNDER A MORE GENERAL MODEL

All the estimators we have been considering are unbiased under a model where the Y's have a common mean μ. If $\hat{T}_0$, $\hat{T}_{BLU}$, or one of the other estimators is used when the correct model requires $E_M(Y_{ij}) = \mathbf{x}_i'\boldsymbol{\beta}$, the estimator of the total will be biased unless the sample satisfies certain balance conditions. The estimator

$$\hat{T} = \sum_{i \in s} (1 + a_i) m_i \bar{Y}_{si}$$

is unbiased under model (8.1.1) when $\Sigma_s (1 + a_i) m_i = M$. However, if $E_M(Y_{ij}) = \mathbf{x}_i'\boldsymbol{\beta}$, then

$$E_M(\hat{T} - T) = \left[\sum_s (1 + a_i) m_i \bar{\mathbf{x}}_{si}' - M\bar{\mathbf{x}}' \right] \boldsymbol{\beta}, \tag{8.5.1}$$

where $\bar{\mathbf{x}}_{si} = \Sigma_{j \in s_i} \mathbf{x}_{ij}/m_i$ and $\mathbf{x}_{ij}$ is the p-vector of auxiliaries associated with unit j in cluster i. The bias will be zero if the sample meets the condition,

$$\sum_s (1 + a_i) m_i \bar{\mathbf{x}}_{si} = M\bar{\mathbf{x}}. \tag{8.5.2}$$

Consider the expansion estimator $\hat{T}_0$ as a representative case. We have $a_i = (M - m)/m$ and the bias is equal to

$$E_M(\hat{T}_0 - T) = M\,[\bar{\mathbf{x}}_s' - \bar{\mathbf{x}}']\,\boldsymbol{\beta},$$

where $\bar{\mathbf{x}}_s = \Sigma_{i \in s} \Sigma_{j \in s_i} \mathbf{x}_{ij}/m$ and $\bar{\mathbf{x}} = \Sigma_{i=1}^{N} \Sigma_{j=1}^{M_i} \mathbf{x}_{ij}/M$.

Thus, the more $\bar{\mathbf{x}}_s$ differs from $\bar{\mathbf{x}}$, the more biased $\hat{T}_0$ will be (unless $\boldsymbol{\beta} = 0$). In *principle*, we can protect the expansion estimator against bias by selecting a sample with simple unweighted balance on the x's. The *practical* difficulty with this is that typically in clustered populations the values of the auxiliary variables for individual units are not known in advance of sampling. In the Labor Force population, for example, we found in Chapter 7 that age, sex, and hours worked per week were good predictors of weekly wages. None of these explanatory variables is likely to be available on individual persons in advance of sampling from a real population.

Because of this, selecting a sample that is balanced in the sense that $\bar{\mathbf{x}}_s = \bar{\mathbf{x}}$ would require special screening procedures that may be quite expensive. After the sample of clusters is selected, all units in the clusters would be screened to determine their values of the auxiliaries. The subsamples of units within clusters would then be selected to achieve $\bar{\mathbf{x}}_s = \bar{\mathbf{x}}$. However, the composition of the sample clusters may make even this elaborate approach unfeasible. If age group is an important auxiliary and there are no persons in the sample clusters over 65 years old, then forcing the proportion of 65+ year-olds in the sample to equal that in the population would be impossible. In other words, the quandary is that balance requires information on the secondary units, prior to sampling the primary units, but we usually need the primary units to arrive at the secondary units.

In this circumstance, what remedies are there? One partial remedy perhaps is to rely on probability sampling to achieve balance in expectation. If n clusters are selected with probability proportional to the size, M_i, of each and an equal probability sample of $\bar{m}$ units is selected within each sample cluster, then the selection probability of unit (ij) is $\pi_{ij} = n\bar{m}/M$. The design-expectation of $\bar{\mathbf{x}}_s$ is then $\bar{\mathbf{x}}$. A similar sort of approach can be used for some of the other members of the class of unbiased estimators (see Exercise 8.14), but probably not in general—in particular, not for the *BLU*. In any case, if, once we have the sample values of x in hand, and find that $\bar{\mathbf{x}}_s \neq \bar{\mathbf{x}}$, then we are still left with a situation in which we must worry about the bias of the estimator $\hat{T}_0$.

In this circumstance, we might try a quick fix. Using the sample data, we can estimate the parameter $\boldsymbol{\beta}$. For unbiasedness, this need not be a sophisticated estimate; ordinary least squares might do, especially if correlations do not run high. If there are clear reasons to suspect heteroscedasticity, we might take that into account. We then can construct an adjusted estimator

$$\hat{T}_{0,\mathrm{adj}} = \hat{T}_0 + M(\bar{\mathbf{x}}' - \bar{\mathbf{x}}_s')\hat{\boldsymbol{\beta}}.$$

More generally, we can use

$$\hat{T}_{\mathrm{adj}} = \hat{T} + \left[M\bar{\mathbf{x}}' - \sum_s (1 + a_i) m_i \bar{\mathbf{x}}_{si}' \right] \hat{\boldsymbol{\beta}},$$

for $\hat{T} = \Sigma_{i \in s}(1 + a_i) m_i \bar{Y}_{si}$. This adjusted (or "calibration") estimator is a more general linear estimator and should suffice for handling bias, at the cost of some noise when the adjustment is unnecessary.

However, we might prefer to use from the start a more general regression estimator tailored to the clustered population, with careful selection of the explanatory variables in the model.

8.6. ESTIMATION UNDER A MORE GENERAL REGRESSION MODEL

In many surveys, particularly of households, the variable(s) of interest depend on characteristics of individual units. For example, income may depend on a person's age, race, sex, profession, educational attainment, and other factors. Considerations like these lead us to the general linear model

$$E_M(\mathbf{Y}) = \mathbf{X}\boldsymbol{\beta}$$
$$\mathrm{var}_M(\mathbf{Y}) = \mathbf{V} \tag{8.6.1}$$

already introduced in Chapter 2.

The optimal predictor under this model was derived in Theorem 2.2.1. As noted previously, the covariance structure in a clustered population will generally be unknown and possibly difficult to estimate. We will here concentrate on finding an estimator of the total that is unbiased under (8.6.1) but not necessarily optimal. Given an unbiased estimator, though, finding a variance estimator that does account for the proper covariance structure is important. In Chapter 9, we will study variance estimators that are consistent under the type of clustered covariance specification in model (8.1.1) and under more general covariance specifications.

If we set $\mathbf{V} = \mathbf{I}$, the *BLU* predictor of the total from Theorem 2.1.1 is

$$\hat{T} = \mathbf{1}_s'\mathbf{Y}_s + \mathbf{1}_r'\mathbf{X}_r\hat{\boldsymbol{\beta}} \tag{8.6.2}$$

with $\hat{\boldsymbol{\beta}} = \mathbf{A}_s^{-1}\mathbf{X}_s'\mathbf{Y}_s$ if $\mathbf{A}_s = \mathbf{X}_s'\mathbf{X}_s$ is nonsingular. If an over-parameterized model is used, then we take $\boldsymbol{\beta}^o = \mathbf{G}\mathbf{X}_s'\mathbf{Y}_s$, where $\mathbf{G}$ is a generalized inverse of $\mathbf{A}_s$ (see Chapter 7). As in expression (7.4.10), the weight vector corresponding to $\hat{T}$, when a g-inverse is used, is

$$\dot{\mathbf{g}}_s = \mathbf{X}_s\mathbf{G}\mathbf{X}_r'\mathbf{1}_r + \mathbf{1}_s, \tag{8.6.3}$$

The estimator in (8.6.2) is unbiased under model (8.6.1), and the error variance is, in general,

$$\mathrm{var}_M(\hat{T} - T) = \mathbf{1}_r'\mathbf{X}_r\mathbf{G}\mathbf{B}_s\mathbf{G}\mathbf{X}_r'\mathbf{1}_r + \mathbf{1}_r'\mathbf{V}_{rr}\mathbf{1}_r - 2\mathbf{1}_r'\mathbf{X}_r\mathbf{G}\mathbf{X}_s'\mathbf{V}_{sr}\mathbf{1}_r, \tag{8.6.4}$$

where $\mathbf{B}_s = \mathbf{X}_s'\mathbf{V}_{ss}\mathbf{X}_s$. The estimator (8.6.2) can also be expressed as

$$\hat{T} = \mathbf{1}_s'\mathbf{Y}_s + \mathbf{T}_{xr}'\hat{\boldsymbol{\beta}},$$

where $\mathbf{T}_{xr} = \Sigma_{i\in s}\Sigma_{j\in r_i}\mathbf{x}_{ij} + \Sigma_{i\in r}\Sigma_{j=1}^{M_i}\mathbf{x}_{ij}$ is the vector of nonsample totals of the auxiliary variables. If the population totals of the auxiliaries, $\mathbf{T}_x =$

$\Sigma_{i=1}^{N} \Sigma_{j=1}^{M_i} \mathbf{x}_{ij}$, are known from the frame, a census, or some other source, then the nonsample total $\mathbf{T}_{xr}$ can be obtained by subtraction.

This type of regression estimator is often extremely useful, especially in household surveys. Frequently, the population comprises groups whose means differ. This corresponds to the special case of (8.6.1), where

$$E_M(Y_{ij}) = \mu_c \tag{8.6.5}$$

when unit ij is in group c ($c = 1, \ldots, C$).

Often group membership is unknown at the time of sampling and can only be determined when the data are collected. A group can cut across clusters and, in design-based sampling, is usually called a *post-stratum* if the membership of a unit in a group is determined *after* the sample is selected. The two-way model with interaction in Section 7.8 illustrated this for an unclustered population. Men and women may, for example, have different average incomes in a population of households. A sample of clusters is selected and households are listed within the sample clusters. Important explanatory variables, like the age and sex of household members, are collected only for individuals who are in the sample. However, the total numbers in the population of males and females in different age groups may be known from a census or from reliable demographic projections.

Under the post-stratification model (8.6.5), the estimator of the total (8.6.2) reduces to

$$\hat{T}_{PS} = \sum_{c=1}^{C} M_c \bar{Y}_{sc},$$

where M_c is the number of units in the population in group c, $\bar{Y}_{sc} = \Sigma_{i\in s} \Sigma_{j\in s_{ic}} Y_{ij}/m_c$, m_c is the number of sample units in group c across all sample clusters, and s_{ic} is the set of sample units in sample cluster i that are also in group c. The variance is given by (8.6.4) with $\mathbf{X}$ an $M \times C$ matrix of 0's and 1's. An element in column c in the matrix is 1 if the corresponding unit is in post-stratum c and 0 if not.

The estimator (8.6.2) can also be evaluated in a cluster sample using both qualitative and quantitative variables, as we did in Chapter 7 for unclustered samples. We illustrate the method using the Labor Force Population first considered in Section 7.9.3. This population is derived from the U.S. Current Population Survey (CPS), which is a clustered sample of households. The clusters we use for illustration here are compact geographic areas composed of about four nearby households, selected in this way to reduce travel costs, with the data being collected by personal interviews at the homes of respondents.

The second and third models in Section 7.9.3 predicted weekly wages based on sex, age, the interaction of sex and age, and the number of hours worked per week. In the next example, we fit the same models using a cluster sample from the Labor Force population to predict the total.

Table 8.6. A Sample of 10 Clusters and 2 Persons per Cluster from the Labor Force Population

Cluster Number	Sample Person within Cluster	Cluster Number	Sample Person within Cluster
4	12	55	218
4	13	55	219
12	39	69	272
12	40	69	274
24	83	84	344
24	84	84	345
25	87	89	364
25	88	89	365
46	181	113	469
46	182	113	471

Example 8.6.1. Estimating total weekly wages in the Labor Force Population. Table 8.6 lists the sample cluster and person numbers in a sample of $n = 10$ clusters from the Labor Force population. The clusters were selected by *srswor*, and from each sample cluster, an *srswor* of $\bar{m} = 2$ persons was selected. Consider predicting the weekly wage for person j in cluster i, who is in the kth level of sex, lth level of age, and who works z_{ij} hours per week, by the model

$$E(Y_{ij}) = \mu + \alpha_k + \beta_l + \gamma z_{ij}, \qquad k = 1, 2; l = 1, \ldots, 4. \qquad (8.6.6)$$

As in Section 7.9.3, the two levels of sex are male and female and the four age categories are 19 and under, 20–24, 25–34, and 35+. Note that we could also write the model, using indicator variables, as

$$E(Y_{ij}) = \mu + \sum_{k=1}^{2} \alpha_k x_{kij} + \sum_{l=1}^{4} \tilde{x}_{lij}\beta_l + \gamma z_{ij},$$

where $x_{kij} = 1$ if person (ij) is the kth sex and 0 if not and $\tilde{x}_{lij} = 1$ if the person is in the lth age group and 0 if not. The matrix $\mathbf{X}_s$ can be constructed from the values of x, $\tilde{x}$, , and z for the sample persons. The extension of model (8.6.6) containing the interaction of age and sex is

$$E(Y_{ij}) = \mu + \alpha_k + \beta_l + (\alpha\beta)_{ij} + \gamma z_{ij}, \qquad k = 1, 2; l = 1, \ldots, 4. \qquad (8.6.7)$$

When specifying models like (8.6.7), it is easy to create a situation where some cross-classified cells are empty. It is important to make simple frequency tabulations from the sample to check this. For the sample in Table 8.6, the numbers of sample persons by age and sex are as follows:

Age Group	Male	Female
19 and under	1	1
20–24	3	2
25–34	5	1
35+	3	4

Note that every combination of age, sex, race need not occur in each sample cluster to allow computation of the estimated total. If, however, even one of the age × sex cells were empty in the entire sample, then the model with interaction could not be fully estimated; this was discussed in Section 7.9.3.

Table 8.7 lists the solutions to the normal equations computed using the S-PLUS function *ginv*. These solutions are quite different from those computed from the systematic sample of $n = 48$ persons used in Section 7.9.3. For model (8.6.5) with no interaction the vector of nonsample auxiliary totals is $\mathbf{T}'_{xr} = (458, 220, 238, 45, 49, 123, 241, 17526)$ while for model (8.9.6) with interaction, $\mathbf{T}'_{xr} = (458, 220, 238, 45, 49, 123, 241, 14, 25, 67, 114, 31, 24, 56, 127, 17526)$. The estimated totals using the two models are 138,555.5 and 139,816.1, slightly lower than the actual population total of 140,818.

The other clustered population in Appendix B is based on a sample of schools and students within schools. The source of the data is the Third

Table 8.7. Parameter Solutions for Models that Include Sex, Age, and Hours Worked as Auxiliaries for Predicting the Total of Weekly Wages in a Cluster Sample from the Labor Force Population

Parameter Solution	Model with No Sex–Age Interaction	Model with Sex–Age Interaction
μ^o	31.81	54.82
α_1^o (male)	78.37	70.36
α_2^o (female)	−46.55	−15.53
β_1^o ($\leqslant 19$)	−82.95	−51.96
β_2^o (20–24)	−65.00	−27.00
β_3^o(25–34)	43.31	20.12
β_4^o (35+)	136.46	113.66
$(\alpha\beta)_{11}^o$ (male × $\leqslant 19$)		−65.49
$(\alpha\beta)_{12}^o$ (male × 20–24)		−88.74
$(\alpha\beta)_{13}^o$ (male × 25–34)		78.20
$(\alpha\beta)_{14}^o$ (male × 35+)		146.38
$(\alpha\beta)_{21}^o$ (female × $\leqslant 19$)		13.53
$(\alpha\beta)_{22}^o$ (female × 20–24)		61.74
$(\alpha\beta)_{23}^o$ (female × 25–34)		−58.08
$(\alpha\beta)_{24}^o$ (female × 35+)		−32.72
$\hat{\gamma}$ (hours worked)	4.62	3.31

International Mathematics and Science Study (TIMSS) conducted in 1993 (Caslyn, Gonzalez, and Frase, 1999). The population, described in the Appendix B.6, contains data for third-grade students in the United States and includes scores on achievement tests in mathematics and science and other student demographic variables. The clusters in this population are the schools which range in size from 7 to 29 students. Math and science achievement scores depend on a number of school and student characteristics, including region of the country, sex, ethnicity, the language spoken at the student's home, the type of community, and the size of the school as measured by enrollment. The coding of these variables is described in more detail in Appendix B.6. The next example illustrates the use of all of these factors in estimating the average test scores for mathematics and science. Thus we have an illustration of using the same set of explanatory variables for estimating population quantities for two different target variables.

Example 8.6.2. Estimating mean test scores in the third-grade population. Table 8.8 lists a sample of 10 schools and 5 students per school. The schools were selected with probability proportional to the number of students per school and the students in each sample school were selected by simple random sampling without replacement.

From this sample, we estimate the mean math and science score per student using a model that includes an intercept and main effects for the qualitative variables, region, sex, ethnicity, language spoken at home, and type of community, plus the quantitative variable, school enrollment. The parameter solutions, computed using a g-inverse, are shown in Table 8.9. (Type of community and ethnicity have been collapsed to 2 and 5 categories compared to the 3 and 6 categories available in the population in Appendix B.6.)

The parameter solutions, $\boldsymbol{\beta}^o = \mathbf{G}\mathbf{X}_s'\mathbf{Y}_s$, being dependent on the Y variable, are different for mathematics and science. However, the weight vector, defined

Table 8.8. A Sample of $n = 10$ Schools and 5 Students per School from the Third-Grade Population

School Number	Sample Students within Cluster
6	90, 93, 96, 104, 109
19	362, 363, 371, 372, 378
28	519, 522, 524, 528, 529
47	848, 852, 856, 858, 870
61	1122, 1123, 1125, 1132, 1142
83	1505, 1507, 1509, 1515, 1517
110	1972, 1973, 1976, 1979, 1980
124	2206, 2209, 2210, 2215, 2216
132	2351, 2358, 2362, 2365, 2367
133	2371, 2373, 2376, 2378, 2387

Table 8.9. Parameter Solutions for the Model that Includes Region, Sex, Ethnicity, Language Spoken at Home, Type of Community, and School Enrollment for Predicting the Mean Mathematics and Science Scores in a Cluster Sample from the Third-Grade Population

Parameter Solution	Math	Science
Intercept	130.92	155.73
Region		
Northeast	62.24	56.48
South	−12.64	−9.85
Central	48.47	57.11
West	32.85	51.99
Community type		
Suburban	84.54	93.99
Non-suburban	46.38	61.74
Sex		
Female	56.12	55.90
Male	74.80	99.83
Language spoken at home same as test language		
Always	69.50	114.08
Sometimes	48.84	36.88
Never	12.57	4.77
Ethnicity		
White	30.55	58.60
Black	−3.70	−22.02
Hispanic	26.35	51.33
Asian	47.89	51.15
Other	29.82	16.66
Enrollment	0.18	0.03

by $\dot{\mathbf{g}}_s = \mathbf{X}_s\mathbf{G}\mathbf{X}_r'\mathbf{1}_r + \mathbf{1}_s$, is the same for both mathematics and science and for any other variable whose mean or total is estimated using the same set of explanatory variables as in Table 8.9. For this sample, the estimated means for the two subjects are computed as $\hat{\mu} = \dot{\mathbf{g}}_s'\mathbf{Y}_s/M$ and are 478.52 for mathematics and 496.13 for science, not far from the population values of 477.70 and 503.53 respectively.

8.7.* ROBUSTNESS AND OPTIMALITY

In Chapters 3 and 4 we studied situations where samples could be selected that would protect the *BLU* predictor against certain kinds of departures from a working model and that would also be optimal if the working model was correct. Analogous results due to Tam (1995) exist for the *BLU* predictor in clustered populations. The practical import appears to be substantially less than those in Chapters 3 and 4, for the reasons given in Section 8.6 above and in the discussion of Example 8.7.1 below. Consider the following model:

$$E_M(\mathbf{Y}) = \mathbf{X}\boldsymbol{\beta}$$
$$\text{cov}_M(Y_{ij}, Y_{i'j'}) = \begin{cases} \sigma_i^2 & i = i', j = j' \\ \sigma_i^2 \rho_i & i = i', j \neq j' \\ 0 & \text{otherwise.} \end{cases} \tag{8.7.1}$$

This is more general than models (8.1.1) and (8.1.3) in allowing the mean to depend on a set of auxiliaries **X**. Let $\hat{T}(\mathbf{X}:\mathbf{V})$ be the *BLU* estimator based on (8.7.1) and assume that **X** has full column rank. For use in the theorem below, define the $M_i \times M_i$ matrix $\tilde{\mathbf{v}}_i = \sigma_i^2 I_{M_i}$ and the $M \times M$ matrix $\tilde{\mathbf{V}} = \text{blkdiag}(\tilde{\mathbf{v}}_i)$. Let $\mathbf{V}_i = \sigma_i^2(1 - \rho_i)\mathbf{I}_{M_i} + \sigma_i^2 \rho_i \mathbf{1}_{M_i}\mathbf{1}'_{M_i}$, so that the population covariance matrix can be written as $\mathbf{V} = \text{blkdiag}(\mathbf{V}_i)$. The theorem, which we have written to closely parallel Theorem 4.2.1, is:

Theorem 8.7.1 (Tam, 1995). Under model (8.7.1), if both $\mathbf{V}\mathbf{1}_M$ and $\tilde{\mathbf{V}}^{1/2}\mathbf{1}_M \in \mathscr{M}(\mathbf{X})$, then

$$\text{var}_M[\hat{T}(\mathbf{X}:\mathbf{V}) - T] \geqslant \left[\frac{1}{m^*}(\mathbf{1}'_M\tilde{\mathbf{V}}^{1/2}\mathbf{1}_M)^2 - \mathbf{1}'_M\mathbf{V}\mathbf{1}_M\right] \tag{8.7.2}$$

where $m^* = \Sigma_{i \in s} m_i/[1 + (m_i - 1)\rho_i]$. The bound is achieved if and only if the sample satisfies

$$\frac{1}{m^*}\mathbf{1}'_s\tilde{\mathbf{V}}_{ss}^{1/2}\mathbf{V}_{ss}^{-1}\mathbf{X}_s = \frac{\mathbf{1}'_M\mathbf{X}}{\mathbf{1}'_M\tilde{\mathbf{V}}^{1/2}\mathbf{1}_M}, \tag{8.7.3}$$

in which case

$$\hat{T}(\mathbf{X}:\mathbf{V}) = \frac{1}{m^*}(\mathbf{1}'_M\tilde{\mathbf{V}}^{1/2}\mathbf{1}_M)(\mathbf{1}'_s\tilde{\mathbf{V}}_{ss}^{1/2}\mathbf{V}_{ss}^{-1}\mathbf{Y}_s). \tag{8.7.4}$$

Proof: The proof is similar to that of Theorem 4.2.1. When $\mathbf{V}\mathbf{1}_M \in \mathscr{M}(\mathbf{X})$, a generalization of Lemma 4.2.1 for the case of a non-diagonal covariance matrix gives $\hat{T}(\mathbf{X}:\mathbf{V}) = \mathbf{1}'_M\mathbf{X}\hat{\boldsymbol{\beta}}$ and $\text{var}_M[\hat{T}(\mathbf{X}:\mathbf{V}) - T] = (\mathbf{1}'_M\mathbf{X}\mathbf{A}_s^{-1}\mathbf{X}'\mathbf{1}_M - \mathbf{1}'_M\mathbf{V}\mathbf{1}_M)$, where $\mathbf{A}_s = \mathbf{X}'_s\mathbf{V}_{ss}^{-1}\mathbf{X}_s$ (see Exercise 8.15). The variance is minimized when $\mathbf{a}'\mathbf{A}_s^{-1}\mathbf{a}$ is minimized, where $\mathbf{a} = \mathbf{X}'\mathbf{1}_M$. Because $\tilde{\mathbf{V}}^{1/2}\mathbf{1}_M \in \mathscr{M}(\mathbf{X})$, there exists a p-vector $\mathbf{c}_1$ such that $\tilde{\mathbf{V}}^{1/2}\mathbf{1}_M = \mathbf{X}\mathbf{c}_1$ and, since $\tilde{\mathbf{V}}^{1/2}$ is diagonal, this ensures that $\tilde{\mathbf{V}}_{ss}^{1/2}\mathbf{1}_s = \mathbf{X}_s\mathbf{c}_1$ for every sample s. From this it follows that

$$\begin{aligned} \mathbf{1}'_s\tilde{\mathbf{V}}_{ss}^{1/2}\mathbf{V}_{ss}^{-1}\tilde{\mathbf{V}}_{ss}^{1/2}\mathbf{1}_s &= \mathbf{c}'_1\mathbf{X}'_s\mathbf{V}_{ss}^{-1}\mathbf{X}_s\mathbf{c}_1 \\ &= \mathbf{c}'_1\mathbf{A}_s\mathbf{c}_1. \end{aligned}$$

Using the form of $\mathbf{V}_{ss(i)}^{-1}$ in Theorem 8.2.2, we also have, after some calculation, $\mathbf{1}_s'\tilde{\mathbf{V}}_{ss}^{1/2}\mathbf{V}_{ss}^{-1}\tilde{\mathbf{V}}_{ss}^{1/2}\mathbf{1}_s = m^*$. Factoring $\mathbf{A}_s$ and applying Schwartz's inequality (Appendix A) gives

$$\begin{aligned} m^*(\mathbf{a}'\mathbf{A}_s^{-1}\mathbf{a}) &= (\mathbf{a}'\mathbf{A}_s^{-1}\mathbf{a})(\mathbf{c}_1'\mathbf{A}_s\mathbf{c}_1) \\ &= (\mathbf{a}'\mathbf{A}_s^{-1/2}\mathbf{A}_s^{-1/2}\mathbf{a})(\mathbf{c}_1'\mathbf{A}_s^{1/2}\mathbf{A}_s^{1/2}\mathbf{c}_1) \\ &\geqslant [(\mathbf{a}'\mathbf{A}_s^{-1/2})(\mathbf{A}_s^{1/2}\mathbf{c}_1)]^2 \\ &= (\mathbf{a}'\mathbf{c}_1)^2 \\ &= (\mathbf{1}_M'\tilde{\mathbf{V}}^{1/2}\mathbf{1}_M)^2. \end{aligned}$$

This establishes the lower bound on the error variance. Equality occurs in (8.7.2) if and only if $\mathbf{a}' = k\mathbf{c}_1'\mathbf{A}_s$ where $k = (\mathbf{1}_M'\tilde{\mathbf{V}}^{1/2}\mathbf{1}_M)/m^*$. This is equivalent to (8.7.3) since $\mathbf{c}_1'\mathbf{A}_s = \mathbf{1}_s'\tilde{\mathbf{V}}_{ss}^{1/2}\mathbf{V}_{ss}^{-1}\mathbf{X}_s$. Now $\mathbf{1}_M'\mathbf{X} = (\mathbf{1}_M'\tilde{\mathbf{V}}^{1/2}\mathbf{1}_M)(\mathbf{1}_s'\tilde{\mathbf{V}}_{ss}^{1/2}\mathbf{V}_{ss}^{-1}\mathbf{X}_s)/m^*$ by (8.7.3), and so, making repeated use of $\mathbf{c}_1'\mathbf{X}_s' = \mathbf{1}_s'\tilde{\mathbf{V}}_{ss}^{1/2}$,

$$\begin{aligned} \hat{T}(\mathbf{X}:\mathbf{V}) &= \mathbf{1}_M'\mathbf{X}\hat{\boldsymbol{\beta}} \\ &= \frac{(\mathbf{1}_M'\tilde{\mathbf{V}}^{1/2}\mathbf{1}_M)}{m^*}\mathbf{1}_s'\tilde{\mathbf{V}}_{ss}^{1/2}\mathbf{V}_{ss}^{-1}\mathbf{X}_s\mathbf{A}_s^{-1}\mathbf{X}_s'\mathbf{V}_{ss}^{-1}\mathbf{Y}_s \\ &= \frac{(\mathbf{1}_M'\tilde{\mathbf{V}}^{1/2}\mathbf{1}_M)}{m^*}\mathbf{c}_1'\mathbf{X}_s'\mathbf{V}_{ss}^{-1}\mathbf{X}_s\mathbf{A}_s^{-1}\mathbf{X}_s'\mathbf{V}_{ss}^{-1}\mathbf{Y}_s \\ &= \frac{1}{m^*}(\mathbf{1}_M'\tilde{\mathbf{V}}^{1/2}\mathbf{1}_M)(\mathbf{1}_s'\tilde{\mathbf{V}}_{ss}^{1/2}\mathbf{V}_{ss}^{-1}\mathbf{Y}_s). \end{aligned}$$

The form of the error variance follows from the fact that $\mathbf{a}'\mathbf{A}_s^{-1}\mathbf{a} = (\mathbf{1}_M'\tilde{\mathbf{V}}^{1/2}\mathbf{1}_M)^2/m^*$ at balance. □

As in Chapter 3, we can rewrite the reduced form of the *BLU* predictor in scalar notation:

$$\hat{T}(\mathbf{X}:\mathbf{V}) = \frac{M\bar{\sigma}}{m^*}\sum_{i\in s}\sigma_i\frac{m_i\bar{Y}_{si}}{v_i}, \tag{8.7.5}$$

where $\bar{\sigma} = \Sigma_{i=1}^N M_i\sigma_i/M$. (Recall that $v(\bar{Y}_{si}) = \sigma_i^2[1 + (m_i - 1)\rho_i]/m_i \equiv v_i/m_i$ is the variance of the sample cluster mean $\bar{Y}_{si}$.) The balance condition (8.7.3) is alternatively

$$\frac{1}{m^*}\sum_{i\in s}\frac{m_i\sigma_i\bar{\mathbf{x}}_{si}}{v_i} = \frac{\Sigma_{i=1}^N M_i\bar{\mathbf{x}}_i}{\Sigma_{i=1}^N M_i\sigma_i}. \tag{8.7.6}$$

Note that in the special case of $\rho_i = \rho = 0$, we have $\tilde{v}_i = \sigma_i^2$, $m^* = m$. In that case, the lower bound on the error variance (8.7.2) reduces to the bound in Theorem 4.2.1, the balance condition (8.7.3) is the same as root($\mathbf{V}$) balance in

(4.2.4), and the estimator of the total becomes $\hat{T}(\mathbf{X}:\mathbf{V}) = m^{-1}(\mathbf{1}'_M\mathbf{V}^{1/2}\mathbf{1}_M)$ $(\mathbf{1}'_s\mathbf{V}_{ss}^{-1/2}\mathbf{Y}_s)$, which is the same as the estimator in Theorem 4.2.1.

Example 8.7.1. Take the special case of model (8.7.1) with $\sigma_i^2 = \sigma^2$ and suppose that an equal number of units $m_i = \bar{m}$ is subsampled from each cluster. By simplifying (8.7.6), the balance condition (8.7.3) is easily shown to be $\bar{\mathbf{x}}_s = \bar{\mathbf{x}}$, that is, simple unweighted balance on the variables in the $\mathbf{X}$ matrix. From (8.7.5) the *BLU* predictor in the balanced case reduces to the expansion estimator $\hat{T}_0 = M\bar{Y}_s$, $\bar{Y}_s = \Sigma_{i\in s}\Sigma_{j\in s_i} Y_{ij}/m$. The balance condition $\bar{\mathbf{x}}_s = \bar{\mathbf{x}}$ seems natural, albeit difficult to achieve in practice. The condition $\mathbf{V}\mathbf{1}_M \in \mathscr{M}(\mathbf{X})$ is awkward. For example, when $\sigma_i^2 = \sigma^2$, we have $\tilde{\mathbf{V}}^{1/2}\mathbf{1}_M = \sigma\mathbf{1}_M$ and

$$\mathbf{V}\mathbf{1}_M = \begin{bmatrix} \sigma^2\{1 + (M_1 - 1)\rho\}\mathbf{1}_{M_1} \\ \vdots \\ \sigma^2\{1 + (M_N - 1)\rho\}\mathbf{1}_{M_N} \end{bmatrix}.$$

Thus, if $\mathbf{V}\mathbf{1}_M = \mathbf{X}\mathbf{c}_2$, we must have $\Sigma_{k=1}^p c_{2k}x_{ijk} = \sigma^2[1 + (M_i - 1)\rho]$ for all $j = 1, \ldots, M_i$. This condition can be forced by having the row corresponding to (cluster i, unit j) include $(1 \;\; M_i)$, but the regression structure for Y is unlikely to include M_i. Since M_i is typically large and may be quite variable, its introduction is likely to add a great deal of extraneous noise into the estimation process. The lower bound on the error variance, achieved in a balanced sample, if it can be selected, is

$$\frac{M^2\sigma^2}{n\bar{m}}[1 + (\bar{m} - 1)\rho] - \sum_{i=1}^{N} M_i\sigma^2[1 + (M_i - 1)\rho].$$

Under standard conditions, the first term is $O(N^2/n)$ while the second is $O(N)$, implying that in samples with a large number of clusters the bound is approximately $M^2\sigma^2[1 + (\bar{m} - 1)\rho]/n\bar{m}$. This can be shown to approximate the error variance of $\hat{T}_0$ when the sampling fraction of units, m/M, and of clusters, n/N, are both small.

The equivalence of the regression estimator to the expansion estimator in balanced sampling has some practical implications. The most important of these is in sample size determination. As we noted above, the lower bound on the error variance of the *BLU* predictor is approximately

$$M^2\sigma^2[1 + (\bar{m} - 1)\rho]/n\bar{m},$$

that is, the approximate error variance of $\hat{T}_0$. Thus, results on sample size determination for $\hat{T}_0$ conceivably could be used as a guide to determine sample sizes in the more general situation. In the following sections, we return to the simple estimators based on the common mean model and give some results on the design of two stage samples.

8.8. EFFICIENT DESIGN FOR THE COMMON MEAN MODEL

Designing a sample survey requires striking a balance between many factors, not all of which lend themselves to mathematical analysis. Availability and mobility of qualified personnel and necessary equipment are often serious limiting factors. Designs for which mathematically simple estimators are appropriate are often desirable because simple estimates having high "face validity" or intuitive appeal are more readily used than ones derived from esoteric mathematical manipulations. Great efficiency under one set of assumptions may entail risk of great inefficiency when those assumptions are violated. These and other considerations must enter into the choice of a sampling plan for a given survey situation.

We first examine how to optimally select clusters. We look at the design problem for the *BLU* estimator under model (8.1.1), and then consider, under greater restrictions, the situation in which $\hat{T}_0$ or one of the other unbiased estimators is used. Next, we take as given the *first* stage units, and inquire about the optimal allocation of the second stage units.

Finally, we examine the optimal relative allocation of first and second stage units under a simple cost model with simplifying assumptions. Studying optimal designs under such simple cost models can generate useful insights. In practice, optimal allocation will be achieved through complicated programming algorithms such as that due to Chromy (1987). All results in this section refer to special cases of model (8.1.1) when a two stage sample is selected. The reader is referred to Royall (1976b) for further results.

8.8.1. Choosing the Set of Sample Clusters for the *BLU* Estimator

For minimizing the variance of $\hat{T}_{BLU}$, when n and m_s are fixed, and $\sigma_i^2 = \sigma^2$ and $\rho_i = \rho > 0$, the optimal first stage sample consists of the n largest clusters. To see this, suppose that some cluster j is in s, with m_j elements to be sampled from cluster j. Suppose a larger cluster l is not in s. The variance will be reduced if cluster j is replaced in the sample by cluster l, with the same number m_j of elements selected from cluster l.

This is apparent from the following expression, obtained by reorganizing (8.2.10),

$$\begin{aligned} \text{var}_M(\hat{T}_{BLU} - T)/\sigma^2 &= (1-\rho)(M-m) \\ &\quad + \rho\left[\sum_r M_i^2 + \sum_s (M_i - m_i)^2 \frac{1-\rho}{1+(m_i-1)\rho}\right] \\ &\quad + \frac{\{M - \Sigma_s m_i[1+\rho(M_i-1)]/[1+(m_i-1)\rho]\}^2}{\Sigma_s m_i/[1+(m_i-1)\rho]}. \end{aligned} \tag{8.8.1}$$

Since $\rho > 0$, the last term on the right-hand side of (8.8.1) becomes smaller when a larger value of M_i is substituted in the sample. Letting $q_j = (1 - \rho)/[1 + (m_j - 1)\rho]$, the difference between the second term of (8.8.1) with cluster j in the sample and with cluster l in the sample is

$$(M_l^2 - M_j^2)(1 - q_j) + 2m_j q_j (M_l - M_j)$$

which is non-negative since $0 < q_j \leqslant 1$ when $\rho > 0$.

For the estimator $\hat{T}_{BLU}$, a given sample of clusters, and a given value of m, the question is open of how the m sampling units should be allocated among the clusters (see remarks in Section 8.1.1). One observation can be noted: when $\sigma_i^2 = \sigma^2$ and $\rho_i = \rho$ with $\rho = 0$ or 1, the error variance does not depend on the allocation rule (see Exercise 8.16).

8.8.2. Choosing the Set of Sample Clusters for the Unbiased Estimators

In this section, we continue to consider the constant parameter model (8.1.3) where the variance and correlation parameters are equal in all clusters. As in the preceding section, we assume $\rho > 0$. When choosing the particular set of sample clusters, we must consider the allocation rule that will be used.

We first consider equal allocation of sample units to clusters. When equal allocation is used, that is, $m_i \equiv \bar{m} = m/n$, we have, by Theorem 8.2.1 that $\hat{T}_p = \hat{T}_0$ and $\hat{T}_H = \hat{T}_1$, so there are only three distinct cases to consider—$\hat{T}_0$, $\hat{T}_1$, and $\hat{T}_R$. In each case, $\text{var}_M(\hat{T} - T) = A + V$, where A is an unvarying component which does not depend on the sizes of the sample clusters, and V varies with the estimator and does depend on the sizes of the sample clusters. The fixed component is

$$A = (1 - \rho)\sigma^2(M - m)M/m + \rho\sigma^2\left(\sum_{i=1}^{N} M_i^2 + M^2/n\right).$$

The second component V is given in Table 8.10.

For $\hat{T}_0$, since we assume that ρ is positive, the variance is minimized by the sample that contains the n largest clusters.

For $\hat{T}_1$, the situation is more complicated. The impact of the second term of V will diminish the more equal in size the M_i are. If ρ is near zero, then, for given M_s, the variance is reduced by taking the M_i as equal as possible. If ρ is near one, then the variance is reduced by taking the M_i as heterogeneous as possible.

For $\hat{T}_R$, make the (typically satisfied) assumption that $M > 2M_s$. Then the second term of V is clearly much larger in magnitude than the corresponding term for $\hat{T}_1$. Furthermore, whatever the value of ρ, the variance will be decreased by taking the M_i as equal as possible. As with the other estimators, we want $\bar{M}_s$ large.

Table 8.10. Varying Component of Variances for Equal Allocation

Estimator	V
$\hat{T}_0 = \hat{T}_p$	$-2\rho\sigma^2 M\bar{M}_s$
$\hat{T}_1 = \hat{T}_H$	$-2\rho\sigma^2 M\bar{M}_s + \sum_s (M_i - \bar{M}_s)^2 \left[\frac{(1-\rho)\sigma^2}{\bar{m}} - \rho\sigma^2\right]$
$\hat{T}_R$	$-2\rho\sigma^2 M\bar{M}_s + \sum_s (M_i - \bar{M}_s)^2 \left[\frac{(1-\rho)\sigma^2}{\bar{m}}\left(\frac{M}{M_s}\right)^2 + \rho\sigma^2\left(\left(\frac{M}{M_s}\right)^2 - 2\left(\left(\frac{M}{M_s}\right)\right)\right)\right]$

We also summarize the results for proportional allocation ($m_i = mM_i/M_s$) when $\sigma_i^2 = \sigma^2$ and $\rho_i = \rho > 0$. Verification is left to Exercise 8.24. With proportional allocation $\hat{T}_0 = \hat{T}_H = \hat{T}_R$ and the error-variance is

$$\begin{aligned}\operatorname{var}_M(\hat{T}_0 - T) &= \operatorname{var}_M(\hat{T}_H - T) = \operatorname{var}_M(\hat{T}_R - T)\\ &= \sum_{i=1}^{N} M_i\sigma^2[1 + (M_i - 1)\rho] + \sigma^2(1-\rho)M(M - 2m)/m\\ &\quad + \frac{\sigma^2\rho}{n}\left(\frac{\Sigma_s (M_i - \bar{M}_s)^2}{n\bar{M}_s^2} + 1\right)M(M - 2M_s).\end{aligned} \tag{8.8.2}$$

The first two terms do not depend on the sample s. When $(M - 2M_s) > 0$ in the last term in (8.8.2), the error-variance of each is small when the sample consists of large clusters of nearly equal size. For $\hat{T}_1$ the components of the error-variance that depend on s are

$$\rho\sigma^2\left[\sum_r M_i^2 + \frac{(M - M_s)^2}{n}\right] + (1-\rho)\sigma^2[\bar{M}_s\bar{M}_s^{(-1)} - 1]\frac{(M - M_s)^2}{m},$$

where $\bar{M}_s^{(-1)} = \Sigma_s M_i^{-1}/n$. For $\hat{T}_p$ the components involving s are

$$(1-\rho)\sigma^2 M^2 \bar{M}_s \bar{M}_s^{(-1)}/m - 2\rho\sigma^2 M\bar{M}_s.$$

In both cases, an efficient sample is one that contains large units of nearly equal size.

8.8.3. Optimal Allocation of Second-Stage Units Given a Fixed Set of First-Stage Sample Units

In cluster sampling, a common practice is to reuse a particular first stage sample for different surveys. Different household surveys, for example, conducted by the same organization often use the same sample of geographic clusters.

This reuse of sample clusters has some major operational advantages, including the ability to maintain a staff of trained interviewers in each cluster. As a result, a relevant question is how to optimally allocate the second-stage sample given that the first stage sample is fixed.

When the sample of clusters s is given, the optimal rules for allocating a fixed number m of observations among the sample clusters are derived for some of the estimators under model (8.1.1). These are shown in Table 8.11. The terms c_i and d_i in the table are defined as $c_i = (1 - 2m/M)(1 - \rho_i)/2\rho_i$, $d_i = 1/(\rho_i \sigma_i^2)$, and $\bar{c}$ and $\bar{d}$ are sample means of c_i and d_i.

We illustrate the derivation of the result for the ratio estimator $\hat{T}_R$. The others are left to Exercise 8.17. Consider equation (8.2.3) when $a_i = M_i N\bar{M}/(m_i n\bar{M}_s) - 1$. We find that

$$\text{var}_M(\hat{T}_R - T) = B + \sum_s (1 - \rho_i)\sigma_i^2 M_i \left(\frac{M_i}{m_i} - 1\right),$$

where B does not involve individual sample values of m_i. This fact and the constraint that $\Sigma_s m_i = m$, lead to the Lagrange function for $\hat{T}_R$

$$\varphi = \sum_s (1 - \rho_i)\sigma_i^2 \frac{M_i^2}{m_i} + \lambda\left(\sum_s m_i - m\right).$$

Differentiating with respect to m_i, equating to 0, and collecting terms gives the estimating equations

$$-\sigma_i^2(1 - \rho_i)\frac{M_i^2}{m_i^2} + \lambda = 0 \quad (i = 1, \ldots, n). \tag{8.8.3}$$

Solving (8.8.3) for m_i leads to the result in Table 8.11.

Table 8.11. Optimal Allocation of Second-Stage Sample Under Model (8.1.1)

Estimator	Optimal m_i Proportional to
$\hat{T}_R$	$M_i \sigma_i \sqrt{1 - \rho_i}$
$\hat{T}_p$	$\sigma_i \sqrt{1 - \rho_i}$
$\hat{T}_1$	$[M_i + (M - M_s)]\sigma_i \sqrt{1 - \rho_i}$
$\hat{T}_0$	$M_i \frac{m}{M} + \left[\bar{c} - \bar{d}\frac{c_i}{d_i} + \left(1 - \frac{M_s}{M}\right)\frac{m}{n}\right]\frac{d_i}{\bar{d}}$

When $\sigma_i^2 = \sigma^2$ and $\rho_i = \rho$, optimal allocation for $\hat{T}_R$ is proportional to cluster size, and for $\hat{T}_p$ it is constant. This same result based on design variances is found in Sukhatme and Sukhatme (1970, p.321). For both $\hat{T}_0$ and $\hat{T}_1$, a weighted average, $w(mM_i/M_s) + (1 - w)(m/n)$, of proportional allocation and constant allocation is optimal (see Exercise 8.18). Note that, when s is fixed, proportional allocation implies $m_i/m = M_i/M_s$. For $\hat{T}_0$ the weight is $w = M_s/M$, the fraction of the population in sample clusters. So, when the first stage sampling fraction is small, constant allocation is suggested for this estimator, while proportional allocation is preferable if this fraction is near unity. For the estimator $\hat{T}_1$ the weight assigned to proportional allocation is $w = M_s/[M_s + n(M - M_s)]$. The optimal rule puts relatively less weight on proportional allocation than the rule for $\hat{T}_0$ does.

It is hard to find explicit optimum allocation formulas for $\hat{T}_H$. Even when $\sigma_i^2 = \sigma^2$ and $\rho_i = \rho$, the optimal m_i are obtained as solutions to a cubic equation and their explicit formulas offer little general insight. Consideration of extreme cases is suggestive, however. If $\rho = 0$, then proportional allocation is optimal for $\hat{T}_H$, while if $\rho = 1$, then equal allocation is optimal. These results and those in Table 8.11 invite the generalization that when $\sigma_i^2 = \sigma^2$ and $\rho_i = \rho$, with $\rho > 0$, optimal allocation for $\hat{T}_H$, as for most other reasonable estimators, is a compromise between constant and proportional allocation.

8.8.4. Optimal First and Second Stage Allocation Considering Costs

A fundamental question to be answered in any sample survey is how many units to select. As in Section 6.2, where we studied stratified populations, the answer depends on the budget and the precision goals of the survey. In sampling from a clustered population, sample size determination is complicated by having to decide on both the number of sample clusters and the number of sample units per cluster. In this section we analyze how to determine the optimal number of sample units at each stage of sampling both when the total cost is fixed and when a relvariance target is fixed.

Suppose that the first stage sample is not fixed, and we want to determine optimal first and second stage sample sizes. We consider the question of choosing n and m in a simple extreme case, using a simple cost model, and under the simplifying assumption $M_i = \bar{M}$, for $i = 1, \ldots, N$. Suppose that the total of the costs that vary with the sample sizes is fixed and is made up of two components:

$$C = c_1 n + c_2 m,$$

where c_1 is the cost per sample cluster, and c_2 is the cost per element sampled. If $\sigma_i^2 = \sigma^2$, $\rho_i = \rho$, and $M_i = \bar{M}$, then Table 8.11 and the discussion that follows it imply that equal allocation is optimal for $\hat{T}_0$, $\hat{T}_1$, $\hat{T}_R$, and $\hat{T}_p$. With equal allocation, all of these estimators, along with $\hat{T}_H$ and the constant parameter $\hat{T}_{BLU}$ are identical to $\hat{T}_0$. Using (8.2.3) with $a_i = (M - m)/m$ and simplifying, the error-variance is then

$$\text{var}_M(\hat{T}_0 - T) = \sigma^2(1-\rho)\frac{M}{m}(M-m) + \sigma^2\rho\frac{M^2}{nN}(N-n) \qquad (8.8.4)$$

(see Exercise 8.19). The following lemma gives the optimal ratio of second stage to first stage sample size.

Lemma 8.8.1. Subject to the total cost constraint, the variance in (8.8.4) is minimized when

$$\frac{m}{n} \equiv \bar{m}_{\text{opt}} = \sqrt{\frac{(1-\rho)c_1}{\rho c_2}}.$$

Proof: The proof is simple and is left to Exercise 8.21. □

Table 8.12 illustrates the effects of different first and second stage costs along with the intracluster correlation ρ on the second stage allocation $\bar{m}$. (Note that for fixed cost C, n decreases as $\bar{m}$ increases.) For a given ratio of c_1 to c_2, the number of units allocated to each cluster decreases with increasing ρ. This agrees with the intuition that as ρ increases less information is obtained per sample unit in a particular cluster so that choosing more clusters and fewer units per cluster is more efficient. The cost ratio c_1/c_2 will depend on the type of survey being done. For example, in a household survey done by personal interview, where the clusters are geographic units and the second stage units are households, various first stage costs like constructing maps, traveling to the clusters, recruiting interviewers, and paying supervisors of fieldwork may be far larger than the per-unit cost of conducting an interview with a household.

Table 8.12. Optimal Second-Stage Sample Sizes $\bar{m}$ for Different Costs and Intracluster Correlations when $\sigma_i^2 = \sigma^2$, $\rho_i = \rho$, $M_i = \bar{M}$, and an Equal Allocation per Cluster Is Used

	ρ				
c_1/c_2	0.01	0.05	0.10	0.20	0.30
1/5	4	2	1	1	1
1	10	4	3	2	2
2	14	6	4	3	2
5	22	10	7	4	3
10	31	14	9	6	5
20	44	19	13	9	7

On the other hand, think of a survey of elementary schools where a school is the first stage unit and the second stage unit is a student. Cluster costs might involve selecting the sample and obtaining cooperation from a school administrator. Suppose that each sample student is interviewed and administered a battery of lengthy standardized tests that must be scored by trained personnel. The second stage costs could conceivably be larger than the cluster costs in this case.

After some rearrangement, expression (8.8.4) can also be written as

$$\text{var}_M(\hat{T}_0 - T) = M^2\left(1 - \frac{m}{M}\right)\frac{\sigma^2}{n\bar{m}}\left\{1 + \left[\left(1 - \frac{m}{M}\right)^{-1}\left(1 - \frac{n}{N}\right)\bar{m} - 1\right]\rho\right\}.$$

If the sampling fraction of units, m/M, and of clusters, n/N, is small, then the error variance is approximately

$$\text{var}_M(\hat{T}_0 - T) \cong M^2\frac{\sigma^2}{n\bar{m}}[1 + (\bar{m} - 1)\rho]. \qquad (8.8.5)$$

The term $M^2\sigma^2/n\bar{m}$ is the variance of $\hat{T}_0$ in a sample of $n\bar{m}$ uncorrelated units. The term $[1 + (\bar{m} - 1)\rho]$ is the amount that the variance is inflated due to units within clusters being correlated. This term is often referred to in the design-based literature as a *design effect* (Kish, 1965). Expression (8.8.5) is a particularly simple one for planning purposes.

Lemma 8.8.1 gave the optimal number of units to select from each sample cluster. The total number of units in the sample will be governed by whether the total survey cost is fixed or whether a target variance is desired for $\hat{T}_0$. The calculations that follow are much the same as we did in Chapter 6 for stratified samples. If the total cost is fixed at some value C^*, Lemma 8.8.2 gives the optimal number of sample clusters.

Lemma 8.8.2. Under model (8.1.3) with $M_i = \bar{M}$, subject to a fixed total cost $C^* = c_1 n + c_2 m$, the number of sample clusters that minimizes the error variance (8.8.4) is

$$n_{\text{opt}} = \frac{C^*}{c_1 + c_2\bar{m}_{\text{opt}}},$$

where $\bar{m}_{\text{opt}}$ is given by Lemma 8.8.1.

Proof: The result comes from substituting $\bar{m}_{\text{opt}} = \sqrt{(1 - \rho)c_1/\rho c_2}$ into the cost function and solving for n. □

On the other hand, if the variance (8.8.4), or equivalently, the relvariance, of $\hat{T}_0$ is fixed at a target value, the optimal number of clusters is given by Lemma 8.8.3.

Lemma 8.8.3. Under model (8.1.3) with $M_i = \bar{M}$, and the relvariance of $\hat{T}_0$ fixed at V_0, the number of sample clusters that minimizes total cost is

$$n_{\text{opt}} = \frac{1}{\bar{m}_{\text{opt}}} \frac{1 + (\bar{m}_{\text{opt}} - 1)\rho}{\dfrac{\mu^2}{\sigma^2} V_0 + \dfrac{1}{N\bar{M}}[1 + (\bar{M} - 1)\rho]},$$

where $\bar{m}_{\text{opt}}$ is given by Lemma 8.8.1.

Proof: The mean of T under the model is $M\mu$. By using (8.8.3), $m = n\bar{m}$, $M = N\bar{M}$, and rearranging, the relvariance is

$$\text{relvar}_M(\hat{T}_0 - T) = \frac{\sigma^2}{\mu^2}\left(\frac{A}{n} - B\right),$$

where $A = [1 + (\bar{m} - 1)\rho]/\bar{m}$ and $B = [1 + (\bar{M} - 1)\rho]/(N\bar{M})$. Setting the relvariance equal to V_0, substituting $\bar{m}_{\text{opt}}$, and solving for n gives the result. □

The above formula for the optimal values of $\bar{m}$ requires an estimate of ρ. If, in addition, we wish to find n to minimize the relvariance, then we must also have the unit relvariance σ^2/μ^2. These parameters can be estimated, as described in Section 8.3, using data collected from a similar population (e.g., the same population at an earlier time period) or from a preliminary sample from the population of interest. The parameters that effect the optimal allocation will differ for different Y variables, and some type of compromise allocation will often be needed. Based on Example 8.3.2 the unit relvariance of the y-variable in the Labor Force population is $\hat{\sigma}^2/\hat{\mu}^2 = 69.77/(49.80)^2 = 0.028$, which is quite small. Larger values are more typical in other populations.

Figure 8.1 is a plot of the relvariance of $\hat{T}_0$ versus $\bar{m}$, the number of sample units allocated to each cluster, for the following combinations of (c_1, c_2): A = (100, 50), B = (100, 100), C = (500, 50), and D = (500, 100). The total cost is fixed at $C^* = 5000$, the unit relvariance is set at $\sigma^2/\mu^2 = 1$, $N = 1000$, and $\bar{M} = 50$. The four panels in the figure are for $\rho = 0.01, 0.05, 0.10$, and 0.20. The dot on each curve is at the optimal value of $\bar{m}$ given by Lemma 8.8.1. For many cost combinations, the optimal is relatively flat in the sense that a wide range of $\bar{m}$'s will yield nearly the minimum relvariance. This is especially true when $\rho = 0.01$ and cost combinations A and C, where the cost per unit within a cluster is cheapest. For combinations B and D, adding more sample units within each cluster is very inefficient, particularly when $\rho = 0.20$, since the cost per unit is high and the additional information gained is limited because of the large intracluster correlation.

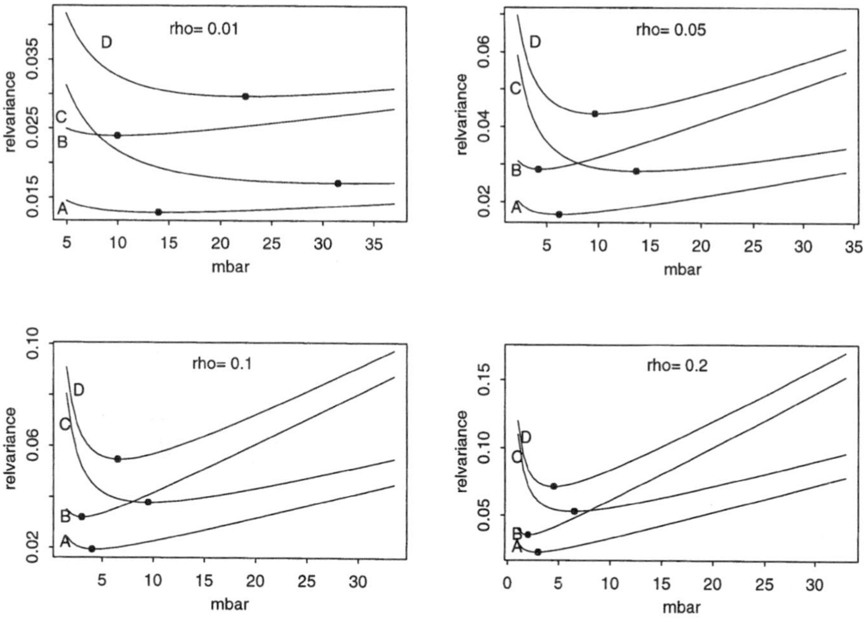

Figure 8.1 Effect of costs and intracluster correlation on the relvariance of the expansion estimator. Curves in each panel are for (c_1, c_2): A = (100,50), B = (100,100), C = (500,50), D = (500,100).

8.9.* ESTIMATION WHEN CLUSTER SIZES ARE UNKNOWN

It often happens that the cluster sizes M_i are actually known only for those clusters in the first stage sample s, but another measure of size x_i is known for all clusters. For example, if the clusters are blocks of dwelling units, x_i might be the number of units in block i at the time of a previous census.

Suppose the sizes M_i can be represented as random variables that depend on the x_i. The probability distributions in earlier sections must now be interpreted as being conditioned on $M_1, \ldots, M_N$, and the *unconditional* expectations and variances with respect to the distribution of the M_i's will be denoted by an asterisk. The following is a useful working model in many populations in which the M_i are proportional to the x_i:

$$E_M^*(M_i) = \beta x_i$$
$$\mathrm{cov}_M^*(M_i, M_{i'}) = \begin{cases} \tau^2 x_i & i = i' \\ 0 & i \neq i'. \end{cases} \tag{8.9.1}$$

For the model conditional on M_i, we again assume that (8.1.1) holds. Denote the vector of cluster sizes as $\mathbf{M} = (M_1, \ldots, M_N)'$. Below, we will denote with an asterisk the unconditional expectations, variances, and covariances, that is, ones taken over both the model for Y_{ij} and M_i.

The expected value of the total for cluster i, using both (8.1.1) and (8.9.1) is $E_M^*[E_M(T_i|\mathbf{M})] = E_M^*(M_i\mu) = \mu\beta x_i$, where the operator E_M without the asterisk continues to denote the conditional expectation given the M's. Because we have a model for the M_i's, and not the M_i's themselves, it is convenient to write a general estimator as a linear combination of estimated cluster totals, $\hat{T}_i = M_i\bar{Y}_{si}$, rather than as a combination of the $\bar{Y}_{si}$, as in earlier sections. In particular, let the general estimator be

$$\hat{T}^* = \sum_{i \in s} \lambda_i^* \hat{T}_i. \tag{8.9.2}$$

The requirement for unbiasedness, $E_M^* E_M(\hat{T}^* - T|\mathbf{M}) = 0$, implies that the λ_i^* must satisfy

$$\sum_s \lambda_i^* x_i/N = \bar{x},$$

where $\bar{x} = \Sigma_{i=1}^N x_i/N$. This condition is met by the ratio $(\lambda_i^* = N\bar{x}/(nx_s))$, regression $(\lambda_i^* = (N/n)[1 + n(x_i - \bar{x}_s)(\bar{x} - \bar{x}_s)/\Sigma_s(x_i - \bar{x}_s)^2])$, and mean of ratios $(\lambda_i^* = N\bar{x}/(nx_i))$ estimators.

To compute an estimator that is more nearly optimal than the three examples above, we first consider an artificial case in which β and μ are known. Following Kelly and Cumberland (1990), suppose that $\mathbf{Z}$ is a $k + p$ random vector with mean $\mathbf{U}$ and covariance matrix $\mathbf{\Sigma}$, and that $\mathbf{Z}$, $\mathbf{U}$, and $\mathbf{\Sigma}$ are partitioned as

$$\mathbf{Z} = \begin{bmatrix} \mathbf{Z}_p \\ \mathbf{Z}_k \end{bmatrix}, \quad \mathbf{U} = \begin{bmatrix} \mathbf{U}_p \\ \mathbf{U}_k \end{bmatrix}, \quad \text{and} \quad \mathbf{\Sigma} = \begin{bmatrix} \mathbf{\Sigma}_{pp} & \mathbf{\Sigma}_{pk} \\ \mathbf{\Sigma}_{kp} & \mathbf{\Sigma}_{kk} \end{bmatrix}.$$

The *BLU* predictor of $\mathbf{Z}_p$ with the form $\mathbf{A} + \mathbf{B}\mathbf{Z}_k$, where $\mathbf{A}$ is $p \times 1$ and $\mathbf{B}$ is $p \times k$, is

$$\hat{\mathbf{Z}}_p = \mathbf{U}_p + \mathbf{\Sigma}_{pk}\mathbf{\Sigma}_{kk}^{-1}(\mathbf{Z}_k - \mathbf{U}_k) \tag{8.9.3}$$

(see Exercise 8.30). For this problem define

$$\mathbf{Z}_p = \begin{bmatrix} \mathbf{Z}_I \\ \mathbf{Z}_{II} \end{bmatrix} \quad \text{and} \quad \mathbf{Z}_k = \begin{bmatrix} \mathbf{Z}_{III} \\ \mathbf{Z}_{IV} \end{bmatrix},$$

where $\mathbf{Z}_I$ is the n-vector of sums of the nonsample Y's in the sample clusters, that is, $Z_{Ii} = \Sigma_{j \in r_i} Y_{ij}$; $\mathbf{Z}_{II}$ is the $(N - n)$-vector of totals for the nonsample clusters with ith element $Z_{IIi} = \Sigma_{j=1}^{M_i} Y_{ij}$, $i \in r$; $\mathbf{Z}_{III}$ is the n-vector of sample totals with $Z_{IIIi} = \Sigma_{j \in s_i} Y_{ij}$, $i \in s$; and $\mathbf{Z}_{IV}$ is the vector of sample cluster sizes with $Z_{IVi} = M_i$, $i \in s$. After the sample is selected, we assume that the sizes of the sample clusters, M_i, are known, although the sizes of the nonsample clusters still are not. Then, $\mathbf{Z}$ has covariance matrix

$$\boldsymbol{\Sigma} = \begin{bmatrix} \mathbf{V}_I & \mathbf{0} & \mathbf{V}_{I,III} & \mathbf{V}_{I,IV} \\ \mathbf{0} & \mathbf{V}_{II} & \mathbf{0} & \mathbf{0} \\ \mathbf{V}_{III,I} & \mathbf{0} & \mathbf{V}_{III} & \mathbf{0} \\ \mathbf{V}_{IV,I} & \mathbf{0} & \mathbf{0} & \mathbf{V}_{IV} \end{bmatrix},$$

where the submatrices $\mathbf{V}_I$, $\mathbf{V}_{I,III}$, and so on, have the obvious, general definitions. We will work out their details below. The unconditional expectations of the various components of $\mathbf{Z}$ are

$$E_M^* E_M(Z_{Ii}|\mathbf{M}) = \mu(\beta x_i - m_i), \quad i \in s; \qquad E_M^* E_M(Z_{IIi}|\mathbf{M}) = \mu\beta x_i, \quad i \in r$$

$$E_M^* E_M(Z_{IIIi}|\mathbf{M}) = \mu m_i, \quad i \in s; \qquad E_M^* E_M(Z_{IVi}|\mathbf{M}) = \beta x_i, \quad i \in s.$$

The $n \times n$ matrix $\mathbf{V}_{I,III}$ is diagonal with elements

$$\begin{aligned} \mathrm{cov}_M^*\left(\sum_{j \in r_i} Y_{ij}, \sum_{j \in s_i} Y_{ij}\right) &= E_M^* E_M\left\{\left[\sum_{j \in r_i} Y_{ij} - E_M^* E_M\left(\sum_{j \in r_i} Y_{ij}\right)\right]\right. \\ &\quad \times \left.\left[\sum_{j \in s_i} Y_{ij} - E_M^* E_M\left(\sum_{j \in s_i} Y_{ij}\right)\right]\right\} \\ &= (\beta x_i - m_i) m_i \rho_i \sigma_i^2, \quad i \in s. \end{aligned}$$

The $n \times n$ matrix $\mathbf{V}_{I,IV}$ is also diagonal with typical element

$$\begin{aligned} \mathrm{cov}_M^*\left(\sum_{j \in r_i} Y_{ij}, M_i\right) &= E_M^* E_M\left[\sum_{j \in r_i} Y_{ij} - E_M^* E_M\left(\sum_{j \in r_i} Y_{ij}\right)\right][M_i - E_M^* E_M(M_i)] \\ &= \mu\tau^2 x_i, \quad i \in s. \end{aligned}$$

$\mathbf{V}_{III}$ is diagonal with elements $m_i v_i$ where $v_i = \sigma_i^2[1 + (m_i - 1)\rho_i]$ as in Theorem 8.1.1. The elements of the diagonal matrix $\mathbf{V}_{IV}$ are $\tau^2 x_i$.

Next, we can apply (8.9.3) with the above definitions to obtain the predictors of the components of $\mathbf{Z}_p$. In particular, we have

$$\hat{\mathbf{B}} = \Sigma_{pk} \Sigma_{kk}^{-1} = \begin{bmatrix} \mathbf{V}_{I,III} & \mathbf{V}_{I,IV} \\ \mathbf{0} & \mathbf{0} \end{bmatrix} \begin{bmatrix} \mathbf{V}_{III}^{-1} & \mathbf{0} \\ \mathbf{0} & \mathbf{V}_{IV}^{-1} \end{bmatrix}$$

and

$$\hat{\mathbf{Z}}_p = \begin{bmatrix} \mu(\beta\mathbf{x}_s - \mathbf{m}_s) \\ \mu\beta\mathbf{x}_r \end{bmatrix} + \hat{\mathbf{B}}\left\{\begin{bmatrix} \mathbf{Z}_{III} \\ \mathbf{M}_s \end{bmatrix} - \begin{bmatrix} \mu\mathbf{m}_s \\ \beta\mathbf{x}_s \end{bmatrix}\right\}, \tag{8.9.4}$$

where $\mathbf{x}_s$ is the $n \times 1$ vector of x_i's for the sample clusters, $\mathbf{x}_r$ is the $(N - n) \times 1$ vector of x_i's for the non-sample clusters, $\mathbf{m}_s$ is the $n \times 1$ vector of sample sizes, m_i, in the sample clusters, and $\mathbf{M}_s$ is the vector of sizes, M_i, for the sample clusters. Thus, from (8.9.4) the predictor for the nonsample total in sample cluster i is

$$\hat{Z}_{Ii} = (M_i - m_i)\mu + (\beta x_i - m_i)w_i(\bar{Y}_{si} - \mu), \quad i \in s, \tag{8.9.5}$$

where $w_i = m_i\rho_i/[1 + (m_i - 1)\rho_i]$ as in Theorem 8.1.1. The predictor for the total in a nonsample cluster is

$$\hat{Z}_{pi} = \mu\beta x_i, \quad i \in r. \tag{8.9.6}$$

Adding the sample total of the Y's, the predictor for the total of the nonsample Y's in the sample clusters, and the predictors of the totals for the nonsample clusters, we get the *BLU* predictor as

$$\hat{T}^*_{BLU} = \sum_{i \in s}\sum_{j \in s_i} Y_{ij} + (M_s - m_s)\mu + \sum_{i \in s}(\beta x_i - m_i)w_i(\bar{Y}_{si} - \mu) + \mu\beta x_r \tag{8.9.7}$$

where $x_r = \Sigma_{i \in r} x_i$. If, instead of its expected value βx_i, we use M_i in the third term, a second estimator, dependent on the parameters μ and β is obtained:

$$\hat{T}^*_{NL} = \sum_{i \in s}\sum_{j \in s_i} Y_{ij} + \sum_{i \in s}(M_i - m_i)[w_i\bar{Y}_{si} + (1 - w_i)\mu] + \mu\beta x_r, \tag{8.9.8}$$

which has an obvious similarity to (8.2.9), the *BLU* predictor when all the cluster sizes are known. $\hat{T}^*_{NL}$ is nonlinear in the random variables in the model since it involves a multiplication in the third term of M_i and $\bar{Y}_{si}$. After some rearrangement, both $\hat{T}^*_{BLU}$ and $\hat{T}^*_{NL}$ can be shown to be in the class defined by (8.9.2).

In the usual case where β and μ are unknown, the estimators $\hat{\beta} = \bar{M}_s/\bar{x}_s$ and $\hat{\mu} = \Sigma_s u_i\bar{Y}_{si}$ with u_i defined in Theorem 8.2.2, can be substituted in either $\hat{T}^*_{BLU}$ or $\hat{T}^*_{NL}$. For the special case of $\rho_i = \rho$ and $\sigma_i^2 = \sigma^2$, the ANOVA estimators described in Section 8.3.1 can also be used in $\hat{T}^*_{BLU}$ or $\hat{T}^*_{NL}$ to estimate unknown parameters. Kelly and Cumberland (1990) report the results of a simulation study using blocks from the Los Angeles, California, area. Clusters were census tracts—a geographic area with about 2000 persons, and second stage units were blocks. The size measures x_i were the 1970 number of blocks per tract while M_i was the number in 1980. The target variable Y was the 1980 block population. They found that $\hat{T}^*_{NL}$ with estimated model parameters had a smaller *mse* than $\hat{T}_p$ with probability proportional to x_i sampling at the first stage and *srswor* at the second. $\hat{T}^*_{NL}$ also had a smaller *mse* than $\hat{T}_R$ with *srswor* at both stages.

8.10. TWO-STAGE SAMPLING IN DESIGN-BASED PRACTICE

Two stage cluster sampling or sampling in even more than two stages is common practice. Sampling techniques for multistage sampling were first developed by design-based practitioners as operationally efficient procedures for data collection. The two key practical reasons for its use are administrative rather than mathematical:

1. A complete list of all units in the population may not be available and producing one would be prohibitively expensive.
2. Field work may be more economical when sample units are clustered.

Hansen, Madow, and Tepping (1983, sec. 4.5) discuss some of these operational issues.

One of the most popular methods of two stage selection is to sample the clusters with probabilities proportional to the number of units, M_i, in each cluster and then to select an equal probability sample of the same size within each sample cluster by simple random or systematic sampling. This yields equal selection probabilities for all units (Section 8.1.2 and Exercise 8.33). In fact, the overall selection probability of a unit in cluster i is

$$\pi_{ij} = \frac{nM_i}{M}\frac{\bar{m}}{M_i} = \frac{n\bar{m}}{M}.$$

The Horvitz-Thompson weight or π-weight is then the same for every sample unit, $\pi_{ij}^{-1} = M/(n\bar{m})$, and the sample is known as *self-weighting* (Hansen, Hurwitz, and Madow, 1953, p. 186) or *epsem* (Kish, 1965, p. 20).

Likewise the scheme in which a simple random sample of clusters is selected and the same sampling fraction f of units is selected from each sample cluster is *epsem* (Section 8.1.2 and Exercise 8.33). This scheme is not usually so practically convenient, since the numbers of sample units from the different clusters can vary considerably. The expected sample size in cluster i is fM_i. If, for example, clusters range in size from 100 to 1000 and the sampling fraction is $f = 0.1$, then the number of sample units ranges from about 10 to 100.

Having the same number of sample units in each cluster is especially advantageous when in-person interviewing is done. Having an equal workload per cluster allows the same number of interviewers to be hired in each cluster.

The sample size $\bar{m}$ can be determined to meet the survey budget and to be the workload that one or more resident interviewers can handle during the allotted survey period. With a procedure where the sample sizes vary, some clusters may have a workload that is too small to keep an interviewer interested and trained, while others may have such a large sample that a sufficient number of good interviewers cannot be hired.

Särndal, Swensson, and Wretman (1992, Chapters 4 and 8) give a comprehensive design-based theory for estimation based on single and multi-stage sampling, with and without auxiliary data. The general regression estimator (GREG) has the same form as in Section 2.7. The GREG is defined as $\hat{T}_{yr} = \hat{T}_{y\pi} + \hat{\mathbf{B}}'(\mathbf{T}_x - \hat{\mathbf{T}}_{x\pi})$, where $\hat{\mathbf{T}}_{x\pi}$ is the π-estimator of the vector of population x-totals, $\mathbf{T}_x$, and $\hat{\mathbf{B}} = \mathbf{A}_{\pi s}^{-1}\mathbf{X}_s'\mathbf{V}_{ss}^{-1}\Pi_y^{-1}\mathbf{y}_s$ with $\mathbf{A}_{\pi s} = \mathbf{X}_s'\mathbf{V}_{ss}^{-1}\Pi_s^{-1}\mathbf{X}_s$, $\mathbf{V}_{ss} = \text{diag}(v_{ij})$ and $\Pi_s = \text{diag}(\pi_{ij})$. The estimator can also be written as a linear combination of the sample y's:

$$\hat{T}_{yr} = \sum_{i \in s} \sum_{j \in s_i} \frac{1}{\pi_{ij}} g_{ij} y_{ij},$$

where $g_{ij} = 1 + (\mathbf{T}_x - \hat{\mathbf{T}}_{x\pi})'\mathbf{A}_{\pi s}^{-1}\mathbf{x}_{ij}/v_{ij}$.

With respect to the distribution generated by two stage random sampling, the GREG is nonlinear due to $\hat{\mathbf{B}}$ and $\hat{\mathbf{B}}\hat{\mathbf{T}}_{x\pi}$ being products of estimators, as it was under single stage sampling. The GREG is model-unbiased under the working model with $E_M(\mathbf{Y}) = \mathbf{X}\boldsymbol{\beta}$ and is approximately design-unbiased when the sample of clusters is large. If the model is mistaken, design-unbiasedness still holds.

Because $\hat{T}_{yr}$ is nonlinear in the design-based approach, its design-variance can only be approximated. The standard technique is to expand $\hat{T}_{yr}$ in a first-order Taylor series and to then compute the design-variance of the resulting approximation. In the case of two stage sampling the asymptotic variance, assuming that each g_{ij} converges to 1, is

$$\text{var}_\pi(\hat{T}_{yr}) \cong \sum_{i=1}^{N} \sum_{i'=1}^{N} (\pi_{1ii'} - \pi_i \pi_{i'}) \frac{e_{i\cdot}}{\pi_i} \frac{e_{i'\cdot}}{\pi_{i'\cdot}} + \sum_{i=1}^{N} \frac{V_i}{\pi_i},$$

where $\pi_{1ii'}$ is the joint selection probability of clusters i and i' in the first-stage sample, $e_{i\cdot} = \Sigma_{j=1}^{M_i}(y_{ij} - \mathbf{x}'\mathbf{B})$, $\mathbf{B} = (\mathbf{X}'\mathbf{V}^{-1}\mathbf{X})^{-1}\mathbf{X}'\mathbf{V}^{-1}\mathbf{y}$, that is, the slope evaluated for the full finite population, and

$$V_i = \sum_{j=1}^{M_i} \sum_{j'=1}^{M_i} (\pi_{jj'|i} - \pi_{j|i}\pi_{j'|i}) \frac{e_{ij}}{\pi_{j|i}} \frac{e_{ij'}}{\pi_{j'|i}}$$

with $e_{ij} = y_{ij} - \mathbf{x}'\mathbf{B}$ and $\pi_{jj'|i}$ is the joint selection probability of units j and j' in cluster i (Särndal, Swensson, and Wretman, 1992, sec. 8.9).

An interesting special case of the GREG is obtained under a stratification working model. If $E_M(Y_{ij}) = \mu_c$ and $\text{var}_M(Y_{ij}) = \sigma_c^2$ for (ij) in group c $(c = 1, \ldots, C)$, then $\mathbf{x}_{ij}$ is $C \times 1$ with a 1 in the position corresponding to the group to which unit (ij) belongs and 0 elsewhere. We have $\mathbf{A}_{\pi s} = \text{diag}(\hat{M}_c/\sigma_c^2)$ with $\hat{M}_c = \Sigma_{i \in s} \Sigma_{j \in s_{ic}} \pi_{ij}^{-1}$, where s_{ic} is the set of sample units in sample cluster i that is in group c. The slope estimator is then $\hat{\mathbf{B}} = (\hat{T}_1/\hat{M}_1, \ldots, \hat{T}_C/\hat{M}_C)'$, and the GREG reduces to

$$\hat{T}_{PS\pi} = \sum_{c=1}^{C} \frac{M_c}{\hat{M}_c} \hat{T}_{c\pi}, \tag{8.10.1}$$

The ratio $\hat{T}_{c\pi}/\hat{M}_c$ is the Hájek estimator of the mean for group c. Recall that we saw this estimator in a simpler form in Section 3.5. Note that (8.10.1) differs from the post-stratified estimator $\hat{T}_{PS} = \Sigma_{c=1}^{C} M_c \bar{Y}_{sc}$ from Section 8.5 that was a special case of the predictor (8.5.2). Selection probabilities, naturally, play no role in $\hat{T}_{PS}$ since it was derived based on a model only. The post-stratified estimator in (8.10.1) and the Horvitz-Thompson estimator, $\hat{T}_{y\pi}$, were studied in some depth using models in Valliant (1993), who noted the importance of using an estimator like $\hat{T}_{PS\pi}$ to avoid model-bias when $E_M(Y_{ij}) = \mu_c$.

As in single-stage sampling, the possibility of model-bias is ignored or is, at most, a secondary consideration in design-based analysis of two stage samples when evaluating estimators. Where from a model-based point of view one estimator is clearly preferable to another for its lack of bias, in design-based theory the one will be judged better than the other based on their respective design variances, but *both* will be acceptable as being (design-)*unbiased*.

For example, consider a Y variable that follows the model $E_M(\mathbf{Y}) = \mathbf{X}\boldsymbol{\beta}$. Suppose that a two-stage probability sample is selected and the π-estimator, $\hat{T}_{y\pi}$, is used. The model-bias of $\hat{T}_{y\pi}$ is

$$E_M(\hat{T}_{y\pi} - T) = (\hat{\mathbf{T}}_{x\pi} - \mathbf{T}_x)\boldsymbol{\beta}.$$

If $\hat{\mathbf{T}}_{x\pi} = \mathbf{T}_x$, this bias will be zero, but in a randomly selected sample, the model-bias may often be substantial. The model-bias averaged over all possible sample, on the other hand, is zero since $E_\pi(\hat{\mathbf{T}}_{x\pi} - \mathbf{T}_x) = \mathbf{0}$. In contrast, the GREG constructed with $E_M(\mathbf{Y}) = \mathbf{X}\boldsymbol{\beta}$ as the working model is model-unbiased and approximately design-unbiased. The superiority of $\hat{T}_{yr}$ over $\hat{T}_{y\pi}$ will show itself as a difference in the design-variances. The design-variance of $\hat{T}_{y\pi}$ will indirectly include a large squared model-bias term, $E_\pi\{[E_M(\hat{T}_{y\pi} - T)]^2\}$ that is not present in the design-variance of the GREG. This is the same point we illustrated for single stage sampling earlier in Section 4.6.

This perspective, in which no design-unbiased (or near unbiased) estimator can be regarded as bad by the very fact that it is unbiased lends a great deal of confidence to survey samplers. One has to be impressed by the straightforwardness and comprehensiveness of the design-based theory, and its ability to circumvent obstacles such as mistaken models. Being, as it were, automatically not wrong, gives samplers a large amount of authority in dealing with physicians, economists, judges, and the like, on statistical issues.

This comprehensiveness of the state of the art for two-stage sampling in design-based theory is in striking contrast to the many open questions and possibilities for improvement we have noted in this chapter. This suggests that the comprehensiveness of the design-based approach is had at a price.

A small sign of this is the simulation study of Section 8.3.4. There, for both the "*pps*" and "*srs*" methods of sampling, $\hat{T}_0$ is the appropriate design-based estimator, because of the *epsem* sampling in both instances. Nonetheless, it does far better overall in the *pps* results, where it hugs the *BLU* estimator, than it does in the *srs* results, where, deviating from the *BLU*, it can perform badly.

Automaticity is bought at the price of lost opportunity.

EXERCISES

8.1 Prove Theorem 8.2.1 and Corollary 8.2.1.

8.2 Show that the formula for $\text{var}_M(\hat{T}_{BLU} - T)$ in Theorem 8.2.3 holds.

8.3 Show that if incorrect values of the parameters σ_i^2 and ρ_i are used in $\hat{T}_{BLU}$ of Theorem 8.2.2, it is still a prediction unbiased estimator of T.

8.4 Prove Corollary 8.2.3.

8.5 Show that $\hat{T}_U = M(\Sigma_s M_i \bar{Y}_{si}/n\bar{M})$ is design-unbiased if simple random samples of clusters and units within clusters are selected but is biased under model (8.1.1). What is $E_p E_M(\hat{T}_U - T)$? Comment on why the expectation over both the design and the model is a misleading criterion here.

8.6 Show that the expectations under model (8.1.3) of the mean squares in Table 8.2 are the values given in the third column of that table.

8.7 The following sample of clusters was selected from the Labor Force population in Appendix B: cluster numbers 8, 45, 49, 52, 53, 59, 65, 83, 84, and 96. Each sample cluster is fully enumerated. (a) Compute the ANOVA estimates of σ^2 and ρ for the variable y. (b) Compute the maximum likelihood estimates of μ, σ^2, and ρ.

8.8 For D_N defined at equation (8.3.5), find necessary and sufficient conditions on the cluster sizes M_i so that $D_N > 1$. For $D_N > 1$, derive the lower bound (8.3.6) on the intracluster correlation when model (8.1.3) holds.

8.9 Derive the error variance given by (8.2.3) for an estimator in the class defined by (8.2.1) under model (8.1.1), that is,

$$\mathrm{var}_M(\hat{T} - T) = v - \sum_s \sigma_i^2 \rho_i (M_i - m_i)^2 + \sum_s \sigma_i^2 \rho_i [m_i a_i - (M_i - m_i)]^2 + \sum_s (1 - \rho_i)\sigma_i^2 m_i a_i^2.$$

8.10 A sample of 10 clusters is selected from the Labor Force population in Appendix B. A sample of $\bar{m} = 2$ persons is selected from each cluster. The particular clusters and persons in the sample are listed in the table below. Using the data for each person for the variable y calculate $\hat{T}_0$, $\hat{T}_R$, $\hat{T}_p$, $\hat{T}_H$, and $\hat{T}_1$.

Cluster	Person	Cluster	Person
10	33	49	191
10	35	49	192
27	95	71	284
27	97	71	285
39	152	79	324
39	153	79	325
45	175	81	336
45	176	81	337
47	183	112	467
47	184	112	468

8.11 Find the determinant of $\mathbf{V}_{ss(i)}$ defined in Section 8.2.2.
(*Note*: $|\mathbf{A} + \mathbf{bb}'| = |\mathbf{A}|(1 + \mathbf{b}'\mathbf{A}^{-1}\mathbf{b})$; see, e.g., Mardia et. al., 1979, sec. A.2.3).

8.12 Show that, if $\rho_i = \rho$, $\sigma_i = \sigma$, and $m_i = \bar{m}$, for $i \in s$, then

$$\hat{T}_{BLU} = \hat{T}_0 + n\bar{w}\,\mathrm{cov}_s(M_i, \bar{Y}_{si}),$$

where

$$\bar{w} = \bar{m}\rho/(1 + (\bar{m} - 1)\rho) \qquad \text{and} \qquad \mathrm{cov}_s(M_i, \bar{Y}_{si}) = \sum_s \bar{Y}_{si}(M_i - \bar{M}_s)/n.$$

8.13 Find $\mathrm{var}_M(\hat{T}_R)$ and $\mathrm{var}_M(\hat{T}_0)$ under model (8.1.1) when $\rho = 0$ and $\sigma_i^2 = k/M_i$. Which is smaller?

8.14 (a) Find balance conditions making $\hat{T}_R$ unbiased under a regression model as in Section 8.5. (b) Do the same for $\hat{T}_p$. (c) Show that if a simple random sample of n clusters is selected along with a simple random sample of m_i units per sample cluster, that the expectation with respect to the sampling design of the bias of $\hat{T}_R$ is zero. (d) Show that if n clusters are selected with probability proportional to M_i, and an equal probability sample of m_i units is selected within sample cluster i, that the expectation with respect to the sampling design of the bias of $\hat{T}_p$ is zero.

8.15 Generalize Lemma 4.2.1 to the case of $\mathbf{V}$ a non-diagnonal matrix, as noted in the proof of Theorem 8.7.1.

8.16 Suppose that model (8.1.3) holds and that the estimator is $\hat{T}_{BLU}$. Show that, for a given sample of clusters, and a given value of m, when $\sigma_i^2 = \sigma^2$ and $\rho_i = \rho$ with $\rho = 0$ or 1, the error variance of $\hat{T}_{BLU}$ does not depend on the allocation rule.

8.17 Derive the optimal second stage sample allocations for $\hat{T}_p$, $\hat{T}_1$, and $\hat{T}_0$ shown in Table 8.11.

8.18 Show that for both $\hat{T}_0$ and $\hat{T}_1$, a weighted average, $w(mM_i/M_s) + (1 - w)(m/n)$, of proportional allocation and constant allocation is optimal when the sample of clusters is fixed and model (8.1.3) holds. What are the weights for $\hat{T}_0$ and $\hat{T}_1$?

8.19 Derive the form of the error variance for $\hat{T}_0$ when model (8.1.3) holds, $M_i = \bar{M}$, and the sample is allocated equally to clusters ($m_i = \bar{m}$):

$$\text{var}_M(\hat{T}_0 - T) = \sigma^2(1 - \rho)\frac{M}{m}(M - m) + \sigma^2\rho\frac{M^2}{nN}(N - n).$$

8.20 Suppose that model (8.1.3) holds. Assume that the estimator $\hat{T}_R$ is used. Instead of proportional allocation, which is optimal, suppose that constant allocation of units to clusters is used. (a) Calculate the efficiency loss defined as $(v_c - v_p)/v_p$, where v_c and v_p represent the variance $\text{var}_M(\hat{T}_R - T)$ under constant and proportional allocation. (b) Show that this efficiency ratio is between 0 and $V_s^2/(1 - m/M)$, where $V_s^2 = \Sigma_s(M_i - \bar{M}_s)^2/(n\bar{M}_s^2)$. (c) Show that the upper bound is achieved at $\rho = 0$.

8.21 Prove Lemma 8.8.1.

8.22 Suppose that the model is (8.1.3), that clusters are of equal size ($M_i = \bar{M}$), and that the same number, $\bar{m}$, of sample units is selected per cluster. Assume that the cost function is $C = c_1 n + c_2 m$ as in Section

8.8.4 and that the estimator is $\hat{T}_0$. Further, suppose that $N = 10{,}000$ and $\bar{M} = 100$. Round the sample sizes requested below to integers. (a) Compute the optimal second stage sample sizes for the following parameter combinations: $(c_1, c_2) = (200, 50)$ and $\rho = 0.02, 0.20$, or 0.50. (b) Suppose that the total budget is $C^* = \$100{,}000$. Determine the number of sample clusters and sample units per cluster that will minimize the error variance of $\hat{T}_0$. Compute the coefficients of variation that will result from this allocation. (c) Suppose that the population unit relvariance of Y is $\sigma^2/\mu^2 = 3$ and that the relvariance target is $V_0^{1/2} = 0.10$. Find the optimal number of sample clusters, sample units per cluster, and the total cost for the combinations of $(c_1, c_2) = (200, 50)$ and $\rho = 0.02, 0.20$, or 0.50. Compute the costs that result from these allocations.

8.23 Derive the components of the error variances in Table 8.10 under model (8.1.3) and equal allocation to clusters, $m_i = \bar{m}$.

8.24 Suppose that model (8.1.3) holds and that $\rho > 0$. Show that for proportional allocation of the second stage sample (i.e., $m_i = nM_i/M$), the error variance of $\hat{T}_0 = \hat{T}_H = \hat{T}_R$ is given by (8.8.2). Show that when $(M - 2M_s) > 0$, the error-variance of each is small when the sample consists of large clusters of nearly equal size, as claimed in Section 8.8.2.

8.25 Suppose that model (8.1.3) holds with $\rho > 0$. Show that for proportional allocation of the second stage sample (i.e., $m_i = nM_i/M$), the error-variances of $\hat{T}_1$ and $\hat{T}_p$ are small when the sample consists of large clusters of nearly equal size, as claimed in Section 8.8.2.

8.26 Under model (8.1.1), the regression of cluster total $T_i = \Sigma_{j=1}^{M_i} Y_{ij}$ on cluster size M_i is a straight line through the origin, $E_M(T_i) = \mu M_i$. Alternatively, suppose that a quadratic regression model is more nearly correct: $E_M(T_i) = \beta_0 + \beta_1 M_i + \beta_2 M_i^2$. Calculate the biases of $\hat{T}_p$ and $\hat{T}_1$ under this polynomial model. For clusters that are balanced on the first two moments of cluster size, that is, $\bar{M}_s = \bar{M}$ and $\bar{M}_s^{(2)} = \bar{M}^{(2)}$, derive relationships between the biases of $\hat{T}_H$ and $\hat{T}_0$ and between the biases of $\hat{T}_1$ and $\hat{T}_p$.

8.27 Show that under the quadratic regression model $E_M(T_i) = \beta_0 + \beta_1 M_i + \beta_2 M_i^2$ if σ_i^2, ρ_i, and m_i are all constant and a balanced sample of clusters is selected ($\bar{M}_s/\bar{M} = 1$), then

$$E_M(\hat{T}_R - T)^2 < E_M(\hat{T}_H - T)^2$$

when $[E_M(\hat{T}_H - T)]^2 > \text{var}_M(\hat{T}_R - T) - \text{var}_M(\hat{T}_H - T)$. Get explicit expressions for $[E_M(\hat{T}_H - T)]^2$ and $\text{var}_M(\hat{T}_R - T) - \text{var}_M(\hat{T}_H - T)$.

8.28 Show that the design-expectation of the model bias of $\hat{T}_p$ under the model $E_M(T_i) = \beta_0 + \beta_1 M_i + \beta_2 M_i^2$ is 0 under a random sampling plan where clusters are selected with probabilities proportional to size, M_i, and a simple random sample is selected within each cluster.

8.29 Show that the design-expectation of the model bias under the model $E_M(T_i) = \beta_0 + \beta_1 M_i + \beta_2 M_i^2$ of $\hat{T}_H$ is approximately eliminated under a random sampling plan where simple random samples are selected at both stages, the first stage sample is large, and the second stage uses a uniform sampling fraction of $f_2 = m/M$.

8.30 Suppose that $\mathbf{Z}$ is a $k + p$ random vector with mean $\mathbf{U}$ and covariance matrix $\boldsymbol{\Sigma}$, and that $\mathbf{Z}$, $\mathbf{U}$, and $\boldsymbol{\Sigma}$ are partitioned as

$$\mathbf{Z} = \begin{bmatrix} \mathbf{Z}_p \\ \mathbf{Z}_k \end{bmatrix}, \quad \mathbf{U} = \begin{bmatrix} \mathbf{U}_p \\ \mathbf{U}_k \end{bmatrix}, \quad \text{and} \quad \boldsymbol{\Sigma} = \begin{bmatrix} \boldsymbol{\Sigma}_{pp} & \boldsymbol{\Sigma}_{pk} \\ \boldsymbol{\Sigma}_{kp} & \boldsymbol{\Sigma}_{kk} \end{bmatrix}.$$

Show that the *BLU* predictor of $\mathbf{Z}_p$ with the form $\mathbf{A} + \mathbf{B}\mathbf{Z}_k$, where $\mathbf{A}$ is $p \times 1$ and $\mathbf{B}$ is $p \times k$, is $\hat{\mathbf{Z}}_p = \mathbf{U}_p + \boldsymbol{\Sigma}_{pk}\boldsymbol{\Sigma}_{kk}^{-1}(\mathbf{Z}_k - \mathbf{U}_k)$.

8.31 Derive the formula for the *BLU* predictor,

$$\hat{T}^*_{BLU} = \sum_{i \in s} \sum_{j \in s_i} Y_{ij} + (M_s - m_s)\mu + \sum_{i \in s} (\beta x_i - m_i) w_i (\bar{Y}_{si} - \mu) + \mu\beta x_r,$$

in (8.9.7).

8.32 Consider the sample, listed below, of 10 schools and 5 students per school from the Third-Grade population in Appendix B.6. (a) Using the data for these students and the same set of explanatory variables as in Example 8.6.2, compute the estimate of the mean mathematics and science scores, $\hat{T}/M$, where $\hat{T}$ is defined by (8.6.2). (b) Next, use only the variables region, community type, and ethnicity plus an intercept and predict the mean mathematics and science scores. In both parts (a) and (b), use the same collapsed categories of community type and ethnicity as in Example 8.6.2.

Sample School	Sample Students
20	386, 389, 395, 401, 402
46	835, 837, 838, 840, 844
50	903, 906, 912, 913, 918
70	1284,1285, 1290, 1293, 1296
74	1368, 1369, 1371, 1374, 1376
85	1533, 1537, 1539, 1542, 1544
92	1635, 1638, 1639, 1640, 1644
109	1937, 1941, 1947, 1957, 1959
114	2046, 2047, 2049, 2057, 2058
126	2242, 2243, 2247, 2250, 2253

8.33 Find the inclusion probabilities of the sample designs (ii.a) and (i.b) of Section 8.1.2, when the second stage is carried out by *srs*. Conclude that both designs are *epsem*.

CHAPTER 9

Robust Variance Estimation in Two-Stage Cluster Sampling

In Chapter 5 we studied variance estimation methods for estimators in single-stage sampling. The goal there was to produce variance estimators that were consistent for the variance under a working model but were also robust to certain departures from that model. The ideas used in that chapter are equally valid in populations for which the cluster models of Chapter 8 apply. As always, we begin with a working model but do not assume that it or any other model is perfectly true. Rather, the model is adopted as a tool for planning and inference. Robustness or insensitivity to changes in the model is a key issue. Thus, we seek variance estimators that are consistent under the working model but that remain consistent under more general models.

9.1. COMMON MEAN MODEL AND A GENERAL CLASS OF VARIANCE ESTIMATORS

The notation used here is the same as in Chapter 8. We begin by estimating the population total $T = \Sigma_{i=1}^{N} \Sigma_{j=1}^{M_i} Y_{ij}$, where the Y_{ij} is a random variable associated with unit j in cluster i. As in the previous chapter, a sample s of n clusters is selected. From sample cluster i a subsample s_i of size m_i is selected from the M_i units. The total number of units in the sample is $m = \Sigma_s m_i$. Initially, we will assume that M_i is known for each of the N clusters in the population. Later, in Section 9.4, we study the case in which a size measure x_i is known for every cluster, but the actual number of units M_i is known only if cluster i is in the first-stage sample.

The working model here is the same as model (8.1.3) and is given by

$$E_M(Y_{ij}) = \mu$$

$$\operatorname{cov}_M(Y_{ij}, Y_{i'j'}) = \begin{cases} \sigma^2 & i = i', j = j' \\ \sigma^2 \rho & i = i', j \neq j' \\ 0 & \text{otherwise.} \end{cases} \tag{9.1.1}$$

The estimators of the total of Y that we consider fall into the general class

$$\hat{T} = \sum_{i \in s} (1 + a_i) m_i \bar{Y}_{si} \tag{9.1.2}$$

where $\bar{Y}_{si} = \Sigma_{j \in s_i} Y_{ij}/m_i$ and a_i is a coefficient that does not depend on the Y's. The estimators of T defined by (9.1.2) are the same as class (8.2.1) in Chapter 8. For later reference, it will also be convenient to write the estimator of T as

$$\hat{T} = \sum_{i \in s} g_i \bar{Y}_{si},$$

where $g_i = (1 + a_i)m_i$. Note that, in the notation of Chapter 2, we would have $\hat{T} = \Sigma_s \Sigma_{s_i} g_{ij} Y_{ij}$, that is, a weighted sum of individual sample Y's. Consequently, if $g_{ij} = g_i/m_i$, we obtain the class of estimators to be studied in this section.

Many of the estimators covered in the last chapter—including $\hat{T}_{BLU}$, $\hat{T}_0$, $\hat{T}_R$, $\hat{T}_p$, $\hat{T}_H$, and $\hat{T}_1$—fall into class (9.1.2). Since model (9.1.1) implies that $E_M(\hat{T} - T) = \mu(\Sigma_{i \in s} g_i - M)$, unbiasedness of estimators in class (9.1.1) requires that

$$\sum_{i \in s} g_i = M, \tag{9.1.3}$$

a condition satisfied by all six estimators mentioned above.

In Chapter 8 we used expression (8.2.3) for comparing the error variances of different estimators. Now, for variance estimation a convenient form of the error variance is given in (9.1.4) below. First, note that under both model (9.1.1) and the more general model in which the variance and correlation parameters, σ_i^2 and ρ_i, differ among the clusters, variables in different clusters are uncorrelated. With $\bar{Y}_i = \Sigma_{j=1}^{M_i} Y_{ij}/M_i$, the estimation error is

$$\hat{T} - T = \sum_s g_i \bar{Y}_{si} - \left(\sum_s M_i \bar{Y}_i + \sum_r M_i \bar{Y}_i \right)$$

A general form of the error variance is

$$\text{var}_M(\hat{T} - T) = B_1 - 2B_2 + B_3, \tag{9.1.4}$$

where $B_1 = \text{var}_M(\hat{T}) = \Sigma_{i \in s} g_i^2 \text{ var}_M(\bar{Y}_{si})$, $B_2 = \Sigma_{i \in s} g_i M_i \text{ cov}_M(\bar{Y}_{si}, \bar{Y}_i)$, and $B_3 = \Sigma_{i=1}^N M_i^2 \text{ var}_M(\bar{Y}_i)$.

As the next theorem shows, the first term of the error variance (9.1.4) is the dominant one, as long as the first-stage sampling fraction of clusters, $f = n/N$, is negligible and certain other population and sample quantities are bounded.

Theorem 9.1.1. If as n, $(N-n) \to \infty$,

(a) $f \to 0$,

(b) $g_i = O(N/n)$, and

(c) M_i, $\text{var}_M(\bar{Y}_{si})$, and $\text{cov}_M(\bar{Y}_{si}, \bar{Y}_i)$ are all $O(1)$ for all i,

then the orders of magnitude of the three variance components, B_1, B_2, and B_3, on the right-hand side of (9.1.4) are N/f, N, and N.

Proof: The proof is simple and left to Exercise 9.1. □

Notice that writing the dominant term of the error variance as $B_1 = \Sigma_{i\in s} g_i^2 \,\text{var}_M(\bar{Y}_{si})$ differs somewhat from the formulation in Section 5.5 for an unclustered population. There, we wrote the estimator of the total as $\hat{T} = \Sigma_s Y_i + \Sigma_s a_i Y_i$, and the dominant term of the error variance was $\Sigma_s a_i^2 v_i$, which was the variance of the estimator of the total for the nonsample units under the working model. In this chapter, it is convenient not to break the estimator of the total into the sum for the sample and an estimator of the sum for the nonsample. Thus, $B_1 = \text{var}_M(\hat{T})$ will be the key term for variance estimation.

Because B_2 and B_3 depend on the variance–covariance structure of units that are not observed, estimators $\hat{B}_2$ and $\hat{B}_3$ of these two terms can be unbiased only under models that relate this structure to that of the sample units. By contrast, B_1 depends only on the variances of sample means, and this enables us to produce an estimator $\hat{B}_1$ that is consistent under models that are much more general than the ones required for estimating B_2 and B_3. Consistent estimation of (9.1.4) is then possible under the more general models when $f \to 0$. This situation parallels the results from Chapter 5, where the variance of the nonsample sum, under some reasonable conditions often met in practice, was a small part of the error variance—a feature that facilitated construction of robust variance estimators.

Under the conditions of Theorem 9.1.1, we can construct a variance estimator that is consistent and approximately unbiased for B_1 under models that are much more general than the working model in (9.1.1). Under the working model all Y_{ij} have the same variance, and all pairs (Y_{ij}, Y_{ik}) within a cluster have the same covariance. Fortunately, we can relax these conditions considerably and still be able to estimate the leading term of (9.1.4). Consider the more general model:

$$\begin{aligned} E_M(Y_{ij}) &= \mu & i &= 1,\ldots,N;\quad j = 1,\ldots,M_i \\ \text{cov}_M(Y_{ij}, Y_{kl}) &= 0, & i &\neq k. \end{aligned} \tag{9.1.5}$$

This model asserts that units in different clusters are uncorrelated, but it imposes no further constraints on the covariance structure within clusters.

Each unit within a cluster may have a different variance, for example, and different pairs of units may have different correlations. Define the residual for cluster i as

$$r_i = \bar{Y}_{si} - \hat{\mu},$$

where $\hat{\mu} = \hat{T}/M$. Under (9.1.5), $E_M(r_i) = 0$ and

$$E_M(r_i^2) = \text{var}_M(\bar{Y}_{si})\left(1 - \frac{2g_i}{M}\right) + \frac{1}{M^2}\sum_{i' \in s} g_{i'}^2 \, \text{var}_M(\bar{Y}_{si}) \qquad (9.1.6)$$

for $i = 1, \ldots, n$.

When $g_i = O(N/n)$ and $M_i = O(1)$, we have $g_i/M = O(n^{-1})$. Thus, under the conditions of Theorem 9.1.1, r_i^2 is an approximately unbiased estimator of $\text{var}_M(\bar{Y}_{si})$ under model (9.1.5). A sandwich variance estimator is then simply

$$v_R = \sum_s g_i^2 r_i^2. \qquad (9.1.7)$$

This estimator is approximately unbiased under either the working model (9.1.1) or the more general model (9.1.5) when the sample of clusters is large and their sampling fraction is small. This version of a sandwich estimator corresponds to v_R in Section 5.5 for unclustered populations, except that, as noted above, the coefficient g_i is appropriate for estimating the full population total rather than the nonsample total as was the case in Section 5.5.

Note that, although the estimator $\hat{\mu}$ in the residual $r_i = \bar{Y}_{si} - \hat{\mu}$ is tied directly to the coefficients g_i used in $\hat{T}$, this is not a critical requirement. We could use another estimator of the form $\hat{\mu} = \Sigma_s u_i \bar{Y}_{si}$ with $\Sigma_s u_i = 1$. The coefficients u_i need only have the property that $u_i = O(n^{-1})$. For example, for the *BLU* predictor under model (9.1.1) we could use $u_i = (m_i/v_i)/\Sigma_s m_i/v_i$ with $v_i = \sigma^2[1 + (m_i - 1)\rho]$. (See also Exercise 9.12.)

9.2.* OTHER VARIANCE ESTIMATORS

In some applications the lower-order variance terms, B_2 and B_3, in (9.1.4) may be too important to be neglected. This section discusses some alternative variance estimators that account for these terms. The first of the alternatives is based on ANOVA estimators of model parameters given in Section 8.3.1. However, this variance estimator is not robust, as will be shown in Section 9.2.1. In Section 9.2.2, we describe modifications of the sandwich estimator that inherit its robustness properties but also include components to estimate the lower-order variance terms.

9.2.1.* Non-Robust ANOVA Estimator

The working model (9.1.1) implies that $\text{var}_M(\bar{Y}_{si}) = \sigma^2[1 + (m_i - 1)\rho]/m_i$ and $\text{cov}_M(\bar{Y}_{si}, \bar{Y}_i) = \sigma^2[1 + (m_i - 1)\rho]/M_i$. Thus we can write the error variance (9.1.4) as a linear function of the parameters $\theta_1 = (1 - \rho)\sigma^2$ and $\theta_2 = \rho\sigma^2$:

$$\text{var}_M(\hat{T} - T) = \sum_s \frac{g_i^2}{m_i}[\theta_1 + m_i\theta_2] - 2\sum_s g_i[\theta_1 + m_i\theta_2] + \sum_{i=1}^{M_i} M_i[\theta_1 + M_i\theta_2]. \tag{9.2.1}$$

We can then estimate (9.2.1) by replacing these quantities by estimators that are unbiased under the working model.

In Section 8.3.1 we examined some analysis-of-variance estimators of model parameters that were unbiased under (9.1.1). These parameter estimates can be used to develop an estimator of the error variance of $\hat{T}$, but the estimator is highly dependent on the working model and is not robust to departures from that model. From Table 8.2 estimators of $\theta_1 = (1 - \rho)\sigma^2$ and $\theta_2 = \rho\sigma^2$ are

$$\hat{\theta}_1 = W \quad \text{and} \quad \hat{\theta}_2 = \frac{B - W}{D}, \tag{9.2.2}$$

where

$$W = \frac{1}{n}\sum_{i\in s}\sum_{j\in s_i}(Y_{ij} - \bar{Y}_{si})^2/(m_i - 1), \qquad B = \sum_{i\in s} m_i(\bar{Y}_{si} - \bar{\bar{Y}}_s)^2/(n - 1),$$

with

$$\bar{\bar{Y}}_s = \sum_s m_i \bar{Y}_{si}/m \quad \text{and} \quad D = \left(m - \sum_{i\in s} m_i^2/m)/(n - 1\right).$$

For illustration, we will consider only the leading term of the variance in (9.1.4). An unbiased estimator of the dominant term of that variance under model (9.1.1) is

$$\hat{B}_1 = \sum_s \frac{g_i^2}{m_i}[\hat{\theta}_1 + m_i\hat{\theta}_2]. \tag{9.2.3}$$

This estimator is not unbiased under the more general model that has σ_i^2 rather than σ^2 and ρ_i instead of ρ:

$$E_M(Y_{ij}) = \mu$$

$$\text{cov}_M(Y_{ij}, Y_{i'j'}) = \begin{cases} \sigma_i^2 & i = i', j = j' \\ \sigma_i^2\rho_i & i = i', j \neq j' \\ 0 & \text{otherwise.} \end{cases} \tag{9.2.4}$$

To find the bias of $\hat{B}_1$ under model (9.2.4), note that

$$E_M(W) = \bar{\theta}_{1s}$$

and

$$E_M(B) = \frac{n\bar{\theta}_{1s} - \bar{\theta}_{1s}^w}{n-1} + \frac{1}{n-1}\left[\sum_s m_i\theta_{2i} - m^{-1}\sum_s m_i^2\theta_{2i}^2\right],$$

where $\theta_{1i} = \sigma_i^2(1-\rho_i)$, $\theta_{2i} = \sigma_i^2\rho_i$, $\bar{\theta}_{1s} = \Sigma_s\theta_{1i}/n$, $\bar{\theta}_{1s}^w = \Sigma_s m_i\theta_{1i}/m$ (see Exercise 9.3). After some calculation, the bias of $\hat{B}_1$ under model (9.2.4) is found to be

$$E_M(\hat{B}_1 - B_1) = \sum_s \frac{g_i^2}{m_i}\{E_M(\hat{\theta}_1 - \theta_{1i}) + m_i E_M(\hat{\theta}_2 - \theta_{2i})\}$$

$$= \sum_s \frac{g_i^2}{m_i}\left\{\bar{\theta}_{1s} - \theta_{1i} + m_i\left[\frac{\bar{\theta}_{1s} - \bar{\theta}_{1s}^w}{(n-1)D} + \frac{1}{(n-1)D}\sum_s m_i\left(1 - \frac{m_i}{m}\right)\theta_{2i} - \theta_{2i}\right]\right\}.$$

If the working model is correct, that is, $\theta_{1i} = \theta_1$ and $\theta_{2i} = \theta_2$, then this bias reduces to 0. If an equal allocation of $m_i = \bar{m}$ units is made to each sample cluster, $\bar{\theta}_{1s} = \bar{\theta}_{1s}^w$ and the bias becomes

$$\sum_s \frac{g_i^2}{\bar{m}}\{\bar{\theta}_{1s} - \theta_{1i} + \bar{m}(\bar{\theta}_{2s} - \theta_{2i})\}. \tag{9.2.5}$$

When $m_i = \bar{m}$, the bias in (9.2.5) will be 0 if the unweighted sample means of the model parameters are equal to the weighted means: $\bar{\theta}_{1s} = \Sigma_s g_i^2\theta_{1i}/\Sigma_s g_i^2$ and $\bar{\theta}_{2s} = \Sigma_s g_i^2\theta_{2i}/\Sigma_s g_i^2$. If g_i is constant, as in $\hat{T}_0$ and $\hat{T}_p$ when $m_i = \bar{m}$, then these unweighted and weighted means will be equal. But, for other estimators like $\hat{T}_{BLU}$, $\hat{T}_R$, $\hat{T}_H$, and $\hat{T}_1$, the coefficients g_i depend on M_i and other quantities that vary among units (see Table 8.1). For those estimators the ANOVA variance estimator will be biased whether or not an equal allocation to sample clusters is used.

9.2.2.* Alternative Robust Variance Estimators

In this section, we describe a robust variance estimator that includes estimators of the lower-order terms in (9.1.4). We can modify the sandwich estimator by first refining the estimator of B_1 and then by devising estimators of the lower-order terms, B_2 and B_3, as in Royall (1986). For the B_2 term in (9.1.4) we will find an estimator that is unbiased and consistent under models more general than (9.1.1), and we will use (9.1.1) only for estimating the third term.

First, we examine a more complicated alternative to the sandwich estimator using a technique devised by Chew (1970). Replacing $E_M(r_i^2)$ by r_i^2 and $\text{var}_M(\bar{Y}_{si})$ by $\hat{v}_i$ in (9.1.6) gives a system of n equations in the unknowns $\hat{v}_i$.

Solving (9.1.6) for $\hat{v}_i$, after the replacements, yields

$$\hat{v}_i = \frac{r_i^2 - A}{1 - 2g_i/M}, \tag{9.2.6}$$

where $A = \Sigma_s g_i^2 \hat{v}_i / M^2$. For (9.2.6) to be well defined, we must have $2g_i/M < 1$ or $g_i < M/2$. Multiplying each side of (9.2.6) by g_i^2/M^2 and summing over s gives A in terms of r_i^2 as

$$A = \frac{\displaystyle\sum_{i\in s} \frac{g_i^2 r_i^2}{M^2(1 - 2g_i/M)}}{1 + \displaystyle\sum_{i\in s} \frac{g_i^2}{M^2(1 - 2g_i/M)}}.$$

The expectation of $\hat{v}_i$ under (9.1.5) is $E_M(\hat{v}_i) = \text{var}_M(\bar{Y}_{si})$, which is easily verified (see Exercise 9.6). Note that even though $\hat{v}_i$ is estimating the positive quantity $\text{var}_M(\bar{Y}_{si})$, it is possible for $\hat{v}_i$ to be negative for some sample clusters. This will mainly be a problem in small- to medium-size cluster samples since as both N and $n \to \infty$, A approaches 0.

Consequently, a bias-robust estimator of the leading term in the error variance (9.1.4) is

$$v_C = \sum_{i\in s} g_i^2 \hat{v}_i. \tag{9.2.7}$$

Note that $O(A) = n^{-1}$, and, in large cluster samples, $\hat{v}_i \cong r_i^2$ implying that (9.2.7) and the sandwich estimator in (9.1.7) are essentially the same. Note also that s must contain $n > 2$ clusters. For $n = 1$, $r_i = \bar{Y}_{si} - \bar{Y}_{si} g_i/M$ vanishes, since condition (9.1.3) for unbiasedness implies that we must have $g_i = M$. For $n = 2$, the term $1 - 2g_i/M$ may fail to be positive. Consider, for example, $\hat{T}_0$ which has $g_i/M = m_i/m$. When $n = 2$, m_i/m will be greater than $\frac{1}{2}$ for one of the sample clusters.

Neither v_R nor v_C alone is adequate if the sampling fraction of clusters, $f = n/N$ is large. In that case, estimators are also needed for the second and third terms of the error variance in (9.1.4), B_2 and B_3.

To estimate $B_2 = \Sigma_{i\in s} g_i M_i \text{cov}_M(\bar{Y}_{si}, \bar{Y}_i)$ in (9.1.4), we need more structure than (9.1.5), but we can still use the more general covariance model in (9.2.4) rather than the restrictive one in the working model (9.1.1). Like (9.1.1), model (9.2.4) says that within a given cluster all Y_{ij} have the same variance and all pairs (Y_{ij}, Y_{ik}) $j \neq k$, have the same covariance. But (9.2.4) allows the variance and covariance to change for every cluster. Under this model $\text{var}_M(\bar{Y}_i) = \text{cov}_M(\bar{Y}_{si}, \bar{Y}_i) = \text{var}_M(\bar{Y}_{si}) - \text{var}_M(\bar{Y}_{si} - \bar{Y}_i) = M_i^{-1}\sigma_i^2[1 + (M_i - 1)\rho_i]$. In addition, if $S_i^2 = \Sigma_{j\in s_i}(Y_{ij} - \bar{Y}_{si})^2/(m_i - 1)$, then $E_M(S_i^2) = \sigma_i^2(1 - \rho_i)$, and the following simple Lemma holds. Note that for S_i^2 to be defined we must have $m_i \geqslant 2$ which can occur only if $M_i \geqslant 2$.

Lemma 9.2.1. Under model (9.2.4) $E_M[M_i^{-1}S_i^2(1-f_i)/f_i] = \text{var}_M(\bar{Y}_{si} - \bar{Y}_i)$, where $f_i = m_i/M_i$.

Proof: The proof is left to Exercise 9.7. □

Thus, for $i \in s$ an estimator of $\text{var}_M(\bar{Y}_i) = \text{cov}_M(\bar{Y}_{si}, \bar{Y}_i)$ is

$$\dot{v}_i = \hat{v}_i - M_i^{-1}S_i^2(1-f_i)/f_i$$

and we can estimate $B_2 = \Sigma_s g_i M_i \,\text{cov}_M(\bar{Y}_{si}, \bar{Y}_i)$ by

$$\sum_s g_i M_i \dot{v}_i. \tag{9.2.8}$$

To estimate the final term in (9.1.4), $\Sigma_1^N \text{var}_M(T_i) = \Sigma_s \text{var}_M(T_i) + \Sigma_r \text{var}_M(T_i)$, we use $\Sigma_s M_i^2 \dot{v}_i$ for the first part, and for the second we must use the full working model (9.1.1), under which

$$\sum_r M_i^2 \,\text{var}_M(\bar{Y}_i) = \theta_1 \sum_r M_i + \theta_2 \sum_r M_i^2. \tag{9.2.9}$$

An unbiased estimator of (9.2.9) is given in Lemma 9.2.2.

Lemma 9.2.2. An unbiased estimator of $\Sigma_r M_i^2 \,\text{var}_M(\bar{Y}_i)$ under model (9.1.1) is

$$\sum_{i \in s}\left(\frac{g_i}{M_i} - 1\right) M_i^2 \dot{v}_i + \left[\sum_{i=1}^N M_i^2 - \sum_s g_i M_i\right]\dot{\theta}_2, \tag{9.2.10}$$

where $\dot{\theta}_2 = \Sigma_s(\hat{v}_i - S_i^2/m_i)/m$ is an unbiased estimator of $\theta_2 = \rho\sigma^2$.

Proof: Since, under model (9.1.1), $E_M(\hat{v}_i) = \sigma^2[1 + (m_i - 1)\rho]/m_i$ and $E_M(S_i^2) = \sigma^2(1-\rho)$, it follows that $E_M(\dot{\theta}_2) = \theta_2$. Using the fact that $\Sigma_s g_i = M$ due to unbiasedness of $\hat{T}$, the expectation of (9.2.10) is

$$\sum_s (g_i\theta_1 + g_i M_i \theta_2) - \sum_s (M_i\theta_1 + M_i^2\theta_2) + \theta_2 \sum_1^N M_i^2 - \theta_2 \sum_s g_i M_i$$
$$= \theta_1 \sum_r M_i + \theta_2 \sum_r M_i^2.$$ □

Note that the estimator $\dot{\theta}_2$ does involve a subtraction and may be negative in a particular sample even though the parameter $\theta_2 = \rho\sigma^2$ is positive when $\rho > 0$.

Combining (9.2.7) for the first term in (9.1.4), (9.2.8) for the second term, and $\Sigma_s M_i^2 \dot{v}_i$ plus (9.2.10) for the third term, we can estimate the error variance (9.1.4) by

$$v_0 = \hat{B}_1 - 2\hat{B}_2 + \hat{B}_3, \tag{9.2.11}$$

where

$$\hat{B}_1 = v_C = \sum_s g_i^2 \hat{v}_i,$$

$$\hat{B}_2 = \sum_s g_i M_i \dot{v}_i,$$

and

$$\hat{B}_3 = \sum_s M_i^2 \dot{v}_i + \sum_{i \in s} \left(\frac{g_i}{M_i} - 1 \right) M_i^2 \dot{v}_i + \left[\sum_{i=1}^N M_i^2 - \sum_s g_i M_i \right] \dot{\theta}_2.$$

By combining terms, the estimator can also be written as

$$v_0 = \sum_s g_i^2 \hat{v}_i - \sum_s g_i M_i \dot{v}_i + \left[\sum_{i=1}^N M_i^2 - \sum_s g_i M_i \right] \dot{\theta}_2. \tag{9.2.12}$$

The estimator v_0 is, by construction, exactly unbiased under the working model (9.1.1) and approximately unbiased under the much more general model (9.1.5). Under appropriate large-sample conditions, similar to the ones employed in Chapter 5, v_0 can also be shown to be consistent, but we will not bother with that formality here.

Alternative estimators that are unbiased under (9.1.1) and asymptotically equivalent to v_0, are obtained simply by replacing θ_1 and θ_2 in (9.2.9) by any unbiased estimators $\hat{\theta}_1$ and $\hat{\theta}_2$. Then we obtain

$$v_w = \hat{B}_1 - 2\hat{B}_2 + \hat{B}_3^*,$$

where $\hat{B}_1$ and $\hat{B}_2$ are defined as above and

$$\hat{B}_3^* = \sum_s M_i^2 \dot{v}_i + \hat{\theta}_1 \sum_r M_i + \hat{\theta}_2 \sum_r M_i^2. \tag{9.2.13}$$

For example, $\hat{\theta}_1$ and $\hat{\theta}_2$ can be obtained by the method of weighted least squares based on the system of equations

$$E_M(\dot{v}_i) = M_i \theta_1 + M_i^2 \theta_2 \quad (i = 1, \ldots, n).$$

Minimizing the sum of squares $\Sigma_s w_i (\dot{v}_i - M_i \theta_1 - M_i^2 \theta_2)^2$ with $w_i = 1/M_i^2$

leads to

$$\hat{\theta}_1 = \left[\left(\sum_s M_i^2\right)\left(n^{-1}\sum_s \dot{v}_i/M_i\right) - \bar{M}_s\left(\sum_s \dot{v}_i\right)\right]\Big/ C \tag{9.2.14}$$

and

$$\hat{\theta}_2 = \left[\left(\sum_s \dot{v}_i\right) - \bar{M}_s\left(\sum_s \dot{v}_i/M_i\right)\right]\Big/ C, \tag{9.2.15}$$

where $C = \Sigma_s (M_i - \bar{M}_s)^2$. The estimator v_w does require that $\Sigma_r M_i$ and $\Sigma_r M_i^2$ be known. The former can be obtained by subtraction if the total number of units in the population M is known. The latter will require that either the size of each nonsample cluster or the population variance among the cluster sizes be known.

9.3. EXAMPLES OF THE VARIANCE ESTIMATORS

Two estimators that fall into the class (9.1.2) are the ratio estimator and the *pps* or mean-of-ratios estimator. As we saw in Section 8.3.4, the ratio estimator can be inefficient in some types of populations. However, examination of the alternative variance estimators for $\hat{T}_R$ and $\hat{T}_p$ provides some useful illustrations of specific formulas and the relative sizes of different terms in the more elaborate estimators from Section 9.2.2.

9.3.1. Ratio Estimator

The ratio estimator $\hat{T}_R = (M/M_s)\,\Sigma_s M_i \bar{Y}_{si}$ has $g_i = M_i M/M_s$. After some simplification, the robust variance estimators can be written as

$$v_R(\hat{T}_R) = \frac{N^2}{n}\left(\frac{\bar{M}}{\bar{M}_s}\right)^2\left(\frac{1}{n}\sum_s M_i^2 r_i^2\right)$$

$$v_C(\hat{T}_R) = \frac{N^2}{n}\left(\frac{\bar{M}}{\bar{M}_s}\right)^2\left(\frac{1}{n}\sum_s M_i^2 \hat{v}_i\right)$$

$$v_0(\hat{T}_R - T) = \frac{N^2}{n}(1-f)\frac{\bar{M}\bar{M}_r}{\bar{M}_s^2}\left(\frac{1}{n}\sum_s M_i^2 \hat{v}_i\right) + \frac{\bar{M}_r}{f\bar{M}_s}\sum_s M_i S_i^2 \frac{(1-f_i)}{f_i}$$
$$+\left(\frac{\bar{M}^{(2)}}{\bar{M}_s^{(2)}} - \frac{\bar{M}}{\bar{M}_s}\right)\frac{1}{f}\sum_s M_i^2\, \dot{\theta}_2,$$

where $\bar{M}^{(2)} = \Sigma_1^N M_i^2/N$, $\bar{M}_s^{(2)} = \Sigma_s M_i^2/n$, $\bar{M}$ and $\bar{M}_s$ are the population and sample means of the M_i, $r_i = \bar{Y}_{si} - \hat{T}_R/M$, and $\hat{v}_i$ is defined by (9.2.6). Notice that the estimator v_0 does explicitly account for the sampling fraction of clusters and units within clusters while v_R and v_C do not. The coefficient of $\dot{\theta}_2$ in $v_0(\hat{T}_R - T)$ vanishes in samples that are balanced on M and M^2. In a sequence of samples approaching balance, that is, $|\bar{M}_s - \bar{M}| \to 0$ and $\bar{M}_s^{(2)} - \bar{M}^{(2)}| \to 0$, the third term of v_0 becomes negligible relative to the second.

We can examine how the general, large sample properties of the robust variance estimators apply in the particular case of the ratio estimator. Under the general model (9.1.5), with $v_i = \text{var}_M(\bar{Y}_{si})$, the actual error-variance is

$$\text{var}_M(\hat{T}_R - T) = \frac{N^2}{n}\left(\frac{\bar{M}}{\bar{M}_s}\right)^2\left(\frac{1}{n}\sum_s M_i^2 v_i\right) - 2N\frac{\bar{M}}{\bar{M}_s}\left(\frac{1}{n}\sum_s M_i \,\text{cov}_M(\bar{Y}_{si}, \bar{Y}_i)\right) + \sum_{i=1}^{N} M_i^2 \,\text{var}_M(\bar{Y}_i).$$

Now, suppose that n and N grow so that $f = n/N \to 0$ and that, as this growth occurs, both the sample and population remain stable in the sense that $\Sigma_s M_i^2 v_i/n$, $N^{-1}\Sigma_{i=1}^N M_i^2 \,\text{var}_M(\bar{Y}_i)$, $\bar{M}_s$, and similar averages all converge to nonzero constants. Since $f \to 0$, $\bar{M}\bar{M}_r/\bar{M}_s^2 \approx \bar{M}^2/\bar{M}_s^2$. Then $\text{var}_M(\hat{T}_R - T)$ and the expectation under (9.1.5) of each of the three variance estimators above equal $(N/f)(\bar{M}/\bar{M}_s)^2 \,\Sigma_s M_i^2 v_i/n + O(N)$, implying that v_R, v_C, and v_0 are all asymptotically unbiased.

When $m_i = M_i$ (single-stage cluster sampling), the cluster total $T_i = M_i\bar{Y}_i$ is observed for all $i \in s$, so that $\hat{T}_R$ is the ordinary ratio estimator based on the sample T_i's with size measure M_i. Because $M_s = m$ with complete enumeration of clusters, the ratio estimator also reduces to the simple expansion estimator, $\hat{T}_0 = (M/m)\,\Sigma_s M_i\bar{Y}_i$. The second term in v_0 is zero and

$$v_0 = \frac{M^2}{m^2}\left(1 - \frac{m}{M}\right)\sum_s M_i^2 \hat{v}_i + \left(\frac{\Sigma_{i=1}^N M_i^2}{\Sigma_s M_i^2} - \frac{N\bar{M}}{n\bar{M}_s}\right)\sum_s M_i^2 \dot{\theta}_2. \qquad (9.2.16)$$

Note that the second term on the right-hand side of (9.2.16) has a lower order than the first and that $\hat{v}_i \cong r_i^2$. If $f = n/N$ is small, then $v_0 \cong (M^2/m^2)\,\Sigma_s M_i^2 r_i^2$, which is the sandwich estimator for the expansion estimator.

The estimators above have been derived under prediction models without reference to the procedure used to select the sample. In design-based practice, $\hat{T}_R$ is commonly suggested when simple random sampling without replacement is used for choosing both the sample clusters and the units within clusters. In this case, the usual design-based estimator (Cochran, 1977, p. 305) is

$$v_c = \frac{N^2(1-f)}{n(n-1)}\sum_s M_i^2 r_i^2 + \frac{1}{f}\sum_s M_i S_i^2 \frac{1-f_i}{f_i}.$$

The dominant term of v_0 differs from v_c by the factor $\bar{M}\bar{M}_r/\bar{M}_s^2$. Thus, for estimating the error-variance when f is small, v_c has a model-bias that can be large in samples where $\bar{M}_r/\bar{M}_s$ is very different from unity. The situation parallels that of the simple ratio estimator in Section 5.8 where the robust variance estimators are approximated by the conventional estimator multiplied by the factor $\bar{x}\bar{x}_r/x_s^2$ (see also Royall and Eberhardt, 1975). The need for the correction factor $\bar{M}\bar{M}_r/\bar{M}_s^2$ was also pointed out by Rustagi (1978).

9.3.2. Mean of Ratios Estimator

As another example, we will consider the $pp(M_i)$ estimator $\hat{T}_p = M\Sigma_s \bar{Y}_{si}/n$ which has $g_i = M/n$. This estimator can also be written in the form of a mean-of-ratios estimator as $\hat{T}_p = M\Sigma_s \hat{T}_i/(nM_i)$ with $\hat{T}_i = M_i\bar{Y}_{si}$. In this case, v_R and v_C have particularly simple forms. Using $1-2g_i/M = (n-2)/n$, the term A in (9.2.6) simplifies to $\Sigma_s r_i^2/[n(n-1)]$ with $r_i = \bar{Y}_{si} - \hat{T}_p/M$. It follows that

$$v_R(\hat{T}_p) = \frac{N^2}{n}\bar{M}^2\left(\frac{1}{n}\sum_s r_i^2\right)$$

$$v_C(\hat{T}_p) = \frac{N^2}{n}\bar{M}^2\left(\frac{1}{n}\sum_s \hat{v}_i\right)$$

$$v_0 = \sum_{i\in s}\frac{M}{n}\left(\frac{M}{n} - M_i\right)\hat{v}_i + \frac{M}{n}\sum_{i\in s} S_i^2\frac{1-f_i}{f_i} + \left[\sum_{i=1}^{N} M_i^2 - N\bar{M}\bar{M}_s\right]\dot{\theta}_2.$$

Note that when the sample weighted balance on M, that is, when $\bar{M}_s = \bar{M}^{(2)}/\bar{M}$ as in Chapters 3 and 4, the final term in v_0 vanishes. Also, in a sequence of samples approaching weighted balance, $|\bar{M}_s - \bar{M}^{(2)}/\bar{M}| \to 0$, the third term in v_0 becomes negligible relative to the second. If clusters are selected with probabilities proportional to their sizes M_i, then the expected value of $\bar{M}_s$ with respect to the sampling design is $\bar{M}^{(2)}/\bar{M}$, so that the design-expectation of the coefficient of $\dot{\theta}_2$ is zero.

In sampling with replacement with probabilities proportional to the M_i, if no cluster is sampled more than once, then the standard design-based variance estimator is

$$v_c = \frac{M^2}{n(n-1)}\sum_s r_i^2$$

(see Särndal, Swensson, and Wretman, 1992, Result 4.5.1, p. 151). Since, again, r_i^2 approximates $\hat{v}_i$, v_c approximates v_R, v_C, and the leading term in v_0. Thus all are asymptotically equivalent as $n \to \infty$ and $f \to 0$.

9.3.3. Numerical Illustrations

In the example below, we illustrate the calculation of some of the alternative variance estimators using $\hat{T}_R$ and $\hat{T}_p$. Recall from the preceding sections that $g_i = M_iM/M_s$ for the ratio estimator and that $g_i = M/n$ for $\hat{T}_p$.

Table 9.1. A Sample of $n = 10$ Clusters From the Labor Force Population

Cluster	Person	Cluster	Person
2	6	56	223
2	7	56	225
20	67	63	246
20	68	63	247
31	111	82	338
31	115	82	339
39	152	85	346
39	153	85	347
42	163	94	387
42	165	94	391

Example 9.2.1. Variance estimation for $\hat{T}_R$ and $\hat{T}_p$. Table 9.1 gives a simple random sample of 10 clusters selected without replacement from the Labor Force population in Appendix B. A simple random sample of $\bar{m} = 2$ persons was selected without replacement from each cluster. In this sample the vector of M_i's for the sample clusters is (3, 3, 5, 2, 3, 4, 2, 2, 4, 7).

Table 9.2 summarizes some of the calculations based on the variable y. The sandwich variance estimates defined by (9.1.7) are $v_R = 1{,}731{,}607$ for $\hat{T}_R$ and $v_R = 1{,}787{,}679$ for $\hat{T}_p$. The adjusted estimator v_C, defined by (9.2.7), is larger than v_R for both the ratio estimator and for $\hat{T}_p$ while the component estimator from (9.2.11) is intermediate between the other two.

The quantities $\hat{v}_i$ and $\dot{v}_i$ estimate $\text{var}_M(\bar{Y}_{si})$ and $\text{var}_M(\bar{Y}_i)$, respectively. Because both $\hat{v}_i$ and $\dot{v}_i$ involve a subtraction, it is possible for each to be negative for particular clusters even though they estimate positive quantities.

Table 9.2. Variance Estimates and Components for Example 9.2.1

	$\hat{T}_R$	$\hat{T}_p$
$\hat{T}$	23,414	23,315
v_R	1,731,607	1,787,679
v_C	1,917,016	1,986,310
v_0	1,752,881	1,795,241
$\hat{B}_1$	1,917,016	1,986,310
$\hat{B}_2$	87,347	104,679
$\hat{B}_3$	10,560	18,290
$\tilde{v}_0$	1,702,393	1,817,887

The vector of v_i's for the ratio estimator is (−10.2, 172.51, 33.43, 231.97, 76.12, 141.29, −10.63, −8.08, 239.66, 3.27) and $\dot{v}_i$ = (−12.08, 165.62, 26.12, 231.97, 75.94, 140.21, −10.63, −8.08, 190.95, −54.33). For $\hat{T}_p$, $\hat{v}_i$ = (−9.06, 173.03, 34.48, 203.4, 78.08, 141.06, −9.00, −7.61, 256.58, 2.62). The vector of $\dot{v}_i$'s for $\hat{T}_p$ is (−10.94, 166.14, 27.17, 203.4, 77.9, 139.98, −9.00, −7.61, 207.87, −54.98). Thus we have a sample in which $\hat{v}_i$ and $\dot{v}_i$ are negative for several sample clusters. Using the formulas following (9.2.11), estimates of the three variance components in v_0 were calculated and are shown separately in the table. Even in this sample of $n = 10$ the first component $\hat{B}_1$ dominates the others. Various ad hoc corrections might be used for the negative $\hat{v}_i$ and $\dot{v}_i$. One is to use $\max(\hat{v}_i, 0)$ and $\max(\dot{v}_i, 0)$ in the component formulas. Employing these two fixes gives adjusted variance estimates of $\tilde{v}_0 = 1{,}702{,}393$ for $\hat{T}_R$ and $\tilde{v}_0 = 1{,}817{,}887$ for $\hat{T}_p$. Notice that, by forcing $\dot{v}_i$ to be nonnegative, we increase the absolute value of $-2\hat{B}_2$, leading to $\tilde{v}_0 < v_0$ for $\hat{T}_R$.

9.4. VARIANCE ESTIMATION FOR AN ESTIMATED TOTAL—UNKNOWN CLUSTER SIZES

In Section 8.9, we studied a situation where the cluster sizes M_i were known only for those clusters in the first-stage sample s, but another measure of size x_i was known for all clusters. We return to that case here to briefly sketch some results on variance estimation.

As in Section 8.9, the sizes M_i are represented as random variables that depend on the x_i. The working model for the M_i is the same as (8.9.1):

$$E_M^*(M_i) = \beta x_i$$
$$\operatorname{cov}_M^*(M_i, M_{i'}) = \begin{cases} \tau^2 x_i & i = i' \\ 0 & i \neq i'. \end{cases} \tag{9.4.1}$$

We again denote with an asterisk the unconditional expectations, variances, and covariances, that is, ones taken over both the model for Y_{ij} and M_i. Thus we have $E_M^*[E_M(T_i|\mathbf{M})] = E_M^*(M_i\mu) = \mu\beta x_i$, where the operator E without the asterisk continues to denote the conditional expectation given the M's. For the model conditional on the M_i's, assume that the working model is (9.1.1).

As in Section 8.9, the general estimator is $\hat{T}^* = \Sigma_{i\in s}\lambda_i^*\hat{T}_i$. The requirement for unbiasedness, $E_M^*E_M(\hat{T}^* - T|\mathbf{M}) = 0$, implies that the λ_i^* must satisfy $\Sigma_s\lambda_i^* x_i/N = \bar{x}$. The error-variance is now

$$\operatorname{var}_M^*(\hat{T}^* - T) = B_1^* - 2B_2^* + B_3^*, \tag{9.4.2}$$

where $B_1^* = \Sigma_{i\in s}\lambda_i^{*2}\operatorname{var}_M^*(\hat{T}_i)$, $B_2^* = \Sigma_{i\in s}\lambda_i^*\operatorname{cov}_M^*(\hat{T}_i, T_i)$, and $B_3^* = \Sigma_{i=1}^N \operatorname{var}_M^*(Y_i)$. Under conditions similar to those in Theorem 9.1.1, the orders of magnitude of B_1^*, B_2^*, and B_3^* are N/f, N, and N.

From the residuals $r_i^* = \hat{T}_i - x_i\hat{T}^*/(N\bar{x})$ we can proceed as before in Section 9.1 to construct a sandwich variance estimator. Similar to expression (9.1.6), we have

$$E_M^*(r_i^{*2}) = \text{var}_M^*(\hat{T}_i)\left(1 - 2\lambda_i^* \frac{x_i}{N\bar{x}}\right) + \left(\frac{x_i}{N\bar{x}}\right)^2 \sum_{i' \in s} \lambda_{i'}^{*2} \, \text{var}_M^*(\hat{T}_{i'}).$$

If $\text{var}_M^*(\hat{T}_i) = O(1)$, $\lambda_i^* = O(N/n)$, and the x_i's are bounded, then $E_M^*(r_i^2) \cong \text{var}_M^*(\hat{T}_i)$ when the sample of clusters is large. This large sample equivalence holds regardless of the internal covariance structure of the clusters. Thus a sandwich estimator of variance is

$$v_R^* = \sum_s \lambda_i^{*2} r_i^{*2}$$

This estimator is robust in the sense of being consistent under models (9.1.5) and (9.4.1) when the sample of clusters is large and the fraction of the clusters in the sample, $f = n/N$, is small.

The approach in Section 9.2.2 can also be used to develop variance estimators that account for the lower order terms in (9.4.2) as described in Royall (1986, sec. 2.3).

9.5. REGRESSION ESTIMATOR

In Section 8.6 we studied a regression estimator of the total that was unbiased under a model where $E_M(\mathbf{Y}) = \mathbf{X}\boldsymbol{\beta}$. We return to that situation here and derive alternative variance estimators that will be consistent under the general variance specification $\text{var}_M(\mathbf{Y}) = \mathbf{V}$. Often, little will be known about $\mathbf{V}$ in a clustered population, and the predictor of the total will be constructed using either $\mathbf{V} = \mathbf{I}$ or $\mathbf{V} = \mathbf{W}^{-1}$ for some diagonal or block diagonal matrix $\mathbf{W}$. The predictor of the total is then

$$\hat{T} = \mathbf{1}_s'\mathbf{Y}_s + \mathbf{1}_r'\mathbf{X}_r\hat{\boldsymbol{\beta}},$$

where $\mathbf{Y}_s$ is the $m \times 1$ vector of Y_{ij}'s for the sample units, and $\mathbf{X}_r$ is the $(M - m) \times p$ matrix of explanatory variables for the nonsample units. If $\mathbf{A}_s = \mathbf{X}_s'\mathbf{W}_s\mathbf{X}_s$ is nonsingular, with $\mathbf{X}_s$ the $m \times p$ matrix of explanatory variables for the sample units, then $\hat{\boldsymbol{\beta}} = \mathbf{A}_s^{-1}\mathbf{X}_s'\mathbf{W}_s\mathbf{Y}$. If an over-parameterized model is used, as in Chapter 7, then $\boldsymbol{\beta}^o = \mathbf{G}\mathbf{X}_s'\mathbf{W}_s\mathbf{Y}_s$ is used, where $\mathbf{G}$ is a generalized inverse of $\mathbf{A}_s$. The weight vector corresponding to $\hat{T} = \mathbf{g}_s'\mathbf{Y}_s$, is either

$$\mathbf{g}_s = \mathbf{W}_s\mathbf{X}_s\mathbf{A}_s^{-1}\mathbf{X}_r'\mathbf{1}_r + \mathbf{1}_s, \qquad (9.5.1)$$

when $\mathbf{A}_s$ is nonsingular, or

$$\mathbf{g}_s = \mathbf{W}_s\mathbf{X}_s\mathbf{G}\mathbf{X}_r'\mathbf{1}_r + \mathbf{1}_s. \tag{9.5.2}$$

when a g-inverse is used. In either case, we can use the same methods to develop robust variance estimators. In the remainder of this section, as a notational shorthand, we will use $\mathbf{G}$ to cover both the singular and nonsingular cases and use $\hat{\beta}$ for the parameter estimator in either case. We also assume that $\mathbf{W}_s = \text{blkdiag}(\mathbf{W}_{si})$ where $\mathbf{W}_{si}$ is the $m_i \times m_i$ weight matrix for sample cluster i.

First, we can write the error variance in three parts as in Section 9.1 and identify the orders of magnitude of each. The estimation error is

$$\hat{T} - T = \sum_s \mathbf{g}_i'\mathbf{Y}_{si} - \left(\sum_s M_i\bar{Y}_i + \sum_r M_i\bar{Y}_i\right),$$

where $\mathbf{g}_i = (g_{i1}, \ldots, g_{im_i})'$ is the part of weight vector for the sample cluster i and $\mathbf{Y}_{si} = (Y_{i1}, \ldots, Y_{im_i})'$ is the data for the sample units from sample cluster i. Similar to expression (9.1.5), we can express the error variance as

$$\text{var}_M(\hat{T} - T) = B_1 - 2B_2 + B_3, \tag{9.5.3}$$

where $B_1 = \Sigma_s \mathbf{g}_i' \text{var}_M(\mathbf{Y}_{si})\mathbf{g}_i$, $B_2 = \Sigma_s \mathbf{g}_i' \text{cov}_M(\mathbf{Y}_{si}, \mathbf{Y}_i)\mathbf{1}_{M_i}$ with $\mathbf{Y}_i = (Y_{i1}, \ldots, Y_{iM_i})'$, and $B_3 = \Sigma_{i=1}^N M_i^2 \text{var}_M(\bar{Y}_i)$. The following result, whose proof is trivial, gives the conditions under which B_1 is the dominant term.

Theorem 9.5.1. If as n, $(N - n) \to \infty$,

(a) $f \to 0$;
(b) $\mathbf{g}_s = O(N/n)$ elementwise; and
(c) $M_i = O(1)$, $\text{var}_M(\mathbf{Y}_{si})$ and $\text{cov}_M(\mathbf{Y}_{si}, \mathbf{Y}_i)$ are each $O(1)$ elementwise for all i,

then the orders of magnitude of the three variance components, B_1, B_2, and B_3, on the right-hand side of (9.5.3) are N/f, N, and N.

In the remainder of Section 9.5, we concentrate on estimating B_1 and neglect the less important terms, B_2 and B_3.

9.5.1. Sandwich Variance Estimator

As in the case of estimation under the common mean model, we can construct a simple, sandwich estimator that is consistent under a reasonably general variance specification. To that end, consider the model:

$$\begin{aligned} E_M(Y_{ij}) &= \mathbf{x}_{ij}'\boldsymbol{\beta} \qquad i = 1, \ldots, N; j = 1, \ldots, M_i \\ \text{cov}_M(Y_{ij}, Y_{kl}) &= 0 \qquad i \neq k. \end{aligned} \tag{9.5.4}$$

This model assumes that the regression specification, $E_M(Y_{ij})$, used to construct $\hat{T}$ is correct. Units in different clusters are assumed to be uncorrelated, but the variance–covariance structure within each cluster is arbitrary. The residual for sample unit ij is

$$r_{ij} = Y_{ij} - \mathbf{x}'_{ij}\hat{\boldsymbol{\beta}},$$

where we allow the estimator of slope to be either $\hat{\boldsymbol{\beta}} = \mathbf{A}_s^{-1}\mathbf{X}'_s\mathbf{W}_s\mathbf{Y}_s$ or $\boldsymbol{\beta}^o = \mathbf{G}\mathbf{X}'_s\mathbf{W}_s\mathbf{Y}_s$. In the remainder of this section, we assume that the elements of $\mathbf{X}$ and $\mathbf{W}$ are bounded.

In Section 9.1, we derived a sandwich estimator using the cluster-level residuals, $r_i = \bar{Y}_{si} - \hat{\mu}$ where $\hat{\mu} = \hat{T}/M$ or another estimator of the common mean. For the regression estimator, we use the vector of residuals for sample cluster i defined as $\mathbf{r}_i = (r_{i1}, \ldots, r_{im_i})'$. The variance estimator we consider is

$$v_R(\hat{T}) = \sum_s (\mathbf{g}'_i\mathbf{r}_i)^2 = \sum_s \mathbf{g}'_i(\mathbf{r}_i\mathbf{r}'_i)\mathbf{g}_i. \tag{9.5.5}$$

This estimator is approximately unbiased under the general model (9.5.4). To show this, first define $\mathbf{X}_{si}$ to be the part of $\mathbf{X}_s$ associated with sample cluster i:

$$\mathbf{X}_{si} = (\mathbf{x}_{i1}, \ldots, \mathbf{x}_{im_i})', \quad j = 1, \ldots, m_i.$$

The vector of sample residuals for cluster i is then

$$\mathbf{r}_i = \mathbf{Y}_{si} - \mathbf{X}_{si}\hat{\boldsymbol{\beta}}.$$

Next, define the hat matrix as

$$\mathbf{H} = \mathbf{X}_s\mathbf{G}\mathbf{X}'_s\mathbf{W}_s = \begin{bmatrix} \mathbf{X}_{s1}\mathbf{G}\mathbf{X}'_{s1}\mathbf{W}_{s1} & \cdots & \mathbf{X}_{s1}\mathbf{G}\mathbf{X}'_{sn}\mathbf{W}_{sn} \\ \vdots & & \vdots \\ \mathbf{X}_{sn}\mathbf{G}\mathbf{X}'_{s1}\mathbf{W}_{s1} & \cdots & \mathbf{X}_{sn}\mathbf{G}\mathbf{X}'_{sn}\mathbf{W}_{sn} \end{bmatrix}$$

and let $\mathbf{H}_{ik} = \mathbf{X}_{si}\mathbf{G}\mathbf{X}'_{sk}\mathbf{W}_{sk}$. As in Chapter 5, $\hat{\mathbf{Y}}_s = \mathbf{H}\mathbf{Y}_s$. In the definition of $\mathbf{H}$, we allow $\mathbf{G}$ to be either $\mathbf{A}_s^{-1}$ or a g-inverse of $\mathbf{A}_s$. The vector of residuals for sample cluster i can then be written as

$$\mathbf{r}_i = \mathbf{Y}_{si} - \sum_{k \in s} \mathbf{H}_{ik}\mathbf{Y}_{sk} = (\mathbf{I}_{m_i} - \mathbf{H}_{ii})\mathbf{Y}_{si} - \sum_{k \neq i} \mathbf{H}_{ik}\mathbf{Y}_{sk}. \tag{9.5.6}$$

Using the fact that $\mathbf{r}_i$ has expectation $\mathbf{0}$ under model (9.5.4), we have

$$\begin{aligned} E_M[(\mathbf{g}'_i\mathbf{r}_i)^2] &= \operatorname{var}(\mathbf{g}'_i\mathbf{r}_i) \\ &= \mathbf{g}'_i\left[(\mathbf{I}_{m_i} - \mathbf{H}_{ii})\operatorname{var}(\mathbf{Y}_{si})(\mathbf{I}_{m_i} - \mathbf{H}_{ii})' + \sum_{k \neq i} \mathbf{H}_{ik}\operatorname{var}(\mathbf{Y}_{sk})\mathbf{H}'_{ik}\right]\mathbf{g}_i. \end{aligned} \tag{9.5.7}$$

If $\mathbf{A}_s = O(n)$ elementwise, $\mathbf{G} = O(n^{-1})$, and the sample sizes m_i are bounded, then $\mathbf{H}_{ik} = O(n^{-1})$. Thus, as $n \to \infty$, $E_M[(\mathbf{g}_i'\mathbf{r}_i)^2] \cong \mathbf{g}_i' \operatorname{var}_M(\mathbf{Y}_{si})\mathbf{g}_i$, and, consequently,

$$E_M(v_R) \cong \sum_s \mathbf{g}_i' \operatorname{var}_M(\mathbf{Y}_{si})\mathbf{g}_i$$

that is, the sandwich estimator v_R is approximately unbiased for B_1, the dominant term in the error variance.

9.5.2. Adjustments to the Sandwich Estimator

If $\mathbf{W}_{si} = k[\operatorname{var}_M(\mathbf{Y}_{si})]^{-1}$ for a constant k, then the expectation of the sandwich estimator simplifies. It is also possible to make an adjustment to v_R to construct an estimator that would be exactly unbiased for B_1 under the working model for which $\operatorname{var}_M(\mathbf{Y}_{si})$ is determined. To see this, let $\mathbf{V}_{ii} = \operatorname{var}_M(\mathbf{Y}_{si})$. We have $\mathbf{H}_{ik} = \mathbf{X}_{si}\mathbf{G}\mathbf{X}_{sk}'\mathbf{V}_{kk}^{-1}$ and, using the fact that $\mathbf{A}_s = \Sigma_{k \in s} \mathbf{X}_{sk}'\mathbf{V}_{kk}^{-1}\mathbf{X}_{sk}$,

$$\begin{aligned}
\sum_{k \in s} \mathbf{H}_{ik}\mathbf{V}_{kk}\mathbf{H}_{ik}' &= \sum_{k \in s} \mathbf{X}_{si}\mathbf{G}\mathbf{X}_{sk}'\mathbf{V}_{kk}^{-1}\mathbf{V}_{kk}\mathbf{V}_{kk}^{-1}\mathbf{X}_{sk}\mathbf{G}'\mathbf{X}_{si}' \\
&= \sum_{k \in s} \mathbf{X}_{si}\mathbf{G}\mathbf{X}_{sk}'\mathbf{V}_{kk}^{-1}\mathbf{X}_{sk}\mathbf{G}'\mathbf{X}_{si}' \\
&= \mathbf{X}_{si}\mathbf{G}\mathbf{A}_s\mathbf{G}'\mathbf{X}_{si}'.
\end{aligned}$$

As in Chapter 7, we will assume that the g-inverse $\mathbf{G}$ is symmetric and satisfies (7.3.5), that is, that $\mathbf{G}\mathbf{A}_s\mathbf{G}' = \mathbf{G}$. The expression above then reduces to

$$\sum_{k \in s} \mathbf{H}_{ik}\mathbf{V}_{kk}\mathbf{H}_{ik}' = \mathbf{X}_{si}\mathbf{G}\mathbf{X}_{si}' = \mathbf{H}_{ii}\mathbf{V}_{ii}. \tag{9.5.8}$$

Using (9.5.8) and the fact that $\mathbf{H}_{ii}\mathbf{V}_{ii} = \mathbf{X}_{si}\mathbf{G}\mathbf{X}_{si}'$ is symmetric, the term in brackets in (9.5.7) can then be written as

$$\begin{aligned}
(\mathbf{I}_{m_i} - \mathbf{H}_{ii})\mathbf{V}_{ii}(\mathbf{I}_{m_i} - \mathbf{H}_{ii})' + \sum_{k \neq i} \mathbf{H}_{ik}\mathbf{V}_{kk}\mathbf{H}_{ik}' &= \mathbf{V}_{ii} - \mathbf{H}_{ii}\mathbf{V}_{ii} - \mathbf{V}_{ii}\mathbf{H}_{ii}' + \sum_{k \in s} \mathbf{H}_{ik}\mathbf{V}_{kk}\mathbf{H}_{ik}' \\
&= \mathbf{V}_{ii} - \mathbf{H}_{ii}\mathbf{V}_{ii} - \mathbf{V}_{ii}\mathbf{H}_{ii}' + \mathbf{H}_{ii}\mathbf{V}_{ii} \\
&= (\mathbf{I}_{m_i} - \mathbf{H}_{ii})\mathbf{V}_{ii}.
\end{aligned}$$

The expectation of the sandwich estimator under the working model then becomes

$$E_M(v_R) = \sum_s \mathbf{g}_i'(\mathbf{I}_{m_i} - \mathbf{H}_{ii})\mathbf{V}_{ii}\mathbf{g}_i.$$

Although the matrix $\mathbf{I}_m - \mathbf{X}_s\mathbf{G}\mathbf{X}_s'$ does not generally have full rank, the matrix $\mathbf{P}_i = \mathbf{I}_{m_i} - \mathbf{H}_{ii}$ often does. When $\mathbf{P}_i^{-1}$ exists, a simple adjustment like the one we used to construct v_D in Section 5.4.1 can be made to form the alternative estimator

$$v_D(\hat{T}) = \sum_s \mathbf{g}_i'\mathbf{P}_i^{-1}(\mathbf{r}_i\mathbf{r}_i')\mathbf{g}_i. \tag{9.5.9}$$

An analog to v_J^* in Section 5.4.1 is

$$v_J^*(\hat{T}) = \sum_s \mathbf{g}_i'\mathbf{P}_i^{-1}(\mathbf{r}_i\mathbf{r}_i')\mathbf{P}_i^{-1}\mathbf{g}_i. \tag{9.5.10}$$

The matrix $\mathbf{P}_i$ may not be invertible when some sample units in a sample cluster have the same values of all auxiliary variables, that is, when some rows of $\mathbf{X}_{si}$ are duplicates. In such clusters, we can set $\mathbf{P}_i^{-1}$ equal to the identity matrix in (9.5.9) and (9.5.10).

Other alternatives to the sandwich estimator can be constructed based on the large sample behavior of $\mathbf{g}_s$ defined by (9.5.1) or (9.5.2). When n is negligible compared to $N - n$, then $\mathbf{g}_s$ is asymptotically equivalent to $\mathbf{a}_s = \mathbf{W}_s\mathbf{X}_s\mathbf{G}\mathbf{X}_r'\mathbf{1}_r$. Three other alternative estimators are then

$$v_R^*(\hat{T}) = \sum_s \mathbf{a}_i'(\mathbf{r}_i\mathbf{r}_i')\mathbf{a}_i, \tag{9.5.11}$$

$$v_D^*(\hat{T}) = \sum_s \mathbf{a}_i'\mathbf{P}_i^{-1}(\mathbf{r}_i\mathbf{r}_i')\mathbf{a}_i, \tag{9.5.12}$$

and

$$v_J^{**}(\hat{T}) = \sum_s \mathbf{a}_i'\mathbf{P}_i^{-1}(\mathbf{r}_i\mathbf{r}_i')\mathbf{P}_i^{-1}\mathbf{a}_i. \tag{9.5.13}$$

By using $\mathbf{a}_i$, the estimator in (9.5.11) has the same form as the sandwich estimator introduced in Section 5.5 for unclustered populations. The estimators in (9.5.10) and (9.5.13) are closely related to the delete-1 jackknife estimator, as we will see in the next section.

9.5.3. Jackknife Estimator

In Chapters 5 and 7 the construction of the jackknife variance estimator began by omitting one sample unit at a time. A jackknife estimator of variance can be formed when the population is clustered by omitting one sample cluster at a time, computing the estimated total from the remaining clusters, and combining the results with an appropriate variance formula. When a sample cluster is deleted, all sample units within the cluster are deleted.

Recall that the estimation error of $\hat{T}$ can be written as $\hat{T} - T = \hat{T}_r - T_r$, where $\hat{T}_r = \mathbf{1}_r'\mathbf{X}_r\boldsymbol{\beta}^o$ and $\boldsymbol{\beta}^o = \mathbf{G}\mathbf{X}_s'\mathbf{W}_s\mathbf{Y}_s$ is a solution to the estimating equations for $\boldsymbol{\beta}$. The predictor of the nonsample total after deleting sample cluster i is $\hat{T}_{r(i)} = \mathbf{1}_r'\mathbf{X}_r\boldsymbol{\beta}_{(i)}^o$ with $\boldsymbol{\beta}_{(i)}^o = (\mathbf{X}_{s(i)}'\mathbf{W}_{s(i)}\mathbf{X}_{s(i)})^-\mathbf{X}_{s(i)}'\mathbf{W}_{s(i)}\mathbf{Y}_{s(i)}$ and the minus superscript denoting a generalized inverse. The subscript notation (i) will be used throughout this section to denote quantities that are computed after omitting sample cluster i. Define the jackknife estimator of the variance of $\hat{T}_r$ to be

$$v_J(\hat{T}_r) = \frac{n-1}{n}\sum_{i\in s}(\hat{T}_{r(i)} - \hat{T}_{r(\cdot)})^2, \tag{9.5.14}$$

where $\hat{T}_{r(\cdot)} = n^{-1}\sum_s \hat{T}_{r(i)}$.

Rather than mechanically deleting a cluster, computing $\hat{T}_{r(i)}$, and cycling through all sample clusters, the following alternative computational form can be used:

$$v_J(\hat{T}_r) = \frac{n-1}{n}\left\{\sum_s(\mathbf{a}_i'\mathbf{P}_i^{-1}\mathbf{r}_i)^2 - n^{-1}\left[\sum_s \mathbf{a}_i'\mathbf{P}_i^{-1}\mathbf{r}_i\right]^2\right\}. \tag{9.5.15}$$

This form, which we derive below, is similar to the ones in Sections 5.4 and 7.10. In the derivation that follows, a particular form of g-inverse, like the one introduced in (7.3.3), is used. However, the end result in (9.5.15) is invariant to the g-inverse used.

To rewrite the jackknife, we will need several intermediate results. First, note that $\mathbf{X}_{s(i)}'\mathbf{W}_{s(i)}\mathbf{X}_{s(i)} = \mathbf{X}_s'\mathbf{W}_s\mathbf{X}_s - \mathbf{X}_{si}'\mathbf{W}_{si}\mathbf{X}_{si}$. To derive a formula for the jackknife analogous to the one in Section 7.10, we need an expression for the g-inverse of $\mathbf{X}_{s(i)}'\mathbf{W}_{s(i)}\mathbf{X}_{s(i)}$, which is the sum of two matrices. Schott (1997, Theorem 5.15) gives the Moore-Penrose g-inverse for a sum of the form $\mathbf{U}'\mathbf{U} + \mathbf{V}'\mathbf{V}$. For the purposes here, a simpler form than the one in Schott is available.

As in Lemma 7.4.1, we rely on the fact that $\mathbf{X}$ can (possibly after rearrangement) be partitioned as $\mathbf{X} = [\mathbf{X}_1 \ \ \mathbf{B}]$, where the columns of $\mathbf{X}_1$ are linearly independent while the columns of $\mathbf{B}$ are linearly dependent on those of $\mathbf{X}_1$. Recall that we also are assuming that $\mathbf{W}_s = \text{blkdiag}(\mathbf{W}_{si})$ with $\mathbf{W}_{si}$ being $m_i \times m_i$, but not necessarily diagonal. The lemma also uses a formula for the regular inverse of the sum of matrices. Specifically, if $\mathbf{A}$, $\mathbf{B}$, $\mathbf{C}$, and $\mathbf{D}$ are conformable matrices, then

$$(\mathbf{A} + \mathbf{CBD})^{-1} = \mathbf{A}^{-1} - \mathbf{A}^{-1}\mathbf{C}(\mathbf{B}^{-1} + \mathbf{D}\mathbf{A}^{-1}\mathbf{C})^{-1}\mathbf{D}\mathbf{A}^{-1} \tag{9.5.16}$$

assuming that $\mathbf{A}^{-1}$ and $\mathbf{B}^{-1}$ exist (Schott, 1997, Theorem 1.7).

Lemma 9.5.1. Let $\mathbf{A}_{s(i)} = \mathbf{X}_{s(i)}'\mathbf{W}_{s(i)}\mathbf{X}_{s(i)}$. Suppose that $\mathbf{X}_s$ and $\mathbf{X}_{si}$ can be partitioned as $\mathbf{X}_s = [\mathbf{X}_{s1} \ \ \mathbf{B}]$ and $\mathbf{X}_{si} = [\mathbf{X}_{si1} \ \ \mathbf{B}_i]$, where $\mathbf{X}_{s1}$ is $m \times q$, $\mathbf{X}_{si1}$ is $m_i \times q$, and $\mathbf{X}_{s1}$ and $\mathbf{X}_{si1}$ have full column rank. The matrix $\mathbf{B}$ *is* $m \times (p-q)$ and $\mathbf{B}_i$ is $m_i \times (p-q)$. Then a g-inverse of $\mathbf{A}_{s(i)}$ is

$$\mathbf{A}_{s(i)}^{-} = \begin{bmatrix} \mathbf{A}_{s1(i)}^{-1} & \mathbf{0} \\ \mathbf{0} & \mathbf{0} \end{bmatrix},$$

where

$$\mathbf{A}_{s1(i)} = \mathbf{X}'_{s1(i)}\mathbf{W}_{s(i)}\mathbf{X}_{s1(i)}, \qquad \mathbf{A}_{s1(i)}^{-1} = \mathbf{A}_{s1}^{-1} + \mathbf{A}_{s1}^{-1}\mathbf{X}'_{si1}\mathbf{W}_{si}\mathbf{P}_i^{-1}\mathbf{X}_{si1}\mathbf{A}_{s1}^{-1},$$

$$\mathbf{A}_{s1} = \mathbf{X}'_{s1}\mathbf{W}_s\mathbf{X}_{s1}, \qquad \text{and} \qquad \mathbf{P}_i = \mathbf{I}_{m_i} - \mathbf{H}_{ii}$$

with $\mathbf{H}_{ii} = \mathbf{X}_{si}\mathbf{G}\mathbf{X}'_{si}W_{si}$ where $\mathbf{G}$ is a g-inverse of $\mathbf{A}_s = \mathbf{X}'_s\mathbf{W}_s\mathbf{X}_s$.

Proof: To simplify notation somewhat, for a matrix $\mathbf{C}'_l$, let $\dot{\mathbf{C}}'_l$ be that matrix multiplied on the right by the square root of the corresponding part of the weight matrix $\mathbf{W}$. For example, $\dot{\mathbf{X}}'_s = \mathbf{X}'_s\mathbf{W}_s^{1/2}$, $\dot{\mathbf{X}}'_{si} = \mathbf{X}'_{si}\mathbf{W}_{si}^{1/2}$, $\dot{\mathbf{X}}'_{s1(i)} = \mathbf{X}'_{s1(i)}\mathbf{W}_{s(i)}^{1/2}$, and $\dot{\mathbf{B}}'_{si1} = \mathbf{B}'_{si1}\mathbf{W}_{si}^{1/2}$. We can write $\mathbf{A}_s = \mathbf{X}'_s\mathbf{W}_s\mathbf{X}_s = \dot{\mathbf{X}}'_s\dot{\mathbf{X}}_s$ as

$$\mathbf{A}_s = \begin{bmatrix} \dot{\mathbf{X}}'_{s1} \\ \dot{\mathbf{B}}' \end{bmatrix} [\dot{\mathbf{X}}_{s1} \quad \dot{\mathbf{B}}] = \begin{bmatrix} \dot{\mathbf{X}}'_{s1}\dot{\mathbf{X}}_{s1} & \dot{\mathbf{X}}'_{s1}\dot{\mathbf{B}} \\ \dot{\mathbf{B}}'\dot{\mathbf{X}}_{s1} & \dot{\mathbf{B}}'\dot{\mathbf{B}} \end{bmatrix}. \tag{9.5.17}$$

We also have

$$\mathbf{X}'_{si}\mathbf{W}_{si}\mathbf{X}_{si} = \begin{bmatrix} \dot{\mathbf{X}}'_{si1}\dot{\mathbf{X}}_{si1} & \dot{\mathbf{X}}'_{si1}\dot{\mathbf{B}}_i \\ \dot{\mathbf{B}}'_i\dot{\mathbf{X}}_{si1} & \dot{\mathbf{B}}'_i\dot{\mathbf{B}}_i \end{bmatrix}. \tag{9.5.18}$$

Subtracting (9.5.18) from (9.5.17), we have an expression for $\mathbf{A}_{s(i)}$:

$$\mathbf{A}_{s(i)} = \begin{bmatrix} \dot{\mathbf{X}}'_{s1}\dot{\mathbf{X}}_{s1} - \dot{\mathbf{X}}'_{si1}\dot{\mathbf{X}}_{si1} & \dot{\mathbf{X}}'_{s1}\dot{\mathbf{B}} - \dot{\mathbf{X}}'_{si1}\dot{\mathbf{B}}_i \\ \dot{\mathbf{B}}'\dot{\mathbf{X}}_{s1} - \dot{\mathbf{B}}'_i\dot{\mathbf{X}}_{si1} & \dot{\mathbf{B}}'\dot{\mathbf{B}} - \dot{\mathbf{B}}'_i\dot{\mathbf{B}}_i \end{bmatrix}. \tag{9.5.19}$$

By rearranging rows and columns, the sample matrix of explanatory variables can be written with the rows for sample cluster i listed last:

$$\mathbf{X}_s = \begin{bmatrix} \mathbf{X}_{s(i)} \\ \mathbf{X}_{si} \end{bmatrix} = \begin{bmatrix} \mathbf{X}_{s1(i)} & \mathbf{B}_{(i)} \\ \mathbf{X}_{si1} & \mathbf{B}_i \end{bmatrix},$$

where $\mathbf{X}_{s1(i)}$ denotes the $(m - m_i) \times q$ sample matrix with full column rank that excludes sample cluster i. The matrix $\mathbf{B}_{(i)}$ consists of the remaining columns of $\mathbf{X}_{s(i)}$ and is $(m - m_i) \times (p - q)$. It follows that an alternative expression for $\mathbf{A}_s$ is

$$\begin{aligned} \mathbf{A}_s &= \begin{bmatrix} \mathbf{X}'_{s1(i)} & \mathbf{X}'_{si1} \\ \mathbf{B}'_{(i)} & \mathbf{B}'_i \end{bmatrix} \begin{bmatrix} \mathbf{W}_{s(i)} & \mathbf{0} \\ \mathbf{0} & \mathbf{W}_{s1} \end{bmatrix} \begin{bmatrix} \mathbf{X}_{s1(i)} & \mathbf{B}_{(i)} \\ \mathbf{X}_{si1} & \mathbf{B}_i \end{bmatrix} \\ &= \begin{bmatrix} \dot{\mathbf{X}}'_{s1(i)}\dot{\mathbf{X}}_{s1(i)} + \dot{\mathbf{X}}'_{si1}\dot{\mathbf{X}}_{si1} & \dot{\mathbf{X}}'_{s1(i)}\dot{\mathbf{B}}_{(i)} + \dot{\mathbf{X}}'_{si1}\dot{\mathbf{B}}_i \\ \dot{\mathbf{B}}'_{(i)}\dot{\mathbf{X}}_{s1(i)} + \dot{\mathbf{B}}'_i\dot{\mathbf{X}}_{si1} & \dot{\mathbf{B}}'_{(i)}\mathbf{B}'_{(i)} + \dot{\mathbf{B}}'_i\dot{\mathbf{B}}_i \end{bmatrix}. \end{aligned} \tag{9.5.20}$$

Subtracting (9.5.18) from (9.5.20) gives another way of writing $\mathbf{A}_{s(i)}$:

$$\mathbf{A}_{s(i)} = \begin{bmatrix} \dot{\mathbf{X}}'_{s1(i)}\dot{\mathbf{X}}_{s1(i)} & \dot{\mathbf{X}}'_{s1(i)}\dot{\mathbf{B}}_{(i)} \\ \dot{\mathbf{B}}'_{(i)}\dot{\mathbf{X}}_{s1(i)} & \dot{\mathbf{B}}'_{(i)}\mathbf{B}'_{(i)} \end{bmatrix}. \tag{9.5.21}$$

By construction, $\dot{\mathbf{X}}'_{s1(i)}\dot{\mathbf{X}}_{s1(i)}$ has a regular inverse. Thus, a g-inverse of $\mathbf{A}_{s(i)}$ is

$$\mathbf{A}^{-}_{s(i)} = \begin{bmatrix} (\dot{\mathbf{X}}'_{s1(i)}\dot{\mathbf{X}}_{s1(i)})^{-1} & \mathbf{0} \\ \mathbf{0} & \mathbf{0} \end{bmatrix},$$

as can be verified directly or by applying Lemma A.10.3.

From (9.5.19) and (9.5.21), $\mathbf{A}_{s1(i)} = \dot{\mathbf{X}}'_{s1(i)}\dot{\mathbf{X}}_{s1(i)} = \dot{\mathbf{X}}'_{s1}\dot{\mathbf{X}}_{s1} - \dot{\mathbf{X}}'_{si1}\dot{\mathbf{X}}_{si1}$. Using

$$\mathbf{G} = \begin{bmatrix} \mathbf{A}^{-1}_{s1} & \mathbf{0} \\ \mathbf{0} & \mathbf{0} \end{bmatrix}, \tag{9.5.22}$$

we have $\mathbf{H}_{ii} = \mathbf{X}_{si}\mathbf{G}\mathbf{X}'_{si}\mathbf{W}_{si} = \mathbf{X}_{si1}\mathbf{A}^{-1}_{s1}\mathbf{X}'_{si1}\mathbf{W}_{si}$. Using this form of $\mathbf{H}_{ii}$ and applying the formula (9.5.16) for the inverse of a sum, with $\mathbf{A} = \dot{\mathbf{X}}'_{s1}\dot{\mathbf{X}}_{s1}$, $\mathbf{B} = \mathbf{I}_{m_i}$, $\mathbf{C} = -\mathbf{X}'_{si1}\mathbf{W}_{si}$, and $\mathbf{D} = \mathbf{X}_{si1}$ yields

$$\mathbf{A}^{-1}_{s1(i)} = \mathbf{A}^{-1}_{s1} + \mathbf{A}^{-1}_{s1}\mathbf{X}'_{si1}\mathbf{W}_{si}\mathbf{P}^{-1}_{i}\mathbf{X}_{si1}\mathbf{A}^{-1}_{s1}$$

as required. □

The solution to the full-sample estimating equations that uses the same type of g-inverse as in Lemma 9.5.1 is $\boldsymbol{\beta}^{o} = \mathbf{G}\mathbf{X}'_{s}\mathbf{W}_{s}\mathbf{Y}_{s}$ with $\mathbf{G}$ defined in (9.2.22). The solution vector $\boldsymbol{\beta}^{o}$ can then be written as

$$\boldsymbol{\beta}^{o} = \begin{bmatrix} \mathbf{A}^{-1}_{s1} & \mathbf{0} \\ \mathbf{0} & \mathbf{0} \end{bmatrix} \begin{bmatrix} \mathbf{X}'_{s1} \\ \mathbf{B}' \end{bmatrix} \mathbf{W}_{s}\mathbf{Y}_{s} = \begin{bmatrix} \mathbf{A}^{-1}_{s1}\mathbf{X}'_{s1}\mathbf{W}_{s}\mathbf{Y}_{s} \\ \mathbf{0} \end{bmatrix}.$$

The predicted values for sample cluster i can be written as $\hat{\mathbf{Y}}_{si} = \mathbf{X}_{si1}\mathbf{A}^{-1}_{s1}\mathbf{X}'_{s1}\mathbf{W}_{s}\mathbf{Y}_{s}$. Using this expression and Lemma 9.5.1, it follows that the solution vector to the normal equations, when cluster i is omitted, is

$$\begin{aligned} \boldsymbol{\beta}^{o}_{(i)} &= \mathbf{A}^{-}_{s(i)}\mathbf{X}'_{s(i)}\mathbf{W}_{s(i)}\mathbf{Y}_{s(i)} \\ &= \begin{bmatrix} \mathbf{A}^{-1}_{s1} + \mathbf{A}^{-1}_{s1}\mathbf{X}'_{si1}\mathbf{W}_{si}\mathbf{P}^{-1}_{i}\mathbf{X}_{si1}\mathbf{A}^{-1}_{s1} & \mathbf{0} \\ \mathbf{0} & \mathbf{0} \end{bmatrix} \left\{ \begin{bmatrix} \mathbf{X}'_{s1} \\ \mathbf{B}' \end{bmatrix} \mathbf{W}_{s}\mathbf{Y}_{s} - \begin{bmatrix} \mathbf{X}'_{si1} \\ \mathbf{B}'_{i} \end{bmatrix} \mathbf{W}_{si}\mathbf{Y}_{si} \right\} \\ &= \begin{bmatrix} \mathbf{A}^{-1}_{s1}\mathbf{X}'_{s1}\mathbf{W}_{s}\mathbf{Y}_{s} \\ \mathbf{0} \end{bmatrix} + \begin{bmatrix} \mathbf{A}^{-1}_{s1}\mathbf{X}'_{si1}\mathbf{W}_{si}(\mathbf{P}^{-1}_{i}\hat{\mathbf{Y}}_{si} - \mathbf{Y}_{si} - \mathbf{P}^{-1}_{i}(\mathbf{I}_{m_i} - \mathbf{P}_{i})\mathbf{Y}_{si}) \\ \mathbf{0} \end{bmatrix} \\ &= \boldsymbol{\beta}^{o} - \begin{bmatrix} \mathbf{A}^{-1}_{s1}\mathbf{X}'_{si1}\mathbf{W}_{si}\mathbf{P}^{-1}_{i}\mathbf{r}_{i} \\ \mathbf{0} \end{bmatrix}, \end{aligned}$$

where $\mathbf{r}_{i} = \mathbf{Y}_{si} - \hat{\mathbf{Y}}_{si}$ is the vector of residuals for sample cluster i.

The matrix of nonsample explanatory variables can be written as $\mathbf{X}_{r} = [\mathbf{X}_{r1}\ \mathbf{B}_{r}]$, where $\mathbf{X}_{r1}$ is $(M - m) \times q$ and $\mathbf{B}_{r}$ is $(M - m) \times q$. After some simplification, we have $\hat{T}_{r(i)} - \hat{T}_{r(\cdot)} = -[\mathbf{a}_i'\mathbf{P}_i^{-1}\mathbf{r}_i - n^{-1}\Sigma_s \mathbf{a}_i'\mathbf{P}_i^{-1}\mathbf{r}_i]$ with

$$\mathbf{a}_i' = \mathbf{1}_r'\mathbf{X}_r\mathbf{G}\mathbf{X}_{si}'\mathbf{W}_{si} = \mathbf{1}_r'\mathbf{X}_{r1}\mathbf{A}_{s1}^{-1}\mathbf{X}_{s1i}'\mathbf{W}_{si}.$$

The jackknife can then be written as

$$v_J(\hat{T}_r) = \frac{n-1}{n}\left\{\sum_s (\mathbf{a}_i'\mathbf{P}_i^{-1}\mathbf{r}_i)^2 - n^{-1}\left[\sum_s (\mathbf{a}_i'\mathbf{P}_i^{-1}\mathbf{r}_i\right]^2\right\}. \qquad (9.5.23)$$

With the form of g-inverse in (9.5.22), we have

$$\mathbf{H}_{ii} = \mathbf{X}_{si1}\mathbf{A}_{s1}^{-1}\mathbf{X}_{si1}'\mathbf{W}_{si} = \mathbf{X}_{si}\mathbf{G}\mathbf{X}_{si}'\mathbf{W}_{si}$$

as noted in the proof of Lemma 9.5.1. Thus the hat matrix is the same as the one used in Section 9.5.1.

Note that if each cluster consisted of a single unit, then $\mathbf{a}_i$ and $\mathbf{r}_i$ are scalars, and $\mathbf{P}_i = 1 - \mathbf{x}_i'\mathbf{G}\mathbf{x}_i w_i$ where $\mathbf{x}_i$ is the vector of explanatory variables for the ith sample unit. If, in addition, $w_i = 1$, then the jackknife reduces to the same form as in (7.10.10). (The generalization of the jackknife in (7.10.10) to the case of general w_i was given in Exercise 7.12). Using methods similar to those in the proof of Theorem 5.4.1, the second term in the braces in (9.5.23) can be shown to converge to 0 in probability. The leading term in the jackknife, $\Sigma_s(\mathbf{a}_i'\mathbf{P}_i^{-1}\mathbf{r}_i)^2$, is equal to the adjusted estimator v_J^{**} and is asymptotically equivalent to v_J^*.

9.6. COMPARISONS OF VARIANCE ESTIMATORS IN A SIMULATION STUDY

The Third Grade population in Appendix B.6 was used in a simulation to compare the performance of the alternative variance estimators for the regression estimator. We selected two sets of 500 samples with $n = 25$ and 50 schools from the population of $N = 135$ clusters. Schools were selected with probabilities proportional to the number of students in each school, using the Hartley-Rao method described in Section 3.4.5. Within each sample school, a simple random sample of five students was selected without replacement.

From each sample the mean scores for mathematics and science were estimated using a regression estimator, $\hat{\mu}_{\text{reg}} = \hat{T}/M$, based on the following 11 explanatory variables: an intercept, categorical variables for sex (female; male), ethnicity (White or Asian; Black, Native American or Other; Hispanic), language spoken at home is the same as the test (Always; Sometimes or Never), and type of community (Suburban; Nonsuburban), and the quantitative variable, school enrollment. In each sample, the seven variance estimators for the regression estimator that were introduced in Section 9.5 were also computed.

Table 9.3 shows some summary results for the regression estimators of the

Table 9.3. Simulation Results for Alternative Variance Estimators for the Linear Regression Estimator

Quantity	Mathematics		Science	
Population Mean	477.70		503.53	
	$n = 25$	$n = 50$	$n = 25$	$n = 50$
Average $\hat{\mu}_\pi$	477.96	477.36	503.56	503.36
rmse($\hat{\mu}_\pi$)	11.07	6.92	12.57	7.69
Average $\hat{\mu}_{\text{reg}}$	477.72	477.45	503.49	503.55
rmse($\hat{\mu}_{\text{reg}}$)	10.17	6.47	10.79	6.56
$\bar{v}_R^{1/2}$	9.81	7.16	10.43	7.55
$(\bar{v}_R^*)^{1/2}$	9.35	6.45	9.95	6.80
$\bar{v}_D^{1/2}$	10.96	7.53	11.59	7.93
$(\bar{v}_D^*)^{1/2}$	10.71	7.15	11.32	7.53
$(\bar{v}_J)^{1/2}$	11.59	7.08	12.16	7.45
$(\bar{v}_J^*)^{1/2}$	12.15	7.86	12.74	8.26
$(\bar{v}_J^{**})^{1/2}$	11.60	7.08	12.16	7.45

Two sets of 500 samples of $n = 25$ and $n = 50$ schools were selected from the Third Grade population. Schools were selected with probabilities proportional to numbers of students and an *srswor* of 5 students was selected per school.

mean mathematics and science scores and for the variance estimators. The regression estimators are nearly unbiased in the 500 samples, as is clear from comparing the population means to the average of the estimates over the samples. As a point of comparison, we also calculated the π-estimator, $\hat{\mu}_\pi = \hat{T}_\pi/M$, in each sample. For the sampling plan here, $\hat{\mu}_\pi = \hat{T}_0/M$, which is the expansion estimator defined in Section 8.2. The root mean square error (*rmse*) of each estimator of the mean was computed as $[\Sigma_{i=1}^{500}(\hat{\mu}_i - \mu)^2/500]^{1/2}$ where $\hat{\mu}_i$ is one of the estimated means, computed from sample i and μ is the population mean. The *rmse* of the regression estimator is about 8.8% smaller than that of the expansion estimator for math (10.17 compared to 11.07) and 16.5% better for science (10.79 vs. 12.57) for samples of size $n = 25$. For $n = 50$, the gains are 7.0% and 17.2%.

The square root of the average value of each variance estimator v for the regression estimator is also shown in Table 9.3 and is denoted as $\bar{v}^{1/2}$. For example, the square root of the average sandwich estimator is shown as $\bar{v}_R^{1/2}$. The pattern among the variance estimators is clear. The jackknife v_J and its variants, v_J^* and v_J^{**}, overestimate the mean square error (mse) on average for both $n = 25$ and $n = 50$. At either sample size the averages of v_J and v_J^{**} are almost equal since the second term of v_J in (9.5.23) is negligible.

The sandwich estimator v_R and the modified version, v_R^*, each underestimate the *mse* for $n = 25$, but for the larger sample size are nearly unbiased or are overestimates. The estimates, v_D, and v_D^*, are overestimates for both sample sizes, and because each involves the adjustment $\mathbf{P}_i^{-1}$, we have $v_D > v_R$ and $v_D^* > v_R^*$ on average.

Table 9.4 summarizes the coverage of 95% confidence intervals constructed from each of the variance estimators. The mathematics and science test scores are themselves nearly normally distributed (see Exercise 9.11). Thus we would expect confidence interval coverage to be good as long as the variance of $\hat{\mu}_{\text{reg}}$ is consistently estimated. When $n = 25$, coverage is at or above the nominal level except for v_R and v_R^*, which have coverage rates of 91.2% to 92.8%. For $n = 50$, coverage is over 95% in all cases except math using v_R^*, where it is 93.2%.

In practice, an *ad hoc* finite population correction (*fpc*), $1 - n/N$, based on the sampling fraction of clusters is sometimes used. In the Third Grade population there are only 135 clusters, so that the *fpc* is substantial, being 0.81 for $n = 25$ and 0.63 for $n = 50$. Multiplying each variance estimate by $1 - n/N$ reduces coverage rates below the desired level in all cases for $n = 50$. We will summarize a few of these results without showing a separate table. The coverage rates for v_R, v_D, and v_J become 90.6, 91.6, and 90.0 for (mathematics, $n = 50$) compared to 96.2, 96.8, and 95.8 with no *fpc*. For ($n = 25$, math) the coverage rates for v_R, v_D, and v_J are 89.6, 92.4, and 93.4 versus 92.4, 95.0, and 95.0 in Table 9.4 without the *fpc*. Thus arbitrary inclusion of an *fpc* does not work well in this experiment, especially at the larger sample size.

EXERCISES

9.1 Prove Theorem 9.1.1.

9.2 Show that $\hat{B}_1 = \Sigma_s(g_i^2/m_i)[\hat{\theta}_1 + m_i\hat{\theta}_2]$, defined by (9.2.3), is a model-unbiased estimator of the dominant term of the error variance in (9.1.4) under the working model (9.1.1).

9.3 Derive the formulas for $E_M(W)$ and $E_M(B)$ in Section 9.2.1 under model (9.2.4).

9.4 Show that

$$E_M(r_i^2) = \text{var}_M(\bar{Y}_{si})\left(1 - \frac{2g_i}{M}\right) + \frac{1}{M^2}\sum_{i' \in s} g_i^2 \text{ var}_M(\bar{Y}_{si'})$$

under model (9.1.5).

Table 9.4. Empirical Coverage of 95% Confidence Intervals using Alternative Variance Estimators for the Linear Regression Estimator

Variance Estimator	Mathematics Lower	Mathematics Middle	Mathematics Upper	Science Lower	Science Middle	Science Upper
$n = 25$						
$v_R^{1/2}$	5.2	92.4	2.4	3.4	92.8	3.8
$(v_R^*)^{1/2}$	5.8	91.6	2.6	4.6	91.2	4.2
$v_D^{1/2}$	3.8	95.0	1.2	2.4	96.2	1.4
$(v_D^*)^{1/2}$	4.2	94.2	1.6	2.4	95.8	1.8
$(v_J)^{1/2}$	3.8	95.0	1.2	2.2	96.6	1.2
$(v_J^*)^{1/2}$	3.4	95.4	1.2	1.8	97.2	1.0
$(v_J^{**})^{1/2}$	3.6	95.2	1.2	2.2	96.6	1.2
$n = 50$						
$v_R^{1/2}$	2.0	96.2	1.8	1.8	97.2	1.0
$(v_R^*)^{1/2}$	3.4	93.2	3.4	2.6	95.2	2.2
$v_D^{1/2}$	2.0	96.8	1.2	1.2	98.2	0.6
$(v_D^*)^{1/2}$	2.0	96.2	1.8	1.8	97.2	1.0
$(v_J)^{1/2}$	2.4	95.8	1.8	2.0	97.0	1.0
$(v_J^*)^{1/2}$	1.8	97.4	0.8	1.0	98.6	0.4
$(v_J^{**})^{1/2}$	2.4	95.8	1.8	2.0	97.0	1.0

Two sets of 500 Samples of $n = 25$ and $n = 50$ schools were selected from the Third Grade population. Schools were selected with probabilities proportional to numbers of students and an *srswor* of 5 students was selected per school.

9.5 Show that under model (9.2.11),

$$\text{var}_M(\bar{Y}_i) = \text{cov}_M(\bar{Y}_{si}, \bar{Y}_i) = \text{var}_M(\bar{Y}_{si}) - \text{var}_M(\bar{Y}_{si} - \bar{Y}_i)$$
$$= M_i^{-1}\sigma_i^2[1 + (M_i - 1)\rho_i].$$

9.6 Show that $E_M(\hat{v}_i) = \text{var}_M(\hat{Y}_i)$ under model (9.1.5) where $\hat{v}_i$ is defined by (9.2.9).

9.7 Prove Lemma 9.2.1, that is, under model (9.2.4)

$$E_M[M_i^{-1}S_i^2(1 - f_i)/f] = \text{var}_M(\bar{Y}_{si} - Y_i), \quad \text{where} \quad S_i^2 = \sum_{j \in s_i} (Y_{ij} - \bar{Y}_{si})^2/(m_i - 1).$$

9.8 Derive weighted least squares estimators for the parameters $\theta_1 = (1 - \rho)\sigma^2$ and $\theta_2 = \rho\sigma^2$ under model (9.2.4) using the fact that $E_M(\dot{v}_i) = M_i\theta_1 + M_i^2\theta_2$ $(i = 1, \ldots, n)$.

9.9 Show that the general formulas for v_R, v_C, and v_0 reduce to (9.2.20), (9.2.21), and (9.2.22) for the ratio estimator, $\hat{T}_R = (M/M_s)\, \Sigma_s M_i \bar{Y}_{si}$.

9.10 The two-stage sample of 10 clusters and $\bar{m} = 2$ persons in the table below was selected from the Labor Force population in Appendix B. Compute $\hat{T}_p$ and $\hat{T}_R$ and the corresponding robust variance estimates v_0, v_1, v_2, and $\tilde{v}_0$ for the variable y. Calculate the quantities $N\hat{B}_1/f$, $-2N\hat{B}_2$, and $N\hat{B}_3$ separately.

Cluster	3	3	7	7	26	26	28	28	31	31
Person	10	11	23	24	91	94	100	101	113	114
Cluster	37	37	42	42	75	75	92	92	113	113
Person	141	147	164	165	299	300	380	381	470	471

9 11 Compute descriptive statistics for the Third Grade population in Appendix B. What statistics do you think would be useful in designing a sample and selecting an estimator for this population? Draw histograms and normal Q–Q plots of the mathematics and science test scores. What will be the effect of these underlying distributions on confidence interval construction when the mean score is estimated from a sample?

9.12 Show that $v_R(\hat{T})$ given by (9.5.5) reduces to an estimator of the form (9.1.7) under the common mean model (9.1.1), in which $\mathbf{X}$ is a single column of ones.

CHAPTER 10

Alternative Variance Estimation Methods

In earlier chapters, we studied methods of estimating the error variance of linear estimators. Samples are also often used to estimate nonlinear functions like ratios of means, odds ratios, or more complicated forms. Section 10.1 discusses how to estimate the variance of nonlinear combinations of linear estimators. This can be done by estimating the variance of a linear approximation to the nonlinear estimator or by applying the jackknife directly to the nonlinear estimator. In Section 10.2 we cover the method of balanced half-sampling or balanced repeated replication, which can be used for either linear or nonlinear estimators. Finally, in Section 10.3 we study a method of smoothing variance estimates known as generalized variance functions.

10.1 ESTIMATING THE VARIANCE OF ESTIMATORS OF NONLINEAR FUNCTIONS

In finite population estimation, we are often interested in relatively complex, nonlinear functions of population totals or means. Two examples are the ratio of the average salary of women to that of men and the ratio of the unemployment rate for white men aged 25-34 to that of Black men in the same age group. In each of these examples, the estimation target is a nonlinear function of population totals, which are themselves linear. The average salary for women, for example, is the sum of the salaries for all women in the population divided by the total number of women. The most straightforward estimator of such a nonlinear function is to substitute linear estimators for each of the linear components of the function. This produces a nonlinear function of the linear estimators. In this section, we cover approximate methods for estimating the variance of such functions. The theory will be kept at an informal level.

As in Chapter 9, consider a clustered population in which a sample of n clusters is selected and a sample of m_i units is selected from sample cluster i. Let j denote a unit within cluster i. Suppose that data are collected on K

variables, $Y_1, \ldots, Y_K$ and that they follow the multivariate model

$$E_M(Y_{kij}) = \mathbf{x}'_{ij}\boldsymbol{\beta}_k \qquad k = 1,\ldots,K;\ i = 1,\ldots,N;\ j = 1,\ldots,M_i$$
$$\text{cov}_M(Y_{kij}, Y_{k'i'j'}) = 0, \qquad i \neq i'. \tag{10.1.1}$$

In words, the mean value of random variable k for unit j in cluster i depends on a set of explanatory variables $\mathbf{x}_{ij}$ that is the same for $Y_1, \ldots, Y_K$, although the slope $\boldsymbol{\beta}_k$ depends on the particular variable. We assume that the random variables observed for units in different clusters are uncorrelated but that within clusters the covariance structure is arbitrary. Different random variables on the same unit can be correlated, a given random variable observed on different units in the same cluster can be correlated, and different random variables observed on different units can be correlated as long as the units are in the same cluster.

Let $\mathbf{T} = (T_1, \ldots, T_K)'$ be the vector of population totals for the K variables, and let $\Psi(\mathbf{T})$ be a differentiable function of the totals. When the K variables follow model (10.1.1), a natural estimator of $\Psi(\mathbf{T})$ is $\Psi(\hat{\mathbf{T}})$ where $\hat{\mathbf{T}} = (\hat{T}_1 \ldots, \hat{T}_K)'$ is the vector of regression estimators with $\hat{T} = \mathbf{g}'_s\mathbf{Y}_s$ and $\mathbf{g}_s = \mathbf{W}_s\mathbf{X}_s\mathbf{G}\mathbf{X}'_r\mathbf{1}_r + \mathbf{1}_s$. The various components of $\mathbf{g}_s$ are defined as in Section 9.4. In this section, we will use $\mathbf{G}$ to denote either a g-inverse or a regular inverse of $\mathbf{X}'_s\mathbf{W}_s\mathbf{X}_s$, whichever is appropriate. Assume that the same weight vector $\mathbf{g}_s$ is used for estimating the total for all K of the Y variables.

The key to developing asymptotic theory for a nonlinear $\Psi(\hat{\mathbf{T}})$ is to be able to approximate it by a linear function of $\hat{\mathbf{T}}$. Thus, assume that the population and the number of sample clusters is large, so that $\Psi(\mathbf{T})$ and $\Psi(\hat{\mathbf{T}})$ can be approximated as

$$M^{-1}\Psi(\mathbf{T}) \cong M^{-1}\Psi(\mathbf{T}_\mu) + M^{-1}\nabla\Psi(\mathbf{T}_\mu)'(\mathbf{T} - \mathbf{T}_\mu) \tag{10.1.2}$$
$$M^{-1}\Psi(\hat{\mathbf{T}}) \cong M^{-1}\Psi(\mathbf{T}_\mu) + M^{-1}\nabla\Psi(\mathbf{T}_\mu)'(\hat{\mathbf{T}} - \mathbf{T}_\mu), \tag{10.1.3}$$

where $\mathbf{T}_\mu$ is the model expectation of $\mathbf{T}$ and $\nabla\Psi(\mathbf{T}_\mu) = (\partial\Psi(\mathbf{T})/\partial T_1, \ldots, \partial\Psi(\mathbf{T})/\partial T_K)$ is the $K \times 1$ vector of first partial derivatives (or gradient) evaluated at $\mathbf{T}_\mu$. Note that we divide by M, the number of units in the population, in (10.1.2) and (10.1.3) because as the population size grows, the totals are unbounded. Thus, $\mathbf{T} - \mathbf{T}_\mu$ and $\hat{\mathbf{T}} - \mathbf{T}_\mu$ may not be small even though $M^{-1}(\mathbf{T} - \mathbf{T}_\mu)$ and $M^{-1}(\hat{\mathbf{T}} - \mathbf{T}_\mu)$ are. Taking the difference between (10.1.3) and (10.1.2), we have

$$M^{-1}[\Psi(\hat{\mathbf{T}}) - \Psi(\mathbf{T})] \cong M^{-1}\nabla\Psi(\mathbf{T}_\mu)'(\hat{\mathbf{T}} - \mathbf{T}). \tag{10.1.4}$$

Dropping the factor in M, the approximate error variance of $\Psi(\hat{\mathbf{T}})$ is then

$$\text{var}_M[\Psi(\hat{\mathbf{T}}) - \Psi(\mathbf{T})] \cong \nabla\Psi(\mathbf{T}_\mu)'\,\text{var}_M(\hat{\mathbf{T}} - \mathbf{T})\nabla\Psi(\mathbf{T}_\mu).$$

When the sampling fraction of clusters is small and the other conditions in Theorem 9.4.1 hold, then $\text{var}_M(\hat{\mathbf{T}} - \mathbf{T}) \approx \text{var}_M(\hat{\mathbf{T}})$ and, consequently,

$$\text{var}_M[\Psi(\hat{\mathbf{T}}) - \Psi(\mathbf{T})] \cong \text{var}_M[\Psi(\hat{\mathbf{T}})] \cong \nabla\Psi(\mathbf{T}_\mu)'\text{var}_M(\hat{\mathbf{T}})\nabla\Psi(\mathbf{T}_\mu). \qquad (10.1.5)$$

A sample estimator can be obtained by substituting a consistent estimator of $\text{var}_M(\hat{\mathbf{T}})$ in (10.1.5) and replacing $\mathbf{T}_\mu$ with $\hat{\mathbf{T}}$. Note that the application here is more elaborate than in Section 9.4 because we have a vector of estimators of totals rather than a single estimator. To estimate the components of the $K \times K$ covariance matrix, $\text{var}_M(\hat{\mathbf{T}})$, we can use the sandwich estimator, v_R, defined in Section 9.5.1, or the variants, v_D, v_J^*, v_R^*, v_D^*, and v_J^{**}, defined in Section 9.5.2. Another choice is the jacknife variance estimator defined in Section 9.5.3. An estimator of the variance of $\hat{T}_k$, using the sandwich estimator in (9.5.5), is

$$v_R(\hat{T}_k) = \sum_s \mathbf{g}_i'\mathbf{r}_{ki}\mathbf{r}_{ki}'\mathbf{g}_i,$$

where g_i is the ith component of $\mathbf{g}_s$ and $\mathbf{r}_{ki} = (r_{kil}, \ldots, r_{kim_i})'$ is the vector of residuals for cluster i with $r_{kij} = Y_{kij} - \mathbf{x}_{ij}'\boldsymbol{\beta}_k^o$ and $\boldsymbol{\beta}^o = \mathbf{G}\mathbf{X}_s'\mathbf{W}_s\mathbf{Y}_s$. The covariance between $\hat{T}_k$ and $\hat{T}_{k'}$ can be estimated by

$$v_R(\hat{T}_k, \hat{T}_{k'}) = \sum_s \mathbf{g}_i'\mathbf{r}_{ki}\mathbf{r}_{k'i}'\mathbf{g}_i.$$

The alternative variance and covariance estimators in Sections 9.5.2 and 9.5.3 are constructed similarly.

Example 10.1.1 Log-odds ratio in a two-way table. Suppose that persons in a population are categorized by sex and as having a chronic health condition, for example, arthritis, or not. The population counts of persons are given by the following two-way table:

Condition	Male	Female
Yes	T_A	T_B
No	T_C	T_D

The odds of having the condition for males is T_A/T_C while for females the odds is T_B/T_D. The logarithm of the ratio of the odds for males to the odds for females is

$$\Psi(T_A, T_B, T_C, T_D) = \log\left(\frac{T_A T_D}{T_B T_C}\right),$$

which is estimated by substituting estimators for the four totals. The gradient of Ψ is $\nabla\Psi(\mathbf{T}_{\boldsymbol{\mu}}) = (T_{\mu A}^{-1}, -T_{\mu B}^{-1}, -T_{\mu C}^{-1}, T_{\mu D}^{-1})$, where $T_{\mu k}$ $(k = A, B, C, D)$ denotes the expected value of a cell total under a model. Assuming that model (10.1.1) holds, we are predicting cell membership, a 0–1 characteristic, with a linear model. Denoting a covariance matrix estimator of $\hat{\mathbf{T}}$ by $v(\hat{\mathbf{T}})$, an estimator of the variance of the log-odds ratio is then

$$v[\Psi(\hat{\mathbf{T}})] = \nabla\Psi(\hat{\mathbf{T}})' v(\hat{\mathbf{T}}) \nabla\Psi(\hat{\mathbf{T}}) = \frac{v_A}{\hat{T}_A^2} + \frac{v_B}{\hat{T}_B^2} + \frac{v_C}{\hat{T}_C^2} + \frac{v_D}{\hat{T}_D^2} - 2\frac{v_{AB}}{\hat{T}_A\hat{T}_B} - 2\frac{v_{AC}}{\hat{T}_A\hat{T}_C}$$
$$- 2\frac{v_{BD}}{\hat{T}_B\hat{T}_D} - 2\frac{v_{CD}}{\hat{T}_C\hat{T}_D} + 2\frac{v_{AD}}{\hat{T}_A\hat{T}_D} + 2\frac{v_{BC}}{\hat{T}_B\hat{T}_C},$$

where, for example, v_A is an estimator of the variance of $\hat{T}_A$ and v_{AC} is an estimator of the covariance of $\hat{T}_A$ and $\hat{T}_C$. The estimator of the variance of the log-odds ratio is, thus, a linear combinatin of estimators of relative variances and relative covariances.

Although algebraic rearrangements are often possible, as in the above example, computer calculations may be just as easily done using the vector–matrix formulation in (10.1.5) with sample estimates substituted where needed.

10.1.1 Variance Estimation for a Ratio of Estimated Totals

In the preceding discussion, we estimated only the dominant term of the approximate error variance of a nonlinear estimator. It is also possible to develop estimators of the lower order variance terms. To illustrate this more detailed approach, consider estimating the ratio of two population quantities, T_y and T_z, that are totals of random variables Y and Z. Examples are (1) the weekly wages per computer scientist with $\hat{T}_y$ being estimated total wages and $\hat{T}_z$ the estimated total number of workers in that occupation and (2) the savings rate for households with $\hat{T}_y$ being the estimated total amount of money saved in a given period of time by all households and $\hat{T}_z$ being their estimated total income.

Assume that the cluster sizes, M_i, are known and consider a working model that is a generalization of (9.1.1) with (Y_{ij}, Z_{ij}) correlated with $(Y_{i'j'}, Z_{i'j'})$ only when the units belong to the same cluster, $i = i'$:

$$E_M\begin{pmatrix} Y_{ij} \\ Z_{ij} \end{pmatrix} = \begin{pmatrix} \mu_y \\ \mu_z \end{pmatrix}$$

$$\text{var}_M\begin{pmatrix} Y_{ij} \\ Z_{ij} \\ Y_{ij'} \\ Z_{ij'} \end{pmatrix} = \begin{pmatrix} \sigma_y^2 & \sigma_{1yz} & \rho_y\sigma_y^2 & \sigma_{2yz} \\ \sigma_{1yz} & \sigma_z^2 & \sigma_{2yz} & \rho_z\sigma_z^2 \\ \rho_y\sigma_y^2 & \sigma_{2yz} & \sigma_y^2 & \sigma_{1yz} \\ \sigma_{2yz} & \rho_z\sigma_z^2 & \sigma_{1yz} & \sigma_z^2 \end{pmatrix}. \tag{10.1.6}$$

Other covariances are zero. Within cluster i there are four covariance terms, one for the covariance of Y and Z values on the same unit, $\text{cov}_M(Y_{ij}, Z_{ij}) = \sigma_{1yz}$, and three for covariances between units, $\text{cov}_M(Y_{ij}, Y_{ij'}) = \rho_y \sigma_y^2$, $\text{cov}_M(Z_{ij}, Z_{ij'}) = \rho_z \sigma_z^2$, and $\text{cov}_M(Y_{ij}, Z_{ij'}) = \sigma_{2yz}$.

Denote the estimator of the ratio by $\hat{T}_y/\hat{T}_z$, where $\hat{T}_y = \Sigma_{i \in s} g_i \bar{Y}_{si}$, $\hat{T}_z = \Sigma_{i \in s} g_i \bar{Z}_{si}$, $\bar{Y}_{si} = \Sigma_{j \in s_i} Y_{ij}/m_i$, and $\bar{Z}_{si} = \Sigma_{j \in s_i} Z_{ij}/m_i$. The coefficients g_i are constructed to make $\hat{T}_y$ and $\hat{T}_z$ unbiased under model (10.1.6). As in Section 9.1, unbiasedness requires $\Sigma_s g_i = M$, the total number of units in the population.

To estimate the error-variance, $\text{var}_M(\hat{T}_y/\hat{T}_z - T_y/T_z)$, we need to account for the fact that in some problems T_y/T_z and $\hat{T}_y/\hat{T}_z$ do not have finite expectations or variances because of tiny but positive probabilities that the denominators vanish. Thus, we state our objectives not in terms of unbiased estimation of $\text{var}_M(\hat{T}_y/\hat{T}_z - T_y/T_z)$. Instead, using the same type of approximations as those given at the beginning of this section, we will find a random variable R, to which the difference is asymptotically equivalent in the sense that $\hat{T}_y/\hat{T}_z - T_y/T_z = R(1 + o_p(1))$, and seek a robust, consistent estimator of $\text{var}_M(R)$.

Denoting the ratio μ_y/μ_z by φ, a first-order Taylor series expansion of $\hat{T}_y/\hat{T}_z = (\hat{T}_y/M)/(\hat{T}_z/M) \equiv \hat{\mu}_y/\hat{\mu}_z$ around φ gives

$$\frac{\hat{\mu}_y}{\hat{\mu}_z} \cong \varphi + \frac{\partial}{\partial \hat{\mu}_y}\left(\frac{\hat{\mu}_y}{\hat{\mu}_z}\right)_{\mu_y, \mu_z} (\hat{\mu}_y - \mu_y) + \frac{\partial}{\partial \hat{\mu}_z}\left(\frac{\hat{\mu}_y}{\hat{\mu}_z}\right)_{\mu_y, \mu_z} (\hat{\mu}_z - \mu_z) = \varphi + \frac{\hat{T}_y - \varphi \hat{T}_z}{M \mu_z}.$$

As before we treat $\hat{T}_y/\hat{T}_z$ as a ratio of means because, as the population size M grows, the totals T_y and T_z are unbounded and, for example, $\hat{T}_y - M\mu_y$ may not be small even though $\hat{\mu}_y - \mu_y$ is. Expanding T_y/T_z around φ gives $T_y/T_z \cong \varphi + (T_y - \varphi T_z)/(M\mu_z)$. Next, define $\bar{D}_{si} = \bar{Y}_{si} - \varphi \bar{Z}_{si}$ and $D_i = T_{yi} - \varphi T_{zi}$ with $T_{yi} = \Sigma_{j=1}^{M_i} Y_{ij}$ and $T_{zi} = \Sigma_{j=1}^{M_1} Z_{ij}$ being the population cluster totals. Taking the difference in the two expansions gives

$$\frac{\hat{T}_y}{\hat{T}_z} - \frac{T_y}{T_z} \cong \frac{\hat{D} - D}{M\mu_z}, \qquad (10.1.7)$$

where $\hat{D} = \Sigma_s g_i \bar{D}_{si}$ and $D = \Sigma_{i=1}^N D_i$. Note that $\hat{D}$ has the same form as $\hat{T}_y$ and $\hat{T}_z$, being a weighted sum of sample cluster means.

Thus we have translated the problem from the difficult one of estimating the error variance of the nonlinear $\hat{T}_y/\hat{T}_z$ to the easier one of estimating the error variance of the linear $\hat{D}/(M\mu_z)$. Proceeding as in the univariate case in Section 9.1, we have

$$\hat{D} - D = \Sigma_s g_i \bar{D}_{si} - \left(\sum_s M_i D_i + \sum_r M_i D_i\right).$$

The variance of $\hat{D} - D$ is equal to

$$\operatorname{var}_M(\hat{D} - D) = \sum_s g_i^2 \operatorname{var}_M(\bar{D}_{si}) - 2 \sum_s g_i M_i \operatorname{cov}_M(\bar{D}_{si}, D_i) + \sum_{i=1}^{N} M_i^2 \operatorname{var}_M(D_i), \tag{10.1.8}$$

analogous to expression (9.1.4). Under conditions similar to those in Theorem 9.1.1 the dominant term of the error variance in (10.1.8) is the first, $\Sigma_s g_i^2 \operatorname{var}_M(\bar{D}_{si})$ (see Exercise 10.1).

Robust variance estimators of (10.1.8) can be constructed using the same steps as in Sections 9.2.2 and 9.2.3. A residual for cluster i is $r_{Di} = \hat{\bar{D}}_{si} - \hat{\hat{D}}/M$, where $\hat{\bar{D}}_{si} = \bar{Y}_{si} - \hat{\varphi}\bar{Z}_{si}$ with $\hat{\varphi} = \hat{\mu}_y/\hat{\mu}_z$ and $\hat{\hat{D}} = \Sigma_s g_i \hat{\bar{D}}_{si}$. But, in this case

$$\begin{aligned} \hat{\hat{D}} &= \sum_s g_i(\bar{Y}_{si} - \hat{\varphi}\bar{Z}_{si}) \\ &= \hat{T}_y - \hat{\varphi}\hat{T}_z, \end{aligned}$$

so that the residual can be rewritten as $r_{Di} = (\bar{Y}_{si} - \hat{\mu}_y) - \hat{\varphi}(\bar{Z}_{si} - \hat{\mu}_z)$. Under a model where $E_M(Y_{ij}, Z_{ij}) = (\mu_y, \mu_z)$ and the number of sample clusters is large so that $E_M(\hat{\varphi}) \cong \varphi$, we have $E_M(r_{Di}) \cong 0$. When units in different clusters are uncorrelated, $E_M(\hat{\bar{D}}_{si}, \hat{\bar{D}}_{si'}) \cong 0$ where $i \neq i'$. Then, similar to expression (9.1.6),

$$E_M(r_{Di}^2) \cong E_M(\hat{\bar{D}}_{si}^2)\left(1 - \frac{2g_i}{M}\right) + \frac{1}{M^2} \sum_{i' \in s} g_{i'}^2 E_M(\hat{\bar{D}}_{si'}^2).$$

Since $E_M(\hat{\bar{D}}_{si}^2) \cong \operatorname{var}_M(\hat{\bar{D}}_{si})$ and we assume that $g_i = O(N/n)$, r_{Di}^2 is approximately unbiased for $\operatorname{var}_M(\hat{\bar{D}}_{si})$.

A sandwich estimator of the dominant term of (10.1.8) is

$$v_R(\hat{D}) = \sum_s g_i^2 r_{Di}^2. \tag{10.1.9}$$

As sketched above, this estimator is approximately unbiased under the working model (10.1.6) and under a more general model where the covariance structure within clusters is unspecified (see Exercise 10.2). Using the approach in Section 9.2.3, a robust estimator of $\operatorname{var}_M(\hat{D} - D)$, that accounts for all three terms in (10.1.8) is given by

$$v_{OD} = \sum_s g_i^2 r_{Di}^2 - \sum_s g_i M_i \dot{v}_{Di} + \left[\sum_{i=1}^{N} M_i^2 - \sum_s g_i M_i\right] \dot{\theta}_{D2}, \tag{10.1.10}$$

where $\dot{v}_{Di} = r_{Di}^2 - M_i^{-1} S_{Di}^2(1 - f_i)/f_i, \quad f_i = m_i/M_i$,

$$S_{Di}^2 = \sum_{j \in s_i} [(Y_{ij} - \bar{Y}_{si}) - \hat{\varphi}(Z_{ij} - \bar{Z}_{si})]^2/(m_i - 1),$$

and $\dot{\theta}_{D2} = \Sigma_s(r_{Di}^2 - S_{Di}^2/m_i)/m$. The error-variance $\text{var}_M(\hat{T}_y/\hat{T}_z - T_y/T_z)$ is then estimated by $v_{0D}/\hat{T}_z^2$. Exercise 10.3(a) is to show that v_{0D} is an approximately unbiased estimator of $\text{var}_M(\hat{D} - D)$ under model (10.1.6). As in Section 9.2.3, an adjustment could be made to the residual r_{Di}^2, to give a statistic $\hat{v}_{Di}$ that is a more refined estimator of the variance of $\bar{D}_{si}$. The adjusted residual $\hat{v}_{Di}$ could then be used in v_{0D} in place of r_{Di}^2 to obtain an estimator analogous to (9.2.12) (see exercise 10.3(b)).

10.1.2 Jackknife and Nonlinear Functions

A jackknife variance estimator for $\Psi(\hat{\mathbf{T}})$ can be constructed directly through the usual algorithm of omitting one first-stage unit, computing the point estimator, and then computing a variance among these delete-one estimators. We sketch the theory in this section for why the jackknife works for nonlinear estimators. Shao and Tu (1995) cover the properties of the jackknife in depth.

First, define the estimator of the total for variable k after deleting cluster i as $\hat{T}_{k(i)} = \Sigma_{i \in s}\Sigma_{j \in s_i} Y_{kij} + \mathbf{1}_r'\mathbf{X}_r\boldsymbol{\beta}_{k(i)}^o$, where $\boldsymbol{\beta}_{k(i)}^o = (\mathbf{X}_{s(i)}'\mathbf{W}_{s(i)}\mathbf{X}_{s(i)})^{-}\mathbf{X}_{s(i)}'\mathbf{W}_{s(i)}\mathbf{Y}_{ks(i)}$. As in Section 9.4.3, the minus superscript denotes a generalized inverse, and the subscript notation (i) denotes quantities that are computed after omitting sample cluster i. Note that $\hat{T}_{k(i)}$ uses a parameter solution for $\boldsymbol{\beta}_k$ that omits cluster i but retains the sum over the full sample of the Y_{kij}'s. The version of the jackknife that we examine is

$$v_J[\Psi(\hat{\mathbf{T}})] = \frac{n-1}{n} \sum_{i \in s} [\Psi(\hat{\mathbf{T}}_{(i)}) - \Psi(\cdot)]^2, \tag{10.1.11}$$

where

$$\Psi(\hat{\mathbf{T}}_{(i)}) = \Psi(\hat{T}_{1(i)}, \ldots, \hat{T}_{K(i)}) \quad \text{and} \quad \Psi(\cdot) = n^{-1} \sum_{i \in s} \Psi(\hat{\mathbf{T}}_{(i)}).$$

We might also center the variance around the full sample estimator $\Psi(\hat{\mathbf{T}})$, rather than the average of the delete-one estimators, and the resulting version of the jackknife will be asymptotically equivalent to (10.1.11).

If the sample of clusters is large, we can expand $\Psi(\hat{\mathbf{T}}_{(i)})$ around the full sample estimator $\Psi(\hat{\mathbf{T}})$ to obtain

$$\Psi(\hat{\mathbf{T}}_{(i)}) \cong \Psi(\hat{\mathbf{T}}) + \nabla\Psi(\hat{\mathbf{T}})'(\hat{\mathbf{T}}_{(i)} - \hat{\mathbf{T}}),$$

from which it follows that

$$\Psi(\hat{\mathbf{T}}_{(i)}) - \Psi(\cdot) \cong \nabla\Psi(\hat{\mathbf{T}})'(\hat{\mathbf{T}}_{(i)} - \hat{\mathbf{T}}_{(\cdot)}), \tag{10.1.12}$$

where $\hat{\mathbf{T}}_{(\cdot)} = n^{-1}\Sigma_s \hat{\mathbf{T}}_{(i)}$. Substituting (10.1.12) into (10.1.11) and rearranging leads to

$$v_J[\Psi(\hat{\mathbf{T}})] \cong \nabla\Psi(\hat{\mathbf{T}})' v_J(\hat{\mathbf{T}}) \nabla\Psi(\hat{\mathbf{T}}), \tag{10.1.13}$$

where $v_J(\hat{\mathbf{T}}) = (n-1)/n \, \Sigma_{i \in s}(\hat{\mathbf{T}}_{(i)} - \hat{\mathbf{T}}_{(\cdot)})(\hat{\mathbf{T}}_{(i)} - \hat{\mathbf{T}}_{(\cdot)})'$ is a jackknife estimator of the $K \times K$ covariance matrix of $\hat{\mathbf{T}}$. Using methods similar to those in Chapter 9, $v_J(\hat{\mathbf{T}})$ can be shown to be consistent and approximately unbiased for $\text{var}_M(\hat{\mathbf{T}})$ under model (10.1.1). The robustness of $v_J(\hat{\mathbf{T}})$ is conveyed to (10.1.11).

Thus we can use the jackknife in either of two ways when estimating the variance of a nonlinear statistic. We can substitute $v_J(\hat{\mathbf{T}})$ and $\nabla\Psi(\hat{\mathbf{T}})$ into (10.1.5) to obtain $v_J^*[\Psi(\hat{\mathbf{T}})] \cong \nabla\Psi(\hat{\mathbf{T}})' v_J(\hat{\mathbf{T}}) \nabla\Psi(\hat{\mathbf{T}})$, or we can use the jackknife directly as in (10.1.11). Either approach will produce a robust estimator of the variance of $\Psi(\hat{\mathbf{T}})$ under the general model (10.1.1). Valliant (1993) and Yung and Rao (1996) study some specific instances of the linearized jackknife, $v_J^*[\Psi(\hat{\mathbf{T}})]$.

10.2 BALANCED HALF-SAMPLE VARIANCE ESTIMATION

Balanced half-sample (*BHS*) or balanced repeated replication (*BRR*) variance estimators, proposed by McCarthy (1966, 1969), are often used in complex surveys because of their applicability to complex estimators and the ease with which they can be programmed. Design-based theory for the method can be found in Krewski and Rao (1981) and Wolter (1984). Note that the type of "balance" discussed in this section differs from the balance on auxiliary variables in earlier chapters.

As for the jackknife, the general idea is to form subsamples from the full sample in a prescribed way, compute an estimate from each subsample, and then compute a variance among the subsample (or replicate) estimates. Assume that the population is stratified and that a sample of $n_h = 2$ primary units is selected from each stratum. The units may be the ones from which data are collected or they may be clusters of other units. There are generalizations of the method to sample sizes other than $n_h = 2$ in Gurney and Jewett (1975) and Wu (1991), but these are seldom used in practice and will not be considered here. A half-sample consists of one of the two sample units from each of the H strata. There are a total of 2^H possible half-samples, but, by clever selection, a much smaller number can be used for variance estimation. A set of J half-samples is defined by the indicators

$$\varsigma_{hi\alpha} = \begin{cases} 1 & \text{if unit } hi \text{ is in half-sample } \alpha \\ 0 & \text{if not} \end{cases}$$

for $i = 1, 2$ and $\alpha = 1, \ldots, J$. Based on the $\varsigma_{hi\alpha}$, define

$$\varsigma_h^{(\alpha)} = 2\varsigma_{h1\alpha} - 1 = \begin{cases} 1 & \text{if unit } h1 \text{ is in half-sample } \alpha \\ -1 & \text{if unit } h2 \text{ is in half-sample } \alpha. \end{cases}$$

Note also that $-\varsigma_h^{(\alpha)} = 2\varsigma_{h2\alpha} - 1$. A set of half-samples is said to be in *full orthogonal balance* if

$$\sum_{\alpha=1}^{J} \varsigma_h^{(\alpha)} = 0, \text{ for all } h \tag{10.2.1}$$

$$\sum_{\alpha=1}^{J} \varsigma_h^{(\alpha)}\varsigma_{h'}^{(\alpha)} = 0 \ (h \neq h') \tag{10.2.2}$$

with a minimal set of half-samples satisfying (10.2.1) and (10.2.2) having $H + 1 \leqslant J \leqslant H + 4$. The number J is the smallest multiple of 4 greater than the number of strata H. Even when H is modest, the size of the minimal set can be considerably smaller than the number of possible half-samples 2^H as the following table shows.

H	J	2^H
5	8	32
10	12	1,024
20	24	1,048,576

Let $\hat{\theta}^{(\alpha)}$ be the estimator, based on half-sample α, with the same form as the full sample estimator $\hat{\theta}$. For an estimator $\hat{\theta}$, a balanced half-sample variance estimator is

$$v_{BHS}(\hat{\theta}) = \sum_{\alpha=1}^{J} (\hat{\theta}^{(\alpha)} - \hat{\theta})^2/J.$$

There are other forms of the *BHS* estimator, noted in Section 10.2.5, but the one above is most commonly used in practice.

10.2.1 Application to the Stratified Expansion Estimator

To illustrate the *BHS* method, consider the model

$$Y_{hi} = \mu_h + \varepsilon_{hi}, \quad \varepsilon_{hi} \sim (0, \sigma_h^2) \tag{10.2.3}$$

with the ε_{hi}'s independent. The *BLU* predictor in that case is the stratified expansion estimator $\hat{T}_0 = \Sigma_h N_h \bar{Y}_{hs}$. When $n_h = 2$, the mean of the units in half-sample α and stratum h is

$$\bar{Y}_{hs}^{(\alpha)} = \varsigma_{h1\alpha} Y_{h1} + \varsigma_{h2\alpha} Y_{h2}$$

The stratified expansion estimator based on half-sample α is $\hat{T}_0^{(\alpha)} = \Sigma_h N_h \bar{Y}_{hs}^{(\alpha)}$. It follows that

$$\begin{aligned}\hat{T}_0^{(\alpha)} - \hat{T}_0 &= \sum_h \frac{N_h}{2} [(2\varsigma_{h1\alpha} - 1) Y_{h1} + (2\varsigma_{h2\alpha} - 1) Y_{h2}] \\ &= \sum_h \frac{N_h}{2} \varsigma_h^{(\alpha)} (Y_{h1} - Y_{h2}). \end{aligned} \tag{10.2.4}$$

Squaring out this difference and summing over half-samples gives

$$\begin{aligned} v_{BHS}(\hat{T}_0) &= J^{-1} \sum_\alpha \sum_h \frac{N_h^2}{4} (\varsigma_h^{(\alpha)})^2 (Y_{h1} - Y_{h2})^2 \\ &\quad + J^{-1} \sum_\alpha \sum_h \sum_{h' \neq h} \frac{N_h N_{h'}}{4} \varsigma_h^{(\alpha)} \varsigma_{h'}^{(\alpha)} (Y_{h1} - Y_{h2})(Y_{h'1} - Y_{h'2}) \\ &= \sum_h \frac{N_h^2}{4} (Y_{h1} - Y_{h2})^2 \end{aligned}$$

with the last equality following from (10.2.2) and the fact that $(\varsigma_h^{(\alpha)})^2 = 1$. When $n_h = 2$, we also have $s_h^2 = (Y_{h1} - Y_{h2})^2/2$ so that

$$v_{BHS}(\hat{T}_0) = \sum_h \frac{N_h^2}{n_h} s_h^2. \tag{10.2.5}$$

In other words, the *BHS* variance estimator equals the standard one that follows from model (10.2.3) when the sampling fraction f_h is 0. Note that the reduction to (10.2.5) required only the orthogonality condition (10.2.2) not (10.2.1). If we also have condition (10.2.1), $\Sigma_{\alpha=1}^J \varsigma_h^{(\alpha)} = 0$, then

$$\begin{aligned} J^{-1} \sum_{\alpha=1}^J \hat{T}^{(\alpha)} &= J^{-1} \sum_{\alpha, h} N_h (\varsigma_{h1\alpha} Y_{h1} + \varsigma_{h2\alpha} Y_{h2}) \\ &= J^{-1} \sum_h N_h \left(Y_{h1} \sum_\alpha \frac{\varsigma_h^{(\alpha)} + 1}{2} + Y_{h2} \sum_\alpha \frac{1 - \varsigma_h^{(\alpha)}}{2} \right) \\ &= \hat{T}_0. \end{aligned}$$

In that case, the average of the *BHS* estimators of the total is also an estimator of the population total.

Recalling earlier results from Section 6.7, $\Sigma_h N_h^2 s_h^2 / n_h$ is still approximately unbiased under model (10.2.3) when $\text{var}_M(\varepsilon_{hi}) = v_{hi}$ instead of σ_h^2. Thus,

$v_{BHS}(\hat{T}_0)$ has the same robustness property. Of course, in the simple case of this section, the standard variance estimator is easy to compute and is robust so that the *BHS* estimator has no advantages. In Section 10.2.3 we consider more complicated situations where the method can be quite useful.

10.2.2 Orthogonal Arrays

In order to implement a *BHS* variance estimator for a survey, we need an orthogonal array of the proper size. Square orthogonal arrays, also known as Hadamard matrices, of size 2, 4, 8, 12, 16, ..., 100 are listed in Wolter (1985). Plackett and Burman (1946) give algorithms for constructing arrays. An array of a given size is not unique since, for example, if A is an orthogonal array, then $-A$ is also. One easy way of generating orthogonal arrays of dimension $2^k \times 2^k$ $(k = 1, 2, \ldots)$ is the following. Let

$$A_2 = \begin{pmatrix} 1 & 1 \\ 1 & -1 \end{pmatrix},$$

a matrix that clearly satisfies (10.2.2), though not (10.2.1). From A_2, an array of dimension 4 can be generated as

$$A_4 = \begin{pmatrix} A_2 & A_2 \\ A_2 & -A_2 \end{pmatrix},$$

and, in general,

$$A_{2^k} = \begin{pmatrix} A_{2^{k-1}} & A_{2^{k-1}} \\ A_{2^{k-1}} & -A_{2^{k-1}} \end{pmatrix}$$

for $k \geqslant 2$. This algorithm can be easily implemented with matrix-oriented software. To illustrate the use of one of the matrices, consider

$$A_8 = \begin{pmatrix} +1 & +1 & +1 & +1 & +1 & +1 & +1 & +1 \\ +1 & -1 & +1 & -1 & +1 & -1 & +1 & -1 \\ +1 & +1 & -1 & -1 & +1 & +1 & -1 & -1 \\ +1 & -1 & -1 & +1 & +1 & -1 & -1 & +1 \\ +1 & +1 & +1 & +1 & -1 & -1 & -1 & -1 \\ +1 & -1 & +1 & -1 & -1 & +1 & -1 & +1 \\ +1 & +1 & -1 & -1 & -1 & -1 & +1 & +1 \\ +1 & -1 & -1 & +1 & -1 & +1 & +1 & -1 \end{pmatrix}$$

generated as described above. Let the rows denote half-samples and the columns strata. $A + 1$ in a column means "select unit 1" for a half-sample while a -1 means "select unit 2." The first half-sample then consists of the first sample unit from each stratum since column 1 of the matrix is all +1's. The second half-sample is the first unit from strata 1, 3, 5, and 7 and the second unit from strata 2, 4, 6, and 8. A_8 satisfies (10.2.2), but not (10.2.1) since the first column consists of all +1's. Columns 2–8, however, do have the property that $\Sigma_{\alpha=1}^{J}\zeta_h^{(\alpha)} = 0$. For $H = 8$ we can have an orthogonal set of half-samples, that is, ones that satisfy (10.2.2), by using A_8. For $H = 7$ we can have a set of half-samples in full orthogonal balance by selecting all 8 rows of A_8 but only the last 7 columns. For $H = 6$, any 6 of columns 2–8 constitute a fully balanced orthogonal set; for $H = 5$, any 5 of columns 2–8 will work; and so on. Fully balanced orthogonal sets for $H = 4$, 5, 6, and, 7 can, thus, be obtained from A_8 with each set having 8 half-samples.

Use of the preceding algorithm and selection of a subset of columns to designate half-samples will generally lead to more than the minimal set. If $H = 46$, for example, the smallest power of 2 greater than 46 is $2^6 = 64$. One way of designating a fully balanced set of orthogonal half-samples is then to generate A_{64}, pick 46 columns from columns 2–64, and use the 64×46 submatrix. The minimal set, on the other hand, has only 48 half-samples. If we select a 48×46 submatrix, excluding the first column of A_{64}, all columns of the submatrix are not necessarily orthogonal. For $H = 46$ it will, consequently, be more economical to begin with the 48×48 orthogonal matrix in Wolter (1985). Note that Wolter's matrices also contain a column of all +1's that should be excluded to obtain a fully balanced set. In the $H = 46$ case, we would begin with the 48×48 Hadamard matrix and extract 46 of columns 2–48 to form a 48×46 matrix.

10.2.3 Extension to Nonlinear Functions

If we were interested only in simple estimators like $\hat{T}_0$, then the more direct methods of Chapter 5 are available, and there would be no advantage to using *BHS*. The usefulness of the method lies in its application to more complex situations such as multi-stage sampling from clustered populations and to more complex estimators. We might, for example, want to estimate the ratio of the unemployment rate for men aged 25–34 to that of women in the same age group using a two-stage sample of households. Section 10.2.4 shows that for certain types of nonlinear estimators, *BHS* provides a consistent variance estimator in clustered populations when the number of strata H is large. Before presenting that result, we give an informal discussion of how *BHS* works for nonlinear estimators under a simpler model.

Suppose that data are collected on p variables, $Y_1, \ldots, Y_p$, and that they follow the multivariate model

$$E_M(Y_{\ell hi}) = \mu_{\ell h}$$

$$\operatorname{cov}_M(Y_{\ell hi}, Y_{\ell' h' i'}) = \begin{cases} \sigma_{\ell\ell' h} & h = h',\ i = i' \\ 0 & \text{otherwise} \end{cases} \qquad (10.2.6)$$

for $\ell = 1, \ldots, p$. In other words, different y variables ℓ and ℓ' for the same unit are correlated but others are not. Let $\mathbf{T} = (T_1, \ldots, T_p)'$ be the vector of population totals for the p variables, and let $\Psi(\mathbf{T})$ be a differentiable function of the totals. As in Section 10.1, a natural estimator of $\Psi(\mathbf{T})$ is $\Psi(\hat{\mathbf{T}})$ where $\hat{\mathbf{T}} = (\hat{T}_{01}, \ldots, \hat{T}_{0p})'$ with

$$\hat{T}_{0\ell} = \sum_h N_h \bar{Y}_{\ell hs}$$

and $\bar{Y}_{\ell hs} = \sum_{i \in s_h} Y_{\ell hi}/n_h$ $(\ell = 1, \ldots, p)$. In vector form $\hat{\mathbf{T}} = N_h \bar{\mathbf{Y}}_{hs}$ where $\bar{\mathbf{Y}}_{hs} = (\bar{Y}_{1hs}, \ldots, \bar{Y}_{phs})'$.

Following the approach in Section 10.1, we develop asymptotic theory for a nonlinear $\Psi(\hat{\mathbf{T}})$ by first approximating it by a linear function of $\hat{\mathbf{T}}$. When $n_h = 2$, the obvious method of achieving large samples is to have a large number, H, of strata. This is, in fact, a fairly common design, and, as noted in chapter 3 can be very similar to systematic sampling. Although the number of strata may be large, there is no requirement that all the $\mu_{\ell h}$'s be distinct in model (10.2.6). When H is large, $\Psi(\hat{\mathbf{T}})$ can be approximated as

$$\Psi(\hat{\mathbf{T}}) \cong \Psi(\mathbf{T}_\mu) + \nabla\Psi(\mathbf{T}_\mu)'(\hat{\mathbf{T}} - \mathbf{T}_\mu), \tag{10.2.7}$$

where $\mathbf{T}_\mu$ is the model expectation of $\mathbf{T}$ and $\nabla\Psi(\mathbf{T}_\mu) = (\partial\Psi(\hat{\mathbf{T}})/\partial\hat{T}_{01}, \ldots, \partial\Psi(\hat{\mathbf{T}})/\partial\hat{T}_{0p})$ is the gradient evaluated at $\mathbf{T}_\mu$.

When $f_h \cong 0$, H is large, and other conditions similar to those in Theorem 5.3.1 hold, then $\text{var}_M[\Psi(\hat{\mathbf{T}}) - \Psi(\mathbf{T})] \cong \text{var}_M[\Psi(\hat{\mathbf{T}})]$. Using (10.2.7),

$$\text{var}_M[\Psi(\hat{T})] \cong \nabla\Psi(\mathbf{T}_\mu)'\text{var}_M(\hat{\mathbf{T}})\nabla\Psi(\mathbf{T}_\mu) \tag{10.2.8}$$

A simple estimator can be obtained by substituting a consistent estimator of $\text{var}_M(\hat{\mathbf{T}})$ in (10.2.8) and replacing $\mathbf{T}_\mu$ with $\hat{\mathbf{T}}$. An estimator for the $(\ell\ell')$th component of $\text{var}_M(\hat{\mathbf{T}})$ is

$$v_{\ell\ell'} = \sum_h \frac{N_h^2}{n_h} s_{\ell\ell' h}$$

with $s_{\ell\ell' h} = \Sigma_{i \in s_h}(Y_{\ell hi} - \bar{Y}_{\ell hs})(Y_{\ell' hi} - \bar{Y}_{\ell' hs})/(n_h - 1)$. Note that, when $n_h = 2$,

$$s_{\ell\ell' h} = (Y_{\ell h1} - Y_{\ell h2})(Y_{\ell' h1} - Y_{\ell' h2})/2. \tag{10.2.9}$$

Letting $\mathbf{s}_h^2$ be the $p \times p$ sample covariance matrix for stratum h with the $(\ell\ell')$th element equal to $s_{\ell\ell' h}$, an estimator of the approximate variance of $\Psi(\hat{\mathbf{T}})$ is

$$v_{\Psi(\hat{\mathbf{T}})} = \nabla\Psi(\hat{\mathbf{T}})'\left(\sum_h \frac{N_h^2}{n_h}\mathbf{s}_h^2\right)\nabla\Psi(\hat{\mathbf{T}}). \tag{10.2.10}$$

The middle matrix in $v_{\Psi(\hat{\mathbf{T}})}$, $\Sigma_h N_h^2 \mathbf{s}_h^2/n_h$, is an unbiased estimator of $\text{var}_M(\hat{\mathbf{T}})$ under model (10.2.6) when $f_h = 0$ and is approximately unbiased under a more general version of (10.2.6) with $\text{cov}_M(Y_{\ell hi}, Y_{\ell' hi}) = v_{\ell\ell' hi}$.

Now, we can show that the *BHS* variance estimator is approximately equal to (10.2.10). Denote the vector of half-sample estimators of the totals as $\hat{\mathbf{T}}^{(\alpha)} = (\hat{T}_{01}^{(\alpha)}, \ldots, \hat{T}_{0p}^{(\alpha)})'$. When H is large, the function of the half-sample estimators, $\Psi(\hat{\mathbf{T}}^{(\alpha)})$, can be expanded around the full-sample estimator $\Psi(\hat{\mathbf{T}})$ as

$$\Psi(\hat{\mathbf{T}}^{(\alpha)}) \cong \Psi(\hat{\mathbf{T}}) + \nabla\Psi(\hat{\mathbf{T}})'(\hat{\mathbf{T}}^{(\alpha)} - \hat{\mathbf{T}}), \tag{10.2.11}$$

where $\nabla\Psi(\hat{\mathbf{T}}) = (\partial\Psi(\hat{\mathbf{T}}^{(\alpha)})/\partial\hat{T}_{01}, \ldots, \partial\Psi(\hat{\mathbf{T}}^{(\alpha)})/\partial\hat{T}_{0p})$ with each of the partial derivatives being evaluated at the full-sample estimator $\hat{\mathbf{T}}$. Next, define $\mathbf{Y}_{hi} = (Y_{1hi}, \ldots, Y_{phi})'$ for $i = 1, 2$, and recall that the vector of full-sample means is $\bar{\mathbf{Y}}_{hs} = (\bar{Y}_{1hs}, \ldots, \bar{Y}_{phs})'$. The corresponding half-sample vector of sample means is $\bar{\mathbf{Y}}_{hs}^{(\alpha)} = (\bar{Y}_{1hs}^{(\alpha)}, \ldots, \bar{Y}_{phs}^{(\alpha)})'$ where $\bar{Y}_{\ell hs}^{(\alpha)} = \varsigma_{h1\alpha} Y_{\ell h1} + \varsigma_{h2\alpha} Y_{\ell h2}$. It follows $\hat{\mathbf{T}}^{(\alpha)} = N_h \bar{\mathbf{Y}}_{hs}^{(\alpha)}$, paralleling the full-sample form, $\hat{\mathbf{T}} = N_h \bar{\mathbf{Y}}_{hs}$. The *BHS* variance estimator is then

$$\begin{aligned}
v_{BHS} &= J^{-1} \sum_\alpha [\Psi(\hat{\mathbf{T}}^{(\alpha)}) - \Psi(\hat{\mathbf{T}})]^2 \\
&\cong J^{-1} \sum_\alpha [\nabla\Psi(\hat{\mathbf{T}})'(\hat{\mathbf{T}}^{(\alpha)} - \hat{\mathbf{T}})]^2 \\
&= J^{-1} \sum_\alpha \nabla\Psi(\hat{\mathbf{T}})' \left[\sum_h N_h(\bar{\mathbf{Y}}_{hs}^{(\alpha)} - \bar{\mathbf{Y}}_{hs})\right]\left[\sum_h N_h(\bar{\mathbf{Y}}_{hs}^{(\alpha)} - \bar{\mathbf{Y}}_{hs})'\right] \nabla\Psi(\hat{\mathbf{T}}).
\end{aligned}$$

Now, define $\Delta_{yh} = \mathbf{Y}_{h1} - \mathbf{Y}_{h2}$. Then, using $\bar{\mathbf{Y}}_{hs}^{(\alpha)} - \bar{\mathbf{Y}}_{hs} = \varsigma_h^{(\alpha)} \Delta_{yh}/2$, we have

$$\begin{aligned}
v_{BHS} &\cong \nabla\Psi(\hat{\mathbf{T}})' \left(\sum_h \frac{N_h^2}{4} \Delta_{yh}\Delta_{yh}'\right) \nabla\Psi(\hat{\mathbf{T}}) \\
&= \nabla\Psi(\hat{\mathbf{T}})' \left(\sum_h \frac{N_h^2}{n_h} \mathbf{s}_h^2\right) \nabla\Psi(\hat{\mathbf{T}}) \\
&= v_{\Psi(\hat{\mathbf{T}})},
\end{aligned}$$

which was defined in (10.2.10).

Thus, by just computing $\Psi(\hat{\mathbf{T}}^{(\alpha)})$ for each half-sample and the *BHS* variance estimator among them, we get an estimator of the variance of the nonlinear Ψ without having to derive a variance formula specific to Ψ.

10.2.4. Two-Stage Sampling

We consider here a stratified version of the common mean model (9.1.5) and return to the univariate case. The population of units is divided into H strata with stratum h containing N_h clusters. Cluster (hi) contains M_{hi} units with the total number of units in stratum h being $M_h = \Sigma_{i=1}^{N_h} M_{hi}$ and the total in the population being $M = \Sigma_{h=1}^{H} M_h$. The total number of clusters in the population is $N = \Sigma_{h=1}^{H} N_h$. Associated with each unit in the population is a random variable Y_{hij} whose finite population total is $T = \Sigma_h \Sigma_{i=1}^{N_h} \Sigma_{j=1}^{M_{hi}} Y_{hij}$. The general model is

$$E_M(Y_{hij}) = \mu_h$$

$$\text{cov}_M(Y_{hij}, Y_{h'i'j'}) = 0 \quad \text{if } (hi) \neq (h'i'), \tag{10.2.12}$$

which is the same as model (9.2.6) but with a subscript for stratum added. Thus we assume that units in a stratum have a common mean and units in different clusters are uncorrelated. We put no further restrictions on the covariances between units in the same cluster.

A two-stage sample is selected from each stratum consisting of $n_h = 2$ sample clusters and a subsample of m_{hi} sample units is selected within sample cluster (hi). The total number of clusters in the sample is $n = \Sigma_h n_h$. The set of sample clusters from stratum h is denoted by s_h and the subsample of units within sample cluster (hi) by s_{hi}.

The general estimator of the total T that we will consider is a stratified version of the one studied in Chapter 9:

$$\hat{T} = \sum_h \sum_{i \in s_h} g_{hi} \bar{Y}_{hsi}, \tag{10.2.13}$$

where g_{hi} is a coefficient that does not depend on the Y's and $\bar{Y}_{hsi} = \Sigma_{j \in s_{hi}} Y_{hij} / m_{hi}$. In order for $\hat{T}$ to be model-unbiased, recall that we want $\Sigma_{i \in s_h} g_{hi} = M_h$. Each g_{hi} may also depend on the particular sample selected. For example, the stratified ratio estimator is $\hat{T}_R = \Sigma_h (M_h / M_{hs}) \Sigma_{i \in s_h} M_{hi} \bar{Y}_{hsi}$ with $M_{hs} = \Sigma_{s_h} M_{hi}$ so that $g_{hi} = (M_h / M_{hs}) M_{hi}$.

The variance estimation theory presented in chapter 9 for the unstratified situation where the number of sample clusters was large can be adapted here. Model (10.2.12) is a stratified version of model (9.1.5). A modification of Theorem 9.1.1, appropriate here, gives conditions under which the dominant term of the error-variance can be identified.

Theorem 10.2.1 If, as $H \to \infty$,

(i) $f = n/N \to 0$,

(ii) $g_{hi} = O(N_h / n_h)$,

(iii) M_{hi}, $\text{var}_M(\bar{Y}_{hsi})$, and $\text{cov}_M(\bar{Y}_{hsi}, \bar{Y}_{hi})$ are $O(1)$ for all h, i, and

(iv) $W_h = N_h / N = O(H^{-1})$,

then

$$\text{var}_M(\hat{T} - T) \approx \text{var}_M(\hat{T}) = \sum_h \sum_{i \in s_h} g_{hi}^2 \, \text{var}_M(\bar{Y}_{hsi}). \tag{10.2.14}$$

The proof is left to Exercise 10.9. □

When the number of strata H is large and the other assumptions in Theorem 10.2.1 hold, the dominant term of the prediction variance is $\text{var}_M(\hat{T})$ just as it was when the sample size of clusters was large and the sampling fraction of clusters was small in Chapter 9. The total number of sample clusters actually is large here, even though we assume $n_h = 2$ in all strata, because $H \to \infty$, and, as in Theorem 9.1.1, we assume that the overall sampling fraction of clusters n/N is negligible.

To evaluate the *BHS* variance estimator and its expectation for the two-stage case, entire clusters are assigned to half-samples. If a particular cluster is in half-sample α, then all units subsampled from that cluster are assigned to α also.

The half-sample estimator of the total is defined as

$$\hat{T}^{(\alpha)} = \sum_h (\varsigma_{h1\alpha} g_{h1}^{(\alpha)} \bar{Y}_{hs1} + \varsigma_{h2\alpha} g_{h2}^{(\alpha)} \bar{Y}_{hs2}).$$

The form of the half-sample term $g_{hi}^{(\alpha)}$ is dictated by the form of $\hat{T}$ and computed as the full sample coefficient would be if the sample size were $n_h = 1$. The α superscript is attached to $g_{hi}^{(\alpha)}$ since the value will differ from the full sample value. Although we do use a superscript α on $g_{hi}^{(\alpha)}$, its value is the same for each half-sample containing unit *hi*. For example, the stratified expansion estimator is $\hat{T}_0 = \Sigma_h (M_h/m_h) \Sigma_s m_{hi} \bar{Y}_{hsi}$ and $g_{hi} = M_h m_{hi}/m_h$ while $g_{hi}^{(\alpha)} = M_h$. We will cover this case in more detail in Example 10.2.1 below.

The difference between the half-sample and full-sample estimators is

$$\hat{T}^{(\alpha)} - \hat{T} = \sum_h \sum_{i \in s_h} (\varsigma_{hi\alpha} g_{hi}^{(\alpha)} - g_{hi}) \bar{Y}_{hsi}.$$

Using the definitions of $\varsigma_{hi\alpha}$ and $\varsigma_h^{(\alpha)}$, we have $\varsigma_{h1\alpha} = (1 + \varsigma_h^{(\alpha)})/2$ and $\varsigma_{h2\alpha} = (1 - \varsigma_h^{(\alpha)})/2$. The difference $\hat{T}^{(\alpha)} - \hat{T}$ can then be written as

$$\hat{T}^{(\alpha)} - \hat{T} = \sum_h \{(\hat{T}_h^{(\alpha)*} - \hat{T}_h) + \tfrac{1}{2} \varsigma_h^{(\alpha)} \Delta_{yh}^{(\alpha)}\},$$

where

$$\hat{T}_h^{(\alpha)*} = \tfrac{1}{2}(g_{h1}^{(\alpha)} \bar{Y}_{hs1} + g_{h2}^{(\alpha)} \bar{Y}_{hs2}),\ \hat{T}_h = \sum_{i \in s_h} g_{hi} \bar{Y}_{hsi}, \text{ and } \Delta_{yh}^{(\alpha)} = g_{h1}^{(\alpha)} \bar{Y}_{hs1} - g_{h2}^{(\alpha)} \bar{Y}_{hs2}.$$

If $\hat{T}^{(\alpha)} - \hat{T}$ is squared out and summed over half-samples, we do not, in general, obtain the kind of tidy reduction that follows expression (10.2.4) using the orthogonality of the $\varsigma_h^{(\alpha)}$'s. On the other hand, if the g_{hi}'s and $g_{hi}^{(\alpha)}$'s have a special form, the reduction does obtain. In particular, suppose that

(*HS*-1) $g_{hi} = g_h$ for all i, and

(*HS*-2) $g_{hi}^{(\alpha)} = 2g_{hi}$.

Note that if *HS*-1 holds, then *HS*-2 implies $g_{hi}^{(\alpha)} = 2g_h$. Since we are restricting ourselves to estimators that are unbiased under model (10.2.12), the condition $\Sigma_{i \in s_h} g_{hi} = M_h$ for unbiasedness implies that $g_h = M_h/2$. Conditions (*HS*-1) and (*HS*-2) restrict the class of estimators to the singleton, $\hat{T}_p = \Sigma_h M_h \Sigma_{s_h} \bar{Y}_{hsi}/n_h$.

When *HS*-1 and *HS*-2 hold, $\hat{T}_h^{(\alpha)*} = \hat{T}_h$, $\Delta_{yh}^{(\alpha)} = 2\Delta_{yh}$ where

$$\Delta_{yh} = g_h(\bar{Y}_{hsi} - \bar{Y}_{hs2})$$

and

$$\hat{T}^{(\alpha)} - \hat{T} = \sum_h \varsigma_h^{(\alpha)} \Delta_{yh}. \tag{10.2.15}$$

Squaring out (10.2.15) and summing over an orthogonal set of half-samples gives the *BHS* estimator as

$$\begin{aligned} v_{BHS} &= J^{-1} \sum_\alpha [\hat{T}^{(\alpha)} - \hat{T}]^2 \\ &= J^{-1} \sum_\alpha \left\{ \sum_h [\varsigma_h^{(\alpha)}]^2 \Delta_{yh}^2 + \sum_{h' \neq h} \varsigma_{h'}^{(\alpha)} \varsigma_h^{(\alpha)} \Delta_{yh'} \Delta_{yh} \right\} \\ &= \sum_h \Delta_{yh}^2. \end{aligned}$$

The expectation under model (10.2.12) is then easily calculated as

$$E_M(v_{BHS}) = \sum_h g_h^2 \sum_{i \in s_h} \mathrm{var}_M(\bar{Y}_{hsi}), \tag{10.2.16}$$

which is the asymptotic variance in Theorem 10.2.1 for $g_{hi} = g_h$.

Example 10.2.1 Expansion estimator: $\hat{T}_0 = (M/m) \Sigma_s m_i \bar{Y}_{si}$. For the full sample $g_{hi} = (M_h/m_h) m_{hi}$. When the (*hi*)th sample cluster is assigned to half-sample α, the number of sample units in the half-sample, $m_h^{(\alpha)}$, is equal to the number in the (*hi*)th cluster, m_{hi}. Thus, $g_{hi}^{(\alpha)} = M_h$ and neither *HS*-1 nor *HS*-2 holds. If $m_{hi} = \bar{m}_h$, an allocation that equalizes workload per cluster, then both *HS*-1 and *HS*-2 hold, and $\hat{T}_0$ reduces to $\hat{T}_p = \Sigma_h M_h \Sigma_{s_h} \bar{Y}_{hsi}/n_h$.

Example 10.2.2 Ratio estimator: $\hat{T}_R = \Sigma_h (M_h/M_{hs}) \Sigma_{i \in s_h} M_{hi} \bar{Y}_{hsi}$. $g_{hi} = (M_h/M_{hs}) M_{hi}$ and $g_{hi}^{(\alpha)} = M_h$. Again, neither *HS*-1 nor *HS*-2 holds. If the clusters are all the same size in a stratum, that is, $M_{hi} = \bar{M}_h$, then $g_{hi} = M_h/2$ and both conditions hold, but in that case $\hat{T}_R = \hat{T}_p$.

As the general discussion and the examples show, the *BHS* variance estimator is more limited than the jackknife in the forms of estimators it can accommodate when model (10.2.12) holds. In fact, the one estimator that satisfies *HS*-1 and *HS*-2, that is, the one having $g_h = M_h/2$, is $\hat{T}_p = \Sigma_h (M_h/2) \Sigma_{s_h} \bar{Y}_{hsi}$. Using arguments presented in section 10.2.3, results for two-stage sampling can be extended to nonlinear combinations if different components are estimated by $\hat{T}_p$.

There are a number of ways of applying *BHS* in practice for which model-based theory has yet to be developed. One common method, when the first-stage units have been selected by systematic sampling, is the following: Sort the first-stage units in the same order as the frame, and then pair adjacent sample units on the sorted list. The sorting characteristics may or may not be related to factors that should be included in a working model. If a regression estimator like the one studied in Section 9.4 is used, an open question is whether the *BHS* variance estimator is consistent. As we saw above, whether the *BHS* variance estimator works depends, in part, on how the coefficients, $g_{hi}^{(\alpha)}$, are constructed.

10.2.5 Other Forms of the BHS Variance Estimator

Two other forms of the *BHS* variance estimator are sometimes used that employ complementary half-samples. The complement of half-sample α is the set of units that are not in that half-sample. When $n_h = 2$, the complement is also obviously a half-sample. The alternative forms are

$$v_{BHS-2}(\hat{\theta}) = \sum_{\alpha=1}^{J} (\hat{\theta}^{(\alpha)} - \bar{\hat{\theta}}^{(\alpha)})^2/J$$

$$v_{BHS-3}(\hat{\theta}) = \sum_{\alpha=1}^{J} (\hat{\theta}^{(\alpha)} - \hat{\theta}_c^{(\alpha)})^2/(4J),$$

$$v_{BHS-4}(\hat{\theta}) = \sum_{\alpha=1}^{J} \{(\hat{\theta}^{(\alpha)} - \hat{\theta})^2 + (\hat{\theta}_c^{(\alpha)} - \hat{\theta})^2\}/(2J),$$

where $\bar{\hat{\theta}}^{(\alpha)} = \Sigma_\alpha \hat{\theta}^{(\alpha)}/J$ and $\hat{\theta}_c^{(\alpha)}$ is the estimator based on the complement of half-sample α. It is an easy exercise to show that, in a set of half-samples in full orthogonal balance, v_{BHS-2}, v_{BHS-3}, and v_{BHS-4} all reduce to (10.2.5) for the stratified expansion estimator in one-stage sampling. The estimators v_{BHS-2} and v_{BHS-3} are sometimes used for nonlinear estimators like those studied in Section 10.2.3. When the number of strata H is large all four *BHS* estimators are asymptotically equivalent so that there is little reason to choose one over another.

An alternative method of variance estimation using balanced half-samples is due to Fay (see Dippo, Fay, and Morganstein, 1984). Rather than eliminating the data for half the sample units, Fay's method uses the data from all sample units but modifies the weights. In the two-stage case considered in Section 10.2.4, the modified, half-sample estimator of the total is

$$\hat{T}^{(\alpha)} = \sum_h [(1 + \varsigma_h^{(\alpha)}\varepsilon) g_{h1} \bar{Y}_{hs1} + (1 - \varsigma_h^{(\alpha)}\varepsilon) g_{h2} \bar{Y}_{hs2}],$$

where $0 < \varepsilon \leqslant 1$. Thus, if unit $(h1)$ is in half-sample α, the weight for that unit is $g_{h1}^{(\alpha)} = (1 + \varsigma_h^{(\alpha)}\varepsilon) g_{h1} = (1 + \varepsilon) g_{h1}$ while the weight for unit $(h2)$ is $g_{h2}^{(\alpha)} = (1 - \varsigma_h^{(\alpha)}\varepsilon) g_{h2} = (1 - \varepsilon) g_{h2}$. If $\varepsilon = 1$, and unit $(h1)$ is in half-sample α, then $g_{h1}^{(\alpha)} = 2g_{h1}$, as in $(HS\text{-}1)$. A modified variance estimator is

$$v_{BHS-F}(\hat{T}) = \frac{1}{J\varepsilon^2} \sum_{\alpha=1}^{J} (\hat{T}^{(\alpha)} - \hat{T})^2.$$

The divisor ε^2 is needed to obtain unbiasedness under the simple model (10.2.3) when $\hat{T}$ is the stratified expansion estimator. When $\varepsilon = 1$, the standard BHS variance estimator is obtained.

The Fay estimator is particularly useful when estimating the variance for domains since it uses all sample units in every half-sample. The standard BHS method, on the other hand, may result in few if any units from a small domain being retained in some half-samples. Judkins (1990) presented a variety of simulation results for the Fay estimator, showing that, in some cases, it could be much more stable than the standard BHS estimator. Rao and Shao (1999) showed that it is design-consistent for the variances of smooth, nonlinear functions and of quantiles. Although no model-based analyses of the Fay estimator have been published, it appears that $(HS\text{-}1)$ and $(HS\text{-}2)$ will again be necessary if v_{BHS-F} is to be approximately model-unbiased.

10.2.6 Partially Balanced Half-Sampling

Partial balancing is a method that is sometimes used to reduce the number of half-sample estimates that must be computed for v_{BHS}. Though computationally expedient, partial balancing can lead to a variance estimator that is inconsistent, as will be demonstrated in this section. Suppose that we have a two-stage sample with $n_h = 2$ but that strata are assigned to groups or superstrata. An attempt may be made to assign the same number of strata to each group, but this is not essential. In a particular group all the sample clusters numbered 1 are associated and assigned as a block to a half-sample. Sample clusters numbered 2 are similarly treated as a block. Figure 10.1 illustrates the grouping of strata and treatment of clusters as blocks.

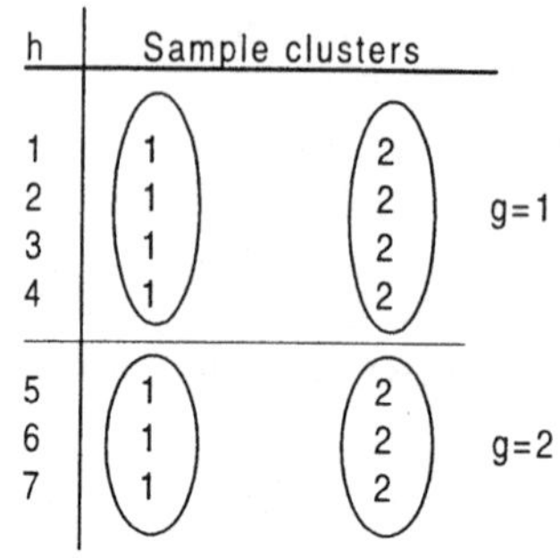

Figure 10.1 An example of grouping strata and treating sample clusters as blocks when partial balancing is used. Circled units are assigned as a block to a half-sample.

If there are $g = 1, \ldots, G$ groups of strata, then the estimator of the total can be written as

$$\hat{T} = \sum_{g=1}^{G} (\hat{T}_{g1} + \hat{T}_{g2}),$$

where $\hat{T}_{gi} = \Sigma_{h \in S_g} g_{hi} \bar{Y}_{hsi}$, $i = 1, 2$ with S_g being the set of strata in group g. The estimator of the total based on half-sample α is

$$\hat{T}^{(\alpha)} = \sum_{g=1}^{G} (\varsigma_{g1\alpha} \hat{T}_{g1}^{(\alpha)} + \varsigma_{g2\alpha} \hat{T}_{g2}^{(\alpha)}),$$

where $\varsigma_{gi\alpha} = 1$ if the units numbered i in group g are in the half-sample and 0 if not, and $\hat{T}_{gi}^{(\alpha)} = \Sigma_{h \in S_g} g_{hi}^{(\alpha)} \bar{Y}_{hsi}$ with $g_{hi}^{(\alpha)}$ computed as it would be for the fully balanced case.

The difference between the grouped half-sample estimator and the full sample estimator is

$$\hat{T}^{(\alpha)} - \hat{T} = \sum_{g=1}^{G} (\varsigma_{g1\alpha} \hat{T}_{g1}^{(\alpha)} - \hat{T}_{g1} + \varsigma_{g2\alpha} \hat{T}_{g2}^{(\alpha)} - \hat{T}_{g2}). \qquad (10.2.17)$$

If $g_{hi}^{(\alpha)} = 2g_{hi}$, that is, *HS*-2 holds, then $\hat{T}_{gi}^{(\alpha)} = 2\hat{T}_{gi}$ and

$$\hat{T}^{(\alpha)} - \hat{T} = \sum_{g} \varsigma_g^{(\alpha)} (\hat{T}_{g1} - \hat{T}_{g2}),$$

where $\varsigma_g^{(\alpha)} = 2\varsigma_{g1\alpha} - 1 = -(2\varsigma_{g2\alpha} - 1)$. With balancing on groups, the grouped *BHS* estimator is

$$v_{GB} = \sum_{g} (\hat{T}_{g1} - \hat{T}_{g2})^2.$$

The expectation of v_{GB}, when HS-1 and HS-2 hold, is easily calculated as

$$E_M(v_{GB}) = \sum_g \sum_{h \in S_g} [g_{h1}^2 v_{h1} + g_{h2}^2 v_{h2}], \qquad (10.2.18)$$

where $v_{hi} = \text{var}_M(\bar{Y}_{hsi})$. This compares to (10.2.16) for the ungrouped case. The grouped BHS estimator is, thus, asymptotically model unbiased.

Even if HS-1 and HS-2 are satisfied, v_{GB} may perform erratically when the number of groups G is not large. Krewski (1978), in a related case, noted the large design-variance of a grouped BHS estimator compared to the standard variance estimator in stratified simple random sampling when the stratified expansion estimator is used. Rao and Shao (1993) proposed a repeatedly grouped balanced half-sample ($RGBHS$) procedure that might be adapted to the partially balanced case. The $RGBHS$ method applies to a case where a large number of units are selected within a stratum and then assigned at random to two groups for variance estimation.

If, as $H \to \infty$, G is fixed, then v_{GB} can be inconsistent in addition to being unstable. An argument similar to one given by Rao and Shao (1993) and Shao (1994) for stratified single-stage sampling will demonstrate this. Let η_g denote the number of strata assigned to group g and suppose that $\min_g(\eta_g) \to \infty$. Under standard conditions,

$$z_g = (\hat{T}_{g1} - \hat{T}_{g2})/\sqrt{D_g} \xrightarrow{d} N(0, 1),$$

where $D_g = \text{var}_M(\hat{T}_{g1} - \hat{T}_{g2}) = \Sigma_{h \in S_g} g_h^2(v_{h1} + v_{h2})$. Since $\text{var}_M(\hat{T} - T) \cong \Sigma_g D_g$,

$$\frac{v_{GB}}{\text{var}_M(\hat{T} - T)} \cong \sum_g \frac{D_g}{\Sigma_{g'} D_{g'}} z_g^2.$$

If $D_g/\Sigma_{g'} D_{g'}$ converges to a constant ω_g, it follows that

$$\frac{v_{GB}}{\text{var}_M(\hat{T} - T)} \to \sum_g \omega_g \chi_g^2, \qquad (10.2.19)$$

where χ_g^2 is a central chi-square random variable with 1 degree of freedom. In other words, rather than converging to 1 as would be the case for a consistent variance estimator, the ratio in (10.2.19) converges to a weighted sum of chi-square random variables. Note that a result similar to (10.2.19) can be obtained if $\eta_g \to \infty$ in only some of the groups. The inconsistency of v_{GB} can manifest itself by $\text{var}_M(v_{GB})$ being large and by the length of confidence intervals being excessively variable. Valliant (1987b, 1996) illustrated some of these problems numerically.

The method of partial balancing may be appealing in some cases because it reduces the number of replicate estimates to compute, but, generally, is a procedure with little to recommend it theoretically unless the number of groups, G, is large.

10.2.7. Design-Based Properties

With some sample designs v_{BHS} may have desirable design-based properties when HS-2 holds, that is, $g_{hi}^{(\alpha)} = 2g_{hi}$, despite having a conditional (model) bias. Define π_{hi} to be the selection probability of unit hi in a sample of $n_h = 2$. If $g_{hi} = M_{hi}/\pi_{hi}$, HS-2 is satisfied when $g_{hi}^{(\alpha)}$ is computed by substituting $\pi'_{hi} = \pi_{hi}/2$ for π_{hi}. In that case, $v_{BHS} = \Sigma_{h,s_h}[M_{hi}\bar{Y}_{hsi}/\pi_{hi} - \hat{T}_h]^2/[n_h(n_h - 1)]$ and v_{BHS} is design-unbiased under with-replacement sampling when $\hat{T}_h$ is design-unbiased. When $g_{hi} = M_{hi}/\pi_{hi}$ and the estimator is a differentiable function of totals defined by (10.2.13), Krewski and Rao (1981) showed that v_{BHS} is design-consistent as $H \to \infty$ and the sampling of clusters is done with replacement. Condition HS-1 is not required for these results. When averaged over the design distribution, the model-bias of $\hat{T}$ turns into a design variance component—a phenomenon that we noted earlier in Chapter 4.

10.3 GENERALIZED VARIANCE FUNCTIONS

In many large scale surveys, particularly ones sponsored by governments, hundreds or even thousands of estimates may be published. Although calculation and publication of a separate estimate of variance might be possible, there are practical reasons why alternative techniques are desirable. By averaging over time or smoothing in some way, variance estimates more stable than point variance estimates can be produced. Presentation of individual sampling errors would double the number of tables in printed reports, increasing computing, printing, and associated personnel costs. Public-use data sets may be available from which many other estimates can be calculated that do not appear in publications. Survey sponsors may, thus, want to provide relatively simple methods of variance estimation that users can implement themselves. Hanson (1978) and Wolter (1985) discuss these reasons in more detail.

These considerations have led to the use of models, sometimes called *generalized variance functions* (GVF's), as a means of approximating standard errors. The relvariance of an estimator $\hat{T}$ of a total T is defined as $\text{var}_M(\hat{T} - T)/[E(T)]^2$. The relvariance can sometimes be written as a function of the expectation of T, for example, $\text{relvar}_M(\hat{T} - T) \approx a + b/E_M(T)$. Given a set of relvariances estimated by direct formulas, as in Chapter 5, by resampling as in the jackknife or balanced half-sampling, or by some other method, parameters of the relvariance model are estimated. The parameter estimates and the estimated total itself are then used to compute a predicted relvariance. Two major surveys in the United States that have used GVF's are the Current Population Survey and the National Health Interview Survey.

10.3.1 Some Theory for *GVF*'s

Though the *GVF* method was initially adopted as an *ad hoc*, practical approach to variance estimation, the method does have a theoretical basis. Consider the simple Bernoulli model for independent random variables Y_i: $E_M(Y_i) = p$ with $\text{var}_M(Y_i) = p(1-p)$. The *BLU* predictor of the total is $\hat{T}_0 = N\bar{Y}_s$. If a sample of size n is selected and the sampling fraction is small, the relvariance is

$$\text{relvar}_M(\hat{T}_0 - T) \approx [E_M(T)]^{-2}\frac{N^2}{n}p(1-p)$$

$$= -\frac{1}{n} + \frac{N}{n}\frac{1}{Np}.$$

Since $E_M(T) = Np$, this relvariance has the form $a + b/E_M(T)$ with $a = -1/n$ and $b = N/n$. In more complicated situations the values of a and b will not be so simple.

Next, consider a more elaborate stratified clustered model for Y_{hij}, a random variable observed for unit j in cluster i within stratum h:

$$E_M(Y_{hij}) = \mu_h$$

$$\text{cov}_M(Y_{hij}, Y_{h'i'j'}) = \begin{cases} \sigma_{hi}^2 & h = h',\ i = i',\ j = j' \\ \sigma_{hi}^2\rho_{hi} & h = h',\ i = i',\ j \neq j' \\ 0 & \text{otherwise.} \end{cases} \qquad (10.3.1)$$

The general class of estimators we consider is again

$$\hat{T} = \sum_h \sum_{s_h} g_{hi}\bar{Y}_{hsi}. \qquad (10.3.2)$$

To study *GVF*'s, we will look at asymptotic properties of the relvariance of $\hat{T}$. The two large sample cases are

1. H fixed and $n_h \to \infty$, and
2. n_h bounded and $H \to \infty$.

In either case, $\text{var}_M(\hat{T} - T) \approx \text{var}_M(\hat{T})$ under appropriate circumstances. Case (1) was covered by Theorem 9.1.1, while case (2) was addressed by Theorem 10.2.1. For both the cases of a large cluster sample within each stratum and a large number of strata we have

$$\text{var}_M(\hat{T} - T) \approx \sum_h \sum_{i \in s_h} g_{hi}^2 v_{hi}/m_{hi},$$

where $v_{hi} = \text{var}_M(\bar{Y}_{hsi})$ and under model (10.3.1)

$$\text{relvar}_M(\hat{T} - T) \cong [E(T)]^{-2}\left[\sum_h \sum_{i \in s_h} g_{hi}^2 \sigma_{hi}^2 [1 + (m_{hi} - 1)\rho_{hi}]/m_{hi}\right]. \tag{10.3.3}$$

When the variance parameter σ_{hi}^2 is a function of the mean μ_h, the approximate relvariance may simplify. In particular, suppose that σ_{hi}^2 is a quadratic function of μ_h:

$$\sigma_{hi}^2 \equiv \sigma_h^2 = \alpha_{1h}\mu_h + \alpha_{2h}\mu_h^2 \tag{10.3.4}$$

with α_{1h} and α_{2h} being constants. Substituting (10.3.4) into (10.3.3) and rearranging gives

$$\begin{aligned} \text{relvar}_M(\hat{T} - T) &\cong \sum_h \pi_h^2 \frac{\alpha_{2h} d_{sh}}{M_h^2} + \left[\sum_h \pi_h \frac{\alpha_{1h} d_{sh}}{M_h}\right] \Big/ E_M(T) \\ &\equiv a + b/E_M(T), \end{aligned} \tag{10.3.5}$$

where $d_{sh} = \Sigma_{i \in s_h} g_{hi}^2 [1 + (m_{hi} - 1)\rho_{hi}]/m_{hi}$, $\pi_h = E_M(T_h)/E_M(T)$, $T_h = \Sigma_{i=1}^{N_h} \Sigma_{j=1}^{M_{hi}} Y_{hij}$, and a and b are defined by the last equality.

An interesting special case is $\hat{T}_p = \Sigma_h (M_h/n_h) \Sigma_{i \in s_h} \bar{Y}_{hsi}$ with $m_{hi} = \bar{m}_h$ and $\rho_{hi} = \rho_h$. In that instance, the coefficients a and b in (10.3.5) reduce to

$$a = \sum_h \pi_h^2 \alpha_{2h} [1 + (\bar{m}_h - 1)\rho_h]/(n_h \bar{m}_h)$$

and

$$b = \sum_h \pi_h \alpha_{1h} M_h [1 + (\bar{m}_h - 1)\rho_h]/(n_h \bar{m}_h). \tag{10.3.6}$$

A few common distributions have the mean-to-variance relationship in (10.3.4). Four examples are:

Example 10.3.1 Binary. $\sigma_h^2 = \mu_h(1 - \mu_h)$ implying that $\alpha_{1h} = 1$ and $\alpha_{2h} = -1$.

Example 10.3.2 Poisson. $\sigma_h^2 = \mu_h$; $\alpha_{1h} = 1$ and $\alpha_{2h} = 0$.

Example 10.3.3 Exponential. $\sigma_h^2 = \mu_h^2$; $\alpha_{1h} = 0$ and $\alpha_{2h} = 1$.

Example 10.3.4. Gamma. $\mu_h = \beta_h/\lambda_h$ for some parameters β_h, λ_h; $\sigma_h^2 = \beta_h/\lambda_h^2 = \mu_h/\lambda_h$, so that $\alpha_{1h} = 1/\lambda_h$ and $\alpha_{2h} = 0$.

Any family of distributions where the variance is a function of the mean will allow $\text{var}_M(\hat{T})$ to be expressed in terms of the stratum means μ_h, although the simple form of the relvariance in (10.3.5) may not obtain. The most frequent use of *GVF*'s is probably for estimators of totals of binary variables—for example, the number of persons unemployed in the last month, the number of persons with arthritis, and the number of families living in poverty.

The model (10.3.1) itself is quite restrictive since it assumes that the mean within stratum h is the same for each unit. Thus, a mean of the form $E_M(Y_{hij}) = \mathbf{x}'_{hij}\boldsymbol{\beta}_h$ is not included unless the explanatory variables are qualitative so that a type of post-stratification model is implied. A more general model having qualitative and quantitative x's will not lead to the form of relvariance in (10.3.5). Note, however, that the strata in model (10.3.1) are not necessarily ones used for sample design. They can be narrowly defined classes of units for which a common mean and variance may be reasonable assumptions.

For a given survey the parameters a and b in (10.3.5) can be estimated by least squares. One of the practical problems involved is selecting a set of estimators of totals to use in the estimation of a and b. The general requirement, as in any regression problem, is that each of the $\hat{T}$'s share the same values of a and b. Examination of the components of a and b permit some rules to be established for variable selection in specific cases. If the estimator is $\hat{T}_p = \Sigma_h(M_h/n_h)\Sigma_{i\in s_h}\bar{Y}_{hsi}$, for example, some jointly sufficient, though not necessary conditions are:

1. Each variable Y_v $(v = 1,\ldots,V)$ follows model (10.3.1) with a common within-stratum variance of the form $\alpha_{1h}\mu_{vh} + \alpha_{2h}\mu_{vh}^2$. In particular, each variable has the same set of coefficients $\{\alpha_{1h}, \alpha_{2h}: h = 1,\ldots,H\}$.
2. $\pi_{vh} = \pi_h$ for all v.
3. $\rho_{vh} = \rho_h$ for all v.
4. The sets of sample sizes $\{n_h, \bar{m}_h: h = 1,\ldots,H\}$ are the same for all v.

Condition (1) is obviously satisfied if α_{1h} and α_{2h} are known constants as in Examples 10.3.1–10.3.3. If the variables are gamma as in Example 10.3.4, they must all have the same λ_h parameters, a considerable restriction. Condition (3) tends to hold for domains that are spread fairly evenly across clusters (see, e.g., Kalton and Blunden, 1973; Kalton, 1977 for some discussion of this from the viewpoint of probability sampling). Both conditions (2) and (3) are likely to be violated for rare domains that are unevenly scattered across clusters. For example, in the United States members of the Hispanic ethnic group are more likely to live in certain large metropolitan areas and in the southwestern part of the country than in other areas. Condition (4) on sample sizes is relevant if variables from several years of a continuing survey are combined.

10.3.2 Estimation of *GVF* Parameters

Because (10.3.5) is simply a regression model that is linear in $1/E_M(T)$, standard methods can be used to estimate a and b. Consider data pairs of the form $(\upsilon_v, \hat{T}_v)$, where for $v = 1,\ldots,V$, $\hat{T}_v$ is the estimator of the total for variable v and υ_v is a consistent estimator of the relvariance of $\hat{T}_v$. One simple approach is to use $\hat{T}_v$ in place of the unknown $E_M(T_v)$ and apply weighted least squares.

The weighted least squares estimators of a and b are

$$\hat{b} = \frac{\Sigma_{v=1}^{V} w_v \upsilon_v (\hat{T}_v^{-1} - \bar{\hat{T}}^{(-1)})}{\Sigma_{v=1}^{V} w_v (\hat{T}_v^{-1} - \bar{\hat{T}}^{(-1)})^2} \quad \text{and} \quad \hat{a} = \bar{\upsilon} - \hat{b}\bar{\hat{T}}^{(-1)},$$

where w_v is a regression weight associated with variable v, $\bar{\hat{T}}^{(-1)} = \Sigma_v w_v \hat{T}_v^{-1} / \Sigma_v w_v$, and $\bar{\upsilon} = \Sigma_v \upsilon_v w_v / \Sigma_v w_v$. The predicted relvariance of $\hat{T}_v$ based on the GVF is

$$\hat{\upsilon}_v = \hat{a} + \hat{b}/\hat{T}_v = \bar{\upsilon} + \hat{b}(\hat{T}_v^{-1} - \bar{\hat{T}}^{(-1)}).$$

If consistent estimators of $\text{var}_M(\hat{T}_v - T_v)$ and $E(\hat{T}_v)$, and hence $\text{relvar}_M(\hat{T}_v - T_v)$ are used, then $\hat{\upsilon}_v$ will be consistent also (see, e.g., Valliant, 1987c). The consistency does not require that $V \to \infty$ as in the usual proof that a regression estimator is consistent. Rather, the result follows mainly from the convergence of the individual relvariance estimators. Any of the consistent variance estimators we have studied in earlier chapters can be used.

The estimators of a and b above are not optimal. For one thing, the relvariance estimators $\hat{\upsilon}_v$ are typically correlated, being based on different Y's for the same set of units. Generalized least squares (GLS) with a full convariance matrix would then be optimal. Estimates of covariances among relvariance estimators are, however, usually difficult to compute and may themselves be unstable. GLS is thus rarely, if ever, used. The weights w_v are typically assigned in a pragmatic way designed to reflect the relative precision of the relvariance estimators. Using the reasoning that larger estimated relvariances themselves have larger variances, a common choice is $w_v = \upsilon_v^{-2}$. Valliant (1987c) reports a simulation study showing that the use of weighted least squares makes an important difference in the precision of GVF's compared to unweighted least squares. That study also confirms that GVF's can be considerably more stable than the point variance estimators.

Estimating a GVF can also be treated as an errors-in-variables problem since the totals $\hat{T}_v$ are sample estimates. A more satisfactory approach to estimation might then be to use measurement error model techniques described in Fuller (1987), though this approach appears not to have been applied in practice.

EXERCISES

10.1 Determine conditions under which the dominant term of the error variance

$$\text{var}_M(\hat{D} - D) = \sum_s g_i^2 \,\text{var}_M(\bar{D}_{si}) - 2\sum_s g_i M_i \,\text{cov}_M(\bar{D}_{si}, D_i) + \sum_{i=1}^{N} M_i^2 \,\text{var}_M(D_i)$$

given in (10.1.8), is the first term.

10.2 Show that the sandwich estimator, $v_R(\hat{D}) = \Sigma_s g_i^2 r_{Di}^2$, defined in (10.1.9), is approximately unbiased under the working model (10.1.6) for $\text{var}_M(\hat{D}) = \Sigma_s g_i^2 \,\text{var}_M(\bar{D}_{si})$ and is also approximately unbiased under a more general model where the covariance structure within clusters is unspecified:

$$E_M(Y_{ij}, Z_{ij}) = (\mu_y, \mu_z)$$

$$\text{cov}_M(Y_{ij}, Y_{i'j'}) = \text{cov}_M(Y_{ij}, Z_{i'j'}) = \text{cov}_M(Z_{ij}, Z_{i'j'}) = 0 \text{ for } i \neq i'.$$

10.3 **(a)** Show that

$$v_{OD} = \sum_s g_i^2 r_{Di}^2 - \sum_s g_i M_i \dot{v}_{Di} + \left[\sum_{i=1}^{N} M_i^2 - \sum_s g_i M_i \right] \dot{\theta}_{D2}$$

defined in (10.1.10) is an approximately unbiased estimator of $\text{var}_M(\hat{D} - D)$ under the working model (10.1.6).

(b) Consider the residual $r_{Di}^* = (\bar{Y}_{si} - \hat{\mu}_y) - \varphi(\bar{Z}_{si} - \hat{\mu}_z)$ that assumes the value of $\varphi = \mu_y/\mu_z$ is known. Derive an adjustment to the squared residual r_{Di}^{*2} such that the adjusted squared residual is an exactly unbiased estimator, under model (10.1.6), of $\text{var}_M(\bar{D}_{si})$ where $\bar{D}_{si} = \bar{Y}_{si} - \varphi\bar{Z}_{si}$. (*Hint:* Use the methods in Section 9.2.3).

10.4 Show that for the stratified expansion estimator, $\hat{T}_0 = \Sigma_h N_h \bar{Y}_{hs}$, the balanced half-sample estimators v_{BHS}, v_{BHS-2}, v_{BHS-3}, and v_{BHS-4} all reduce to $\Sigma_h N_h^2 s_h^2/n_h$. Further, show that if the half-sample totals are computed using $Y_{hi}^* = \sqrt{1 - f_h}\, Y_{hi}$ rather than Y_{hi}, then all three variance estimators are equal to $\Sigma_s N_h^2(1 - f_h) s_h^2/n_h$.

10.5 Show that $v_{BRR-2} \leqslant v_{BRR}$ for any estimator $\hat{\theta}$.

10.6 Does the balanced half-sample method apply to $\hat{T}_{BLU}$ under the model

$$E_M(Y_{hij}) = \mathbf{x}_{hij}'\boldsymbol{\beta}_h, \ \text{var}_M(Y_{hij}) = v_{hij} \text{ when } n_h = 2\text{? Why or why not?}$$

10.7 Suppose that a stratified sample of small business establishments is selected from a population with $H = 6$ and for each stratum $n_h = 2$ and $N_h = 50$. The sample data for the number of employees Y in each establishment are:

h	1	2	3	4	5	6
$i = 1$	13	12	9	6	9	5
$i = 2$	5	15	11	14	20	9

Compute the stratified expansion estimate $\hat{T}_{y0} = \Sigma_h N_h \bar{y}_{hs}$, v_{BHS}, and v_0. Use columns 2–7 of the 8 × 8 Hadamard matrix in Section 10.2.2. List each of the half-sample estimates of $\hat{T}_{y0}$.

10.8 In the sample of businesses in the previous problem, suppose that the following numbers of employees z were found to be over 50 years of age:

h	1	2	3	4	5	6
$i = 1$	3	12	9	3	9	5
$i = 2$	3	2	8	4	4	4

Using stratified expansion estimators, $\hat{T}_{y0}$ and $\hat{T}_{z0}$, estimate the proportion p of employees in the population who are over 50 years of age as the estimated total over 50 divided by the estimated grand total of employees. Using the BHS method, estimate the variance of $\hat{p} = \hat{T}_{y0}/\hat{T}_{z0}$ as $v_{BHS} = \Sigma_\alpha[\hat{p}^{(\alpha)} - \hat{p}]^2/J$ with $\hat{p}^{(\alpha)}$ being the proportion based on half-sample α.

10.9 Prove Theorem 10.2.1.

CHAPTER 11

Special Topics and Open Questions

Estimation in finite population sampling includes a number of specialized topics that we review in this chapter. These subjects are not as much a part of the mainstream of survey sampling as the ones in earlier chapters but are fertile areas for research. Several of the topics covered here also integrate research that is ongoing in the rest of statistics into the study of finite populations.

Section 11.1 reviews some of the methods that have been proposed for handling outliers when estimating means and totals. Practitioners have long recognized that unusual data points may foul the properties of estimates and have applied various ad hoc remedies. The first section describes how some of the more formal outlier-robust procedures can be applied in finite population estimation.

Earlier chapters dealt mainly with linear models, but many variables, particularly binary ones, are usually better described by nonlinear models like the logistic. Section 11.2 discusses the estimation of totals of categorical variables under such models. Nonparametric estimation has seen rapid advances in the rest of statistics in recent years, but relatively little has been done to apply these methods to survey sampling. Section 11.3 describes the limited amount of research that has been done in this area and suggests some directions for future work.

An area of substantial importance is the estimation of cumulative distribution functions (CDFs) and quantiles. Standard estimators can have extremely bad conditional biases if useful auxiliary information is available but is not used. Section 11.4 covers estimation of CDFs and quantiles when the response variable Y follows a regression model and also describes the techniques that are appropriate when simpler population structures hold.

The last section of this chapter gives a brief introduction to small area estimation. This topic has been the subject of a large amount of research due to the common desire to wring as much information from a given data set as possible. This section can serve as an elementary introduction to the large and often difficult literature on small area estimation.

11.1 ESTIMATION IN THE PRESENCE OF OUTLIERS

Guarding against model failure has been one of our major themes. In this section we examine the situation—common in the statistical literature on robustness outside of sampling—where the working model holds for most of the population but a small percentage of units are "contaminated" by following a model whose mean or variance is far removed from that of the core model. These contaminated units are known as *outliers*. When trying to make inferences about model parameters, it has long been recognized that the performance of standard estimators, like the sample mean, is highly vulnerable to outliers resulting from contaminated distributions (see, e.g., Tukey, 1960).

Survey samples often contain outliers—points whose values seem to have little in common with the rest of the sample. Economic data on expenditures, revenues, or prices can be extremely skewed and a few sample observations may be far larger than the others. The possibility that a handful of unusual points can have a substantial effect on an estimated slope is illustrated in Figure 11.1. Borrowing an example from Rousseeuw (1984), we generated 45 observations that followed the model $Y_i = 2x_i + \varepsilon_i\sqrt{x_i}$ with x_i uniformly distributed between 1 and 7 and $\varepsilon_i \sim N(0, 1)$. Those points are shown as dots in the figure. Five additional points were distributed as bivariate normal with mean $(\mu_x, \mu_y) = (5, 25)$, standard deviation $(\sigma_x, \sigma_y) = (1, 4)$, and correlation 0. Those points are the circles in Figure 11.1. The solid line is the fitted

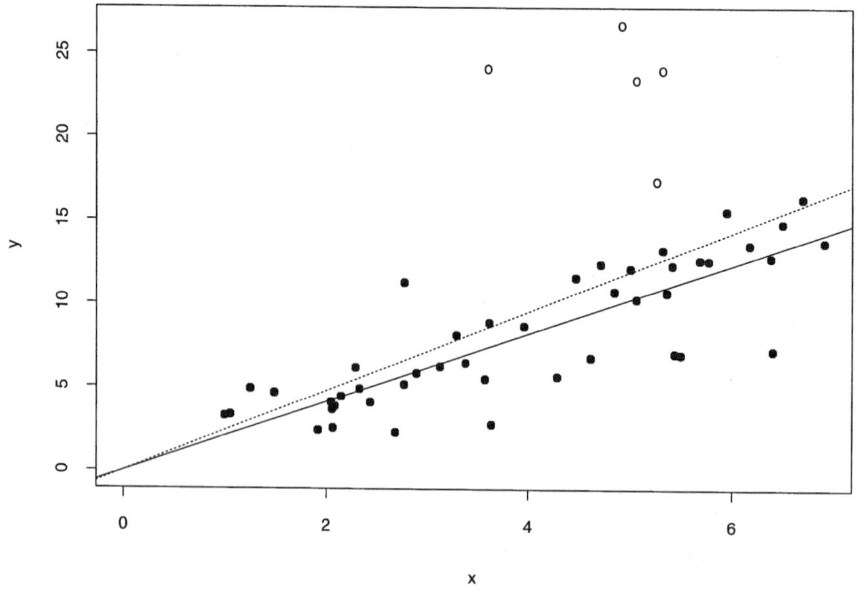

Figure 11.1 An illustration of the effect of outliers on an estimate of slope.

straight-line for the model $Y_i = \beta x_i + \varepsilon_i \sqrt{x_i}$, $\varepsilon_i \sim (0, \sigma^2)$ using only the 45 dot points, which has $\hat{\beta} = 2.05$. The dashed line is the fitted line from all 50 points, which produces $\hat{\beta} = 2.38$—about 16% higher than the estimate based on 45 points.

Clearly, the five unusual points, which are 10% of the sample, have an effect on the estimate of slope. If, for example, this slope is then used in a ratio estimate for the population total, there will be a major difference in the value of $\hat{T}_R$, depending on how the five outliers are treated. How those points should be handled depends on whether there are other points like them (and in like proportions) in the nonsample portion of the population—something that may not be known with any assurance.

There may be many reasons why specific units are different from the majority, but a general classification of outliers given by Chambers (1986) is useful:

- Nonrepresentative
  - The value for a sample unit is incorrect (e.g., coding error), or
  - The value is unique to a particular population unit.
- Representative
  - The value associated with a sample unit is correctly recorded.
  - The value is not unique (there may be others like it in the nonsample part of the population).

Nonrepresentative values that are simply the result of coding or other data processing errors should be resolved as part of data editing. An extreme value that is correct may also be a one-of-a-kind occurrence that should be flagged as such for special handling during estimation. In a survey of establishments in a metropolitan area, for example, a single large manufacturer may have 5,000 employees. No other establishment may approach its size, and its data may not be useful for predicting any totals for the nonsample. Deciding whether such a unit is unique may require nonstatistical investigation, but this extra work is part of good survey practice.

Representative outliers are the more interesting and difficult ones to deal with statistically. If the finite population quantity of interest is the total of all the population elements, including possible outliers in the nonsample, the sample outliers have some information relevant to inference. The question is how much weight or "influence" to allow them to have when constructing an estimate. Survey practitioners routinely employ *ad hoc* methods, like assigning a weight of 1 to large units, to combat outliers, but the theory for the outlier problem is considerably less well developed than for other areas we have studied. Chambers and Kokic (1993) and Lee (1995) survey much of the literature for outlier-robust finite population estimation. Material in this section is derived largely from those reviews and from Chambers (1986).

Outliers are typically thought of as having large data values, but with the linear estimators that we have studied most often in this book, influential units can occur in another way also. As introduced in Chapter 2, the *BLU* predictor of a total under a model and many other estimators can be written as weighted sum of the target Y variable:

$$\hat{T} = \sum_{i \in s} w_i Y_i.$$

An outlier could be a point with a large Y value, a large weight w_i, or both. Whether a particular point is an outlier or not also depends on whether an estimate is for the full population or a domain. A particular product $w_i Y_i$ may not be particularly large compared to the sample sum at high levels of aggregation but may be the major part of an estimate for a subgroup.

When the estimator is linear and outliers are felt to be a problem, two traditional approaches have been suggested for guarding against unusual points:

1. *Modify weights:* Reduce the values of w_i associated with the sample outliers but leave the Y values untouched; and
2. *Modify values:* Change the Y's for the sample outliers, leaving the weights unchanged.

The sampling literature on these approaches is limited even though both are reasonably common in survey practice. Hidiroglou and Srinath (1981) suggest post-stratifying the outliers into a special stratum with lower weights than they might normally have. Kish (1965) introduced a "surprise stratum" strategy that would apply to a continuing survey. An outlier would be placed in the surprise stratum and its Y modified by being replaced by an average of the unit's previous, more typical values. Winsorized estimators, that is, one that truncate large Y's at a cutoff value were studied by Ernst (1980), Rivest (1994), and Searls (1966).

11.1.1 Gross Error Model

The situation we consider here is one in which the sample outliers are representative, that is, there are units like them in the nonsample, but those units constitute a small percentage of the population. We begin with a mixture model

$$Y_i = (1 - \delta_i)(\mu_1 + \varepsilon_{1i}) + \delta_i(\mu_2 + \varepsilon_{2i}), \tag{11.1.1}$$

where the errors are uncorrelated and $\varepsilon_{ki} \sim (0, \sigma_k^2)$, $(k = 1, 2)$. The indicator δ_i is a 0–1 variable denoting whether a unit is an outlier or not with probability $\pi_i = \Pr(\delta_i = 1)$. Typically, the probability π_i of being an outlier is small, and

$\mu_2 \gg \mu_1$. It may also often be the case that $\sigma_2 \gg \sigma_1$. A key feature of the outlier problem is that we cannot identify which units belong to the "normal" part of the population ($\delta_i = 0$) or the "outlier" part ($\delta_i = 1$) in advance of sampling. After a sample is selected, we may be able to recognize sample outliers, but we still will not know which units in the nonsample are outliers. This eliminates the possibility of post-stratifying at the estimation stage. As we saw in Chapter 5 in the analysis of the Counties70 population, the most problematic situations occur where there are outliers in the population but few if any in the sample.

Suppose, for the moment, that we take no special precautions for dealing with outliers and assume that all units have a common mean μ_1 and known variance σ_1^2. The least squares estimator for μ_1 is the value that minimizes

$$\sum_{i \in s} \left(\frac{Y_i - \mu_1}{\sigma_1} \right)^2. \tag{11.1.2}$$

Differentiating with respect to μ_1 and setting equal to 0, gives the estimating equation

$$\sum_{i \in s} \left(\frac{Y_i - \mu_1}{\sigma_1} \right) = 0. \tag{11.1.3}$$

The solution, which is $\bar{Y}_s$ in this simple case, will clearly be affected by large values of the standardized residual $r_i = (Y_i - \mu_1)/\sigma_1$.

One method of generating a robust alternative estimator is to fortify the distance function (11.1.2) or the estimating equation (11.1.3) against undesirably influential points. The method called M-estimation (for maximum-likelihood type) was introduced by Huber (1964, 1981) for that purpose. An M-estimate $\tilde{\mu}_1$ of the mean μ_1 is defined as the solution to

$$\sum_{i \in s} \psi \left(\frac{Y_i - \tilde{\mu}_1}{\tilde{\sigma}_1} \right) = 0, \tag{11.1.4}$$

where $\tilde{\sigma}_1$ is a robust estimate of the scale parameter σ_1 from the non-outlier part of the sample. Examples of ψ-functions are given below. The ψ-functions are closely related to *influence functions* as defined by Hampel, Ronchetti, Rousseeuw, and Stahel (1986). As a convenience, we will loosely use the terms "influence function" and "ψ-function" interchangeably in this section.

In some cases, the estimating equation (11.1.4) can come from differentiating a distance function $\Sigma_s \rho(Y_i; \mu_1)$, but (11.1.4) can be defined even when ψ is not the derivative of any real-valued function. Estimators defined by estimating equations like (11.1.4) are thus more general than ones defined as the solutions of minimization problems.

The role of the ψ-function is to curb the influence of large residuals on the estimate $\tilde{\mu}_1$. Note that if $r_i = (Y_i - \mu_1)/\sigma_1$ is used instead of $\tilde{r}_i = (Y_i - \tilde{\mu}_1)/\tilde{\sigma}_1$

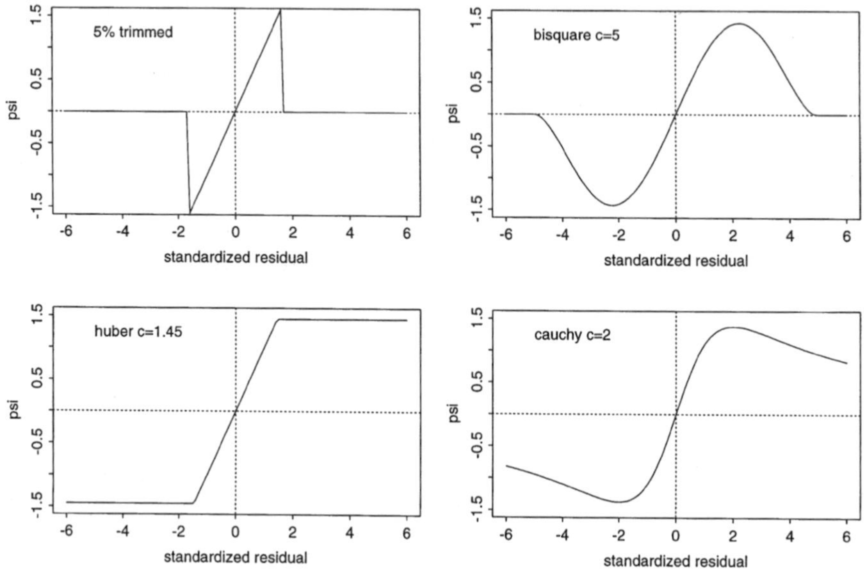

Figure 11.2 Graphs of the 5% trimmed mean, bisquare ($c = 5$), Huber ($c = 1.45$), and Cauchy ($c = 2$) ψ-functions.

and if $\psi(r_i) = r_i$, then (11.1.4) is just the standard estimating function (11.1.3). The ψ-function for ordinary least squares, thus, places no bound at all on the effect of large residuals. But, to be efficient under the working model, it is desirable that $\psi(r)/r \to 1$ as $r \to 0$.

Figure 11.2 shows four examples of ψ-functions that bound the influence of large residuals. One of the simplest methods is to delete or *trim* a proportion α at each end of the sample distribution, retaining only the central $(1 - 2\alpha)n$ observations. Letting r be a standardized residual, the ψ-function for this type of trimmed mean is

$$\psi(r) = \begin{cases} r & |r| \leqslant c \\ 0 & |r| > c, \end{cases} \tag{11.1.5}$$

where the "tuning constant" c is selected based on the amount of trimming desired. The bisquare function is defined as

$$\psi(r) = \begin{cases} r\left(1 - \dfrac{r^2}{c^2}\right)^2 & |r| \leqslant c \\ 0 & |r| > c, \end{cases} \tag{11.1.6}$$

where c is a tuning constant often set to 5, the value used in the figure. A variation on these is to truncate large Y values to some fixed value c rather than deleting the observation entirely. The Huber ψ-function is

$$\psi(r) = \begin{cases} r & |r| \leqslant c \\ c\,\text{sgn}(r) & |r| > c \end{cases} \tag{11.1.7}$$

with c being a different tuning constant and $\text{sgn}(r) = 1$ if r is positive and -1 if negative. The commonly recommended value of $c = 1.45$ is used in the figure. Another choice for ψ is the Cauchy function

$$\psi(r) = \frac{r}{c^2 + r^2} \tag{11.1.8}$$

with $c = 2$ in the figure. For purposes of comparison, the Cauchy function in Figure 11.2 has been multiplied by a constant to make the vertical scale similar to those in the other panels of the figure. Note that multiplication of $\psi(\cdot)$ by a constant does not affect the solution to (11.1.4).

Each of these ψ-functions is called *skew-symmetric*, meaning $\psi(-r) = -\psi(r)$. All have the feature of reducing the effect of large standardized residuals, but the extent of the reduction varies among the functions. The trimmed ψ-function is equal to r in a range on either side of 0, but eliminates residuals outside the range. The bisquare is roughly equal to r for $-\sqrt{c} \leqslant r \leqslant \sqrt{c}$ (here $c = 5$) and descends to 0 outside that range. Both the trimmed and bisquare functions have been called *redescending* (Hampel et al., 1986) since each drops to 0 beyond a certain point, thereby discarding observations with residuals beyond a cutoff. The Huber function in the figure is exactly equal to r in $[-c, c]$ with $c = 1.45$ and truncates standardized residuals to c in absolute value outside that range. The Cauchy assigns the smallest penalty to large residuals by very gradually tapering to 0 as the residuals become larger in absolute value.

Because, in practice, the scale factor σ_1 is unknown unless there are past data from the same or a very similar population, an estimate must be used. Two popular, outlier-resistant choices are the median absolute deviation (MAD) and the interquartile range (IQR) estimators:

$$IQR = [Y_{(3n/4)} - Y_{(n/4)}]/1.35, \quad \text{and}$$

$$MAD = \underset{i \in s}{\text{median}}\,(|Y_i - \text{median}(Y_i)|)/0.6745,$$

where $Y_{(i)}$ is the ith sample order statistic. When $Y \sim N(\mu, \sigma^2)$,

$$[Y_{(3n/4)} - Y_{(n/4)}] \to \sigma[\Phi^{-1}(0.75) - \Phi^{-1}(0.25)] \approx 1.35\sigma \quad \text{and}$$

$$\underset{i \in s}{\text{median}}\,(|Y_i - \text{median}(Y_i)|) \to \text{median}(|Y - \mu|) \approx 0.6745\sigma,$$

which explains the origins of the constant divisors in the *IQR* and *MAD* estimators.

Now, consider the special case of the mixture model (11.1.1) when $\mu_2 = \mu_1$ given by

$$Y_i = \mu_1 + (1 - \delta_i)\varepsilon_{1i} + \delta_i \varepsilon_{2i} \tag{11.1.9}$$

with $\varepsilon_{ji} \sim N(0, \sigma_j^2)$, $j = 1,2$, and $\sigma_2 = k\sigma_1$. This is the gross error model suggested by Tukey (1960). When all units have a common mean μ_1, one estimate of the population total is

$$\hat{T}_{\text{robust}} = \sum_{i \in s} Y_i + (N - n)\tilde{\mu}_1 \tag{11.1.10}$$

with $\tilde{\mu}_1$ a solution of (11.1.4). We conducted a simulation study to compare the expansion estimator $\hat{T}_0 = N\bar{Y}_s$ with robust alternatives computed from (11.1.10) using the ψ-functions for the 5% trimmed mean (i.e., the mean after eliminating the lowest 5% and highest 5% of the ordered Y's), bisquare ($c = 5$), and Huber ($c = 1.45$) ψ-functions. Each of these outlier-resistant alternatives is available in S-Plus™ (Statistical Sciences, Inc., 1995). The scale factor σ_1 was estimated by the *MAD*. Because of the nature of the ψ-functions, $\tilde{\mu}_1$ must generally be computed iteratively—a topic we do not cover here. Hampel et al. (1986) describe algorithms for several choices of ψ.

Table 11.1 gives the results of simulations comparing the root mean square errors (*rmse*) of the several estimators. Populations were generated with $N = 1000$, $\mu_1 = \mu_2 = 50$, $\sigma_1 = 2$, and $k = 5$ for each of six different levels of contamination. For each level of contamination, 1,000 populations were generated and a simple random sample of size $n = 50$ was selected without replacement from each. Regardless of the amount of contamination, all

Table 11.1 Results for the Expansion, 5% Trimmed, Bisquare, and Huber Estimators of the Population Total from 1,000 Simple Random Samples of Size 50

	Average Over Samples				*rmse*	$rmse(\hat{T})/rmse(\hat{T}_0)$		
Contamination π	$\hat{T}_0$	$\hat{T}_{\text{trim}}$ $\alpha=.05$	$\hat{T}_{\text{bisquare}}$ $c=5$	$\hat{T}_{\text{huber}}$ $c=1.45$	$\hat{T}_0$	$\hat{T}_{\text{trim}}$ $\alpha=.05$	$\hat{T}_{\text{bisquare}}$ $c=5$	$\hat{T}_{\text{huber}}$ $c=1.45$
0	50,011	50,012	50,010	50,009	272.65	1.02	1.04	1.03
0.01	49,989	49,989	49,990	49,990	313,60	0.92	0.94	0.93
0.05	49,988	49,994	49,994	49,992	409.41	0.77	0.75	0.76
0.10	50,039	50,028	50,015	50,020	519.04	0.73	0.63	0.65
0.15	49,995	50,001	50,004	50,004	592.19	0.76	0.59	0.63
0.20	49,956	49,965	49,985	49,977	658.41	0.79	0.56	0.60

The population was generated from the gross error model in (11.1.9) with $\mu_1 = \mu_2 = 50$, $\sigma_1 = 2$, and $\sigma_2 = 5\sigma_1$

estimators are approximately unbiased for the population total since $E(T) = 50{,}000$. With no contamination, $\hat{T}_0$ is *BLU* and is 2–4% more efficient than the *M*-estimators; so, there would be minimal loss in using the robust estimators instead of $\hat{T}_0$. With only 1% contamination, the *M*-estimators are 6–8% more efficient and are progressively better as the percentage contamination increases. By the time δ reaches 20%, the bisquare estimator has a root mean square error (*rmse*) only 56% of that of the expansion estimator. Trimming is the least efficient approach among the robust alternatives, at least with the particular choices of tuning constants here used. Overall outlier robust estimation is an improvement on the *BLU* estimator in this experiment.

If, however, the contaminated units do not have the same mean as the other units, the robust estimators can be biased and have large *rmse*'s. Suppose that $\tilde{\mu}_1$ is unbiased for μ_1. Under model (11.1.1) with $\pi_i = \pi$, the bias of $\hat{T}_{\text{robust}}$ is

$$E_M(\hat{T}_{\text{robust}} - T) = (N - n)\pi(\mu_1 - \mu_2).$$

Table 11.2 shows the outcomes of a second set of simulations where $\mu_1 = 50$, $\mu_2 = 70$, and the other parameters are the same as in the Table 11.1 simulations. Note that there is quite a bit of overlap between the $N(50, 2)$ and $N(70, 10)$ distributions so that how to post-stratify into two groups would not be obvious. With 1% contamination, the *M*-estimators are 15–18% more efficient than $\hat{T}_0$, but with higher levels of contamination the outlier-robust alternatives fall apart. At 20% contamination the *rmse* of $\hat{T}_{\text{bisquare}}$ is 2.99 times the *rmse* of $\hat{T}_0$. This is predominantly due to the squared bias component, since the bias of $\hat{T}_{\text{bisquare}}$ is 3654 and the corresponding root mean square error about 3700. Even with only 5% contamination, the *rmse*'s of $\hat{T}_{\text{trim}}$, $\hat{T}_{\text{bisquare}}$, and $\hat{T}_{\text{huber}}$ are 12%, 42%, and 22% larger than that of the expansion estimator. There are basically two strata in the population but the *M*-estimators discount the information obtained on the second stratum.

One approach to trying to correct the bias of the alternative estimators (11.1.1) is to increase the tuning constants c in the bisquare and Huber functions. Larger values of c will lead to less heavy discounting of large residuals. We reran the simulation reported in Table 11.2 with $c = 10$ for the bisquare and 3 for the Huber function. Results for 5% and 20% contamination were the following:

		Average over samples		$rmse(\hat{T})/rmse(\hat{T}_0)$	
Contamination π	Avg. T	$\hat{T}_{\text{bisquare}}$ $c = 10$	$\hat{T}_{\text{huber}}$ $c = 3$	$\hat{T}_{\text{bisquare}}$ $c = 10$	$\hat{T}_{\text{huber}}$ $c = 3$
0.05	51,002	50,147	50,528	1.27	0.92
0.20	53,999	51,127	52,814	2.33	1.32

Table 11.1.2 Results for the Expansion, 5% Trimmed, Bisquare, and Huber Estimators of the Population Total from 1,000 Simple Random Samples of Size 50

		Average Over Samples				rmse	$rmse(\hat{T})/rmse(\hat{T}_0)$		
Contamination π	Avg. T	T_0	$\hat{T}_{\text{trim}}$ $\alpha = .05$	$\hat{T}_{\text{bisquare}}$ $c = 5$	$\hat{T}_{\text{huber}}$ $c = 1.45$	$\hat{T}_0$	$\hat{T}_{\text{trim}}$ $\alpha = .05$	$\hat{T}_{\text{bisquare}}$ $c = 5$	$\hat{T}_{\text{huber}}$ $c = 1.45$
0.00	50,000	50,011	50,012	50,010	50,009	272.65	1.02	1.04	1.03
0.01	50,205	50,194	50,047	50,016	50,045	400.66	0.82	0.85	0.82
0.05	51,002	50,992	50,409	50,061	50,224	700.11	1.12	1.42	1.22
0.10	52,001	51,973	51,208	50,128	50,479	913.30	1.25	2.09	1.72
0.15	52,995	53,064	52,265	50,228	50,836	1126.41	1.17	2.48	1.97
0.20	53,999	53,966	53,186	50,345	51,209	1233.20	1.20	2.99	2.32

The population was generated from the gross error model in (11.1.9) with $\mu_1 = 50$, $\mu_2 = 70$, $\sigma_1 = 2$, and $\sigma_2 = 5\sigma_1$

At 5% contamination the Huber estimator now has a smaller *rmse* than the bisquare or the expansion estimator. But, with 20% contamination both $\hat{T}_{\text{bisquare}}$ and $\hat{T}_{\text{huber}}$ are still less efficient than $\hat{T}_0$.

Thus, when the degree of contamination is small, say, 1–3%, these alternative estimators can be very useful. However, blind application of the robust estimators can be very inefficient in some populations.

We can try to correct the bias of the alternative estimators under (11.1.1) more directly, but not with very impressive success. In the case where the mixture probabilities are all equal ($\pi_i = \pi$) and $E_M(\tilde{\mu}_1) \cong \mu_1$, note that $E_M(\hat{T}_{\text{robust}} - T) = -(N-n)E_M(\bar{r}_s)$ where $\bar{r}_s = \Sigma_s(Y_i - \tilde{\mu}_1)/n$ (see Exercise 11.1). This suggests a bias-corrected estimator defined as

$$\hat{T} = \sum_s Y_i + (N-n)(\tilde{\mu}_1 + \bar{r}_s).$$

However, it is easy to show that this is just the expansion estimator $\hat{T}_0$, which we know from Table 11.1 can be highly inefficient. A more elaborate version of bias correction will be described in the next section where a regression model is considered.

11.1.2 Simple Regression Model

In this section we consider again the straight-line model through the origin:

$$Y_i = \beta x_i + \varepsilon_i, \varepsilon_i \sim (0, v_i) \tag{11.1.11}$$

with the errors independent. The *BLU* predictor of the total $T = \Sigma_{i=1}^N Y_i$ is

$$\hat{T} = \sum_{i \in s} Y_i + \hat{\beta} \sum_{i \in r} x_i$$

with $\hat{\beta} = \Sigma_s(Y_i x_i/v_i)/\Sigma_s(x_i^2/v_i)$. The least squares estimate $\hat{\beta}$ is sensitive to sample outliers as Figure 11.1 illustrates and a start toward robust estimation is to replace $\hat{\beta}$ with a more outlier robust version.

Our real target, however, is the population total T and we want to derive a predictor $\tilde{T}$ that is outlier-robust in the sense that the prediction error $\tilde{T} - T$ is minimally affected by sample outliers. As motivation for an alternative estimator defined below, note that the *BLU* predictor can also be written as

$$\begin{aligned} \hat{T} &= \sum_s Y_i + \beta \sum_r x_i + (\hat{\beta} - \beta) \sum_r x_i \\ &= \sum_s Y_i + \beta \sum_r x_i + \sum_s u_i \frac{Y_i - \beta x_i}{\sqrt{v_i}}, \end{aligned} \tag{11.1.12}$$

where

$$u_i = \frac{x_i}{\sqrt{v_i}} \frac{\Sigma_r x_i}{\Sigma_s x_i^2 / v_i}.$$

The least squares estimator of the total is, thus, the sum of three terms—(1) the sample sum, (2) the expected value of the nonsample total, and (3) the difference between an estimator of the nonsample total and its expectation. Large residuals in the third term on the right-hand side of (11.1.12) can have an extreme effect on the estimated total.

The decomposition above suggests that $\hat{T}$ might be made more outlier-robust by substituting a resistant estimator for $\hat{\beta}$ and by somehow limiting the influence of outliers on the third term in (11.1.2). Chambers (1986) proposed

$$\tilde{T}_1 = \sum_s Y_i + \tilde{\beta} \sum_r x_i + \sum_s u_i \psi(\tilde{r}_i), \qquad (11.1.13)$$

where $\tilde{r}_i = (Y_i - \tilde{\beta} x_i)/\sqrt{v_i}$, $\tilde{\beta}$ is some resistant estimator of β, and $\psi(\cdot)$ is a bounded influence function, possibly different from the one used to estimate β. A simpler choice, analogous to estimator (11.1.10) of the last section, is

$$\tilde{T}_2 = \sum_s Y_i + \tilde{\beta} \sum_r x_i,$$

which was suggested by Lee (1991). A variant was studied by Gwet and Rivest (1992).

When estimating the slope β, there are three types of points that are a concern—(1) points with outliers, that is, points with y values a substantial distance from the line $y = \beta x$, (2) high leverage points, that is, x-outliers, and (3) points that are both. Figure 11.3 is a plot of a sample of 20 points from the Hospitals population with three artificial points added. The point A is clearly an outlier. Point B is a high leverage point; it may have a substantial effect on the estimated slope, even though it is not an outlier. Point C is a high leverage outlier since its x-value is large and it has a large residual.

Four fitted lines are shown on the figure. The solid line is the plot of $y = \hat{\beta} x$ with $\hat{\beta} = \bar{y}_s / \bar{x}_s$ being the *BLU* estimate under $M(0, 1{:}x)$ using the 20 sample points but not A, B, and C. The upper dotted line is the fit when point A is added to the 20 while the lower dashed line is estimated from the 20 points plus C. The dashed line closest to the solid line is based on a bisquare M-estimate of β (discussed below) from the 20 points plus C. A bisquare fit (not shown) using the 20 sample points plus A was also almost identical to the one drawn in the figure. The least squares lines are pulled up or down depending on whether A or C is in the sample, but the M-estimate with a bisquare ψ-function is only slightly effected by A or C.

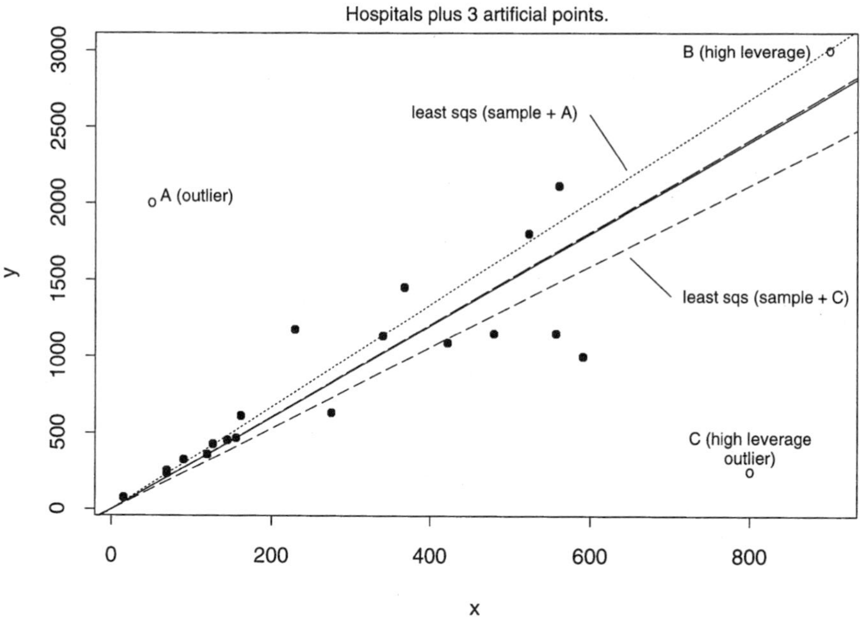

Figure 11.3 Outliers, leverage points, and different estimates of slope. Twenty sample points from Hospitals plus 3 artificial points.

The standard least squares estimate of β is the solution to the estimating equation $\Sigma_s x_i(Y_i - \beta x_i)/v_i = 0$, and a corresponding M-estimator version is

$$\sum_s \frac{x_i}{\sqrt{v_i}} \phi(\tilde{r}_i) = 0, \tag{11.1.14}$$

where ϕ is an influence function, perhaps different from the function ψ. Equation (11.1.14) is the type that was solved for the bisquare estimate in Figure 11.3. Expression (11.1.14) is still somewhat vulnerable to points with large values of x, that is, high leverage points. This problem can be combated with generalized M-estimators or GM-estimators (Hampel et al., 1986, pp. 315–323) that are solutions to equations of the form

$$\sum_s w(x_i)\phi\{\tilde{r}_i u(x_i)\} = 0$$

with $w(\cdot)$ and $u(\cdot)$ being positive weight functions selected to down-weight leverage points.

Corresponding to the regression model (11.1.11), a version of the mixture model in this case is

$$Y_i = (1 - \delta_i) Y_{1i} + \delta_i Y_{2i}, \tag{11.1.15}$$

where $Y_{ki} = \beta_k x_i + \varepsilon_{ki}$ $(k = 1, 2)$ with the errors being uncorrelated, $\varepsilon_{ki} \sim (0, v_{ki})$, and $\pi = \Pr(\delta_i = 1)$. Below, the v_{ki} are treated as if known, although in practice the iterative solution to (11.1.14) will incorporate estimates of the variance parameters. As in the gross error model (11.1.1), we have in mind a case where π is small. Chambers (1986) gives asymptotic properties of $\tilde{T}_1$ under (11.1.15), some of which we sketch here. Carroll and Ruppert (1988) cover asymptotics for M-estimators in the general regression case.

Suppose that β_M is the large sample solution to (11.1.14), equivalent to $E_M \Sigma_s x_i \phi(\tilde{r}_i)/\sqrt{v_{1i}} = 0$. Using v_{1i} in the estimating equation implies that β_M is estimated in a way that downplays the effect of the $\delta_i = 0$ cases. Since

$$E_M[\psi(\tilde{r}_i)] = E_M[\psi(\tilde{r}_i)|\delta_i = 1]\Pr(\delta_i = 1) + E_M[\psi(\tilde{r}_i)|\delta_i = 0]\Pr(\delta_i = 0),$$

the asymptotic bias of $\tilde{T}_1$ is

$$E_M(\tilde{T}_1 - T) \approx (N - n)\bar{x}_r\{(1 - \pi)[B_1 - (\beta_1 - \beta_M)] + \pi[B_1 - (\beta_2 - \beta_M)]\}, \tag{11.1.16}$$

where $B_k = \Sigma_s z_i E_M[\psi(\tilde{r}_{ki})]/\Sigma_s z_i^2$ with $z_i = x_i/\sqrt{v_{1i}}$. Similarly, the asymptotic bias of $\tilde{T}_2$ is

$$E_M(\tilde{T}_2 - T) \approx -(N - n)\bar{x}_r\{(1 - \pi)(\beta_1 - \beta_M) + \pi(\beta_2 - \beta_M)\}. \tag{11.1.17}$$

In the case where all units follow the core expected value model

$$Y_i = \beta_1 x_i + (1 - \delta_i)\varepsilon_{1i} + \delta_i \varepsilon_{2i}, \tag{11.1.18}$$

the bias of $\tilde{T}_2$ vanishes as long as the solution to $\Sigma_s x_i \phi(\tilde{r}_i)/\sqrt{v_{1i}} = 0$ converges to β_1. If ψ is skew-symmetric, then $\tilde{T}_1$ is also approximately unbiased. Both $\tilde{T}_1$ and $\tilde{T}_2$ can have much smaller variances than the *BLU* estimator under (11.1.15) when $v_{2i} \gg v_{1i}$ since they both discount the effect of large residuals.

The difficulty with $\tilde{T}_1$ and $\tilde{T}_2$ under the more general model (11.1.15) is the same as with $\hat{T}_{\text{robust}}$ defined in (11.1.10). Both the outlier-resistant choices can be severely biased. Moreover, for reasons detailed in Chambers (1986), there is a tradeoff between bias and variance. In $\tilde{T}_1$, choosing $\psi(r) = r$ produces an asymptotically unbiased estimator (see Exercise 11.2) but increases the variance compared to using a bounded influence function. The third term in $\tilde{T}_1$ is needed for unbiasedness when $\psi(r) = r$ and can thus be viewed as a kind of bias correction term. Choosing $\psi(r) = 0$ makes $\tilde{T}_1 = \tilde{T}_2$ and leads to a small variance but if β_1 and β_2 are much different, the bias can be substantial. In short, the resistant alternatives can be extremely effective if (11.1.18) is the correct model, but if not, the risk in using them can be considerable. Two examples will show the extremes of what can happen.

Example 11.1.1 A sample of $n = 20$ units from Counties70 in Appendix B consists of these unit numbers:

18	32	69	109
116	123	177	182
203	220	231	235
238	243	247	256
260	261	272	296

Suppose that, in addition to these 20, we have two other sample points: $(x_1, y_1) = (30{,}000, 5{,}000)$ and $(x_2, y_2) = (50{,}000, 10{,}000)$. These are not part of Counties70 but are injected here for the sake of illustration. Figure 11.4 is a plot of the full Counties70 population with the 20 real sample points shown as circles and nonsample points as dots. The two artificial outliers are triangles. The solid line is the least squares fit of $M(0, 1{:}x)$ while the dashed lines are based on the bisquare ($c = 4.685$, an alternative choice of tuning constant to the $c = 5$ used earlier) and Huber ($c = 1.345$, another common alternative) estimates of β in $M(0, 1{:}x)$. If we take the population totals to be the Counties70 total plus the two injected outliers, then $T_y = 11{,}258{,}111$ and $T_x = 2{,}795{,}075$. The ratio estimate from the sample of 22 is $\hat{T}_R = 8{,}514{,}508$ which is about a 24% under-estimate. The resistant

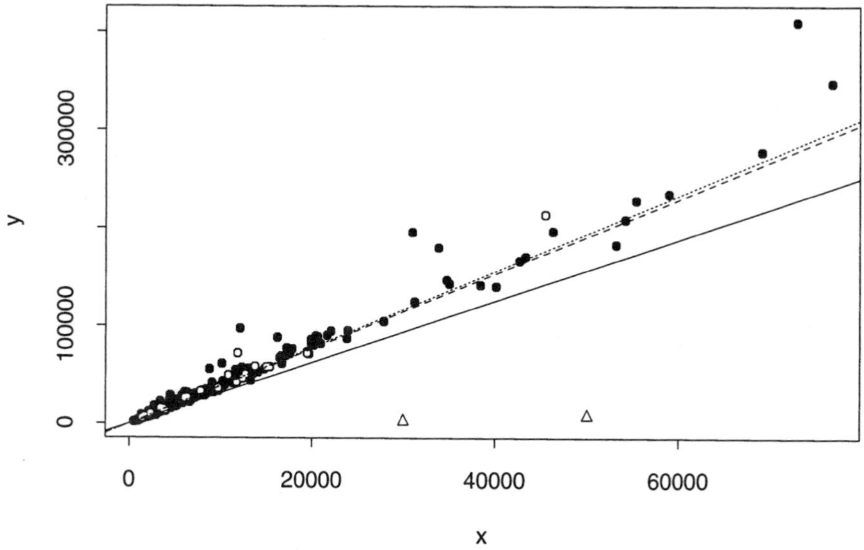

Figure 11.4 A sample of $n = 22$ (20 real units + 2 artificial outliers) from Counties70 and the nonsample points plotted together. Nonsample units are dots; real sample units are circles; two artificial units are triangles.

estimates are

	$\tilde{T}_1$	$\tilde{T}_2$
bisquare ($c = 4.685$)	10,674,684	10,558,581
Huber ($c = 1.345$)	10,540,699	10,393,614

These are each low but within 8% of the population total. For $\tilde{T}_1$ the same ψ-function was used for the estimate $\tilde{\beta}$ and for the third term in (11.1.13).

Example 11.1.2 Now, to show how ambiguous the situation with outliers can be, suppose that in Example 11.1.1 there are actually 20 points in the population equal to $(x_1, y_1) = (30{,}000, 5{,}000)$ and 20 equal to $(x_2, y_2) = (50{,}000, 10{,}000)$, but we have the same sample of 22 units. The two outliers in the sample are thus representative of others in the nonsample. The population totals become $T_y = 11{,}543{,}111$ and $T_x = 4{,}315{,}075$. The ratio estimator is 12,696,811 and the resistant estimators are

	$\tilde{T}_1$	$\tilde{T}_2$
bisquare ($c = 4.685$)	16,652,862	16,466,191
Huber ($c = 1.345$)	16,437,441	16,200,959

The estimators $\hat{T}_R$, $\tilde{T}_1$, and $\tilde{T}_2$ each retain their same estimates of slope as in Example 11.1.1 since the sample is the same, but the nonsample part of the population has many more points with small y's. The robust estimates of β lead to severe overpredictions for those points and overestimates of the total of y.

In Section 5.7.2 we saw that Counties70 contains eight points that are above the population regression line that can be considered outliers. The presence or absence of those points in the sample has a large effect on t-statistics even when the minimal estimator $\hat{T}(x^{1/2}, x:x)$ is used with the jackknife variance estimator, and a weighted balanced sample is selected. Example 11.1.2 is a less subtle illustration of the same type of problem, that is, the unseen, nonsample outliers can degrade the quality of inferences more than ones in the sample.

11.1.3 Areas for Research

Variance estimation for the outlier resistant estimators of totals is an open problem. Lee (1991) discusses method-of-moments types of estimators of the asymptotic variance of $\tilde{T}_2$ and gives some limited empirical results. Another possibility is resampling. Chambers and Kokic (1993) used a version of the bootstrap to calculate confidence intervals with $\tilde{T}_1$ in two populations. They used the bisquare ψ-function with $c = 4.685$ to derive the estimator $\tilde{\beta}$, and three different Huber functions (based on different values of the tuning constant

c) were used as the ψ-function in (11.1.13). In the population with fewer outliers, the coverage probability of 95% confidence intervals was close to the nominal level for $\tilde{T}_1$ and the bootstrap. In the population with a substantial number of outliers, Chambers and Kokic found that the ratio estimator together with v_H, defined in Section 5.1.2, had coverage levels as good or better than $\tilde{T}_1$ and the bootstrap. However, all were 5–15% below the nominal 95% level.

As we have noted in other contexts, confidence intervals based on biased estimators will be centered in the wrong spot and will not, typically, have nominal coverage probabilities. Even though the outlier robust alternatives may sometimes have good mean square error performance, there remains the problem of how to adequately correct for the bias when constructing confidence intervals.

Outliers will also affect the more general regression estimators studied throughout this book. There has been a substantial amount of research in this area in the rest of statistics that has not been transferred to finite population estimation.

11.2 NONLINEAR MODELS

Virtually all the previous material in this book has involved the estimation of means and totals of quantitative variables in situations where a linear model is a reasonable working specification. A natural situation where a nonlinear model may be appropriate is that of estimating the proportion of units or total number of units that have a particular characteristic. For example, we might want to estimate the proportion of business establishments that have laid-off workers in the first quarter of a year or the number of farms that operated at a deficit during the previous year. In such cases we are measuring a 0–1 variable Y on each unit in the population—either the unit has the characteristic or not. The expected value specification for the target variable Y gives the probability that a unit has the characteristic.

The standard practice in surveys is to estimate the total of a binary random variable with a linear estimator of the form $\hat{T} = \Sigma_s w_i Y_i$. Such an estimator will be unbiased if a linear model holds for Y_i, for example, $E_M(Y_i) = \beta_0 + \beta_1 x_i$. A problem with a linear model for a binary variable in the presence of auxiliaries is that the expectation is not confined to $[0, 1]$ as a legitimate probability should be. For units whose probability of having the characteristic is not near 0 or 1, a linear model may be a reasonable description. But, there are many alternatives that appropriately restrict the expectation. For example, in the case of a single auxiliary, three are

$$\text{Exponential: } E_M(Y_i) = 1 - \exp(-\beta x_i),$$

$$\text{Logistic: } E_M(Y_i) = \{1 + \exp[-(\beta_0 + \beta_1 x_i)]\}^{-1}, \qquad \text{and}$$

$$\text{Complementary log-log: } E_M(Y_i) = 1 - \exp[-\exp(\beta_0 + \beta_1 x_i)].$$

Thorough treatments of nonlinear regression can be found in Bates and Watts (1988) and Seber and Wild (1989).

Estimators of totals and means can be developed under nonlinear models that are very similar in appearance to the *BLU* predictors introduced in Chapter 2. Suppose, as usual, that the population vector of target values is $\mathbf{Y} = (Y_1, \ldots, Y_N)'$, where each Y can be continuous or binary. The vector of expected values of $\mathbf{Y}$ will be denoted as $\mathbf{f}(\boldsymbol{\beta}) = (f(\mathbf{x}_1; \boldsymbol{\beta}), \ldots, f(\mathbf{x}_N; \boldsymbol{\beta}))$, where f is a nonlinear function of the components of the $N \times p$ parameter vector $\boldsymbol{\beta}$ and $\mathbf{x}_i$ is a p-vector of auxiliaries associated with unit i. The $N \times N$ covariance matrix for $\mathbf{Y}$ is $\mathbf{V}$. As in earlier chapters, after a sample is selected, the population can be split into the sample s and the nonsample r. The full specification of the model for $\mathbf{Y}$ is

$$E_M(\mathbf{Y}) = \mathbf{f}(\boldsymbol{\beta}) = [\mathbf{f}_s(\boldsymbol{\beta})', \mathbf{f}_r(\boldsymbol{\beta})']'$$

$$\text{var}_M(\mathbf{Y}) = \mathbf{V} = \begin{bmatrix} \mathbf{V}_{ss} & \mathbf{V}_{sr} \\ \mathbf{V}_{rs} & \mathbf{V}_{rr} \end{bmatrix}, \tag{11.2.1}$$

where $\mathbf{f}(\boldsymbol{\beta})$ and $\mathbf{V}$ are decomposed in the obvious way. We also need the vectors and matrices of first partial derivatives defined by

$$\mathbf{F}_i(\boldsymbol{\beta}) = \left[\frac{\partial f(\mathbf{x}_i; \boldsymbol{\beta})}{\partial \beta_1}, \ldots, \frac{\partial f(\mathbf{x}_i; \boldsymbol{\beta})}{\partial \beta_p}\right]' \quad \text{for } i = 1, \ldots, N, \qquad \text{and}$$

$$\begin{aligned} \mathbf{F}(\boldsymbol{\beta}) &= [\mathbf{F}_1(\boldsymbol{\beta}), \ldots, \mathbf{F}_N(\boldsymbol{\beta})]' \\ &= [\mathbf{F}_s(\boldsymbol{\beta})', \mathbf{F}_r(\boldsymbol{\beta})']' \end{aligned}$$

where $\mathbf{F}_s(\boldsymbol{\beta})$ is the $n \times p$ matrix of first partial derivatives for the sample units and $\mathbf{F}_r(\boldsymbol{\beta})$ is the $(N-n) \times p$ matrix of partials for the nonsample units. In much of this section the argument $\boldsymbol{\beta}$ will be suppressed in $\mathbf{F}_i(\boldsymbol{\beta})$, $\mathbf{F}_s(\boldsymbol{\beta})$, and $\mathbf{F}_r(\boldsymbol{\beta})$ for compactness of notation.

If $\boldsymbol{\beta}$ were known, the *BLU* predictor of T is simply the sample sum of the Y's plus the *BLU* predictor of the nonsample sum as stated in the following theorem.

Theorem 11.2.1 Under model (11.2.1) with $\boldsymbol{\beta}$ known, among linear estimators of the form $\hat{T} = \mathbf{g}_s'\mathbf{Y}_s$ satisfying $E_M(\hat{T} - T) = 0$, the error variance $E_M(\hat{T} - T)^2$ is minimized by

$$\hat{T}^* = \sum_{i \in s} Y_i + \mathbf{1}_r'[\mathbf{f}_r(\boldsymbol{\beta}) + \mathbf{V}_{rs}\mathbf{V}_{ss}^{-1}(\mathbf{Y}_s - \mathbf{f}_s(\boldsymbol{\beta}))],$$

where $\mathbf{1}_r$ is an $(N-n)$-vector of all 1's.

Proof: The proof is left to Exercise 11.6.

When $\boldsymbol{\beta}$ is unknown, an estimator must be used. The standard estimator of $\boldsymbol{\beta}$ in a nonlinear regression problem is obtained by generalized least squares (*GLS*). The *GLS* estimator is the value $\hat{\boldsymbol{\beta}}$ that minimizes the sum of squares

$$[\mathbf{Y}_s - \mathbf{f}_s(\boldsymbol{\beta})]' \mathbf{V}_{ss}^{-1} [\mathbf{Y}_s - \mathbf{f}_s(\boldsymbol{\beta})]. \tag{11.2.2}$$

Differentiating (11.2.2) with respect to $\boldsymbol{\beta}$ and setting the result to 0 leads to this set of p estimating equations in the p unknowns $\beta_1, \ldots, \beta_p$:

$$\mathbf{F}_s(\boldsymbol{\beta})' \mathbf{V}_{ss}^{-1} [\mathbf{Y}_s - \mathbf{f}_s(\boldsymbol{\beta})] = 0 \tag{11.2.3}$$

The value of $\boldsymbol{\beta}$ that solves this set of equations is the estimate $\hat{\boldsymbol{\beta}}$. Except in some special cases, the justification for the *GLS* estimator is the intuitive reasonableness of least squares. The estimator $\hat{\boldsymbol{\beta}}$ is generally biased but under some reasonable conditions is consistent. If $\mathbf{Y}$ is multivariate normal, then the *GLS* estimator is also the maximum likelihood estimator (*MLE*) which has good large sample properties. Jennrich (1969) showed that, when the Y's are independent with equal variances, the *GLS* estimator of $\hat{\boldsymbol{\beta}}$ is asymptotically normal with mean $\boldsymbol{\beta}$ and is a consistent estimator of $\boldsymbol{\beta}$.

Because a closed-form solution for $\hat{\boldsymbol{\beta}}$ in (11.2.3) cannot be found, iterative methods are used (see, e.g., Agresti, 1990, secs. 4.7.2, 13.1.3). Denote the sum of squares in (11.2.2) by $Q(\boldsymbol{\beta})$, the left-hand side of (11.2.3) by $q(\boldsymbol{\beta})$ and let $\mathbf{H}$ be the $p \times p$ matrix of second derivatives having entries $\partial^2 Q / \partial\beta_a \partial\beta_b$. Let $\mathbf{q}^{(t)}$ and $\mathbf{H}^{(t)}$ be the values of $q(\boldsymbol{\beta})$ and $\mathbf{H}$ evaluated at $\boldsymbol{\beta}^{(t)}$, the tth guess for $\hat{\boldsymbol{\beta}}$. At step t in the iterative process, the sum of squares, $Q(\boldsymbol{\beta})$, is approximated near $\boldsymbol{\beta}^{(t)}$ by a second order Taylor series expansion:

$$Q^{(t)}(\boldsymbol{\beta}) \cong Q(\boldsymbol{\beta}^{(t)}) + \mathbf{q}^{(t)\prime}(\boldsymbol{\beta} - \boldsymbol{\beta}^{(t)}) + \tfrac{1}{2}(\boldsymbol{\beta} - \boldsymbol{\beta}^{(t)})' \mathbf{H}^{(t)} (\boldsymbol{\beta} - \boldsymbol{\beta}^{(t)}).$$

Solving $\partial Q^{(t)} / \partial \boldsymbol{\beta} = \mathbf{q}^{(t)} + \mathbf{H}^{(t)}(\boldsymbol{\beta} - \boldsymbol{\beta}^{(t)}) = \mathbf{0}$ for $\boldsymbol{\beta}$ yields the iteration equations

$$\boldsymbol{\beta}^{(t+1)} = \boldsymbol{\beta}^{(t)} - [\mathbf{H}^{(t)}]^{-1} \mathbf{q}^{(t)}, \tag{11.2.4}$$

assuming $\mathbf{H}^{(t)}$ has an inverse. Given an initial value for $t = 0$, the set of p equations in (11.2.4) is solved for $\boldsymbol{\beta}^{(1)}$, $\boldsymbol{\beta}^{(1)}$ is used to solve for $\boldsymbol{\beta}^{(2)}$, and so on, until the solution changes minimally between steps or until a specified number of iterations have been done. The iterative technique defined by (11.2.4) is known as the Newton-Raphson method.

When estimating the population total, the obvious candidate comes from substituting $\hat{\boldsymbol{\beta}}$ into the estimator from Theorem 11.2.1, giving

$$\hat{T} = \sum_{i \in s} Y_i + \mathbf{1}_r' \{\mathbf{f}_r(\hat{\boldsymbol{\beta}}) + \mathbf{V}_{rs} \mathbf{V}_{ss}^{-1} [\mathbf{Y}_s - \mathbf{f}_s(\hat{\boldsymbol{\beta}})]\}. \tag{11.2.5}$$

Notice that if $\mathbf{f}(\boldsymbol{\beta}) = \mathbf{X}\boldsymbol{\beta}$, then (11.2.5) is

$$\hat{T} = \sum_{i \in s} Y_i + \mathbf{1}_r'[\mathbf{X}_r\hat{\boldsymbol{\beta}} + \mathbf{V}_{rs}\mathbf{V}_{ss}^{-1}(\mathbf{Y}_s - \mathbf{X}_s\hat{\boldsymbol{\beta}})],$$

which is just the *BLU* predictor under the general linear model in Theorem 2.2.1. An approximation to the error variance of $\hat{T}$ can be had by analogy to result (2.2.2) for the linear case:

$$\begin{aligned} \text{var}_M(\hat{T} - T) \cong\; & \mathbf{1}_r'(\mathbf{F}_r - \mathbf{V}_{rs}\mathbf{V}_{ss}^{-1}\mathbf{F}_s)(\mathbf{F}_s'\mathbf{V}_{ss}^{-1}\mathbf{F}_s)^{-1}(\mathbf{F}_r - \mathbf{V}_{rs}\mathbf{V}_{ss}^{-1}\mathbf{F}_s)'\mathbf{1}_r \\ & + \mathbf{1}_r'(\mathbf{V}_{rr} - \mathbf{V}_{rs}\mathbf{V}_{ss}^{-1}\mathbf{V}_{sr})\mathbf{1}_r. \end{aligned} \tag{11.2.6}$$

The approximation comes from first using (11.2.5) to write out the variance in general, and then applying standard Taylor series approximations to compute variance and covariance components as necessary. For instance, $\text{var}_M[\mathbf{f}_r(\hat{\boldsymbol{\beta}})] \cong \mathbf{F}_r(\mathbf{F}_s'\mathbf{V}_{ss}^{-1}\mathbf{F}_s)^{-1}\mathbf{F}_r'$ using $\mathbf{f}_r(\hat{\boldsymbol{\beta}}) \cong \mathbf{f}_r(\boldsymbol{\beta}) + \mathbf{F}_r(\hat{\boldsymbol{\beta}} - \boldsymbol{\beta})$. Valliant (1985) covers some of the details.

11.2.1 Model for Bernoulli Random Variables

Although a nonlinear model may apply to many situations, the archetypal case is a 0–1 variable whose probability of being 1 depends on various pieces of auxiliary information. Suppose that $Y_1, \ldots, Y_N$ are independent Bernoulli random variables with

$$\begin{aligned} E_M(Y_i) &= f(\mathbf{x}_i; \boldsymbol{\beta}) \\ \text{var}_M(Y_i) &= f(\mathbf{x}_i; \boldsymbol{\beta})[1 - f(\mathbf{x}_i; \boldsymbol{\beta})] \end{aligned} \tag{11.2.7}$$

where $0 \leqslant f(\mathbf{x}_i; \boldsymbol{\beta}) \leqslant 1$. The covariance matrix under model (11.2.7) is automatically unknown since we assume that the parameter $\boldsymbol{\beta}$ is unknown. When the Y's are independent, the estimator of the total reduces to

$$\hat{T} = \sum_{i \in s} Y_i + \mathbf{1}_r'\mathbf{f}_r(\hat{\boldsymbol{\beta}}). \tag{11.2.8}$$

Under model (11.2.7), we can write down the likelihood and compute the *MLE* of $\boldsymbol{\beta}$. The log-likelihood function for $\boldsymbol{\beta}$ is

$$L(\boldsymbol{\beta}) = \sum_{i \in s} [Y_i \log(f_i) + (1 - Y_i)\log(1 - f_i)]$$

with $f_i \equiv f(\mathbf{x}_i; \boldsymbol{\beta})$. The derivative of $L(\boldsymbol{\beta})$ with respect to $\boldsymbol{\beta}$ is $\mathbf{F}_s(\boldsymbol{\beta})'\mathbf{V}_{ss}^{-1}[\mathbf{Y}_s - \mathbf{f}_s(\boldsymbol{\beta})]$ with $\mathbf{V}_{ss} = diag[f_i(1 - f_i)]$ (see Exercise 11.8). Thus the maximum likelihood estimating equations have the same form as those obtained by *GLS* in (11.2.3).

The estimate of $\boldsymbol{\beta}$ must also be solved for iteratively. Since we have a likelihood to work with, the Fisher scoring algorithm can be used (see, e.g., Agresti, 1990, sec. 13.1.3). The formula for Fisher scoring at iteration step t is

$$\boldsymbol{\beta}^{(t+1)} = \boldsymbol{\beta}^{(t)} + [\mathbf{I}^{(t)}]^{-1}\mathbf{q}^{(t)}, \tag{11.2.9}$$

where $\mathbf{I} = -E_M[\partial^2 L(\boldsymbol{\beta})/\partial\beta_a\partial\beta_b]$ is the expected value of the $p \times p$ matrix of second derivatives of the log-likelihood, that is, the Fisher information matrix.

If there exists a function g such that $g[f(\mathbf{x}_i;\boldsymbol{\beta})] = \mathbf{x}_i'\boldsymbol{\beta}$, then the Newton-Raphson and Fisher scoring algorithms are exactly the same (Nelder and Wedderburn, 1972 or Agresti, 1990, sec. 13.1.4). The function g is referred to as the *canonical link*. In the logistic model, for example, where $f(\mathbf{x}_i;\boldsymbol{\beta}) = [1 + \exp(-\mathbf{x}_i'\boldsymbol{\beta})]^{-1}$, the canonical link is the logit function defined as

$$\text{logit}[f(\mathbf{x}_i;\boldsymbol{\beta})] = \log\left(\frac{f_i}{1-f_i}\right) = \mathbf{x}_i'\boldsymbol{\beta}.$$

For the complementary log-log function with $f(\mathbf{x}_i;\boldsymbol{\beta}) = 1 - \exp[-\exp(\mathbf{x}_i'\boldsymbol{\beta})]$. the canonical link is

$$\log[-\log(1 - f(\mathbf{x}_i;\boldsymbol{\beta}))] = \mathbf{x}_i'\boldsymbol{\beta}.$$

The variance of $\hat{T}$ in (11.2.8), when observations are independent, is the special case of (11.2.6) given by

$$\text{var}_M(\hat{T} - T) \cong \mathbf{1}_r'\mathbf{F}_r(\mathbf{F}_s'\mathbf{V}_{ss}^{-1}\mathbf{F}_s)^{-1}\mathbf{F}_r'\mathbf{1}_r + \mathbf{1}_r'\mathbf{V}_{rr}\mathbf{1}_r, \tag{11.2.10}$$

which can estimated by

$$v_{(\hat{T}-T)} = \mathbf{1}_r'\mathbf{F}_r(\hat{\boldsymbol{\beta}})\,[\mathbf{F}_s(\hat{\boldsymbol{\beta}})'\,\hat{\mathbf{V}}_{ss}^{-1}\mathbf{F}_s(\hat{\boldsymbol{\beta}})]^{-1}\mathbf{F}_r(\hat{\boldsymbol{\beta}})'\mathbf{1}_r + \mathbf{1}_r'\hat{\mathbf{V}}_{rr}\mathbf{1}_r, \tag{11.2.11}$$

where $\hat{\mathbf{V}}_{ss} = diag[\mathbf{f}_s(\hat{\boldsymbol{\beta}})(1 - \mathbf{f}_s(\hat{\boldsymbol{\beta}}))]$ and $\hat{\mathbf{V}}_{rr} = diag[\mathbf{f}_r(\hat{\boldsymbol{\beta}})(1 - \mathbf{f}_r(\hat{\boldsymbol{\beta}}))]$. Valliant (1985) gave conditions under which, as $N, n \to \infty$ and $n/N \to 0$, $\hat{T} - T$ is asymptotically normal and $v_{(\hat{T}-T)}$ is consistent in the sense that $v_{(\hat{T}-T)}/\text{var}_M(\hat{T} - T) \xrightarrow{p} 1$.

Though nonlinear models like the logistic and exponential models have found wide application in data analysis, particularly for epidemiological data, their use in estimation of totals or means in sample surveys has been minimal due in part to their being operationally cumbersome. Unlike many of the estimators we have studied, $\hat{T} = \Sigma_{i\in s}Y_i + \mathbf{1}_r'\mathbf{f}_r(\hat{\boldsymbol{\beta}})$ generally requires that the predictions for nonsample units be explicitly summed. Each component of $\mathbf{f}_r(\hat{\boldsymbol{\beta}})$, that is, $f(\mathbf{x}_i;\hat{\boldsymbol{\beta}})$ for $i \in r$, must be evaluated separately.

Recall that the *BLU* predictor under a linear model can be written as $\hat{T} = \Sigma_{i\in s}g_iY_i$ with g_i being a weight that does not depend on the Y's (see Section

2.4). When the vector of weights is $\mathbf{g}_s = \mathbf{X}_s(\mathbf{X}_s'\mathbf{X}_s)^{-1}\mathbf{X}_r'\mathbf{1}_r + \mathbf{1}_s$, for example, we only need to know the totals of the explanatory variables, $\mathbf{1}_r'\mathbf{X}_r = \Sigma_r \mathbf{x}_i$, for the nonsample units. These nonsample totals may be available from the frame or by subtracting the sample totals from separate population census totals. An explicit sum over the nonsample units is avoided, which is an attractive practical advantage. Thus, estimating totals of 0–1 variables using linear, rather than nonlinear, estimators is common practice.

11.2.2 Areas for Research

Surveys are often used to estimate rare characteristics, like the prevalence in a population of the number of persons who have a particular type of disability or who have a certain chronic health condition. Rare characteristics are ones for which linear models and linear estimators may be especially poor. Little, if any, research has been done on how much improvement can be made by using nonlinear estimators or whether it is feasible to use the estimators.

Robust variance estimation for the class of estimators given by (11.2.8) is a largely unexplored topic with the exception of Valliant (1986). The approach to robust variance estimation used in Chapters 5, 6, and 9 is to guard against departures from the working model's variance specification in cases where the expected value is correctly specified. In model (11.2.7), when the variance is misspecified, the mean will be also. Thus, having a well-fitting model is critical. Some work on regression diagnostics for logistic and other models estimated from survey data has been done by Korn and Graubard (1999) but much remains to be done. No work at all appears to have been done on whether balancing of some type has a role in robustness for Bernoulli models or other nonlinear models. The biostatistical literature on generalized estimating equations and accompanying variance estimators (e.g., Liang and Zeger, 1986; Liang, Zeger, and Qaqish, 1992; Zeger and Liang, 1986) may be particularly useful if applied to finite population estimation.

11.3 NONPARAMETRIC ESTIMATION OF TOTALS

In Chapters 3 and 4 we emphasized the achievement of bias-robustness of estimators of totals through the use of different types of balanced sampling. Often samples are selected with no attempt made to achieve balance and robust estimators must be constructed using other approaches. Another way of guarding against the failure of the working model is the use of nonparametric regression. Eubank (1988) and Hardle (1990) cover the general theory of nonparametric regression. Kuo (1988) first suggested using nonparametric regression for finite population estimation in the context of cumulative distribution function estimation, which we take up in Section 11.4. Nonparametric regression can also be used in the estimation of a total, either as an alternative to parametric methods (Section 11.3.1) or in combination with them (Section 11.3.2).

11.3.1 Nonparametric Regression for Totals

Suppose that Y has an ill-defined but smooth relationship to a single auxiliary variable x, that is,

$$Y_i = m(x_i) + \varepsilon_i, \tag{11.3.1}$$

where $m(x)$ is a continuous function (often assumed for theoretical reasons to be twice differentiable) and the errors ε_i are independent with mean 0. The variance of the ε_i may depend on x. Assume that the x's are sorted in ascending order so that $x_1 \leqslant x_2 \leqslant \cdots \leqslant x_N$. Since the population total can be written as $T = \Sigma_{i \in s} Y_i + \Sigma_{i \in r} Y_i$, a natural estimator of the total is

$$\hat{T}_{np} = \sum_{i \in s} Y_i + \hat{T}_{np,r}, \tag{11.3.2}$$

where $\hat{T}_{np,r} = \Sigma_{i \in r} \hat{m}(x_i)$ with $\hat{m}(x_i)$ an estimate of $E_M(Y_i) = m(x_i)$. Such an estimate is achieved using techniques of nonparametric regression, one version of which, kernel estimation, originating in work of Nadarya (1964) and Watson (1964), is as follows:

Let $K(\cdot)$ be a symmetric density function, for example, the standard normal density, $K(u) = (2\pi)^{-1/2} \exp(-u^2/2)$, the uniform density $K(u) = I\{-1 \leqslant u \leqslant 1\}/2$, or the biweight $K(u) = 15(1 - u^2)^2/16$. For a given scaling factor or "bandwidth" $b > 0$, define $K_b(u) = K(u/b)$. Then a nonparametric estimator of $m(x)$ is

$$\hat{m}(x) = \sum_{j \in s} w_j(x) Y_j, \tag{11.3.3}$$

where the weight associated with sample unit j is $w_j(x) = K_b(x_j - x)/\Sigma_s K_b(x_j - x)$. Note that this weight depends on the particular x-value for which the prediction is made.

This method is known as *fixed bandwidth kernel smoothing* with $K(\cdot)$ being the kernel. A prediction is made at a given value of x by taking a weighted average of the Y's for nearby units. The weights given to different sample units depend both on the form of the kernel and the size of the bandwidth. If, for example, $K_b(u) = \exp(-u^2/2b)$ and b very large, each sample Y will get a weight approaching $1/n$ and the predictor at each value of x is near the sample mean $\bar{Y}_s$. As $b \to 0$, almost all weight is put on units whose x_i's are close to the x for which a prediction is needed. As b becomes small, the nonparametric prediction in (11.3.3) becomes more unstable with relatively few observations getting most of the weight.

Using (11.3.3) the estimator of the nonsample sum is

$$\hat{T}_{np,r} = \sum_{i \in r} \sum_{j \in s} w_j(x_i) Y_j$$
$$= \sum_{j \in s} \tilde{w}_j Y_j, \tag{11.3.4}$$

where $\tilde{w}_j = \Sigma_{i \in r} w_j(x_i)$. We note that $\hat{T}_{np,r}$ is a linear estimator, and that $\Sigma_{j \in s} \tilde{w}_j = N - n$.

Dorfman (1994) gave conditions under which the relative bias of $\hat{T}_{np}$ is $O_p(b^3 + \sqrt{b/n})$. The bias thus goes to 0 in large samples as long as $b \to 0$. He also showed that the error variance of $\hat{T}_{np}/N$ is $O_p(n^{-1})$ as long as $b \to 0$ and $nb \to \infty$. The implication of this theory is that the bandwidth should be small, consistent with yielding a window around each nonsample x containing enough sample points to allow reasonable prediction. Standard methods of selecting the bandwidth b based on meeting an optimality criterion like minimizing integrated squared error, $\int [\hat{m}(t) - m(t)]\, dt$, will give a bandwidth too large to minimize $E_M(\hat{T}_r - T_r)^2$.

Chambers, Dorfman, and Wehrly (1993), abbreviated as CDW below, suggested that b be selected to minimize $E_M(\hat{T}_r - T_r)^2$ under a working model. In practice, this may have to be done by defining a grid of potential bandwidths and choosing the value on the grid that yields the smallest value of an estimate of $E_M(\hat{T}_r - T_r)^2$.

CDW noted two practical problems in computing $\hat{T}_r$ in (11.3.4). The first is the predictor's instability when the sample distribution of the x's is highly skewed. A fixed bandwidth kernel estimator has, by definition, a constant width smoothing window. Thus, the method works best when the sample x's are evenly spread around every point where a prediction is needed. One alternative is to work with transformed x data whose sample values are approximately evenly spaced. Even if a transformation of x is possible, a second problem is that the sample x's may not be near many of the nonsample ones for which predictions are needed. This sparse data problem is similar to the "boundary effects" noted by Gasser and Müller (1979) in nonparametric regression, that is, biases arising when attempting to estimate $m(\cdot)$ close to the edge of the x data range. An adaptive bandwidth b may be needed to cope with gaps in the x data.

An important alternative to kernel estimation is local linear regression (see, e.g., Fan and Gibjels (1992), Ruppert and Wand (1994)). Breidt and Opsomer (1999), in a design-based framework, get local linear estimation based estimators that asymptotically achieve the Godambe-Joshi lower bound for anticipated variance under specified conditions.

Dorfman (1994) gives estimators of $\text{var}(\hat{T}_{np,r} - T_r)$ based on nonparametric regression estimates of $\text{var}(Y_i)$, and Breidt and Opsomer (1999) give variances, based on design-based methodology. An unexplored alternative is the use of replication methods.

11.3.2 Nonparametric Calibration Estimation

Nonparametric regression can be used in conjunction with standard parametric regression methods for estimating totals, with possible gains in efficiency and bias reduction. The basic idea, due to Chambers, Dorfman, and Wehrly (1993) (CDW), extends to estimation of $T = \Sigma_{i=1}^{N} h(Y_i)$, totals of functions of the variable of interest. As usual, we let $T = T_s + T_r$, with $T_s = \Sigma_s h(Y_j)$ and $T_r = \Sigma_r h(Y_i)$.

Let $Y_i = p(x_i) + \varepsilon_i$ be a working parametric model, and let $\hat{h}_i$ be estimates of the components $h(Y_i)$, based on the working model. (If $h(Y_i) = Y_i$, then $\hat{h}_i$ will usually be the *BLU* estimator $\hat{p}(x_i)$ gotten by substituting estimates of the parameters into $p(x_i)$; Section 11.4 treats of the situation where $h(Y_i) = I\{Y_i \leqslant t\}$, an indicator function.) The parametrically based estimator of T is $\hat{T} = \Sigma_s h(Y_i) + \Sigma_r \hat{h}_i \equiv T_s + \hat{T}_r$.

Suppose, in fact, a model M holds in which $E_M[h(Y_i)] = m(x_i)$ (replacing Y_i by $h(Y_i)$ in (11.3.1)). Let $\delta(x_i) = m(x_i) - E_M(\hat{h}_i)$. Then $E_M(T_r) = E_M(\hat{T}_r) + \Sigma_r \delta(x_i)$. The second term on right represents the bias in $\hat{T}_r$ due to model misspecification (and possibly other sources.)

For sample points, we have the residuals

$$r_i = h(Y_i) - \hat{h}_i,$$

which have expectation $E_M(r_i) = \delta(x_i)$. For a given non-sample point i, we can estimate $\delta(x_i)$ by nonparametric regression: $\hat{\delta}(x_i) = \Sigma_{j \in s} w_j(x_i) r_j$, and arrive at the bias-adjusted *calibrated nonparametric predictor*

$$\hat{T}_{CDW} = \hat{T} + \sum_{j \in s} \tilde{w}_j r_j,$$

with $w_j(x_i)$ and $\tilde{w}_j$ defined as in Section 11.3.1. In the case where $h(Y_i) = Y_i$, and the parameter estimates of $p(x)$ are linear functions of the Y, $\hat{T}_{CDW}$ will be linear in the sample Y's.

Theoretical properties of this estimator have not been studied except in the special case of $h(Y_i) = I\{Y_i \leqslant t\}$ (see Section 11.4.6). However, the following small simulation study on the Hospitals population ($N = 393$) suggests the possible effectiveness of this technique.

One thousand samples of size $n = 64$ were selected by *srswor*. The total of the Y values was estimated in each sample, using the ratio estimator $\hat{T}_R$, with the working model $Y_i = \beta x_i + x_i^{1/2} \varepsilon_i$, $\varepsilon_i \sim (0, \sigma^2)$. Also recorded were the vector of sample residuals $\mathbf{r}_s = \mathbf{y}_s - \hat{\beta}\mathbf{x}_s$. Weights $\tilde{w}_i$, for $i \in s$, were calculated by the following steps:

1. The x-values were transformed to $x^{1/4}$. Under this transformation, the size of the larger gaps in the x's seemed roughly equal. The range of $x^{1/4}$ was [1.78, 5.60], so the average distance between x's was 0.0097 in the population and 0.06 in a "typical" sample.

2. In each sample, for each value $x_i^{1/4}$ in the corresponding non-sample, the distance δ_i was calculated to the fourth least distant sample value $x_j^{1/4}$, $j \in s$. For example, the 330th unit in Hospitals has $x_i^{1/4} = 4.73$; if the fourth roots of sample x's were, in order, ..., 4.52, 4.54, 4.66, 4.68, 4.78, 4.81, 4.82, 4.83, ..., then $\delta_i = 0.10$. For one of the more extreme samples—one having larger gaps—a window of width $b^* = 0.6$ covered 97% of these gaps, while $b^* = 0.3$ covered 92%. We focused on these two bandwidths, using a variable bandwidth $b = \max(b^*, \delta_i)$ at each non-sample point, to guarantee at least four sample points contributing to the estimate at any non-sample point.

3. The kernel used was the bi-weight $K(u) = \frac{15}{16}(1-u^2)^2 I\{|u| \leqslant 1\}$. Then, for the particular bandwidth b, $w_j(x_i) = K_b(x_j^{1/4} - x_i^{1/4})/\Sigma_{j' \in s} K_b(x_{j'}^{1/4} - x_i^{1/4})$, $i \in r$, $j \in s$, and $\tilde{w}_j$ was defined by (11.3.4).

The nonparametric calibration estimators $\hat{T}_{CDW,b} = \hat{T}_R + \Sigma_{j \in s} \tilde{w}_j r_j$ were calculated for $b^* = 0.3$ and $b^* = 0.6$, for each sample, with the adjustments to bandwidth described above.

Figure 11.5 depicts the biases $\hat{T} - T$ of the ratio estimator and the CDW estimator at the two nominal bandwidths 0.3 and 0.6, smoothed against the sample means $\bar{x}_s$, using the Royall and Cumberland (1981a) device of plotting mean conditional bias, that is, with samples sorted by $\bar{x}_s$, mean bias is plotted for each of 20 groups of 50 samples. Note that $T = 320{,}159$, so that the relative

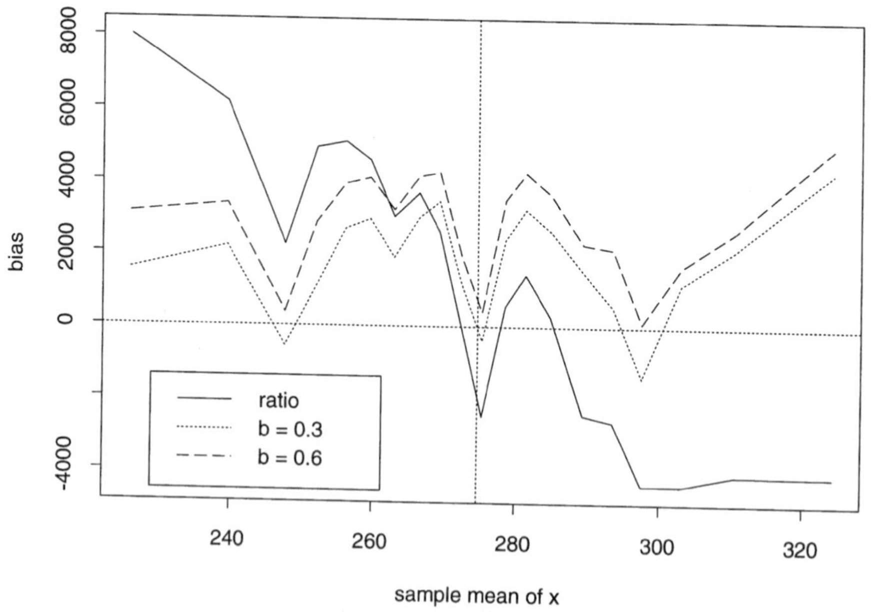

Figure 11.5 Biases of the ratio estimator and the CDW estimator at the two nominal bandwidths (0.3 and 0.6) plotted versus $\bar{x}_s$ for 500 *srswor* samples of $n = 64$. A vertical line is drawn at the population mean $\bar{x}$.

bias of the estimators is not great. The population mean of the x's is $\bar{x} = 274.7$. As we would anticipate, the bias of the ratio estimator trends from being too high to being too low. Except for the anomalous behavior at the very high $\bar{x}$'s, the nonparametric calibration estimators maintain roughly the same level of bias across $\bar{x}_s$, and it is typically less than that of the ratio estimator. Across all samples, the *mean square errors* for $\hat{T}_R$, $\hat{T}_{CDW,0.3}$, $\hat{T}_{CDW,0.6}$ were, respectively, 135.6, 123.0, and 118.6 times 10^6, so globally in terms of mean square error the longer bandwidth did better.

The original *parametric* estimator need not be based on the *BLU*. Chambers (1996a) carried out a large empirical study on establishment data, investigating the nonparametric calibration of ridge regression based estimators of total using multiple auxiliary data, with results that were encouraging for nonparametric calibration.

Variance estimation and confidence interval estimation have yet to be explored for the calibration estimators.

11.4 DISTRIBUTION FUNCTION AND QUANTILE ESTIMATION

Throughout this text, we have focused on the goal of estimating totals, means, and ratios. For some analyses, a more useful description of a population is the cumulative distribution function (CDF) of the variable in question, in which, for selected values t, we determine the proportion of units having a value less than or equal to t. Sedransk and Sedransk (1979), for instance, used estimated CDFs to compare patient care in radiation therapy facilities in a national survey of cancer patient medical records. Specific features of a distribution, like the median and the interquartile range, may also be estimated. The U.S. Bureau of Labor Statistics, for example, regularly publishes estimates of median earnings for wage and salary workers using data from the Current Population Survey of households. Study of the income distribution and its changes over time has always been important in economics. In the United States in 1993 a family of four needed an annual income of about $15,000 to be above the "poverty threshold," a figure that varies depending on family size and economic conditions (U.S. Department of Labor, 1995). In 1969 an estimated 9.6% of families were living in poverty in the United States but in 1993 the percentage was 12.6, according to the same report. For these and similar quantities, it is useful to estimate a CDF.

Figure 11.6 is a plot of the Cancer population from Appendix B in which units are counties, Y is breast cancer mortality in 1950–1969 for white females and x is the adult female population in 1960. The population size $N = 301$ is known. If we want the proportion of counties with the number of deaths less than or equal to 50, this corresponds to counting all the observations below the line drawn at $Y = 50$. When a sample is selected, we will know whether each sample Y is below the line. Estimating the population CDF is, then, equivalent to estimating the number of nonsample Y's below $t = 50$.

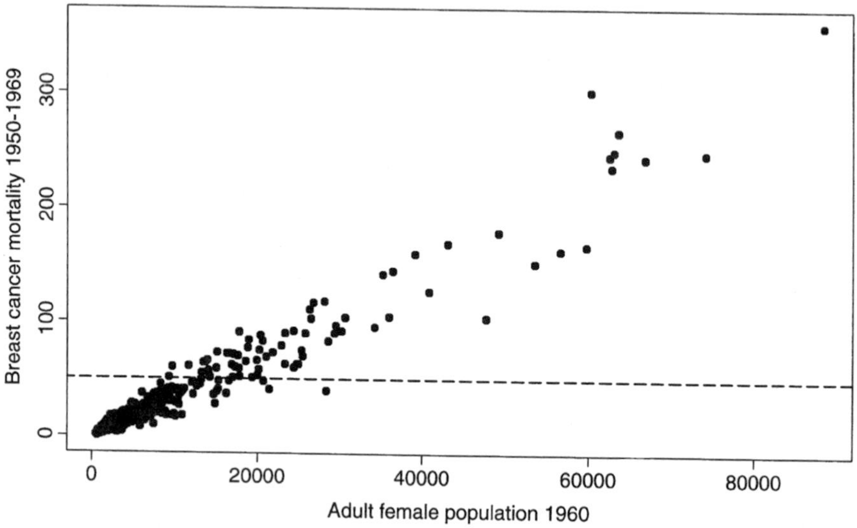

Figure 11.6 Plot of the Cancer population with an example value at which to evaluate the CDF.

Formally, we define the finite population distribution function $F_N(\cdot)$ of a variable Y, evaluated at some point t, as

$$F_N(t) = \frac{1}{N} \sum_{i=1}^{N} I\{Y_i \leqslant t\}, \tag{11.4.1}$$

where $I\{Y_i \leqslant t\}$ is the indicator function,

$$I\{Y_i \leqslant t\} = \begin{cases} 1 & Y_i \leqslant t \\ 0 & Y_i > t \end{cases}.$$

The CDF is a type of finite population total, not of the Y's themselves, but of a function of the Y's, $I\{Y_i \leqslant t\}$. The corresponding finite population α-quantile of Y is

$$q_\alpha = \inf\{t; F_N(t) \geqslant \alpha\}, \tag{11.4.2}$$

that is, the smallest value of t for which at least $100\alpha\%$ of the Y's are less than or equal to that value. If the inverse of $F_N(\cdot)$ is well defined, the α-quantile can also be expressed as $q_\alpha = F_N^{-1}(\alpha)$.

The finite population CDF is a step function as pictured in Figure 11.7. In panel (a) the finite population of y's consists of the numbers $\{1, 3, 6, 11, 12\}$.

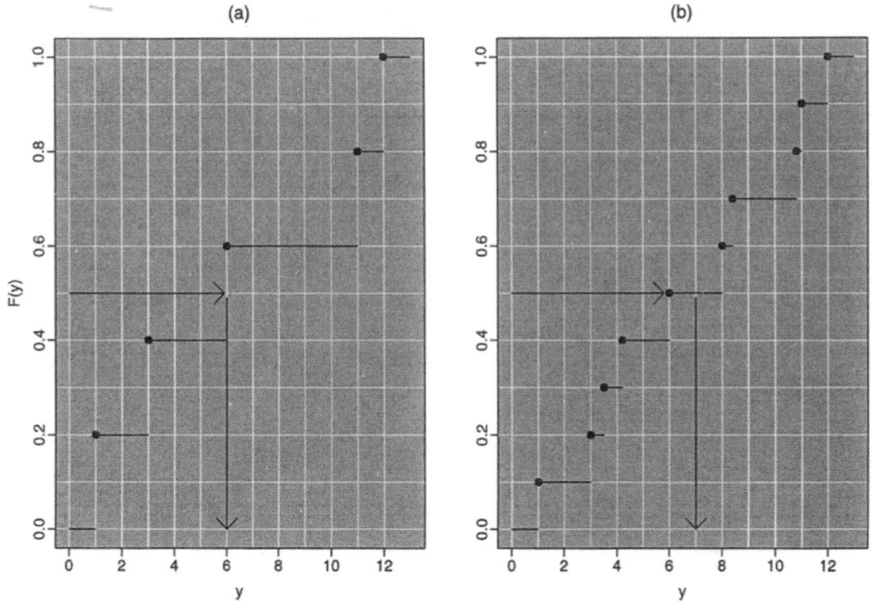

Figure 11.7 Example cumulative distribution functions and the medians of the distributions.

At each point, $F_N(\cdot)$ has a jump of size $1/N = 1/5$. We have drawn the graphs with a dot where an actual data point occurs, followed by a horizontal line to the value where another data value is observed, at which point a jump is shown. Often a vertical line is drawn at each jump to create a stair-step type of picture, but these are omitted in Figure 11.7 to emphasize the discontinuities at the jump points. The median, that is, the 0.5-quantile or 50th percentile, in (a) is $F_N^{-1}(0.5) = 6$, corresponding to the $[N/2]$ order statistic, where "[ ]" means the smallest integer greater than or equal to the number in the brackets. In panel (b) the population of y's is $\{1, 3, 3.5, 4.2, 6, 8, 8.4, 10.8, 11, 12\}$. In that case, $q_{0.5}$ defined by (11.4.2) is again 6, being the smallest value of y for which $F_N(y) \geqslant 0.5$. But, if we consider $F_N^{-1}(0.5)$, the inverse CDF maps to any of the values $6 \leqslant y < 8$. The value of the median is traditionally taken to be $(y_{(5)} + y_{(6)})/2 = 7$. This computation is consistent with assuming that $F_N(\cdot)$ is the realization of some continuous, underlying CDF F and that $(y_{(5)} + y_{(6)})/2$ is a "smoothed-out" estimate of the median of F. As in other parts of this book, we take our estimation target to be the finite population quantity $F_N(\cdot)$, rather than the superpopulation function $F(\cdot)$.

In the remainder of Section 11.4, we discuss the estimation of the cumulative distribution function and corresponding quantiles when models of various complexity describe the population. The description here has benefited from the presentation in Chambers (1996b).

11.4.1 Estimation Under Homogeneous and Stratified Models

The simplest model to consider is one in which the probability of having a value less than or equal to t is the same for all units and all units are uncorrelated. The model we consider is then

$$E_M[I\{Y_i \leqslant t\}] = \Pr(Y_i \leqslant t) = p(t)$$
$$\text{var}_M[I\{Y_i \leqslant t\}] = p(t)\,[1 - p(t)] \tag{11.4.3}$$

and Y_i and Y_j are uncorrelated when $i \neq j$. In this simple situation, the best linear unbiased estimator of $F_N(t)$, the population total of the 0–1 variables $I\{Y_i \leqslant t\}$, is the sample proportion of units that have Y values less than or equal to t,

$$\hat{F}_N(t) = \frac{1}{n} \sum_{i \in s} I\{Y_i \leqslant t\} \equiv p_s(t).$$

The error variance of $F_N(t)$ is

$$\text{var}_M[\hat{F}_N(t) - F_N(t)] = \frac{(1-f)}{n} p(t)\,[1 - p(t)]$$

for which an unbiased estimator is

$$v[\hat{F}_N(t) - F_N(t)] = \frac{(1-f)}{n-1} p_s(t)[1 - p_s(t)].$$

These results are the familiar ones for estimating a proportion that were introduced earlier in Example 2.3.5.

Note that if the underlying model on the Y's is $Y_i = \mu + \varepsilon_i$, where the errors have a common distribution, $\varepsilon_i \sim G(\cdot)$, then (11.4.3) follows with $p(t) = G(t - \mu)$.

An estimator of a quantile can be obtained by inverting the estimator of the cumulative distribution function. Given a value of α, an estimator of q_α is defined by

$$\hat{q}_\alpha = \inf\{t; \hat{F}_N(t) \geqslant \alpha\}. \tag{11.4.4}$$

Both the population CDF, $F_N(t)$, and its estimator, $\hat{F}_N(t)$, are step functions as shown in Figure 11.7. The estimator $\hat{F}_N(t)$ has jumps only at the sample Y values. Thus (11.4.4) does not have a unique solution except when t is equal to Y_i for one of the sample units. A smoothing operator is often used to make $\hat{F}_N(t)$ continuous and monotone increasing on an interval (a, b) with $a < n^{-1}$ and $b > (n-1)/n$. After applying the smoother, a unique solution for an estimated quantile will exist.

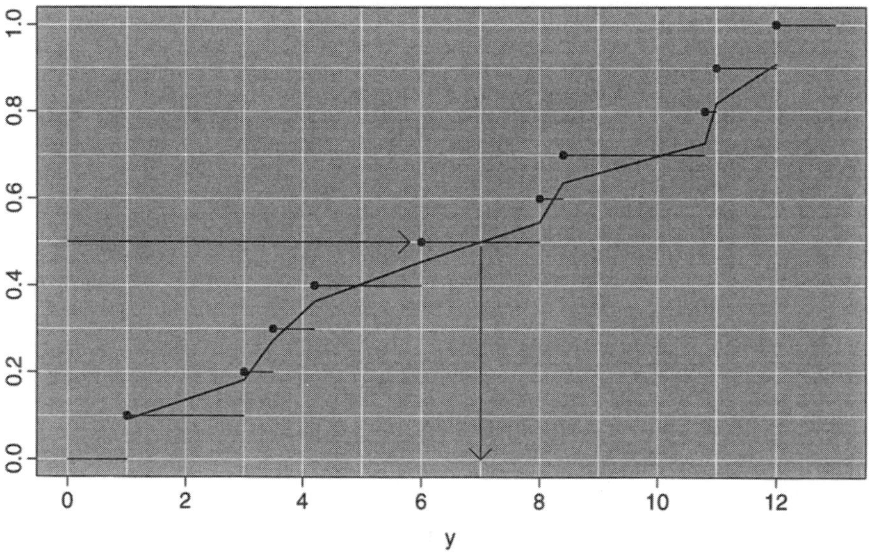

Figure 11.8 Example of a smoothed cumulative distribution function.

One simple, smoothed version of $\hat{F}_N(t)$ is found in Hogg and Tanis (1983). Let $Y_{(i)}$ be the ith sample order statistic. For $Y_{(i)} < t < Y_{(i+1)}$, the value of the smoothed CDF estimator is

$$\hat{F}_N^*(t) = \frac{1}{n+1}\left[i + \frac{t - Y_{(i)}}{Y_{(i+1)} - Y_{(i)}}\right]. \tag{11.4.5}$$

Figure 11.8 is a graph of the CDF from panel (b) of Figure 11.7 with the smoothed CDF superimposed on the step function. The unique median based on the smoothed $\hat{F}_N^*(t)$ is 7, which is the average of $y_{(5)}$ and $y_{(6)}$. As the figure makes clear, $\hat{F}_N^*(t)$ is not defined outside the range of the observed data, that is, for $t < y_{(1)}$ or $t > y_{(n)}$.

Woodruff (1952) suggested a simple way of getting a confidence interval for a quantile q_α. The general idea is to put a confidence interval around $F(q_\alpha)$ and then invert the endpoints to get an interval estimate for q_α. In particular, we find points $q_{L\alpha}$ and $q_{U\alpha}$ such that

$$\begin{aligned}\hat{F}_N(q_{L\alpha}) &= \alpha - z_{1-\gamma/2}\sqrt{v[\hat{F}_N(\hat{q}_\alpha) - F_N(q_\alpha)]}\\ \hat{F}_N(q_{U\alpha}) &= \alpha + z_{1-\gamma/2}\sqrt{v[\hat{F}_N(\hat{q}_\alpha) - F_N(q_\alpha)]}\end{aligned},$$

where $z_{1-\gamma/2}$ is the $(1-\gamma/2)$-percentile from the standard normal distribution. A $100(1-\gamma)\%$ confidence interval for q_α is then $[q_{L\alpha}, q_{U\alpha}]$. Francisco and Fuller (1991) gave design-based theory showing that the Woodruff method gives correct coverage rates asymptotically, and also introduced a more complicated "test inversion" procedure for constructing a confidence interval that also yields nominal coverage rates in large samples. In that procedure the confidence interval for q_α is

$$\left\{t: \frac{|\hat{F}_N(t) - \alpha|}{\sqrt{v[\hat{F}_N(t) - F_N(t)]}} \leqslant z_{1-\gamma/2}\right\}.$$

This is the set of t's that would not be rejected in a test of the hypothesis $H_0: F_N(t) = \alpha$. The Francisco-Fuller theory is design-based, but their approach, which is akin to calibration in a regression setting (cf. Carroll and Ruppert, 1988, sec. 2.9.3) is appropriate to a model-based setting also.

The extension of the CDF estimator to a stratified population is straightforward. Suppose that $h = 1, \ldots, H$ index the strata and that for units $i = 1, \ldots, N_h$ in stratum h the model is

$$E_M[I\{Y_{hi} \leqslant t\}] = \Pr(Y_{hi} \leqslant t) = p_h(t)$$
$$\text{var}_M[I\{Y_{hi} \leqslant t\}] = p_h(t)[1 - p_h(t)] \qquad (11.4.6)$$

with all Y_{hi}'s uncorrelated. The best linear unbiased estimator of $F_N(t)$ is

$$\hat{F}_N(t) = \frac{1}{N} \sum_h N_h p_{hs}(t), \qquad (11.4.7)$$

where $p_{hs}(t) = n_h^{-1} \Sigma_{i \in s_h} I\{Y_{hi} \leqslant t\}$. Derivation of the error variance and its unbiased estimator is left to Exercise 11.11.

Since the estimator $p_{hs}(t)$ is the mean of indicator functions, the stratified estimator (11.4.7) is a weighted sum of stratum-level step functions. Each of the stratum-level CDF's can be smoothed as in (11.4.5) to obtain $\hat{F}_h^*(t)$, say, and the estimate at a given point t is then the weighted sum of the smoothed CDF's:

$$\hat{F}_N^*(t) = \frac{1}{N} \sum_h N_h \hat{F}_h^*(t).$$

Either the Woodruff method or the Francisco-Fuller approach can be applied to the stratified estimator to construct a confidence interval for a quantile.

11.4.2 Estimation of $F_N(\cdot)$ Under a Regression Model

We will study the estimation of $F_N(\cdot)$ under the linear regression model

$$Y_i = \mathbf{x}_i'\boldsymbol{\beta} + \varepsilon_i, \; i = 1, \ldots, N, \qquad (11.4.8)$$

where $\mathbf{x}_i' = (x_{i1}, \ldots, x_{ip})$ is a vector of p auxiliaries, and the ε_i's are independent errors with $E_M(\varepsilon_i) = 0$ and $\text{var}_M(\varepsilon_i) = v_i$. Note that the finite population CDF can be written as the sum over the sample plus the nonsample:

$$F_N(t) = \frac{1}{N}\left[nF_s(t) + (N-n)F_r(t)\right], \tag{11.4.9}$$

where $F_s(t) = \Sigma_s I\{Y_i \leqslant t\}/n$ and $F_r(t) = \Sigma_r I\{Y_i \leqslant t\}/(N-n)$. The first term in brackets on the right-hand side of (11.4.9), being the sum over the sample indicator functions, is known, and the task is to estimate the second term, $F_r(t)$. One straightforward approach would be to estimate $F_N(t)$ as

$$\tilde{F}(t) = \frac{1}{N}\left[nF_s(t) + (N-n)\tilde{F}_r(t)\right]$$

where

$$\tilde{F}_r(t) = \Sigma_r I\{\hat{Y}_i \leqslant t\}/(N-n)$$

with

$$\hat{Y}_i = \mathbf{x}_i'\hat{\boldsymbol{\beta}} \quad \text{and} \quad \hat{\boldsymbol{\beta}} = (\mathbf{X}_s'\mathbf{V}_{ss}^{-1}\mathbf{X}_s)^{-1}\mathbf{X}_s'\mathbf{V}_{ss}^{-1}\mathbf{Y}_s,$$

the standard least squares estimator from Section 2.2. (Assume, for simplicity, that $\mathbf{X}_s'\mathbf{V}_{ss}^{-1}\mathbf{X}_s$ has full rank.) However, the estimator $\tilde{F}(t)$ is biased (see Exercise 11.13).

A better approach, due to Chambers and Dunstan (1986), is to build an estimator from the model's residuals. Let $G_i(t) = \Pr(Y_i \leqslant t)$ and $G(\cdot)$ be the distribution function of the normalized errors $\varepsilon_i/\sqrt{v_i}$. The expected value of $F_r(t)$ is then

$$\begin{aligned} E_M\left[\sum_{i\in r} I\{Y_i \leqslant t\}\right] &= \sum_{i\in r} \Pr(Y_i \leqslant t) \\ &= \sum_{i\in r} \Pr\left(\frac{\varepsilon_i}{\sqrt{v_i}} \leqslant \frac{t - \mathbf{x}_i'\boldsymbol{\beta}}{\sqrt{v_i}}\right) \\ &= \sum_{i\in r} G((t - \mathbf{x}_i'\boldsymbol{\beta})/\sqrt{v_i}). \end{aligned}$$

A natural estimator of $E_M(F_r(t))$ is

$$\hat{F}_r(t) = \frac{1}{N-n}\sum_{i\in r} \hat{G}_i(t) \tag{11.4.10}$$

with $\hat{G}_i(t) = \Sigma_{j\in s} I\{r_j \leqslant (t - \mathbf{x}_i'\hat{\boldsymbol{\beta}})/\sqrt{v_i}\}/n$, where $r_j = (Y_j - \mathbf{x}_j'\hat{\boldsymbol{\beta}})/\sqrt{v_j}$. To compute $\hat{G}_i(t)$ for each nonsample point, $i \in r$, we determine the sample proportion of standardized residuals that are less than or equal to $((t - \mathbf{x}_i'\hat{\boldsymbol{\beta}})/\sqrt{v_i}$. Notice that this requires that we know the individual x's for all nonsample points.

The event $r_j \leqslant (t - \mathbf{x}_i'\hat{\boldsymbol{\beta}})/\sqrt{v_i}$ is equivalent to

$$\frac{\varepsilon_j}{\sqrt{v_j}} \leqslant t_i + \left(\frac{\mathbf{x}_j'}{\sqrt{v_j}} - \frac{\mathbf{x}_i'}{\sqrt{v_i}}\right)(\hat{\boldsymbol{\beta}} - \boldsymbol{\beta}) \cong t_i + \left(\frac{\mathbf{x}_j'}{\sqrt{v_j}} - \frac{\mathbf{x}_i'}{\sqrt{v_i}}\right)(\boldsymbol{\beta}_{(i)} - \boldsymbol{\beta}),$$

where $t_i = (t - \mathbf{x}_i'\boldsymbol{\beta})/\sqrt{v_i}$ and $\hat{\boldsymbol{\beta}}_{(i)}$ is the *BLU* estimator of $\boldsymbol{\beta}$ omitting sample unit i. Now, suppose that $G(t)$ is differentiable at all values of t with derivative $g(t)$. Then using a first-order linear approximation at the point t_i,

$$\begin{aligned} E_M(I\{r_j \leqslant (t - \mathbf{x}_i'\hat{\boldsymbol{\beta}})/\sqrt{v_i}\}) &\cong E_M\left\{G\left(t_i + \left[\frac{\mathbf{x}_j'}{\sqrt{v_j}} - \frac{\mathbf{x}_i'}{\sqrt{v_i}}\right][\hat{\boldsymbol{\beta}}_{(i)} - \boldsymbol{\beta}]\right)\right\} \\ &\cong G(t_i) + g(t_i)\left(\frac{\mathbf{x}_j'}{\sqrt{v_j}} - \frac{\mathbf{x}_i'}{\sqrt{v_i}}\right)E_M(\hat{\boldsymbol{\beta}}_{(i)} - \boldsymbol{\beta}). \end{aligned} \tag{11.4.11}$$

Under model (11.4.8) it follows that $E_M(\hat{G}_i(t)) \cong G_i(t) = G((t - \mathbf{x}_i'\boldsymbol{\beta})/\sqrt{v_i})$ as long as $\hat{\boldsymbol{\beta}}$ is consistent for $\boldsymbol{\beta}$. Thus an approximately unbiased estimator of the expectation of $F_N(t)$ is

$$\hat{F}_{CD}(t) = \frac{1}{N}[nF_s(t) + (N - n)\hat{F}_r(t)]. \tag{11.4.12}$$

See Chambers, Dorfman, and Hall (1992) for more detailed analysis. A corresponding quantile estimator is simply

$$\hat{q}_\alpha = \inf\{t: \hat{F}_{CD}(t) \geqslant \alpha\}. \tag{11.4.13}$$

Example 11.4.1 The units in the table below are a systematic, equal probability sample of $n = 10$ from the Cancer population in Appendix B. The population was sorted in ascending order on x, the adult female population in 1960 before sampling. When there is a strong relationship between Y and x, as is the case here, a systematic sample has the clear advantage of spreading the sample units over the likely range of Y. The population CDF, computed from all $N = 301$ units, and the sample estimate $\hat{F}_{CD}(t)$ are plotted in the upper half of Figure 11.9. For clarity, we have connected the points in the step functions in the figure. $\hat{F}_{CD}(t)$ was calculated under the model $M(1, 1\!:\!x)$. To compute $\hat{F}_{CD}(t)$ we need an estimate of the variance parameter σ^2 since it enters into $\hat{G}_i(t)$ for each nonsample unit through $v_i = \sigma^2 x_i$. For this example, we used the standard estimator $\hat{\sigma}^2 = \mathbf{e}_s' \, diag(x_i^{-1})\mathbf{e}_s/(n - 1)$ where $\mathbf{e}_s = \mathbf{Y}_s - \mathbf{X}_s\hat{\boldsymbol{\beta}}$.

The estimate $\hat{F}_{CD}(t)$ was computed at 20 values of t equal to the 5, 10, 15,..., 95, and 100% quantiles in the full Cancer population. Note that even with a

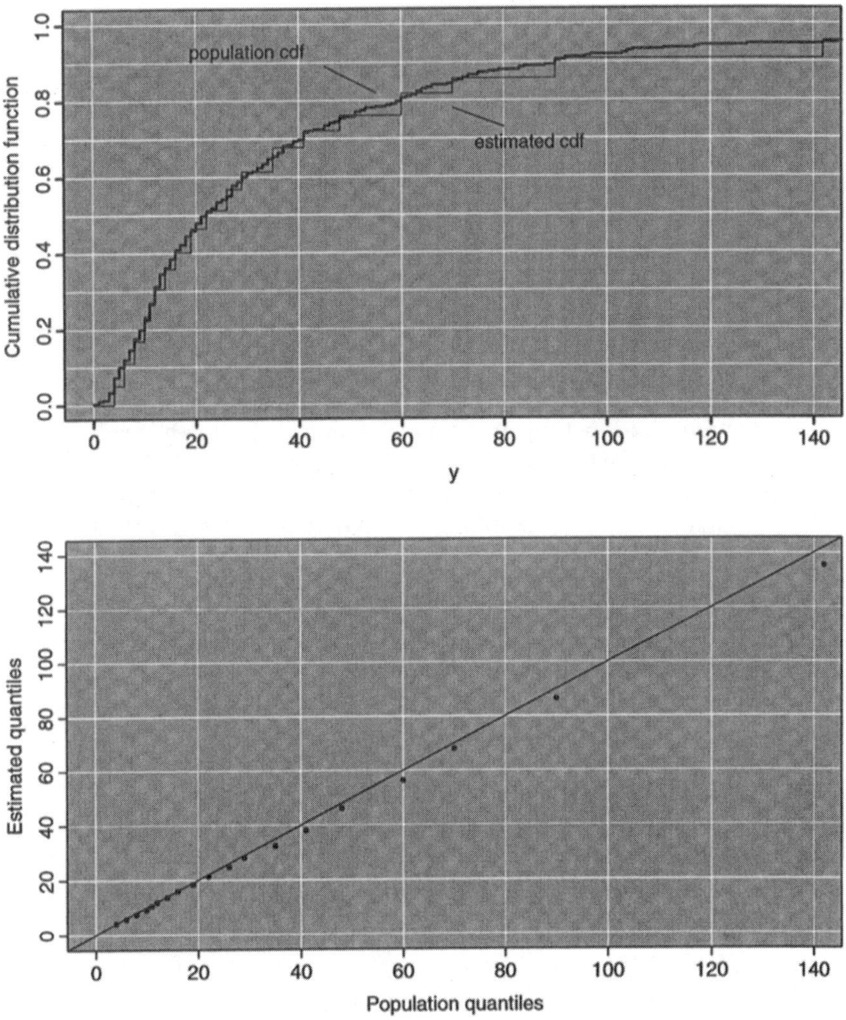

Figure 11.9 Cumulative distribution function for the Cancer population and an estimated CDF plus quantile-quantile plot.

sample of 10 we can evaluate the CDF at many different choices of t—not just the 20 used in the figure. Recall that for homogeneous or stratified populations in Section 11.4.1, $\hat{F}_N(t)$ had a jump point at each sample value. When regression model (11.4.8) holds, we have much more flexibility. For example, we could use a set of t's equal to the sample Y values and each predicted nonsample value, $\hat{Y}_i = \mathbf{x}_i'\hat{\boldsymbol{\beta}}$. Thus we can estimate the value of the CDF for every unit in the population as long as we know the auxiliary $\mathbf{x}_i$ for all population units.

Unit No.	x	y
11	1,114	5
41	2,154	4
71	2,793	12
102	3,733	12
132	5,108	12
162	7,031	29
192	9,215	34
222	13,206	55
252	19,818	66
282	34,109	96

As in Section 11.4.1, getting an estimate of a quantile like the median or the first or third quartile based on (11.4.12) can be somewhat cumbersome computationally since $\hat{F}_{CD}$ is a step function without a neat inverse. One method, similar to that for smoothing the CDF as in (11.4.5) is the following. For a given quantile q_α:

1. Evaluate $\hat{F}_{CD}(t)$ at a set of ordered values $t_1 < t_2 < \cdots < t_K$.
2. Find k such that $\alpha_1 = \hat{F}_{CD}(t_{k-1}) < \alpha \leqslant \hat{F}_{CD}(t_k) = \alpha_2$.
3. Compute the interpolated value

$$\hat{q}_\alpha = t_k - [t_k - t_{k-1}]\frac{\alpha_2 - \alpha}{\alpha_2 - \alpha_1}.$$

$\hat{F}_{CD}(\hat{q}_\alpha)$ can then be checked for closeness to α. If necessary, another interpolation can be performed between $\hat{q}_\alpha$ and either t_k or t_{k-1}. In Example 11.4.1, $t_{11} = 22$ and $t_{10} = 19$ with $\hat{F}_{CD}(22) = 0.513$ and $\hat{F}_{CD}(19) = 0.464$. Interpolating as above gives the estimated median as $\hat{q}_{0.50} = 21.224$. Similar computations lead to $\hat{q}_{0.25} = 10.521$ and $\hat{q}_{0.75} = 46.131$. These estimates compare to the actual population quartiles of 11, 22, and 48. Note that the Y variable in the Cancer population—breast cancer mortality—is naturally measured as an integer. Reporting quantiles to several decimals is, thus, specious accuracy but does illustrate the general interpolation procedure described above. Korn and Graubard (1999) and Korn, Midthune, and Graubard (1997) give alternative methods of interpolation.

The lower half of Figure 11.9 is a quantile–quantile plot of the estimated 5, 10, 15, ..., 95% quantiles versus the same quantiles for the full Cancer population. This so-called "q-q" plot (see, e.g., Cleveland, 1993) is usually more informative when comparing two distributions than the CDF plot in the upper panel of Figure 11.9. In this case, the q–q plot confirms the impression from the CDF plot that the sample estimates are close to the population quantities for the smaller quantiles but are slight underestimates for larger values of Y.

11.4.3 Large Sample Properties

Because both $\hat{F}_{CD}(t)$ and $F_N(t)$ are nonlinear in Y, large sample approximations are needed to study their properties. As noted in the last section, $\hat{F}_{CD}(t)$ is approximately unbiased under the working model. The prediction error, $\hat{F}_{CD}(t) - F_N(t)$, can be written as

$$\begin{aligned}\hat{F}_{CD}(t) - \hat{F}_N(t) &= (1-f)[\hat{F}_r(t) - F_r(t)] \\ &= (1-f)\{\hat{F}_t(t) - E_M(F_r(t)) - [F_r(t) - E_M(F_r(t))]\}. \quad (11.4.14)\end{aligned}$$

We can get an expression for $\text{var}_M[\hat{F}_{CD}(t) - F_N(t)]$ through a linear approximation to (11.4.14). Using (11.4.11), we have

$$\begin{aligned}\hat{F}_r(t) &= \frac{1}{N-n}\sum_{i\in r}\hat{G}_i(t) \\ &\cong \frac{1}{N-n}\sum_{i\in r}G_i(t) + \frac{1}{(N-n)n}\sum_{i\in r}g(t_i)\sum_{j\in s}\left(\frac{\mathbf{x}_j'}{\sqrt{v_j}} - \frac{\mathbf{x}_i'}{\sqrt{v_i}}\right)(\hat{\boldsymbol{\beta}} - \boldsymbol{\beta}).\end{aligned}$$

Since $E_M(F_r(t)) = \Sigma_{i\in r}G_i(t)/(N-n)$, it follows that

$$\begin{aligned}\hat{F}_r(t) - E_M(F_r(t)) &= \frac{1}{(N-n)n}\sum_{i\in r}g(t_i)\sum_{j\in s}\left(\frac{\mathbf{x}_j'}{\sqrt{v_j}} - \frac{\mathbf{x}_i'}{\sqrt{v_i}}\right)(\hat{\boldsymbol{\beta}} - \boldsymbol{\beta}) \\ &= \mathbf{d}'(\hat{\boldsymbol{\beta}} - \boldsymbol{\beta}), \quad (11.4.15)\end{aligned}$$

where $\mathbf{d}$ is the $p \times 1$ vector with kth element

$$\frac{1}{(N-n)n}\sum_{i\in r}g(t_i)\sum_{j\in s}\left(\frac{x_{jk}}{\sqrt{v_j}} - \frac{x_{ik}}{\sqrt{v_i}}\right), \qquad k = 1,\ldots,p.$$

From (11.4.15) the approximate variance of $\hat{F}_r(t)$ is

$$\text{var}_M[\hat{F}_r(t)] \cong \mathbf{d}'\,\text{var}_M(\hat{\boldsymbol{\beta}})\mathbf{d}. \quad (11.4.16)$$

For a fixed value of t, the nonsample population CDF, $F_r(t) = \Sigma_r I\{Y_i \leqslant t\}/(N-n)$, is the sum of uncorrelated Bernoulli random variables under model (11.4.8). Thus

$$\text{var}_M[F_r(t) - E_M(F_r(t))] = \text{var}_M[F_r(t)] = \frac{1}{(N-n)^2}\sum_{i\in r}G_i(t)[1 - G_i(t)]. \quad (11.4.17)$$

Since the sample and nonsample Y's are independent, the error variance is approximately

$$\text{var}_M[\hat{F}_{CD}(t) - F_N(t)] \cong (1-f)^2\left\{\mathbf{d}'\,\text{var}_M(\hat{\boldsymbol{\beta}})\mathbf{d} + \frac{1}{(N-n)^2}\sum_{i\in r} G_i(t)\,[1 - G_i(t)]\right\}. \tag{11.4.18}$$

Under some reasonable conditions, the elements of $\mathbf{d}$ will be $O(1)$, $\text{var}_M(\hat{\boldsymbol{\beta}}) = O(n^{-1})$ elementwise, and $\text{var}_M[F_r(t) = O[(N-n)^{-1}]$, so that, as $f \to 0$, the dominant term of the error variance is $(1-f)^2\,[\mathbf{d}'\,\text{var}_M(\hat{\boldsymbol{\beta}})\mathbf{d}]$. Chambers and Dunstan (1986) give conditions under which $\hat{F}_{CD}(t) - F_N(t)$ is asymptotically normal with mean 0 and with variance given by (11.4.18).

11.4.4 The Effect of Model Misspecification

The definition of $\hat{F}_{CD}(t)$ depends on both the assumption for the mean and the variance of Y_i. If model (11.4.8) is misspecified, $\hat{F}_{CD}(t)$ can be biased just as an estimator of the total might be. Suppose that the correct model is

$$Y_i = \mu(\mathbf{x}_i) + \varepsilon_i, \qquad i = 1, \ldots, N, \tag{11.4.19}$$

where μ is a scalar function of the vector $\mathbf{x}_i$ and $\varepsilon_i \sim (0, a_i)$ and independent. The nonsample estimator $\hat{F}_r(t)$ is built around $\hat{G}_i(t) = \Sigma_{j\in s} I\{r_j \leqslant (t - \mathbf{x}_i'\hat{\boldsymbol{\beta}})/\sqrt{v_i}\}/n$. Suppose that in large samples $(\hat{\boldsymbol{\beta}} - \boldsymbol{\beta}^*) = o_p(1)$, where $\boldsymbol{\beta}^*$ is a constant to which $\hat{\boldsymbol{\beta}}$ converges, and let $\mu_i = \mu(\mathbf{x}_i)$. We have $r_j \leqslant (t - \mathbf{x}_i'\hat{\boldsymbol{\beta}})/\sqrt{v_i}$ if and only if

$$\frac{Y_j - \mu_j}{\sqrt{a_j}} \leqslant \left[\frac{t - \mathbf{x}_i'\hat{\boldsymbol{\beta}}}{\sqrt{v_i}} + \frac{\mathbf{x}_j'\hat{\boldsymbol{\beta}}}{\sqrt{v_j}}\right]\sqrt{\frac{v_j}{a_j}} - \frac{\mu_j}{\sqrt{a_j}}.$$

Adding and subtracting $\sqrt{v_j}\mathbf{x}_i'\boldsymbol{\beta}^*/\sqrt{a_j v_i}$ and $\mathbf{x}_j'\boldsymbol{\beta}^*/\sqrt{a_j}$ from the right-hand side and doing some rearranging, we have

$$\begin{aligned}\frac{Y_j - \mu_j}{\sqrt{a_j}} &\leqslant \sqrt{\frac{v_j a_i}{a_j v_i}}\,\frac{t - \mathbf{x}_i'\hat{\boldsymbol{\beta}}^*}{\sqrt{a_i}} + \frac{\mathbf{x}_j'\boldsymbol{\beta}^* - \mu_j}{\sqrt{a_j}} + \sqrt{\frac{v_j}{a_j}}\left(\frac{\mathbf{x}_j'}{\sqrt{v_j}} - \frac{\mathbf{x}_i'}{\sqrt{v_i}}\right)(\hat{\boldsymbol{\beta}} - \boldsymbol{\beta}^*)\\ &= k_{ji}t_{ai}(\boldsymbol{\beta}^*) + \Delta_j + o_p(1),\end{aligned} \tag{11.4.20}$$

where

$$t_{ai}(\boldsymbol{\beta}^*) = (t - \mathbf{x}_i'\boldsymbol{\beta}^*)/\sqrt{a_i}, \qquad k_{ji} = \sqrt{v_j a_i/(a_j v_i)},$$

and

$$\Delta_j = (\mathbf{x}_j'\boldsymbol{\beta}^* - \mu_j)/\sqrt{a_j}.$$

Thus, the event that $r_j \leqslant (t - \mathbf{x}_i'\hat{\boldsymbol{\beta}})/\sqrt{v_i}$, which is the core of $\hat{G}_i(t)$, is equivalent to the event that $(Y_j - \mu_j)/\sqrt{a_j}$ is less than or equal to the right-hand side of (11.4.20). Consequently,

$$E_M[\hat{F}_{CD}(t) - F_N(t)] \cong \frac{1}{N-n}\sum_{i\in r}\left\{\left[\frac{1}{n}\sum_{j\in s} G(k_{ji}t_{ai}(\boldsymbol{\beta}^*) + \Delta_j)\right] - G(t_{ai}(\mu))\right\}, \tag{11.4.21}$$

where $t_{ai}(\mu) = (t - \mu_i)/\sqrt{a_i}$. So, $\hat{F}_{CD}(t)$ has a bias that depends on misspecification of both the mean and variance in the model.

If the mean is properly specified, that is, $\mu(\mathbf{x}_i) = \mathbf{x}_i'\boldsymbol{\beta}$, then $\Delta_j = 0$ and (11.4.21) becomes

$$E_M[\hat{F}_{CD}(t) - F_N(t)] \cong \frac{1}{N-n}\sum_{i\in r}\left\{\left[\frac{1}{n}\sum_{j\in s} G(k_{ji}t_{ai}(\boldsymbol{\beta}))\right] - G(t_{ai}(\boldsymbol{\beta}))\right\}, \tag{11.4.22}$$

which is not zero, in general. In other words, misspecifying the variance alone is enough to bias the CDF estimator. This contrasts to the case of estimating a total T where a wrong variance specification in a linear model might lead to inefficiency but not to bias in the estimator of T. To examine this point further, suppose that the working variance is not too badly in error so that k_{ji} is near 1. Expanding $G(k_{ji}t_{ai}(\boldsymbol{\beta}))$ in a first-order Taylor series about $t_{ai}(\boldsymbol{\beta})$ gives

$$G(k_{ji}t_{ai}) \cong G(t_{ai}) + g(t_{ai})(k_{ji} - 1)t_{ai},$$

where we abbreviate $t_{ai}(\boldsymbol{\beta})$ as t_{ai}. An interesting case is $v_i/a_i = x_i^d$ with $-1 \leqslant d \leqslant 1$. (For example, if $v_i = \sigma^2 x_i$ and $a_i = \sigma^2 x_i^{3/2}$, then $d = -1/2$.) Then the approximate bias is

$$E_M[\hat{F}_{CD}(t) - \hat{F}_N(t)] \cong \frac{1}{N-n}\sum_{i\in r}\left\{g(t_{ai})t_{ai}\left[\frac{\bar{x}_s^{(d/2)}}{x_i^{d/2}} - 1\right]\right\}, \tag{11.4.23}$$

where $\bar{x}_s^{(d/2)} = \Sigma_s x_i^{d/2}/n$. Since $t_{ai}/x_i^{d/2} = (t - \mathbf{x}_i'\boldsymbol{\beta})/\sqrt{v_i} \equiv t_{vi}$, expression (11.4.23) will be 0 if this balance condition holds:

$$\bar{x}_s^{(d/2)} = \sum_{i\in r} u_i x_i^{d/2} \tag{11.2.24}$$

with $u_i = g(t_{ai})t_{vi}/\Sigma_{i\in r}g(t_{ai})t_{vi}$.

The balance condition in (11.2.24) equates a sample moment to a weighted nonsample moment, reminiscent of the overbalance condition in Scott, Brewer, and Ho (1978). Since the weights u_i involve the density function g, this type of balance cannot be easily achieved in practice.

Suppose we had a simple balanced sample, with $\bar{x}_s^{(d/2)} = \bar{x}_r^{(d/2)}$ and suppose that $t = \bar{\mathbf{x}}_r\boldsymbol{\beta}$. Then (11.4.23) reduces to

$$E_M[\hat{F}_{CD}(t) - F_N(t)] \cong \frac{1}{N-n}\sum_{i\in r}\left\{g\left(\frac{(\bar{\mathbf{x}}_r - \mathbf{x}_i)'\boldsymbol{\beta}}{\sqrt{a_i}}\right)\frac{(\bar{\mathbf{x}}_r - \mathbf{x}_i)'\boldsymbol{\beta}}{\sqrt{v_i}}\,[\bar{x}_r^{(d/2)} - x_i^{d/2}]\right\}. \tag{11.4.25}$$

If β is a positive scalar and all x's are positive, then (11.4.25) is strictly positive when $d > 0$ and strictly negative when $d < 0$ (see Exercise 11.15). That is, in balanced samples, for values of t near the mean of Y's, $\hat{F}_{CD}(t)$ will tend to be positively biased when v_i overstates the true variance and negatively biased when it understates the variance, even when $E_M(Y_i)$ is correctly specified.

Some device other than balance is required for dealing with such biases.

11.4.5 Design-Based Approaches

The simplest design-based estimator uses no auxiliary data and is

$$\hat{F}_\pi(t) = \frac{\sum_{i\in s}\pi_i^{-1}I\{Y_i \leqslant t\}}{\sum_{i\in s}\pi_i^{-1}}, \tag{11.4.26}$$

where π_i is the selection probability of unit i. When the design uses simple random sampling, (11.4.26) reduces to $p_s(t)$, the estimator for homogeneous populations in Section 11.4.1. However, if Y is related to an auxiliary x, $\hat{F}_\pi(t)$ can have a substantial conditional bias, depending on the configuration of the x's in the sample. For t equal to the median and the third quartile in populations similar to Hospitals, Cancer, and Counties70, Chambers and Dunstan (1986) and various other authors have empirically demonstrated biases of this design-based estimator similar to ones we observed earlier in Figure 3.3.

Rao, Kovar, and Mantel (1990) introduced several alternatives to $\hat{F}_\pi(t)$ and $\hat{F}_{CD}(t)$, including the following, which is motivated by model (11.4.8):

$$\hat{F}_{RKM}(t) = \frac{1}{N}\left[\sum_{j\in s}\pi_j^{-1}I\{Y_j \leqslant t\} + \sum_{i=1}^{N}\hat{G}_{\pi i}(t) - \sum_{j\in s}\pi_j^{-1}\hat{G}_{\pi ic}(t)\right],$$

where

$$\hat{G}_{\pi i}(t) = \frac{\sum_{j\in s}\pi_j^{-1}I\{r_j \leqslant (t - \mathbf{x}_i'\hat{\boldsymbol{\beta}})/\sqrt{v_i}\}}{\sum_{j\in s}\pi_j^{-1}} \quad \text{and}$$

$$\hat{G}_{\pi ic}(t) = \frac{\sum_{j\in s}(\pi_i/\pi_{ij})I\{r_j \leqslant (t - \mathbf{x}_i'\hat{\boldsymbol{\beta}})/\sqrt{v_i}\}}{\sum_{j\in s}(\pi_i/\pi_{ij})}$$

with π_{ij} being the joint selection probability of units i and j. The ratio π_{ij}/π_i is the conditional probability that unit j is in the sample given that unit i has been selected. This estimator is asymptotically design-unbiased and model-unbiased under the regression model (11.4.8) (see Exercise 11.16). A closely related estimator was proposed by Godambe (1989).

Dorfman (1993) noted that $\hat{F}_{RKM}(\cdot)$ was a step toward a model robust CDF estimator. Replacing $\hat{G}_{\pi i}(t)$ and $\hat{G}_{\pi ic}(t)$ by $\hat{G}_i(t)$ used in (11.4.10), we have

$$\hat{F}_{RKM^*}(t) = \frac{1}{N}\left[\sum_{j\in s} \pi_j^{-1} I\{Y_j \leqslant t\} + \sum_{i=1}^{N} \hat{G}_i(t) - \sum_{j\in s} \pi_j^{-1}\hat{G}_i(t)\right]$$

$$= \hat{F}_{CD}(t) + \frac{1}{N}\sum_{j\in s} (\pi_j^{-1} - 1)[I\{Y_j \leqslant t\} - \hat{G}_j(t)],$$

and it was shown that, if the π_j^{-1} properly reflect the number of population units near the sample point j with respect to the true underlying model M, then $E_M(\hat{F}_{RKM^*}(t) - F_N(t)) \cong 0$. A detailed simulation study concluded that $\hat{F}_{RKM}(\cdot)$ and $\hat{F}_{RKM^*}(\cdot)$ were less sensitive to model misspecification and outliers than $\hat{F}_{CD}(\cdot)$, and that, on the other hand, if routine model diagnostics are used to sharpen the model, then $\hat{F}_{CD}(\cdot)$ would generally be most efficient, often by a wide margin.

Chambers, Dorfman, and Hall (1992) compare $\hat{F}_\pi(\cdot)$, $\hat{F}_{RKM}(\cdot)$, and $\hat{F}_{CD}(\cdot)$ theoretically under a homoscedastic linear model. One surprising result is that under certain (pathological) conditions, the (model-based) variance of $\hat{F}_{CD}(\cdot)$ can exceed that of $\hat{F}_\pi(\cdot)$, even when the working linear model holds. Wang and Dorfman (1996) give an alternative estimator having lower mean square error than $\hat{F}_{RKM}(\cdot)$ or $\hat{F}_{CD}(\cdot)$ under the homoscedastic linear model on Y.

Another design-based approach, when an auxiliary x is available, is to post-stratify the sample based on x, that is, sort the sample based on x and divide it into H groups. A post-stratified CDF estimator is

$$\hat{F}_{PS}(t) = \frac{1}{N}\sum_{h=1}^{H} N_h \hat{F}_h(t),$$

where N_h is the number of population units in post-stratum h (which must be known), and $\hat{F}_h(t) = \Sigma_{i\in s_h}\pi_i^{-1} I_{\{Y_i \leqslant t\}}/\Sigma_{i\in s_h}\pi_i^{-1}$. This estimator, in effect, fits a common mean within each stratum. When Y is related to x, and the post-strata are numerous and narrow, this estimator can be quite effective, as demonstrated by Silva and Skinner (1995).

11.4.6 Nonparametric Regression-Based Estimators

A model-based approach to estimating CDF's using nonparametric regression with an auxiliary variable x, was introduced by Kuo (1988), who directly smooths the sample indicators $I\{Y_j \leqslant t\}$ for $j \in s$. The estimator of $\Pr(Y_i \leqslant t)$, $i \in r$, is

$$\tilde{G}_i(t) = \sum_{j \in s} w_i(x_j) I\{Y_j \leqslant t\},$$

where $w_i(x_j)$ are weights based on a kernel density, as in Section 11.3.1, or on a nearest neighbor approach, as in Kuo (1988). Then

$$\hat{F}_{KUO}(t) = \frac{1}{N} [nF_s(t) + (N - n) \hat{F}_{KUO,r}(t)],$$

where

$$\hat{F}_{KUO,r}(t) = (N - n)^{-1} \sum_{i \in r} \tilde{G}_i(t) = (N - n)^{-1} \sum_{j \in s} \tilde{w}_j I\{Y_j \leqslant t\},$$

with $\tilde{w}_j = \Sigma_{i \in r} w_j(x_i)$ as in Section 11.3.1. Underlying this estimator is the model

$$h(Y_i) \equiv I\{Y_i \leqslant t\} = m(x_i) + \varepsilon_i,$$

that is, it is assumed that $h(Y_i)$ follows model (11.3.1).

Suppose that Y itself follows model (11.3.1). An estimator patterned after the Chambers-Dunstan estimator $\hat{F}_{CD}(t)$, but replacing the BLU estimator with a nonparametric regression estimator, was introduced by Dorfman and Hall (1993) for the case where the errors follow a common distribution function, $G(t) = \Pr(\varepsilon \leqslant t)$, and, in particular, have a common variance.

As in Section 11.3.1, let $m(x)$ be estimated by $\hat{m}(x) = \Sigma_{j \in s} w_j(x) Y_j$, with the residual for sample unit j being $\hat{\varepsilon}_j = Y_j - \hat{m}(x_j)$. Then an estimator of $\Pr(Y_i \leqslant t) = G(t - m(x_i))$, for $i \in r$, is

$$\ddot{G}_i(t) = \frac{1}{n} \sum_{j \in s} I\{\hat{\varepsilon}_j \leqslant (t - \hat{m}(x_i))\},$$

yielding the nonparametric regression based estimator

$$\hat{F}_{CDnp}(t) = \frac{1}{N} [nF_s(t) + (N - n) \hat{F}_{CDnp,r}(t)], \tag{11.4.27}$$

where $\hat{F}_{CDnp,r}(t) = (N - n)^{-1} \Sigma_{i \in r} \ddot{G}_i(t)$.

In the case where there is a reasonable working parametric model, Chambers, Dorfman, and Wehrly (1993) introduced the nonparametric calibration estimator, discussed in general terms in Section 11.3.2. For the CDF their estimator takes the form

$$\hat{F}_{CDW}(t) = \frac{1}{N}\left[nF_s(t) + \sum_{i\in r}\hat{G}_i(t) + \sum_{j\in s}\tilde{w}_j[I\{Y_j \leqslant t\} - \hat{G}_j(t)]\right]$$

$$= \hat{F}_{CD}(t) + \frac{1}{N}\sum_{j\in s}\tilde{w}_j[I\{Y_j \leqslant t\} - \hat{G}_j(t)],$$

where $\hat{G}_j(t)$ is as at (11.4.10) and $\hat{F}_{CD}(t) = N^{-1}[nF_s(t) + \Sigma_{i\in r}\hat{G}_i(t)]$ as in (11.4.12). The bias-calibrated CDF estimator is thus the sum of the working model estimator, $\hat{F}_{CD}(t)$, and a residual correction similar to the bias-calibrated estimator of the total in (11.3.5). The relationship between $\hat{F}_{CDW}(t)$ and $\hat{F}_{RKM^*}(t)$ or $\hat{F}_{RKM}(t)$ is clear; the only major difference is in the weights applied to the residuals: $\hat{F}_{RKM}(t)$ uses $(\pi_j^{-1} - 1)$, which *anticipates* the relation between the sample and non-sample units, and $\hat{F}_{CDW}(t)$ uses $\hat{w}_j$ which reflects this relation as it is in the actual sample at hand.

Dorfman and Hall (1993) consider the estimators $\hat{F}_{CD}(t)$, $\hat{F}_{RKM}(t)$, $\hat{F}_{CDnp}(t)$, $\hat{F}_{KUO}(t)$, and $\hat{F}_{CDW}(t)$ under the nonparametric regression model (11.3.1) on Y with homoscedastic errors. The working model for $\hat{F}_{CD}(t)$, $\hat{F}_{RKM}(t)$, and $\hat{F}_{CDW}(t)$ was the simple linear model $Y_i = \alpha + \beta x_i + \varepsilon_i$, with $\varepsilon_i \sim (0, \sigma^2)$. Biases of these three estimators are $O(n^{-1})$, if the working model is correct. If it is not, then the biases are of the order of a constant for $\hat{F}_{CD}(t)$ (very poor, i.e., the bias squared swamps the variance), of order $O(n^{-1/2})$ for $\hat{F}_{RKM}(t)$ (poor, i.e., the bias squared is on a par with the variance), and of order $O[(bn)^{-1} + b^2]$ for $\hat{F}_{KUO}(t)$, $\hat{F}_{CDnp}(t)$ and $\hat{F}_{CDW}(t)$. In the last case, if the bandwidth has the form $b = Cn^{-c}$ for some constant C and for $1/4 < c < 1/2$ (a very mild condition), then the bias is $o(n^{-1/2})$, and is a negligible part of the mean square error.

Overall, the nonparametric calibration estimator $\hat{F}_{CDW}(t)$ seems the most versatile, having bias of the same order as the best whether the working model holds or not. The straightforward estimator $\hat{F}_{KUO}(t)$ is appealing for its simplicity, and might be preferable when available working models cannot be trusted.

The above possibilities are not exhaustive. Dorfman and Hall (1993) and Kuk (1993) offer other alternatives.

11.4.7 Some Open Questions

No research appears to have been done on the estimation of the variance of any of the nonparametric CDF estimators or on confidence interval construction. How to use multiple explanatory variables in predicting Y and, thus, $F_N(t)$ is largely unexplored. Little, if any, work, has been done on the important issue of quantile estimation via nonparametric CDF estimators.

Replication of some sort, like the bootstrap or balanced repeated replication, may be a possibility for variance estimation and confidence interval construction, particularly since analytic derivation of variances for nonparametric CDF and quantile estimators may be difficult. Dorfman and Valliant (1993) presented empirical comparisons of methods of confidence interval construction in which BRR performed better than the Woodruff or Francisco-Fuller methods. Rao and Shao (1993) and Shao and Wu (1992) gave some design-based theory showing that BRR provides a consistent variance estimator of quantiles. This work needs to be extended to quantiles derived from parametric and nonparametric regression methods.

11.5 SMALL AREA ESTIMATION

In some applications there is a need to make estimates for population subgroups where sample sizes are small or, perhaps, even zero. This type of problem is usually referred to as *small area estimation* since geographic areas were one of the first applications of the methodology described in this section. More generally, estimates are made for domains based on demographic and socioeconomic variables like age, ethnic group, or local government area.

Samples are designed within time, cost, and other constraints that restrict the number of estimates that can reliably be made using standard methods. However, demand for additional, detailed estimates has led to the development of ways to try to wring more information from samples. Governments are probably the biggest users of these methods since small area statistics are often used in formulating policies and programs and in distributing funds to local jurisdictions.

The statistical methods often involve the use of data for larger domains to construct estimates for smaller ones. There are a variety of model-based methods for doing this including Bayes, empirical Bayes, James-Stein estimators, and the prediction estimators that we will cover here. The literature on this subject is voluminous, and we will hardly scratch the surface of the subject. Platek et al. (1987) and Schaible (1996) are collections that cover a number of applications and some theory. These books contain long lists of references that interested readers may consult. Ghosh and Rao (1994) review much of the theory that has been developed for small area problems and give many additional references. One application of small area estimation has been to adjust for undercoverage of some demographic groups in the decennial census of the United States (see, e.g., Ericksen and Kadane, 1985, Ericksen, Kadane, and Tukey, 1989). This particular use proved to be extremely controversial (Freedman and Navidi, 1986) because of the unavailability of data to validate models—a common situation in small area problems.

We concentrate on the problem of estimating the total of Y over a specified domain d within a larger finite population, following Royall (1979) and Steinberg (1979). The sample s from the larger population may be far from ideal for our problem. The sample may have been selected for some other purpose, and it might contain few, if any, units from domain d.

Theorem 2.2.1 provides the basis for finding optimal predictors under models that are appropriate to the small area problem. We begin by considering cases in which the population is cross-classified. Each unit falls into one of $c = 1, \ldots, C$ classes and $d = 1, \ldots, D$ domains. A domain might be the adult population of the state of Idaho and a class could be all males in the labor force who are 25–44 years old. Let the sample from cell (c, d) be s_{cd} and the set of nonsample units be r_{cd}. As usual, the domain total T_d can be broken into the sum for the sample units and nonsample units:

$$T_d = \sum_{c=1}^{C} \sum_{i \in s_{cd}} Y_i + \sum_{c=1}^{C} \sum_{i \in r_{cd}} Y_i.$$

The number of units in the population in cell (c, d) is N_{cd} and the number of sample units is n_{cd}. A distinct possibility is that s_{cd} is empty and $n_{cd} = 0$, especially if no special control can be exercised over the distribution of sample units among the cells.

11.5.1 Estimation when Cell Means Are Unrelated

A relatively simple model under which units in a cell have a common mean and different units are correlated within a cell is

Model A:

$$E_M(Y_i) = \mu_{cd} \qquad i \text{ in } (c, d)$$

$$\operatorname{cov}_M(Y_i, Y_{i'}) = \begin{cases} \sigma_{cd}^2 & i = i',\ i \text{ in } (c, d) \\ \sigma_{cd}^2 \rho_{cd} & i \neq i',\ i \text{ and } i' \text{ in } (c, d) \\ 0 & \text{otherwise.} \end{cases} \tag{11.5.1}$$

With this model the cell means are unrelated and an application of Theorem 2.2.1 shows that the *BLU* predictor of T_d is the simple post-stratified estimator

$$\hat{T}_d^{(A)} = \sum_c \hat{T}_{cd}^{(A)},$$

where $\hat{T}_{cd}^{(A)} = N_{cd} \bar{Y}_{s_{cd}}$ is the expansion estimator with $\bar{Y}_{s_{cd}} = \Sigma_{s_{cd}} Y_i / n_{cd}$ (see Exercise 11.17). Each cell total is estimated in $\hat{T}_d^{(A)}$ using only data from the cell itself—N_{cd} and $\bar{Y}_{s_{cd}}$. The error variance $\hat{T}_d^{(A)}$ is

$$\operatorname{var}_M(\hat{T}_d^{(A)} - T_d) = \sum_c N_{cd}^2 (1 - f_{cd}) \sigma_{cd}^2 (1 - \rho_{cd}).$$

An unbiased estimator of the error variance is obtained when $\sigma_{cd}^2(1 - \rho_{cd})$ is replaced by the estimator

$$v_{cd} = \sum_{i \in s_{cd}} (Y_i - \bar{Y}_{s_{cd}})^2 / (n_{cd} - 1)$$

(see Exercise 11.17).

If all of the C cells are not represented in the sample, then $\hat{T}_d^{(A)}$ cannot be computed. When s_{cd} is empty, we cannot estimate μ_{cd} unless this parameter is related in some way to parameters in cells that were sampled. This is the unfortunate and unavoidable fact that makes small area estimation difficult. We must either draw an adequate sample from cell (c, d) or we must rely on whatever assumptions are required for estimating T_{cd} from observations in other cells. To the extent that each cell is unique, we will be frustrated in all efforts to provide estimates for cells where only small samples are available. If there are similarities and regularities among the cells, we can use observations from some cells to make inferences about others. These "similarities and regularities" are just the relationships that we express through models like the ones that follow. We emphasize that because of the very data sparseness, verification of the model for the domain of application, use of balanced samples, and so forth, will often be difficult or impossible, rendering our inferences tenuous.

11.5.2 Cell Means Determined by Class but Uncorrelated

A simple model under which unbiased estimation of T_d is possible even when some classes are not represented in the sample is this:

Model B:

$$E_M(Y_i) = \mu_c \qquad i \text{ in } (c, d)$$

$$\text{cov}_M(Y_i, Y_{i'}) = \begin{cases} \sigma_{cd}^2 & i = i',\ i \text{ in } (c, d) \\ \sigma_{cd}^2 \rho_{cd} & i \neq i',\ i \text{ and } i' \text{ in } (c, d) \\ 0 & \text{otherwise.} \end{cases} \qquad (11.5.2)$$

Each class is a distinct population and the class/domain cells are like the clusters in Chapters 8 and 9. The Y's are sometimes conceived of as being generated by a two-stage process in which the class c cell means μ_c are themselves realized values of uncorrelated random variables having mean μ_c and variance τ_c^2 and, given μ_c, the Y_i in cell (c, d) are exchangeable in the sense of having the same mean μ_c, the same variance $\tilde{\sigma}_{cd}^2$, and the same covariance $\tilde{\sigma}_{cd}^2 \tilde{\rho}_{cd}$. Then, it follows that $\sigma_{cd}^2 = \tau_c^2 + \tilde{\sigma}_{cd}^2$ and $\sigma_{cd}^2 \rho_{cd} = \tau_c^2 + \tilde{\sigma}_{cd}^2 \tilde{\rho}_{cd}$.

Model B says that there is a common expected value for all units in a class regardless of the domain. But, it recognizes through ρ_{cd} that units within the same class/domain cell are more alike than class c units that are not in the same domain. Assume that every class c is represented in the sample although some classes may have no sample units from domain d. In other words, n_{cd} may be 0 but $n_c = \Sigma_{d=1}^{D} n_{cd} > 0$ for all c. Denote the variance of the sample mean for cell (c, d) when $n_{cd} > 0$ as

$$v_{cd} = \text{var}_M(\bar{Y}_{s_{cd}}) = \frac{\sigma^2_{cd}}{n_{cd}}[1 + (n_{cd} - 1)\rho_{cd}].$$

If $n_{cd} = 0$, we define $\Sigma_{s_{cd}} Y_i = \bar{Y}_{s_{cd}} = v_{cd}^{-1} = 0$. Because the class/domain cells are treated as clusters in Model B, Theorem 8.2.2 can be used to find the *BLU* predictor for cell (c, d) as

$$\hat{T}^{(B)}_{cd} = \sum_{s_{cd}} Y_i + (N_{cd} - n_{cd})[w_{cd}\bar{Y}_{s_{cd}} + (1 - w_{cd})\hat{\mu}_c],$$

where $w_{cd} = n_{cd}\rho_{cd}/[1 + (n_{cd} - 1)\rho_{cd}]$ and $\hat{\mu}_c = \Sigma_j u_{cj}\bar{Y}_{s_{cj}}$ with u_{cj} defined for all sampled cells as

$$u_{cj} = v_{cj}^{-1} \Big/ \sum_{j=1}^{D} v_{cj}^{-1}.$$

The *BLU* predictor of the domain total is obtained by summing the cell estimators:

$$\hat{T}^{(B)}_d = \sum_c \hat{T}^{(B)}_{cd}.$$

In contrast to $\hat{T}^{(A)}_{cd} = N_{cd}\bar{Y}_{s_{cd}}$ above, $\hat{T}^{(B)}_{cd}$ uses data from all the domains because the cell mean μ_c depends only on class. Suppose, for example, that the variable Y is annual income, that class c is all males in the labor force who are 25–44 years old and the domains are the states in the northwestern region of the United States. The estimator $\hat{T}^{(B)}_{cd}$ of the total income of those males in the state of Idaho then uses data from Montana, Washington, Oregon, and any other states in the northwest region. This use of data from domains outside the particular one of interest is sometimes colloquially referred to as "borrowing strength" from the other domains. (Of course, if the model assumptions are wrong, one might equally well speak of "borrowing weakness.")

Before considering the error variance, we introduce a variation on the problem that may be instructive. Suppose that Model B applies and that we have, in addition to the sample, a supplementary estimate $\dot{\mu}_c$ of the class mean μ_c for $c = 1, \ldots, C$. Consider linear estimators of the form

$$\hat{T}_d = \sum_c a_{cd}\bar{Y}_{s_{cd}} + \sum_c b_{cd}\dot{\mu}_c.$$

It is assumed that $a_{cd} = 0$ when $n_{cd} = 0$. Notice that the *BLU* predictor $\hat{T}^{(B)}_d$ can be written in this form if we allow the "supplementary" estimate to be from the sample: $\dot{\mu}_c = \hat{\mu}_c$. If $\dot{\mu}_c$ is unbiased for μ_c under Model B, then $E_M(\hat{T}_d - T_d) =$

$\Sigma_c(a_{cd} + b_{cd} - N_{cd})\mu_c$. For $\hat{T}_d$ to be unbiased, we must have $b_{cd} = N_{cd} - a_{cd}$. Consequently,

$$\hat{T}_d = \sum_c a_{cd}(\bar{Y}_{scd} - \hat{\mu}_c) + \sum_c N_{cd}\hat{\mu}_c. \tag{11.5.3}$$

Reparameterizing, let $\theta_{cd} = (a_{cd} - n_{cd})/(N_{cd} - n_{cd})$. Note that $\theta_{cd} = 0$ when $n_{cd} = 0$, since $a_{cd} = 0$ in that circumstance. Unbiasedness implies that (11.5.3) can be rewritten as

$$\hat{T}_d = \sum_c n_{cd}\bar{Y}_{scd} + \sum_c (N_{cd} - n_{cd})[\theta_{cd}\bar{Y}_{scd} + (1 - \theta_{cd})\hat{\mu}_c] \tag{11.5.4}$$

for some constants θ_{cd}, $c = 1, \ldots, C$. If $\hat{\mu}_c$ is uncorrelated with Y-values in classes other than c, then the optimal θ_{cd}'s for $n_{cd} > 0$ are given in the following Lemma.

Lemma 11.5.1 The set of θ_{cd}'s that minimize the error variance of (11.5.4) under Model B, assuming that $\hat{\mu}_c$ is uncorrelated with Y-values in classes other than c, are, for $c = 1, \ldots, C$,

$$\theta^*_{cd} = \frac{\text{cov}_M(\bar{Y}_{scd} - \hat{\mu}_c,\ \bar{Y}_{rcd} - \hat{\mu}_c)}{\text{var}_M(\bar{Y}_{scd} - \hat{\mu}_c)} = \text{cor}_M(\bar{Y}_{scd} - \hat{\mu}_c,\ \bar{Y}_{rcd} - \hat{\mu}_c)\sqrt{\frac{\text{var}_M(\bar{Y}_{rcd} - \hat{\mu}_c)}{\text{var}_M(\bar{Y}_{scd} - \hat{\mu}_c)}} \tag{11.5.5}$$

where cor stands for correlation. The error variance using the optimal θ_{cd}'s is

$$\text{var}_M(\hat{T}_d - T_d) = \sum_c (N_{cd} - n_{cd})^2 v_{\bar{Y}_{rcd}}(1 - r^2_{cd}) \tag{11.5.6}$$

where $v_{\bar{Y}_{rcd}} = \text{var}_M(\bar{Y}_{rcd} - \hat{\mu}_c)$ and $r_{cd} = \text{cor}_M(\bar{Y}_{scd} - \hat{\mu}_c,\ \bar{Y}_{rcd} - \hat{\mu}_c)$ for $n_{cd} > 0$ and $r_{cd} = 0$ when $n_{cd} = 0$.

Proof: The general form of the error variance of $\hat{T}_d$ is

$$\begin{aligned}\text{var}_M(\hat{T}_d - T_d) = &\sum_c (N_{cd} - n_{cd})^2\theta^2_{cd}\,\text{var}_M(\bar{Y}_{scd} - \hat{\mu}_c)\\ &+ \sum_c (N_{cd} - n_{cd})^2\,\text{var}_M(\bar{Y}_{rcd} - \hat{\mu}_c)\\ &- 2\sum_c (N_{cd} - n_{cd})\theta_{cd}\,\text{cov}_M(\bar{Y}_{scd} - \hat{\mu}_c,\ \bar{Y}_{rcd} - \hat{\mu}_c).\end{aligned} \tag{11.5.7}$$

Taking the derivative of (11.5.7) with respect to θ_{cd}, setting to 0, and solving for θ_{cd} gives the optima. Substituting the optimal values into (11.5.7) and rearranging yields (11.5.6). □

Note that if μ_c is known and is used in (11.5.4), the optimal weights θ^*_{cd} reduce to $\theta^*_{cd} = n_{cd}\rho_{cd}/[1 + (n_{cd} - 1)\rho_{cd}] = w_{cd}$, the same ones that were used in the *BLU* predictor $\hat{T}^{(B)}_{cd}$ where μ_c was estimated from the sample. In the special case of $\rho_{cd} = 0$, all weight is given to the known value μ_c, as we would expect. When μ_c is known and the optimal θ^*_{cd}'s are used, the error variance (11.5.6) under the more general Model B can be rewritten. Noting that $\rho_{cd} = w_{cd}/[w_{cd} + n_{cd}(1 - w_{cd})]$, we have after some algebra

$$\text{var}_M(\hat{T}_d - T_d) = \sum_c \frac{N^2_{cd}}{n_{cd}}(1 - f_{cd})\sigma^2_{cd}(1 - \rho_{cd})[1 - (1 - f_{cd})(1 - w_{cd})]. \quad (11.5.8)$$

If the sample contains no units from some cell (c, d), that is, $n_{cd} = 0$, (11.5.8) is ill-defined. In such a case, we predict T_{cd} by $N_{cd}\mu_c$ and note that $\text{var}_M(N_{cd}\mu_c - T_{cd}) = \text{var}_M(T_{cd}) = N_{cd}\sigma^2_{cd}[1 + (N_{cd} - 1)\rho_{cd}]$. Expression (11.5.8) then will apply if we set $w_{cd}/n_{cd} = N^{-1}_{cd} + \rho_{cd}/(1 - \rho_{cd})$, so that the (c, d) summand becomes $\text{var}_M(T_{cd})$.

When no supplementary estimates are available for μ_c, the *BLU* predictor under Model B is $\hat{T}^{(B)}_d$ with error variance given by the following lemma.

Lemma 11.5.2. Under Model B the error variance of $\hat{T}^{(B)}_{cd} = \Sigma_c \hat{T}^{(B)}_{cd}$, with

$$\hat{T}^{(B)}_{cd} = \sum_{s_{cd}} Y_i + (N_{cd} - n_{cd})[w_{cd}\bar{Y}_{s_{cd}} + (1 - w_{cd})\hat{\mu}_c],$$

is

$$\text{var}_M(\hat{T}^{(B)}_d - T^{(B)}_d) = \sum_c \frac{N^2_{cd}}{n_{cd}}(1 - f_{cd})\sigma^2_{cd}(1 - \rho_{cd})$$
$$\times [1 - (1 - f_{cd})(1 - w_{cd})(1 - u_{cd})]. \quad (11.5.9)$$

Proof: The error variance of the estimator for class c is

$$\text{var}_M(\hat{T}^{(B)}_{cd} - T^{(B)}_{cd}) = M^2_{cd}\left\{v_{cd}[w_{cd} + (1 - w_{cd})u_{cd}]^2 + (1 - w_{cd})^2 \sum_{j \neq d} u^2_{cj}v_{cj}\right\}$$
$$+ \text{var}_M\left(\sum_{r_{cd}} Y_i\right) - 2M_{cd}[w_{cd} + (1 - w_{cd})u_{cd}]\text{cov}_M\left[\bar{Y}_{s_{cd}}, \sum_{r_{cd}} Y_i\right],$$

where $M_{cd} = N_{cd} - n_{cd}$. Next, add and subtract $(1 - w_{cd})^2 u^2_{cd}v_{cd}$ within the braces and use $v_{cd} = \sigma^2_{cd}\rho_{cd}/w_{cd}$ and $\Sigma^D_{j=1} u^2_{cj}v_{cj} = \mu_{cd}v_{cd}$ to obtain

$$\text{var}_M(\hat{T}^{(B)}_{cd} - T^{(B)}_{cd})$$
$$= M^2_{cd}\sigma^2_{cd}\left\{\frac{(1 - w_{cd})^2}{w_{cd}}\rho_{cd}u_{cd} + M^{-1}_{cd}[1 + (M_{cd} - 1)\rho_{cd}] - \rho_{cd}w_{cd}\right\}.$$

Noting that $M_{cd} = N_{cd}(1 - f_{cd})$ and $\rho_{cd}/(1 - \rho_{cd}) = w_{cd}/[n_{cd}(1 - w_{cd})]$ and doing some rearranging produces (11.5.9). □

The error variance (11.5.9) has an interesting interpretation in light of the earlier work in this section. Under Model B the post-stratified estimator $\hat{T}_d^{(A)}$ is unbiased with error variance $\Sigma_c N_{cd}^2(1 - f_{cd})\sigma_{cd}^2(1 - \rho_{cd})$, which is expression (11.5.9), excluding the part in the brackets. When μ_c is known, the error variance of the *BLU* predictor is given by (11.5.8), which differs from (11.5.9) by the factor $1 - u_{cd}$ in the brackets. The ratio of the error variance of the *BLU* predictor $\hat{T}_{cd}^{(B)}$ of the (c, d) cell total to the variance of the post-stratified estimator is

$$\frac{\text{var}_M(\hat{T}_{cd}^{(B)} - T_{cd}^{(B)})}{\text{var}_M(\hat{T}_{cd}^{(A)} - T_{cd}^{(A)})} = 1 - (1 - f_{cd})(1 - w_{cd})(1 - \mu_{cd}). \qquad (11.5.10)$$

This ratio is less than or equal to 1, implying that the *BLU* predictor is potentially more precise than the post-stratified estimator, depending on the sizes of f_{cd}, w_{cd}, and u_{cd}. It can be shown that the right-hand side of (11.5.10) is bounded above by $3\max(f_{cd}, w_{cd}, u_{cd})$ (see Exercise 11.23).

When there are many domains, the *BLU* predictor of the cell (c, d) total can be a substantial improvement over the post-stratified estimator because the *BLU* uses data from all domains in its estimator of μ_c while the post-stratified estimator uses data only from cell (c, d). If, for example, the sampling fractions, f_{cd}, and the intracell correlations, ρ_{cd}, are small, and the variances v_{cd} are about the same for each domain, then $3\max(f_{cd}, w_{cd}, u_{cd})$ will be approximately $3/D$. As the number of domains increases so does the advantage of the *BLU* predictor, assuming that Model B is correct.

11.5.3 Synthetic and Composite Estimators

An estimator that is sometimes used when Model B seems reasonable is the *synthetic estimator*

$$\hat{T}_d^{(\text{syn})} = \sum_c N_{cd}\hat{\mu}_c,$$

where $\hat{\mu}_c = \Sigma_{j=1}^D \ell_{cj}\bar{Y}_{scd}$ is a weighted average of the cell means and $\Sigma_{j=1}^D \ell_{cj} = 1$. Gonzalez, Placek, and Scott (1996) give examples of synthetic estimation used for some health-related characteristics like percentage of births by Cesarean section and percentage of infants that were jaundiced within individual States.

The synthetic estimator can be written as $\hat{T}_d^{(\text{syn})} = \Sigma_c n_{cd}\hat{\mu}_c + \Sigma_c(N_{cd} - n_{cd})\hat{\mu}_c$. In this form it is apparent that the (known) sample sum $\Sigma_c n_{cd}\bar{Y}_{scd}$ is, in effect, being estimated by $\Sigma_c n_{cd}\hat{\mu}_c$. An obvious modification of $\hat{T}_d^{(\text{syn})}$ is to avoid estimating what we already know and to use

$$\hat{T}_d^{(\text{mod})} = \sum_c n_{cd}\bar{Y}_{scd} + \sum_c (N_{cd} - n_{cd})\hat{\mu}_c.$$

In many situations, the modified estimator will be more precise than the synthetic (see Exercise 11.19).

Returning to $\hat{T}_d^{(B)}$, notice that if $\sigma_{cd}^2 = \sigma^2$, the *BLU* predictor does not depend on that constant since the σ^2's enter the estimator only through the weights u_{cj}. If the ρ_{cd}'s are also set to a constant ρ, then a family of estimators, $\hat{T}_d^{(\rho)}$, is generated. We readily see that $\hat{T}_d^{(0)} = \hat{T}_d^{(\text{mod})}$ and $\hat{T}_d^{(1)} = \hat{T}_d^{(A)}$. The estimator $\hat{T}_d^{(\rho)}$ with $0 < \rho < 1$ represents a compromise between the synthetic and the post-stratified estimator.

Another way of striking a compromise between $\hat{T}_d^{(\text{mod})}$ and $\hat{T}_d^{(A)}$ is simply to take a weighted average in each cell (c, d):

$$\begin{aligned}\hat{T}_{cd}^{(w)} &= w_{cd}\hat{T}_{cd}^{(A)} + (1 - w_{cd})\,\hat{T}_{cd}^{(\text{mod})} \\ &= n_{cd}\bar{Y}_{s_{cd}} + (N_{cd} - n_{cd})\left[w_{cd}\bar{Y}_{s_{cd}} + (1 - w_{cd})\sum_j \frac{n_{cj}}{n_c}\bar{Y}_{s_{cj}}\right]\end{aligned} \tag{11.5.11}$$

and estimate T_d by $\Sigma_d \hat{T}_{cd}^{(w)}$.

Weighted averages of this sort are referred to as *composite estimators* (see, for example, Ghosh and Rao, 1994). The optimal weights w_{cd} can be found from an application of the following result whose proof is left to Exercise 11.20.

Lemma 11.5.3 Suppose that X and Y are unbiased predictors of a random variable Z. Consider composite predictors of the form $\hat{Z} = \alpha X + (1 - \alpha) Y$ where $0 \leqslant \alpha \leqslant 1$. The value of α that minimizes the error variance of $\hat{Z}$ is $\alpha_{\text{opt}} = \text{cov}(X - Y, Z - Y)/\text{var}(X - Y)$.

Consider the special case of Model B with $\sigma_{cd}^2 = \sigma^2$ and $\rho_{cd} = \rho$. To apply Lemma 11.5.3, we equate X with $\bar{Y}_{s_{cd}}$, Y with $\Sigma_j n_{cj}\bar{Y}_{s_{cj}}/n_c$, and Z with $\bar{Y}_{r_{cd}}$, leading to

$$\text{cov}(X - Y, Z - Y) = \rho\sigma^2\left(1 - \frac{n_{cd}}{n_c}\right) - \frac{n_{cd}}{n_c}v_{cd} + \sum_j \left(\frac{n_{cj}}{n_c}\right)^2 v_{cj}, \qquad \text{and}$$

$$\text{var}(X - Y) = v_{cd}\left(1 - \frac{n_{cd}}{n_c}\right)^2 + \sum_{j \neq d}\left(\frac{n_{cj}}{n_c}\right)^2 v_{cj}.$$

Next, note that $n_{cd}v_{cd}/n_c = (1 - \rho)\sigma^2/n_c + \rho\sigma^2 n_{cd}/n_c$, which results in

$$\text{cov}(X - Y, Z - Y) = \rho\sigma^2\left[\left(1 - \frac{n_{cd}}{n_c}\right)^2 + \sum_j\left(\frac{n_{cj}}{n_c}\right)^2\right] \qquad \text{and} \tag{11.5.12}$$

$$\text{var}(X - Y) = \frac{1 - \rho}{n_{cd}}\left(1 - \frac{n_{cd}}{n_{cd}}\right) + \rho\left(1 - \frac{n_{cd}}{n_c}\right)^2 + \rho\sum_{j \neq d}\left(\frac{n_{cj}}{n_c}\right)^2.$$

The optimum value of w_{cd} is then

$$w_{cd}^* = 1/(1 + r_{cd}),$$

where

$$r_{cd} = \frac{1-\rho}{n_{cd}}\left(1 - \frac{n_{cd}}{n_c}\right) \Big/ \operatorname{cov}(X - Y, Z - Y)$$

and $\operatorname{cov}(X - Y, Z - Y)$ is defined by (11.5.12).

For a given value of ρ, the composite estimator $\hat{T}_d^{(w)}$ is closely related to $\hat{T}_d^{(\rho)}$. In both cases increasing either n_{cd} or ρ gives relatively more weight to the cell mean $\bar{Y}_{s_{cd}}$ in estimating the total T_{cd}. As in the cluster sampling case in Chapter 8, this is logical since the more alike units within a cell are, as measured by ρ, the less useful units from other cells will be in estimating T_{cd}. When $\rho = 0$, $\hat{T}_d^{(w)} = \hat{T}_d^{(\rho)} = \hat{T}_d^{(\mathrm{mod})}$, while when $\rho = 1$, $\hat{T}_d^{(w)} = \hat{T}_d^{(\rho)} = \hat{T}_d^{(A)}$. For intermediate values of ρ, the main difference between $\hat{T}_d^{(w)}$ and $\hat{T}_d^{(\rho)}$ is in their respective estimates of μ_c:

$$\sum_j \frac{n_{cj}}{n_c} \bar{Y}_{s_{cj}} \quad \text{and} \quad \frac{\Sigma_j \bar{Y}_{s_{cj}}/[1 + (1-\rho)/(n_{cj}\rho)]}{\Sigma_j 1/[1 + (1-\rho)/(n_{cj}\rho)]}$$

(see Exercise 11.24). The estimate of μ_c used in $\hat{T}_d^{(w)}$ gives the cell mean $\bar{Y}_{s_{cj}}$ a weight proportional to the sample size n_{cj}, while the estimate used in $\hat{T}_d^{(\rho)}$ gives the cell means more nearly equal weights. For this reason $\hat{T}_d^{(\rho)}$ will be less likely to be dominated by cells with unusually large sample sizes.

11.5.4 Using Auxiliary Data

Although the simple case of a common mean within a class is an interesting beginning, auxiliary data are often available that can be used in estimation. Suppose that for each unit in cell (c, d) we have a $p \times 1$ vector of auxiliaries $\mathbf{x}_{cdi}$ and that the following model holds:

$$E_M(Y_i) = \mathbf{x}'_{cdi}\boldsymbol{\beta}_c \qquad i \text{ in } (c, d)$$

$$\operatorname{cov}_M(Y_i, Y_{i'}) = \begin{cases} \sigma_{cd}^2 & i = i',\ i \text{ in } (c, d) \\ \sigma_{cd}^2 \rho_{cd} & i \neq i',\ i \text{ and } i' \text{ in } (c, d) \\ 0 & \text{otherwise,} \end{cases} \tag{11.5.13}$$

where $\boldsymbol{\beta}_c = (\beta_{c1}, \ldots, \beta_{cp})'$. Our estimation target is again the domain d total, $T_d = \Sigma_{c=1}^{C} \Sigma_{i \in U_{cd}} Y_i$, where U_{cd} is the universe of units in cell (c, d).

The $p \times 1$ parameter $\boldsymbol{\beta}_c$ depends on the class but not the domain. Thus we will be able to use data from all the domains in estimating $\boldsymbol{\beta}_c$ even though our target is T_d for a particular one of the domains. An application of Theorem 2.2.1 will again produce the *BLU* predictor. Let $\mathbf{Y}_{scd}$ be the $n_{cd} \times 1$ vector of sample units from cell (c, d), $\mathbf{Y}_{rcd}$ be the $(N_{cd} - n_{cd}) \times 1$ vector of nonsample units, $\mathbf{X}_{scd}$ be the $n_{cd} \times p$ matrix of auxiliaries for the sample units in the cell, and $\boldsymbol{\beta} = (\boldsymbol{\beta}'_1, \ldots, \boldsymbol{\beta}'_C)'$ be the $Cp \times 1$ vector of parameters for all classes together. The $n_d \times Cp$ matrix of auxiliaries for the domain d sample is then

$$\mathbf{X}_{sd} = \begin{bmatrix} \mathbf{X}_{s1d} & \mathbf{0} & \cdots & \mathbf{0} \\ \mathbf{0} & \mathbf{X}_{s2d} & \cdots & \mathbf{0} \\ \vdots & \vdots & \ddots & \mathbf{0} \\ \mathbf{0} & \mathbf{0} & \cdots & \mathbf{X}_{sCd} \end{bmatrix},$$

where the $\mathbf{0}$'s are appropriately sized matrices of all zeroes. The full sample $n \times Cp$ matrix of auxiliaries is

$$\mathbf{X}_s = \begin{bmatrix} \mathbf{X}_{s1} \\ \mathbf{X}_{s2} \\ \vdots \\ \mathbf{X}_{sD} \end{bmatrix}.$$

The $n_{cd} \times n_{cd}$ covariance matrix for $\mathbf{Y}_{scd}$ is $\mathbf{V}_{ss(cd)} = \sigma^2_{cd}(1 - \rho_{cd})\mathbf{I}_{n_{cd}} + \sigma^2_{cd}\rho_{cd}\mathbf{1}_{n_{cd}}\mathbf{1}'_{n_{cd}}$ where $\mathbf{I}_{n_{cd}}$ is the $n_{cd} \times n_{cd}$ identity matrix and $\mathbf{1}_{n_{cd}}$ is an $n_{cd} \times 1$ matrix of 1's. The inverse of $\mathbf{V}_{ss(cd)}$ is

$$\mathbf{V}^{-1}_{ss(cd)} = \frac{1}{\sigma^2_{cd}(1 - \rho_{cd})[1 + (n_{cd} - 1)\rho_{cd}]}$$
$$\times \begin{bmatrix} 1 + (n_{cd} - 2)\rho_{cd} & -\rho_{cd} & \cdots & -\rho_{cd} \\ -\rho_{cd} & 1 + (n_{cd} - 2)\rho_{cd} & & -\rho_{cd} \\ \vdots & & \ddots & \\ -\rho_{cd} & -\rho_{cd} & \cdots & 1 + (n_{cd} - 2)\rho_{cd} \end{bmatrix},$$

similar to the case of cluster sampling in Theorem 8.2.2. The covariance matrix for $(\mathbf{Y}_{rcd}, \mathbf{Y}_{scd})$ is $\mathbf{V}_{rs(cd)} = \sigma^2_{cd}\rho_{cd}\mathbf{1}_{N_{cd}-n_{cd}}\mathbf{1}'_{n_{cd}}$.

The domain total T_d is a linear combination to which Theorem 2.2.1 applies, and we must evaluate the various components needed to compute the *BLU* predictor. The vector $\boldsymbol{\gamma}$ in the theorem is 1 in places corresponding to Y's in domain d and 0 elsewhere. The *BLU* estimator of $\boldsymbol{\beta}$ is $\hat{\boldsymbol{\beta}} = (\hat{\boldsymbol{\beta}}'_1, \ldots, \hat{\boldsymbol{\beta}}'_C)'$, where

$$\hat{\boldsymbol{\beta}}_c = \sum_{d=1}^{D} (\mathbf{X}'_{scd}\mathbf{V}^{-1}_{ss(cd)}\mathbf{X}_{scd})^{-1} \sum_{d=1}^{D} \mathbf{X}'_{scd}\mathbf{V}^{-1}_{ss(cd)}\mathbf{Y}_{scd}.$$

Further, we have $\gamma_r' \mathbf{V}_{rs}\mathbf{V}_{ss}^{-1}(\mathbf{Y}_s - \mathbf{X}_s\hat{\boldsymbol{\beta}}) = \Sigma_{c=1}^{C} \mathbf{1}'_{N_{cd}-n_{cd}} \mathbf{V}_{rs(cd)}\mathbf{V}_{ss(cd)}^{-1}(\mathbf{Y}_{scd} - \mathbf{X}_{scd}\hat{\boldsymbol{\beta}}_c)$ and, using the forms of $\mathbf{V}_{rs(cd)}$ and $\mathbf{V}_{ss(cd)}^{-1}$, $\mathbf{1}'_{N_{cd}-n_{cd}} \mathbf{V}_{rs(cd)}\mathbf{V}_{ss(cd)}^{-1}(\mathbf{Y}_{scd} - \mathbf{X}_{scd}\hat{\boldsymbol{\beta}}_c) = (N_{cd} - n_{cd}) n_{cd}\rho_{cd}[1 + (n_{cd} - 1)\rho_{cd}]^{-1}(\bar{Y}_{s_{cd}} - \bar{\mathbf{X}}'_{scd}\hat{\boldsymbol{\beta}}_c)$ where $\bar{\mathbf{X}}_{s_{cd}} = \Sigma_{s_{cd}}\mathbf{x}_{cdi}/n_{cd}$. Putting the various results together, we have

$$\hat{T}_d = \sum_{c=1}^{C} \{n_{cd}\bar{Y}_{s_{cd}} + (N_{cd} - n_{cd})[\bar{\mathbf{X}}'_{r_{cd}}\hat{\boldsymbol{\beta}}_c + w_{cd}(\bar{Y}_{s_{cd}} - \bar{\mathbf{X}}'_{s_{cd}}\hat{\boldsymbol{\beta}}_c)]\} \qquad (11.5.14)$$

with $\bar{\mathbf{X}}_{r_{cd}} = \Sigma_{r_{cd}}\mathbf{x}_{cdi}/(N_{cd} - n_{cd})$ and w_{cd} as defined in Section 11.5.2. In the special case of $\mathbf{x}_{cdi} = \mathbf{1}$ for all units, it is easy to verify that (11.5.14) reduces to the *BLU* predictor $\hat{T}_d^{(B)} = \Sigma_c \hat{T}_{cd}^{(B)}$ defined in Section 11.5.2.

When sampling fraction is small in each cell (c, d), that is, $n_{cd}/N_{cd} \to 0$, then the *BLU* predictor becomes

$$\hat{T}_d = \sum_{c=1}^{C} N_{cd}\{w_{cd}[\bar{Y}_{s_{cd}} + (\bar{\mathbf{X}}_{cd} - \bar{\mathbf{X}}_{s_{cd}})'\hat{\boldsymbol{\beta}}_c] + (1 - w_{cd})\bar{\mathbf{X}}'_{cd}\hat{\boldsymbol{\beta}}_c\},$$

where $\bar{\mathbf{X}}_{cd}$ is the population mean vector in the cell. The sum in the braces is a weighted average of a regression estimator, $\bar{Y}_{s_{cd}} + (\bar{\mathbf{X}}_{cd} - \bar{\mathbf{X}}_{s_{cd}})'\hat{\boldsymbol{\beta}}_c$, and a term, $\bar{\mathbf{X}}'_{cd}\hat{\boldsymbol{\beta}}_c$, reminiscent of a synthetic estimator.

11.5.5 Auxiliary Data at the Cell Level

In some applications auxiliary data will be available only at the domain level (as opposed to the lower unit level), and supplemental estimates are available for the cell totals. A prominent example is the case of adjustment for undercoverage in the U.S. 1980 decennial census. A separate Post-Enumeration Program (PEP) was done after the census was taken in which intensive efforts were made in a large sample of compact geographic areas to locate and count members of demographic groups most likely to be missed in the regular census. The PEP provided the supplemental small area estimates. Ericksen and Kadane (1985) and Freedman and Navidi (1986) describe this application in detail.

Let the vector of auxiliaries for cell (c, d) be $\mathbf{x}'_{cd} = (x_{cd1}, \ldots, x_{cdp})$ and suppose that the cell total follows the model

$$T_{cd} = \mathbf{x}'_{cd}\boldsymbol{\beta}_c + \varepsilon_{1cd},$$

where the errors are distributed as $\varepsilon_{1cd} \sim (0, \sigma_{1cd}^2)$ and are independent. Suppose further that we have unbiased predictors $\hat{T}_{cd}$ of each cell total and that

$$\hat{T}_{cd} = T_{cd} + \varepsilon_{2cd}$$

$\varepsilon_{2cd} \sim (0, \sigma^2_{2cd})$ and independent. Since the slope is common across the domains, all of $\hat{T}_{cd}$ $(d = 1, \ldots, D)$ can be used in estimating $\boldsymbol{\beta}_c$.

Among the estimators $\hat{T}_d = \Sigma^C_{c=1} \Sigma^D_{d=1} \lambda_{cd} \hat{T}_{cd}$, we want to find the set of λ_{cd} (c $= 1, \ldots, C$) that yields an unbiased estimator and minimizes the variance of $\hat{T}_d$. Unbiasedness implies that $E_M(\hat{T}_d - T_d) = \Sigma_c[(\Sigma^D_{j=1} \lambda_{cj} \mathbf{x}'_{cj}) - \mathbf{x}'_{cd}]\boldsymbol{\beta}_c$ for all values of $\boldsymbol{\beta}_c$. Thus

$$\sum_j \lambda_{cj} \mathbf{x}_{cj} = \mathbf{x}_{cd}. \tag{11.5.15}$$

The prediction variance of $\hat{T}_d$ is

$$\text{var}_M(\hat{T}_d - T_d) = \sum_c \left[\sum_{j \neq d} \lambda^2_{cj} \sigma^2_{\cdot cj} + (\lambda_{cd} - 1)^2 \sigma^2_{1cd} + \lambda^2_{cd} \sigma^2_{2cd} \right], \tag{11.5.16}$$

where $\sigma^2_{\cdot cd} = \sigma^2_{1cd} + \sigma^2_{2cd}$. To minimize this variance it is sufficient to minimize each term in the sum over c. To that end, define the Lagrange function

$$\phi = \left[\sum_{j \neq d} \lambda^2_{cj} \sigma^2_{\cdot cj} + (\lambda_{cd} - 1)^2 \sigma^2_{1cd} + \lambda^2_{cd} \sigma^2_{2cd} \right] + \boldsymbol{\eta}' \left(\sum_d \lambda_{cd} \mathbf{x}_{cd} - \mathbf{x}_{cd} \right).$$

where $\boldsymbol{\eta}$ is a $p \times 1$ Lagrange multiplier. We have two sets of estimating equations—one for λ_{cd} and one for λ_{cj} $(j \neq d)$:

$$\frac{\partial \phi}{\partial \lambda_{cd}} = 2(\lambda_{cd} - 1)\sigma^2_{1cd} + 2\lambda_{cd}\sigma^2_{2cd} + \boldsymbol{\eta}' \mathbf{x}_{cd} = 0 \tag{11.5.17}$$

$$\frac{\partial \phi}{\partial \lambda_{cj}} = 2\lambda_{cj}\sigma^2_{\cdot cj} + \boldsymbol{\eta}' \mathbf{x}_{cj} = 0 \ (j \neq d). \tag{11.5.18}$$

Dividing each equation by 2, multiplying each by its corresponding $\mathbf{x}'_{cd}/\sigma^2_{\cdot cd}$, and summing over d yields

$$\sum_d \lambda_{cd} \mathbf{x}'_{cd} - \gamma_{cd} \mathbf{x}'_{cd} + \frac{\boldsymbol{\eta}'}{2} \sum_d \frac{\mathbf{x}_{cd} \mathbf{x}'_{cd}}{\sigma^2_{\cdot cd}} = 0,$$

where $\gamma_{cd} = \sigma^2_{1cd}/\sigma^2_{\cdot cd}$. Using the constraint (11.5.15), we have

$$\frac{\boldsymbol{\eta}'}{2} = (\gamma_{cd} - 1)\mathbf{x}'_{cd} \mathbf{A}^{-1}_{cd},$$

where $\mathbf{A}_{cd} = \Sigma_d \mathbf{x}_{cd} \mathbf{x}'_{cd}/\sigma^2_{\cdot cd}$. Substituting into (11.5.17) and (11.5.18), we obtain

$$\lambda_{cd} = \gamma_{cd} + (1 - \gamma_{cd}) \mathbf{x}'_{cd} \mathbf{A}^{-1}_{cd} \mathbf{x}_{cd}/\sigma^2_{\cdot cd} \tag{11.5.19}$$

$$\lambda_{cj} = (1 - \gamma_{cd}) \mathbf{x}'_{cd} \mathbf{A}^{-1}_{cd} \mathbf{x}_{cj}/\sigma^2_{\cdot cj} \ (j \neq d). \tag{11.5.20}$$

Finally, the *BLU* predictor is

$$\hat{T}_d = \sum_{c=1}^{C} \sum_{d=1}^{D} \lambda_{cd} \hat{T}_{cd} = \sum_{c} [\gamma_{cd} \hat{T}_{cd} + (1 - \gamma_{cd}) \mathbf{x}'_{cd} \hat{\boldsymbol{\beta}}_c] \tag{11.5.21}$$

with $\hat{\boldsymbol{\beta}}_c = \mathbf{A}_{cd}^{-1} \Sigma_d \mathbf{x}_{cd} \hat{T}_{cd} / \sigma^2_{\cdot cd}$. Within each class c, (11.5.21) is a composite of the direct estimator and the regression synthetic estimator $\mathbf{x}'_{cd} \hat{\boldsymbol{\beta}}_c$.

Example 11.5.1 For the 1980 U.S. Census Ericksen and Kadane's (1985) cells were 66 areas of three types: states like Alaska and Wyoming; central cities like New York City, Los Angeles, or Chicago; and states apart from central cities like New York State apart from New York City, or California apart from Los Angeles, San Diego, and San Francisco. T_{cd} was the true census undercount of the total population in the area, and $\hat{T}_{cd}$ was the PEP estimate of the undercount. The true undercount was modeled as

$$T_{cd} = \beta_0 + \beta_1 x_{cd1} + \beta_2 x_{cd2} + \beta_3 x_{cd3} + \varepsilon_{1cd}$$

with x_{cd1} = percent minority in the area, x_{cd2} = crime rate in the area, x_{cd3} = percent of the area population that the Census Bureau attempted to count using a personal interview by a enumerator in the regular census. Some counties were enumerated by a mail survey in the census. The general thinking behind this model was that minority groups and persons in high crime areas were more likely to be undercounted.

11.5.6 Need for a Small Area Estimation Canon

Despite the large amount of literature on the subject, small area estimation is generally regarded with unease by the survey practitioner. This is not surprising, if the practitioner assumes a design-based paradigm, in which models play an incidental role. Thus it is often those outside the sampling profession who are most eager to make use of small area methods.

Also, from the standpoint of the presuppositions of this text, that unease is warranted, not because of the use of models, but because of their *misuse*. A running theme throughout this text is the need to guard inference against model failure, either by the adroit use of model-robust methods such as balanced sampling, or by careful and adequate model verification. It must be acknowledged, however, that model verification in the sampling context is a subject in its infancy.

Nowhere is verification more needed, and more difficult, than in the circumstance of small area estimation. An important component of such verification is cross-validation on small domains for which data are available. This involves estimating the model parameters using data outside of the domain and comparing predicted results on the domain with actual sample values. The degree and type of cross-validation will depend on the amount and sort of data available within the domain. In the case of a domain totally lacking data, we need to investigate the validity of *a priori* justifications for applying the model, even when it is well verified on available data. Such justifications need to be published along with any estimates. We need also to accompany estimates with well-grounded variance estimates. Much work needs to be done for developing a generally accepted canon of model verification and sound variance estimation for small area estimation.

EXERCISES

11.1 Suppose that the mixture probabilities π_i are all equal to π in model (11.1.1).

(a) Show that the bias of $\hat{T}_{\text{robust}} = \Sigma_{i\in s} Y_i + (N-n)\Sigma_{i\in r}\tilde{\mu}_1$ is

$$E_M(\hat{T}_{\text{robust}} - T) = -(N-n)\,E_M(\bar{r}_s) \quad \text{where} \quad \bar{r}_s = \Sigma_s(Y_i - \tilde{\mu})/n.$$

(b) Use (a) to suggest a bias-corrected estimator. What does this estimator simplify to? Is the result outlier-robust?

11.2 Show that $\tilde{T}_1 = \Sigma_s Y_i + \tilde{\beta}\Sigma_r x_i + \Sigma_s u_i \psi(\tilde{r}_i)$ with $u_i = (x_i/\sqrt{v_i})(\Sigma_r x_i/\Sigma_s x_i^2/v_i)$ reduces to a type of ratio estimator when $\psi(r) = r$. Show that the asymptotic bias of $\tilde{T}_1$ is 0 under model (11.1.15).

11.3 The list below gives a sample of $n = 20$ from a population suspected of following the gross error model (11.1.9). (a) Compute the estimates of the finite population total defined by $\hat{T}_{\text{robust}} = \Sigma_{i\in s} Y_i + (N-n)\tilde{\mu}_1$ using the bisquare ($c = 5$) and Huber ($c = 1.45$) influence functions. (b) Compute the expansion estimate.

9.80	8.60	10.57	11.52	8.03
8.24	10.86	9.65	12.89	8.56
13.45	11.41	8.93	10.72	13.48
12.57	9.34	8.93	9.15	10.23
15.48	30.29	27.14	21.17	28.72

11.4 The following unit numbers are a sample from the Counties70 population in the Appendix:

18	32	69	109
116	123	177	182
203	220	231	235
238	243	247	256
260	261	272	296

(a) Compute the ratio estimator $\hat{T}_R = N\bar{Y}_s\bar{x}/\bar{x}_s$.

(b) Compute the two outlier-resistant estimators $\tilde{T}_1$ and $\tilde{T}_2$ defined in Section 9.3.2. Calculate two versions of each using (1) the bisquare with $c = 4.685$ to estimate β and the same bisquare for the ψ function in $\tilde{T}_1$, and (2) the Huber function with $c = 1.345$ for both $\hat{\beta}$ and ψ. Use $v_i = \sigma^2 x_i$ as in $M(0, 1{:}x)$ and estimate σ with the median absolute deviation of the fitted residuals as described in Section 11.1.1.

(c) Plot x and y from the sample and the full population. Include on the plot the fitted least squares line for the model $M(0, 1{:}x)$ and the bisquare and Huber estimates $\hat{\beta}$.

11.5 Suppose that the sample from Counties70 is the same 20 units plus two others: $(x_1, y_1) = (20{,}000,\ 300{,}000)$ and $(x_2, y_2) = (30{,}000,\ 350{,}000)$. Assume that the population totals of y and x are equal to the Counties70 totals plus the values for these two additional points, that is, $T_y = 11{,}893{,}111$, and $T_x = 2{,}765{,}075$. Repeat parts (a), (b), and (c) of the last problem. What effect do the two outliers have on the different estimators?

11.6 Prove Theorem 11.2.1 which says that if the parameter $\boldsymbol{\beta}$ is known in nonlinear model (11.2.1), then the *BLU* predictor of the population total T is

$$\hat{T}^* = \sum_{i \in s} Y_i + \mathbf{1}_r'[\mathbf{f}_r(\boldsymbol{\beta}) + \mathbf{V}_{rs}\mathbf{V}_{ss}^{-1}(\mathbf{Y}_s - \mathbf{f}_s(\boldsymbol{\beta}))].$$

11.7 Derive the approximation (11.2.6) to the error variance of the estimator of the total under the nonlinear model (11.2.1):

$$\begin{aligned} \mathrm{var}_M(\hat{T} - T) \cong\ & \mathbf{1}_r'(\mathbf{F}_r - \mathbf{V}_{rs}\mathbf{V}_{ss}^{-1}\mathbf{F}_s)(\mathbf{F}_s'\mathbf{V}_{ss}^{-1}\mathbf{F}_s)^{-1}(\mathbf{F}_r - \mathbf{V}_{rs}\mathbf{V}_{ss}^{-1}\mathbf{F}_s)'\mathbf{1}_r \\ & + \mathbf{1}_r'(\mathbf{V}_{rr} - \mathbf{V}_{rs}\mathbf{V}_{ss}^{-1}\mathbf{V}_{sr})\mathbf{1}_r. \end{aligned}$$

11.8 Suppose that model (11.2.7) holds. (a) Write the likelihood function for $\boldsymbol{\beta}$ and show that the log-likelihood is $L(\boldsymbol{\beta}) = \Sigma_{i \in s}[Y_i \log(f_i) + (1 - Y_i) \times \log(1 - f_i)]$. (b) Show that the maximum likelihood estimator of $\boldsymbol{\beta}$ is the solution to $\mathbf{F}_s(\boldsymbol{\beta})'\mathbf{V}_{ss}^{-1}[\mathbf{Y}_s - \mathbf{f}_s(\boldsymbol{\beta})] = \mathbf{0}$.

11.9 Plot the weights $w(u_i) = K_b(u_i)/\Sigma_i K_b(u_i)$ for u_i between -3 and 3 in increments of 0.10 for $K_b(u) = \exp(-u^2/2b)$ and for $b = 5, 3, 1, 0.5$. What is the effect of the bandwidth b on the weight $w(u_i)$?

11.10 Suppose that $Y_i, \ldots, Y_N$ are independent and follow model (11.4.3). (a) Show the best linear unbiased estimator of $F_N(t)$ is $\hat{F}_N(t) =$

$(1/n)\,\Sigma_{i\in s} I\{Y_i \leqslant t\} \equiv p_s(t)$. (b) Show that the error variance of $\hat{F}_N(t)$ is $\mathrm{var}_M[\hat{F}_N(t) - F_N(t)] = (1-f)/n \times p(t)\,[1-p(t)]$ and that an unbiased variance estimator is $v[\hat{F}_N(t)] = (1-f)/(n-1)\,p_s(t)\,[1-p_s(t)]$.

11.11 Suppose that $Y_1, \ldots, Y_N$ are independent and follow model (11.4.4). (a) Show the best linear unbiased estimator of $\hat{F}_N(t)$ is $\hat{F}_N(t) = \Sigma_h N_h p_{hs}(t)/N$. (b) Derive the error variance of $\hat{F}_N(t)$ and an unbiased estimator of the error variance.

11.12 Suppose that the Y's are independent and follow the stratified model $E_M(Y_{hi}) = \mu_h$ and $\mathrm{var}_M(Y_{hi}) = \sigma_h^2$. A stratified sample of size n_h is selected from stratum h with the total sample being $n = \Sigma_{h=1}^{H} n_h$.

(a) What simpler form does the CDF estimator $\hat{F}_{CD}(t) = (1/N)[nF_s(t) + (N-n)\,\hat{F}_r(t)]$ reduce to?

(b) If we define the weight for each sample unit from stratum h to be N_h/n_h, that is, the weight associated with the stratified expansion estimator which is BLU, show that the following algorithm for estimating a quantile q_α is equivalent to finding $\hat{q}_\alpha = \inf\{t: F_N(t) \geqslant \alpha\}$ where $\hat{F}_N(t)$ is defined by (11.4.7):

- **(i)** Sort the sample y's in ascending order disregarding strata membership
- **(ii)** Cumulate the weights, $w_i = N_h/n_h$ for $i \in s_h$, associated with the sorted y's to form the sequence $C_j = \Sigma_{i\leqslant j; i\in s} w_i$, $j = 1, \ldots, n$
- **(iii)** Estimate q_α as the first y whose cumulant C_j is greater than or equal to $N\alpha$.

11.13 Consider the CDF estimator $\tilde{F}(t) = (1/N)[nF_s(t) + (N-n)\tilde{F}_r(t)]$ where

$$\tilde{F}_r(t) = \sum_r I\{\hat{Y}_i \leqslant t\}/(N-n) \text{ with } \hat{Y}_i = \mathbf{x}_i'\hat{\boldsymbol{\beta}} \text{ and}$$

$$\hat{\boldsymbol{\beta}} = (\mathbf{X}_s'\mathbf{V}_{ss}^{-1}\mathbf{X}_s)^{-1}\mathbf{X}_s'\mathbf{V}_{ss}^{-1}\mathbf{Y}_s.$$

Show that this estimator is biased under model (11.4.3). (*Hint:* $E_M(I\{\hat{Y}_i \leqslant t\}) \neq \Pr(Y_i \leqslant t)$).

11.14 Use the following sample of units from the Cancer population to

(a) calculate $\hat{F}_{CD}(t)$ at the values of t given by the sample y's and the predicted y's based on the model $M(1, 1{:}x)$. Plot the result as a step function.

(b) Estimate the 25th, 50th, and 75th percentiles of the population.

(c) Compute the population quartiles for comparison. Note that this problem will require a computer program to be written.

Unit	x	y
4	681	4
23	1635	6
75	2929	14
101	3706	7
196	9445	38
235	15204	48
275	28024	118
278	29254	90
288	42997	169
294	60161	302

11.15 Show that expression (11.4.25) for the bias of $\hat{F}_{CD}(t)$ is strictly positive if $d > 0$ and strictly negative if $d < 0$ assuming that β is a positive scalar and all x's are positive.

11.16 Show that $\hat{F}_{RKM}(t)$ is asymptotically design-unbiased and asymptotically model-unbiased under the regression model (11.4.8).

11.17 **(a)** Show that under Model A of Section 11.4

$$E_M(Y_i) = \mu_{cd} \qquad i \text{ in } (c, d)$$

$$\operatorname{cov}_M(Y_i, Y_{i'}) = \begin{cases} \sigma_{cd}^2 & i = i',\ i \text{ in } (c, d) \\ \sigma_{cd}^2 \rho_{cd} & i \neq i',\ i \text{ and } i' \text{ in } (c, d) \\ 0 & \text{otherwise} \end{cases}$$

the *BLU* predictor is $\hat{T}_d^{(A)} = \Sigma_c \hat{T}_{cd}^{(A)}$ where $\hat{T}_{cd}^{(A)} = N_{cd} \bar{Y}_{s_{cd}}$ and $\bar{Y}_{s_{cd}} = \Sigma_{s_{cd}} Y_i / n_{cd}$.

(b) Show that the error variance of $\hat{T}_d^{(A)}$ is

$$\operatorname{var}_M(\hat{T}_{cd}^{(A)} - T_d) = \sum_c N_{cd}^2 (1 - f_{cd}) \sigma_{cd}^2 (1 - \rho_{cd}).$$

(c) Show that $v_{cd} = \Sigma_{i \in s_{cd}} (Y_i - \bar{Y}_{scd})^2 / (n_{cd} - 1)$ is an unbiased estimator of $\sigma_{cd}^2 (1 - \rho_{cd})$ under the model in part (a) above.

11.18 Find the values of θ_{cd} $(c = 1, \ldots, C)$ in

$$\hat{T}_d = \sum_c n_{cd} \bar{Y}_{s_{cd}} + \sum_c (N_{cd} - n_{cd}) [\theta_{cd} \bar{Y}_{s_{cd}} + (1 - \theta_{cd}) \hat{\mu}_c]$$

that minimize the error variance under Model B of Section 9.6. Assume that $\hat{\mu}_c$ is uncorrelated with y-values in classes other than c.

11.19 Show that,

(a) Show that, if $(\hat{\mu}_c - \bar{Y}_{s_{cd}})$ and $(\hat{\mu}_c - \bar{Y}_{r_{cd}})$ are positively correlated, then the modified synthetic estimator $\hat{T}_d^{(\mathrm{mod})} = \Sigma_c n_{cd} \bar{Y}_{s_{cd}} + \Sigma_c (N_{cd} - n_{cd}) \hat{\mu}_c$ has smaller error variance under Model B than does the synthetic estimator $\hat{T}_d^{(\mathrm{syn})} = \Sigma_c N_{cd} \hat{\mu}_c$. Assume that $\hat{\mu}_c = \Sigma_{j=1}^{D} \ell_{cj} \bar{Y}_{s_{cj}}$ and $\Sigma_{j=1}^{D} \ell_{cj} = 1$.

(b) Show that, if $\hat{\mu}_c = \Sigma_{j=1}^{D} n_{cj} \bar{Y}_{s_{cj}} / n_c$, $\sigma_{cd}^2 = \sigma^2$, and $\rho_{cd} = \rho$, then the error variance of $\hat{T}_d^{(\mathrm{mod})}$ is always smaller than that of $\hat{T}_d^{(\mathrm{syn})}$.

11.20 Suppose that X and Y are unbiased predictors of a random variable Z. Consider composite predictors of the form $\hat{Z} = \alpha X + (1 - \alpha) Y$ where $0 \leqslant \alpha \leqslant 1$.

(a) What value of α is optimal?

(b) For what range of values of α does the composite estimator have a smaller error variance than either X or Y?

11.21 Verify that the *BLU* predictor

$$\hat{T}_d = \sum_{c=1}^{C} \{ n_{cd} \bar{Y}_{s_{cd}} + (N_{cd} - n_{cd}) [\bar{\mathbf{X}}'_{r_{cd}} \hat{\boldsymbol{\beta}}_c + w_{cd} (\bar{Y}_{s_{cd}} - \bar{\mathbf{X}}'_{s_{cd}} \hat{\boldsymbol{\beta}}_c)] \}$$

given by (11.5.14) reduces to $\hat{T}_d^{(B)} = \Sigma_c \hat{T}_{cd}^{(B)}$ in the special case of $\mathbf{x}_{cdi} = \mathbf{1}$ for all units.

11.22 Suppose we have a population composed of $d = 1, \ldots, D$ states and $c = 1, \ldots, C_d$ counties in state d. Suppose that it is desired to use the general type of model for both counties and states:

$$T_j = \mathbf{x}'_j \boldsymbol{\beta} + \varepsilon_{1j},\ \varepsilon_{1j} \sim (0, \sigma^2),$$

where j indexes either counties or states and the errors are independent. Show that this model cannot hold both for the smaller units (counties) and the larger units (states).

11.23 The ratio of the error variance of the *BLU* predictor under Model B to the error variance of the post-stratified estimator was given in expression (11.5.10) as

$$\frac{\mathrm{var}_M(\hat{T}_{cd}^{(B)} - T_{cd}^{(B)})}{\mathrm{var}_M(\hat{T}_{cd}^{(A)} - T_{cd}^{(A)})} = 1 - (1 - f_{cd})(1 - w_{cd})(1 - u_{cd}).$$

Show that this ratio is bounded above by $3 \max(f_{cd}, w_{cd}, u_{cd})$. (*Hint:* Show that $1 - (1 - f_{cd})(1 - w_{cd})(1 - u_{cd}) \leqslant 1 - (1 - F)^3 \leqslant 3F$ where $F = \max(f_{cd}, w_{cd}, u_{cd})$).

11.24 Show that the estimator of μ_c in $\hat{T}_d^{(\rho)}$ in Section 11.5.3 can be written as

$$\hat{\mu}_c = \frac{\Sigma_j \bar{Y}_{scj}/[1 + (1 - \rho)/(n_{cj}\rho)]}{\Sigma_j 1/[1 + (1 - \rho)/(n_{cj}\rho)]}.$$

What can you say about the relative size of the weights on the cell means based on this form?

APPENDIX A

Some Basic Tools

This appendix lays out some basic notational conventions and other results that are used in various chapters of this book. The presentation is intended to be informal and to serve as a quick reference for readers who already have some familiarity with the ideas. More detailed discussion is available in many sources at different levels of formality, for example, Bickel and Doksum (1977), Rao (1973), or Serfling (1981).

A.1. Orders of Magnitude, $O(\cdot)$ and $o(\cdot)$

The symbols $O(\cdot)$ and $o(\cdot)$ are called "big oh" and "little oh," respectively, and denote ways of comparing the magnitudes of two functions $u(x)$ and $v(x)$ as the argument x tends to a limit L. The notation $u(x) = O(v(x))$ as $x \to L$ means that $|u(x)/v(x)|$ remains bounded as $x \to L$. The notation $u(x) = o(v(x))$ as $x \to L$ denotes that

$$\lim_{x \to L} \frac{u(x)}{v(x)} = 0.$$

Probabilistic versions of these order of magnitude relations are given in Section A.3.

A.2. Convergence in Probability and in Distribution

We say that the sequence of random variables $\{Z_n\}$ *converges in probability* to the random variable Z and write $Z_n \xrightarrow{p} Z$ if $P(|Z_n - Z| \geqslant \varepsilon) \to 0$ as $n \to \infty$ for every $\varepsilon > 0$.

Let F denote the distribution function for a random variable Y, that is $F_Y(t) = P(Y \leqslant t)$ for all $t \in R$. The sequence of random variables $\{Z_n\}$ is said to *converge in distribution or in law* to the random variable Z if $F_{z_n}(t) \to F_z(t)$ for every point t such that F_z is continuous at t as $n \to \infty$. Convergence in distribution is denoted by $Z_n \xrightarrow{d} Z$.

A.3. Probabilistic Orders of Magnitude, $O_p(\cdot)$ and $o_p(\cdot)$

A sequence of random variables $\{Z_n\}$, with respective distribution functions $\{F_n\}$, is said to be *bounded in probability* if for every $\varepsilon > 0$ there exist M_ε and n_ε such that

$$F_n(M_\varepsilon) - F_n(-M_\varepsilon) > 1 - \varepsilon \quad \text{for all } n > n_\varepsilon.$$

This condition is denoted $Z_n = O_p(1)$. It is a fact that $Z_n \xrightarrow{d} Z \Rightarrow Z_n = O_p(1)$.

For two sequences of random variables $\{Z_n\}$ and $\{Y_n\}$, the notation $Z_n = O_p(Y_n)$ means that the sequence $Z_n/Y_n = O_p(1)$. The notation $Z_n = o_p(Y_n)$ means that $Z_n/Y_n \xrightarrow{p} 0$.

A.4. Chebyshev's Inequality

If Z is any random variable, then

$$P[|Z| \geqslant \varepsilon] \leqslant \frac{E(Z^2)}{\varepsilon^2},$$

where $\varepsilon > 0$. If, for example, we have a sequence of estimators $\{Z_n\}$, each of which has expectation $E(Z_n) = \theta$ and has a variance that approaches 0 as $n \to \infty$, then applying Chebyshev's inequality to the sequence $Z_n - \theta$ shows that $Z_n \xrightarrow{p} \theta$.

A.5. Cauchy-Schwarz Inequality

For any two random variables X_1 and X_2 such that $E(X_1^2) < \infty$ and $E(X_2^2) < \infty$

$$|E(X_1X_2)| \leqslant \sqrt{E(X_1^2)E(X_2^2)}.$$

Equality holds if and only if one of X_1 and X_2 equals 0 or $X_1 = aX_2$ for some constant a.

A non-probabilistic version is the following. Suppose that $\mathbf{X} = (x_1, \ldots, x_n)'$ and $\mathbf{Y} = (y_1, \ldots, y_n)'$ are two non-random vectors. Then,

$$|\mathbf{X}'\mathbf{Y}| \leqslant \sqrt{(\mathbf{X}'\mathbf{X})(\mathbf{Y}'\mathbf{Y})}.$$

A.6. Slutsky's Theorem

If $Z_n \xrightarrow{d} Z$ and $U_n \xrightarrow{p} u_0$ (a constant), then

(a) $Z_n + U_n \xrightarrow{d} Z + u_0$
(b) $U_n Z_n \xrightarrow{d} u_0 Z$.

A.7. Taylor's Theorem

Taylor's Theorem is used in the text for large sample analyses in which a function of a statistic, like an estimated mean, is approximately linear when the statistic is near its expectation or near the value to which it converges.

A.7.1. Univariate Version

Let the function g have a finite nth derivative $g^{(n)}$ everywhere in the open interval (a, b) and $(n-1)$th derivative $g^{(n-1)}$ continuous in the closed interval $[a, b]$. Let $x \in [a, b]$. For each point $y \in [a, b]$, $y \neq x$, there exists a point z interior to the interval joining x and y such that

$$g(y) = g(x) + \sum_{k=1}^{n-1} \frac{g^{(k)}(x)}{k!}(y - x)^k + \frac{g^{(n)}(z)}{n!}(y - x)^n.$$

A.7.2. Multivariate Version

Let the function g defined on R^m possess continuous partial derivatives of order n at each point of an open set $A \subset R^m$. Let $x \in A$. For each point $\mathbf{y}$, $\mathbf{y} \neq \mathbf{x}$, such that the line segment $L(\mathbf{x}, \mathbf{y})$ joining $\mathbf{x}$ and $\mathbf{y}$ lies in A, there exists a point $\mathbf{z}$ in the interior of $L(\mathbf{x}, \mathbf{y})$ such that

$$g(\mathbf{y}) = g(\mathbf{x}) + \sum_{k=1}^{n-1} \frac{1}{k!} \sum_{i_k=1}^{m} \cdots \sum_{i_1=1}^{m} \frac{\partial^k g(t_1, \ldots, t_m)}{\partial t_{i_1} \cdots \partial t_{i_k}}\bigg|_{\mathbf{t}=\mathbf{x}} \prod_{j=1}^{k} (y_{i_j} - x_{i_j})$$
$$+ \frac{1}{n!} \sum_{i_1=1}^{m} \cdots \sum_{i_n=1}^{m} \frac{\partial^n g(t_1, \ldots, t_m)}{\partial t_{i_1} \cdots \partial t_{i_n}}\bigg|_{\mathbf{t}=\mathbf{z}} \prod_{j=1}^{n} (y_{i_j} - x_{i_j}).$$

For example, suppose that $\mathbf{y}$ is a vector of estimators whose expectation is the vector $\boldsymbol{\theta}$. Under the conditions in the multivariate version of Taylor's Theorem, there is a point $\mathbf{z}$ on the line segment $L(\boldsymbol{\theta}, \mathbf{y})$ such that

$$g(\mathbf{y}) = g(\boldsymbol{\theta}) + \nabla g(\boldsymbol{\theta})'(\mathbf{y} - \boldsymbol{\theta}) + \tfrac{1}{2}(\mathbf{y} - \boldsymbol{\theta})'\nabla g^{(2)}(\mathbf{z})(\mathbf{y} - \boldsymbol{\theta}),$$

where $\nabla g(\boldsymbol{\theta})$ is the m-vector of first partial derivatives of $g(\mathbf{t})$ evaluated at the point $\mathbf{t} = \boldsymbol{\theta}$ and $\nabla g^{(2)}(\mathbf{z})$ is the $m \times m$ matrix of second partial derivatives of $g(\mathbf{t})$ evaluated at $\mathbf{t} = \mathbf{z}$.

A.8. Central Limit Theorems for Independent, not Identically Distributed Random Variables

Theorem A.8.1 (Eicker, 1963). Let $\mathbf{E}$ be the set of sequences of independent random variables $\varepsilon_1, \varepsilon_2, \ldots$ having means zero, finite variances $\psi_1, \psi_2, \ldots$, and distribution functions $F_1 F_2, \ldots$ with all F_i belonging to a family $\mathbf{F}$. Define $\{a_{nk}: n = 1, 2, \ldots; k = 1, 2, \ldots\}$ to be a double sequence of real constants, and $\{k_n\}$ to be a sequence of positive integers such that $a_{nk_n} \neq 0$ and $a_{nk} = 0$ for

$k > k_n$ ($n = 1, 2, \ldots$). Let $B_n^2 = \Sigma_{k=1}^{k_n} a_{nk}^2 \psi_k$ and $\zeta_n = B_n^{-1} \Sigma_{k=1}^{k_n} a_{nk} \varepsilon_k$. Then, for $\Phi(x)$ the standard normal distribution function,

$$\sup_{\varepsilon \in \mathbf{E}, -\infty < x < \infty} |P(\zeta_n < x) - \Phi(x)| \to 0 \quad \text{and} \quad \sup_{\varepsilon \in \mathbf{E}} \max_{k=1,\ldots,k_n} P(|B_n^{-1} a_{nk} \varepsilon_k| > \delta) \to 0$$

for every $\delta > 0$ if and only if

(i) $\max_{k=1,\ldots,k_n} a_{nk}^2 / \Sigma_{k=1}^{k_n} a_{nk}^2 \to 0$ as $n \to \infty$,

(ii) $\sup_{G \in \mathbf{F}} \int_{|z|>c} z^2 \, dG(z) \to 0$ *as* $c \to \infty$, and

(iii) $\inf_{G \in \mathbf{F}} \int z^2 \, dG(z) > 0$.

In words, the interpretation of the theorem is that the standardized statistic ζ_n is asymptotically normal if and only if (i) the known elements a_{nk}^2 each make an asymptotically negligible contribution to their sum $\Sigma_{k=1}^{k_n} a_{nk}^2$, (ii) the distribution functions generating the random variables ε_i are not too thick tailed, and (iii) the set of possible variances have a positive lower bound.

A typical application of this theorem for the purposes of this text is to define the circumstances under which an estimator of a total is asymptotically normal. For example, suppose that the errors are defined by the model specification $Y_i = \mathbf{x}_i' \boldsymbol{\beta} + \varepsilon_i$, where $\boldsymbol{\beta}$ is a $p \times 1$ parameter vector and $\mathbf{x}_i$ is a $p \times 1$ vector of auxiliaries. When the estimator of the total has the form $\hat{T} = \Sigma_s g_i Y_i$, the estimation error can be written as $\hat{T} - T = \Sigma_s a_i \varepsilon_i - \Sigma_r \varepsilon_i$, where $a_i = g_i - 1$. Under conditions given in Theorem 2.5.1, $\Sigma_s a_i \varepsilon_i / \sqrt{\text{var}_M(\hat{T} - T)}$ is normally distributed in large samples and $\Sigma_r \varepsilon_i / \sqrt{\text{var}_M(\hat{T} - T)}$ converges in probability to 0.

The next theorem also relates to large sample normality and is used in the proof of Theorem 5.3.1 where the sandwich variance estimator is proved to be consistent. This theorem is a restatement of Theorem 4 of Section 28 in Gnedenko and Kolmogorov (1968).

Theorem A.8.2 (Gnedenko and Kolmogorov, 1968). Consider the double sequence of row-independent random variables ξ_{nk}, $k = 1, 2, \ldots, n = 1, 2, \ldots$ with means 0 and with the property that $\sup_{1 \leqslant k \leqslant k_n} P(|\xi_{nk}| > \varepsilon) \to 0$, as $n \to \infty$. Suppose $\text{var}(\zeta_n) = 1$ where $\zeta_n = \Sigma_{k=1}^{n_k} \xi_{nk}$. Then $|P(\zeta_n < x) - \Phi(x)| \to 0$, for all x if and only if $P(|\eta_n^2 - 1| > \varepsilon) \to 0$, for all $\varepsilon > 0$, where $\eta_n^2 = \Sigma_{k=1}^{k_n} \xi_{nk}^2$.

A.9. Central Limit Theorem for Simple Random Sampling

Asymptotic results can be derived for finite populations using only a probability sampling plan to generate the statistical distribution. The theorem below gives conditions under which the sample mean has an asymptotic normal distribution generated by repeated sampling from a fixed finite population. The artifice used for large sample theory is to conceive of the finite population as being embedded in a sequence of populations that is growing ever larger. No assumption about an underlying superpopulation model is made.

Let $U = \{1, 2, \ldots, N\}$ denote a finite population of N units, each of which has associated with it a real number Y_i. A simple random sample of n units is selected without replacement from U, and the population mean $\bar{Y} = \Sigma_{i=1}^{N} Y_i/N$ is estimated by the sample mean $\bar{y} = \Sigma_{i \in s} Y_i/n$. The finite population variance of the Y_i's is $S_Y^2 = \Sigma_{i=1}^{N}(Y_i - \bar{Y})^2/(N-1)$. For the asymptotic formulation, we embed the finite population in a sequence of populations $\{U_v\}$ indexed by v where n_v and N_v increase without bound as $v \to \infty$.

Theorem A.9.1 (Hájek, 1960). Suppose that $n_v \to \infty$ and $(N_v - n_v) \to \infty$ as $v \to \infty$. Then, under simple random sampling without replacement

$$\frac{\sqrt{n_v}(\bar{y}_v - \bar{Y}_v)}{\sqrt{1 - f_v}\, S_{vY}} \xrightarrow{d} N(0, 1) \quad \text{as } v \to \infty$$

if and only if $\{Y_{vj}\}$ satisfies the Lindeberg-Hájek condition

$$\lim_{v \to \infty} \sum_{T_v(\delta)} \frac{(Y_{vj} - \bar{Y}_v)^2}{(N_v - 1)S_{vY}^2} = 0 \quad \text{for any } \delta > 0,$$

where $T_v(\delta)$ is the set of units in U_v for which $|Y_{vj} - \bar{Y}_v|/\sqrt{1 - f_v}\, S_{vY} > \delta\sqrt{n_v}$.

A.10. Generalized Inverse of a Matrix

A *generalized inverse* (or g-inverse) of a matrix $\mathbf{A}$ is defined to be any matrix $\mathbf{G}$ that satisfies the equation

$$\mathbf{AGA} = \mathbf{A}. \tag{A.10.1}$$

A g-inverse of $\mathbf{A}$ is often denoted by $\mathbf{A}^-$, using an obvious analogy to the notation $\mathbf{A}^{-1}$. The matrix $\mathbf{A}$ does not have to be either symmetric or square but for the applications in this book will be both since we are interested in matrices with the form $\mathbf{A} = \mathbf{X}'\mathbf{V}^{-1}\mathbf{X}$. The following two lemmas are taken from Searle (1971, secs. 1.3, 1.5, and 1.7).

Lemma A.10.1. Let $\mathbf{X}$, $\mathbf{P}$, and $\mathbf{Q}$ be real valued matrices with dimensions $n \times p$, $q \times p$, and $q \times p$. Then

(i) $\mathbf{X}'\mathbf{X} = \mathbf{0}$ implies $\mathbf{X} = \mathbf{0}$.

(ii) $\mathbf{PX}'\mathbf{X} = \mathbf{QX}'\mathbf{X}$ implies $\mathbf{PX}' = \mathbf{QX}'$.

Proof: The first is true because $\mathbf{X}'\mathbf{X} = \mathbf{0}$ implies that the sum of squares of the elements in each row is zero and hence the elements themselves are all 0. To prove (ii), note that

$$(\mathbf{PX'X} - \mathbf{QX'X})(\mathbf{P} - \mathbf{Q})' = (\mathbf{PX'} - \mathbf{QX'})(\mathbf{PX'} - \mathbf{QX'})' = \mathbf{0}.$$

By (i), $\mathbf{PX'} = \mathbf{QX'}$. □

Lemma A.10.2. When $\mathbf{G}$ is a g-inverse of $\mathbf{X'X}$, then

(i) $\mathbf{G'}$ is also a g-inverse of $\mathbf{X'X}$.
(ii) $\mathbf{XGX'X} = \mathbf{X}$, that is, $\mathbf{GX'}$ is a g-inverse of $\mathbf{X}$.
(iii) $\mathbf{XGX'}$ is invariant to $\mathbf{G}$, that is, $\mathbf{XGX'} = \mathbf{XFX'}$ for any two g-inverses $\mathbf{F}$ and $\mathbf{G}$.
(iv) $\mathbf{XGX'}$ is symmetric whether $\mathbf{G}$ is or not.

Proof: By definition, $\mathbf{G}$ satisfies $\mathbf{X'XGX'X} = \mathbf{X'X}$. Transposing gives $\mathbf{X'XG'X'X} = \mathbf{X'X}$, proving (i). Applying Lemma A.10.1(ii) with $\mathbf{P} = \mathbf{X'XG'}$ and $\mathbf{Q} = \mathbf{I}$ implies that $\mathbf{X'XG'X'} = \mathbf{X'}$. Transposing again, we obtain (ii). To prove (iii), suppose that $\mathbf{F}$ is some other g-inverse different from $\mathbf{G}$. Then (i) means that $\mathbf{XGX'X} = \mathbf{XFX'X}$. By Lemma A.10.1(ii), $\mathbf{XGX'} = \mathbf{XFX'}$, that is, $\mathbf{XGX'}$ is the same for all g-inverses of $\mathbf{X'X}$. To prove (iv), note that by (i) $\mathbf{G}$ and $\mathbf{G'}$ are both g-inverses. A.10.2(iii) then implies that $\mathbf{XGX'} = \mathbf{XG'X'} = (\mathbf{XGX'})'$, as desired. □

A generalized inverse of a matrix can be constructed using the *singular value decomposition* (SVD) of the matrix, as described below.

A real-valued $p \times q$ matrix $\mathbf{A}$ can be decomposed into the product of an orthogonal $p \times q$ matrix $\mathbf{U}$, a diagonal $q \times q$ matrix $\mathbf{D}$, and another $q \times q$ orthogonal matrix $\mathbf{V}$:

$$\mathbf{A} = \mathbf{UDV'}. \tag{A.10.2}$$

The matrices $\mathbf{U}$ and $\mathbf{V}$ are orthogonal in the sense that $\mathbf{U'U} = \mathbf{V'V} = \mathbf{I}_q$ where $\mathbf{I}_q$ is the $q \times q$ identity matrix. The elements of the diagonal matrix $\mathbf{D}$ are called the *singular values* of $\mathbf{A}$. The squares of the singular values of $\mathbf{A}$ are the eigenvalues of $\mathbf{A'A}$. Expression (A.10.2) is called the *singular value decomposition* (SVD) of $\mathbf{A}$. If $\mathbf{A}$ is symmetric, then $\mathbf{U} = \mathbf{V}$ and the square root of $\mathbf{A}$ can be constructed as

$$\mathbf{A}^{1/2} = \mathbf{UD}^{1/2}\mathbf{U'}, \tag{A.10.3}$$

where $\mathbf{D}^{1/2}$ is the diagonal matrix of the square roots of the diagonal elements of $\mathbf{D}$.

Using the SVD, a g-inverse of $\mathbf{A}$ is $\mathbf{G} = \mathbf{VD}^{-1}\mathbf{U'}$. To see that this $\mathbf{G}$ satisfies

(A.10.1), note that, by the orthogonality of $\mathbf{U}$ and $\mathbf{V}$,

$$\begin{aligned}\mathbf{AGA} &= \mathbf{UDV}'(\mathbf{VD}^{-1}\mathbf{U}')\mathbf{UDV}'\\ &= \mathbf{UDV}'\\ &= \mathbf{A}.\end{aligned}$$

When any of the elements of $\mathbf{D}$ are zero, that is, $\mathbf{A}'\mathbf{A}$ has one or more zero eigenvalues, then $\mathbf{G}$ is modified as follows. Any zeros are omitted from $\mathbf{D}$ as are the corresponding columns of $\mathbf{U}$ and $\mathbf{V}$. Denote the matrices after the deletions have been made as $\mathbf{V}^*$, $\mathbf{D}^*$, and $\mathbf{U}^*$. The g-inverse is then computed as $\mathbf{G} = \mathbf{V}^*\mathbf{D}^{*-1}\mathbf{U}^{*\prime}$.

Lemma A.10.3 for a generalized inverse of a partitioned matrix will be useful when studying models that involve both qualitative and quantitative auxiliary variables in Chapter 7. The Lemma refers to a partitioned matrix $\mathbf{M}$ with the form

$$\mathbf{M} = \begin{bmatrix}\mathbf{X}'\\ \mathbf{Z}'\end{bmatrix}[\mathbf{X} \quad \mathbf{Z}] \equiv \begin{bmatrix}\mathbf{A} & \mathbf{B}\\ \mathbf{B}' & \mathbf{D}\end{bmatrix}, \text{ say.}$$

Lemma A.10.3. A generalized inverse of $\mathbf{M}$ is

$$\begin{aligned}\mathbf{M}^- &= \begin{bmatrix}\mathbf{A}^- + \mathbf{A}^-\mathbf{B}\mathbf{Q}^-\mathbf{B}'\mathbf{A}^- & -\mathbf{A}^-\mathbf{B}\mathbf{Q}^-\\ -\mathbf{Q}^-\mathbf{B}'\mathbf{A}^- & \mathbf{Q}^-\end{bmatrix}\\ &= \begin{bmatrix}\mathbf{A}^- & \mathbf{0}\\ \mathbf{0} & \mathbf{0}\end{bmatrix} + \begin{bmatrix}-\mathbf{A}^-\mathbf{B}\\ \mathbf{I}\end{bmatrix}\mathbf{Q}^-[-\mathbf{B}'\mathbf{A}^- \quad \mathbf{I}],\end{aligned}$$

where $\mathbf{A}^-$ is a g-inverse of $\mathbf{A}$, $\mathbf{Q} = \mathbf{D} - \mathbf{B}'\mathbf{A}^-\mathbf{B}$, and $\mathbf{Q}^-$ is a g-inverse of $\mathbf{Q}$.

The proof of this lemma, using Lemma A.10.2, is left to the exercises after Chapter 7.

APPENDIX B

Datasets

This appendix contains several populations that are used for the examples and homework problems in the text. All of these populations are available at the John Wiley worldwide web site, ftp://ftp.wiley.com/public/sci_tech_med/finite_populations. The populations might also be used for further research on the topics discussed in this book. Four of the populations—Cancer, Hospitals, Counties60, and Counties70—were studied by Royall and Cumberland (1981a, b). The source of each population and some characteristics of each are described in Tables B.1 and B.2. These populations have been used subsequently by a number of authors in studies of ratio and regression estimation, variance estimation, and confidence interval coverage.

The last two populations are clustered and can be used in studying various topics in Chapters 7–11. The fifth population is a smaller version of a clustered population used by Valliant (1987c) and contains demographic and economic variables from a labor force survey described in more detail in Section B.5. The sixth population is from the Third International Mathematics and Science Study (TIMSS). The TIMSS population is described in Section B.6.

Table B.2. Population Statistics for the Unclustered Populations

Population	N	$\bar{x}$	T	$\dfrac{\Sigma_1^N (x_i - \bar{x})^2}{\bar{x}^2(N-1)}$	$\rho(x, y)$
Cancer	301	11,288	11,994	1.49	0.967
Hospitals	393	274.7	320,159	0.60	0.910
Counties60	304	8,931	10,006,523	1.69	0.998
Counties70	304	8,931	11,243,111	1.69	0.982

Table B.1. Unclustered Populations Used in the Text and Exercises

Population	Description	Source	x	y	N
Cancer	Counties in North Carolina, South Carolina, and Georgia with 1960 white female population $<100{,}000$	x: U.S. Census y: Mason and McKay (1974)	Adult female population, 1960	Breast cancer mortality, 1950–1969 (white females)	301
Hospitals	National sample of short-stay hospitals with fewer than 1000 beds	Herson (1976). From National Center for Health Statistics Hospital Discharge Survey (January 1968)	Number of beds	Number of patients discharged	393
Counties60	Counties in North Carolina, South Carolina, and Georgia with $<100{,}000$ households in 1960	U.S. Census	Number of households 1960	Population, excluding residents of group quarters, 1960	304
Counties70	Counties in North Carolina, South Carolina, and Georgia with $<100{,}000$ households in 1960	U.S. Census	Number of households 1960	Population, excluding residents of group quarters, 1970	304

Appendix B.1. Cancer Population

Unit No.	y	x	Unit No.	y	x
1	1	445	51	8	2261
2	0	599	52	6	2317
3	3	677	53	8	2333
4	4	681	54	16	2393
5	3	746	55	10	2404
6	4	869	56	4	2419
7	1	950	57	11	2462
8	5	976	58	10	2467
9	5	1096	59	11	2477
10	5	1098	60	9	2483
11	5	1114	61	11	2511
12	7	1125	62	14	2591
13	5	1236	63	6	2624
14	6	1285	64	8	2690
15	3	1291	65	12	2731
16	3	1318	66	15	2735
17	2	1323	67	9	2736
18	8	1327	68	13	2747
19	9	1438	69	18	2782
20	7	1479	70	15	2783
21	4	1536	71	12	2793
22	6	1598	72	11	2891
23	6	1635	73	12	2894
24	11	1667	74	12	2906
25	4	1696	75	14	2929
26	7	1792	76	12	2935
27	7	1795	77	3	2962
28	4	1808	78	5	3054
29	6	1838	79	7	3112
30	13	1838	80	9	3118
31	3	1847	81	11	3185
32	8	1933	82	14	3217
33	8	1959	83	18	3236
34	4	1990	84	11	3290
35	9	2003	85	11	3314
36	10	2070	86	4	3316
37	7	2091	87	13	3401
38	8	2099	88	10	3409
39	5	2104	89	10	3426
40	11	2147	90	9	3470
41	4	2154	91	11	3488
42	12	2163	92	12	3511
43	11	2172	93	4	3549
44	9	2174	94	16	3571
45	13	2183	95	20	3578
46	17	2193	96	5	3620
47	11	2210	97	15	3654
48	10	2212	98	15	3680
49	4	2236	99	12	3683
50	4	2245	100	7	3688

Unit No.	y	x	Unit No.	y	x
201	29	9994	251	51	19274
202	17	10033	252	66	19818
203	29	10049	253	53	19906
204	41	10144	254	58	20065
205	31	10303	255	75	20140
206	35	10416	256	88	20268
207	27	10461	257	83	20539
208	37	10670	258	48	20636
209	18	10844	259	69	20969
210	41	10875	260	41	21353
211	39	10890	261	73	21757
212	41	11105	262	79	22811
213	61	11622	263	63	23245
214	46	12038	264	90	23258
215	47	12173	265	92	24296
216	36	12181	266	60	24351
217	43	12608	267	63	24692
218	45	12775	268	63	24896
219	46	12915	269	75	25275
220	45	13021	270	70	25405
221	49	13142	271	90	25715
222	55	13206	272	111	26245
223	64	13407	273	103	26408
224	64	13647	274	117	26691
225	66	13870	275	118	28024
226	57	13989	276	40	28270
227	53	14089	277	83	28477
228	51	14197	278	90	29254
229	36	14620	279	97	29422
230	28	14816	280	92	30125
231	59	14952	281	104	30538
232	39	15039	282	96	34109
233	73	15049	283	142	35112
234	41	15179	284	105	35876
235	48	15204	285	145	36307
236	37	16161	286	160	39023
237	72	16239	287	127	40756
238	72	16427	288	169	42997
239	48	16462	289	104	47672
240	62	16793	290	179	49126
241	51	16925	291	152	53464
242	71	17027	292	163	56529
243	60	17201	293	167	59634
244	70	17526	294	302	60161
245	59	17666	295	246	62398
246	91	17692	296	236	62652
247	52	17742	297	250	62931
248	65	18482	298	267	63476
249	77	18731	299	244	66674
250	84	18835	300	248	74005
			301	360	88456

Appendix B.2. Hospitals Population

Unit No.	y	x	Unit No.	y	x
1	57	10	51	115	56
2	64	14	52	220	57
3	41	15	53	247	57
4	76	15	54	297	59
5	35	16	55	308	61
6	56	18	56	58	61
7	42	19	57	91	62
8	48	19	58	182	62
9	23	20	59	242	63
10	90	20	60	222	63
11	22	20	61	240	64
12	170	24	62	225	64
13	120	24	63	239	65
14	36	25	64	231	65
15	49	25	65	255	67
16	14	25	66	321	67
17	92	25	67	215	67
18	64	25	68	259	68
19	103	25	69	233	69
20	100	26	70	253	69
21	98	26	71	209	70
22	89	27	72	216	70
23	85	28	73	315	70
24	109	29	74	233	70
25	118	30	75	258	70
26	79	30	76	244	70
27	52	32	77	200	73
28	125	32	78	297	74
29	66	34	79	297	80
30	100	35	80	301	80
31	75	35	81	266	81
32	108	37	82	270	86
33	95	38	83	310	88
34	78	38	84	326	90
35	153	39	85	120	91
36	124	40	86	243	95
37	87	40	87	243	95
38	121	40	88	377	96
39	213	41	89	228	96
40	210	43	90	308	98
41	141	43	91	346	99
42	174	47	92	444	100
43	81	48	93	362	100
44	173	49	94	383	100
45	260	50	95	318	100
46	186	50	96	373	100
47	296	50	97	414	100
48	87	50	98	265	100
49	229	50	99	227	102
50	194	50	100	371	102

Unit No.	y	x	Unit No.	y	x
101	518	103	151	778	156
102	309	104	152	487	159
103	311	106	153	637	160
104	327	108	154	590	160
105	389	110	155	402	160
106	439	110	156	611	161
107	368	110	157	690	163
108	298	110	158	360	165
109	273	111	159	662	169
100	498	111	160	689	170
111	95	111	161	665	170
112	594	113	162	592	175
113	440	116	163	446	178
114	134	118	164	84	180
115	345	119	165	479	180
116	360	119	166	531	180
117	431	120	167	573	181
118	373	120	168	481	184
119	414	120	169	652	184
120	467	121	170	635	185
121	534	122	171	1011	187
122	416	123	172	713	188
123	231	125	173	625	192
124	427	126	174	504	193
125	535	127	175	744	195
126	323	128	176	586	196
127	577	128	177	695	200
128	707	129	178	697	204
129	426	130	179	670	204
130	381	134	180	622	205
131	411	135	181	703	206
132	362	135	182	814	207
133	505	137	183	726	207
134	592	138	184	670	210
135	244	139	185	918	214
136	355	141	186	726	214
137	322	142	187	590	224
138	384	143	188	587	224
139	453	144	189	558	224
140	828	145	190	1186	225
141	475	145	191	410	227
142	337	145	192	732	227
143	283	145	193	955	228
144	470	150	194	1175	229
145	621	150	195	439	229
146	376	151	196	931	231
147	543	151	197	684	233
148	538	152	198	669	235
149	486	154	199	629	235
150	467	155	200	925	235

Unit No.	y	x	Unit No.	y	x
201	610	241	251	985	309
202	601	242	252	1042	310
203	858	244	253	1226	310
204	490	244	254	912	312
205	1084	247	255	1016	313
206	1028	247	256	944	318
207	928	248	257	876	322
208	810	252	258	1232	324
209	995	252	259	1049	325
210	956	254	260	1210	327
211	1160	255	261	946	327
212	705	256	262	471	330
213	974	257	263	872	330
214	1076	260	264	1400	332
215	788	261	265	1425	338
216	795	261	266	885	339
217	811	263	267	1133	340
218	1009	264	268	1097	340
219	609	265	269	1166	347
220	1106	268	270	906	347
221	773	268	271	1219	348
222	884	269	272	1173	350
223	887	269	273	1098	352
224	951	270	274	876	354
225	956	273	275	915	357
226	1201	275	276	976	358
227	1063	275	277	1029	359
228	632	275	278	889	361
229	852	276	279	966	365
230	754	279	280	956	365
231	861	279	281	1766	365
232	767	282	282	1225	366
233	456	284	283	1453	367
234	1007	285	284	1413	368
235	941	286	285	1156	370
236	1097	287	286	787	373
237	233	289	287	1137	374
238	824	291	288	1231	374
239	764	295	289	896	378
240	842	297	290	1009	380
241	539	297	291	1150	385
242	778	300	292	1272	386
243	557	300	293	1373	390
244	958	302	294	1389	391
245	715	303	295	926	393
246	1036	304	296	1060	394
247	1153	307	297	2190	400
248	855	307	298	1219	400
249	935	307	299	1095	400
250	1031	308	300	1634	401

Unit No.	y	x	Unit No.	y	x
301	1719	408	351	1645	551
302	989	411	352	1478	556
303	808	411	353	1152	558
304	1369	417	354	2116	562
305	1040	418	355	1828	566
306	1315	419	356	1789	573
307	1089	422	357	1509	577
308	1347	425	358	1415	579
309	1632	437	359	1583	583
310	1346	437	360	1326	584
311	1370	438	361	999	591
312	1105	445	362	1648	592
313	1705	450	363	2154	600
314	1584	451	364	1785	606
315	1948	461	365	1218	606
316	1617	463	366	1463	613
317	1665	467	367	2240	625
318	1012	469	368	1684	631
319	1322	470	369	1606	635
320	1239	472	370	1620	650
321	1258	474	371	2150	652
322	1835	478	372	1376	658
323	1534	479	373	1707	670
324	1149	480	374	1504	684
325	1390	490	375	1893	712
326	1126	492	376	2089	712
327	1355	493	377	2058	719
328	1301	496	378	1283	760
329	1657	498	379	2844	774
330	1785	500	380	2171	785
331	1744	500	381	1239	816
332	1669	505	382	1706	817
333	1527	506	383	2766	829
334	2031	509	384	1715	830
335	2051	510	385	2135	838
336	834	517	386	1624	857
337	1232	523	387	2818	860
338	1350	524	388	2700	904
339	1805	524	389	1678	918
340	1420	530	390	1394	936
341	2034	534	391	1894	937
342	1418	536	392	1765	957
343	1522	538	393	2268	986
344	1386	540			
345	1376	541			
346	1093	543			
347	1780	543			
348	1547	549			
349	986	550			
350	1287	550			

Appendix B.3. Counties 60 Population

Unit No.	y	x	Unit No.	y	x
1	1876	482	51	7666	1964
2	2425	559	52	7864	1971
3	2670	748	53	7342	1985
4	3247	753	54	8354	1989
5	3240	830	55	8167	2095
6	3323	897	56	8420	2107
7	3588	965	57	7870	2119
8	3874	990	58	8260	2130
9	4536	1072	59	8338	2159
10	4111	1143	60	8569	2180
11	4514	1161	61	7650	2198
12	4546	1182	62	8030	2215
13	5419	1290	63	8374	2219
14	5290	1307	64	9910	2234
15	5458	1313	65	8902	2240
16	5320	1337	66	8125	2263
17	5090	1343	67	8386	2279
18	5810	1367	68	9236	2279
19	6065	1404	69	8889	2294
20	5905	1412	70	9143	2383
21	5323	1446	71	8994	2388
22	5575	1450	72	8851	2422
23	5863	1457	73	9204	2422
24	5513	1494	74	9071	2430
25	6181	1522	75	8853	2462
26	6184	1529	76	10014	2494
27	5761	1532	77	9635	2505
28	5849	1559	78	9837	2519
29	6100	1591	79	9913	2530
30	6088	1630	80	10974	2533
31	6414	1633	81	9267	2553
32	6364	1651	82	10143	2626
33	6284	1681	83	10221	2694
34	6518	1695	84	10382	2702
35	7097	1710	85	10434	2758
36	6497	1734	86	10107	2764
37	6491	1765	87	11101	2796
38	6793	1791	88	11327	2857
39	6750	1799	89	11132	2892
40	6619	1800	90	10886	2902
41	6656	1806	91	12222	2926
42	5792	1826	92	12207	2957
43	7328	1837	93	11460	2968
44	7927	1839	94	11684	2978
45	7041	1850	95	11439	2985
46	7327	1852	96	10856	3060
47	6874	1857	97	11246	3089
48	6536	1861	98	11587	3152
49	6920	1880	99	11759	3199
50	7280	1916	100	11970	3218

Unit No.	y	x	Unit No.	y	x
101	13006	3239	151	17197	4861
102	13044	3245	152	18265	4884
103	13348	3254	153	17766	4887
104	12687	3265	154	20605	4907
105	12128	3312	155	18815	4910
106	13393	3354	156	19545	4942
107	12559	3366	157	17766	4976
108	11276	3390	158	17686	4982
109	13189	3423	159	20114	5014
110	13120	3453	160	19325	5045
111	15703	3466	161	18130	5110
112	13353	3491	162	18451	5137
113	13078	3581	163	20397	5151
114	13095	3587	164	19632	5242
115	13954	3662	165	22059	5365
116	16637	3694	166	19214	5530
117	13611	3714	167	20367	5531
118	13887	3734	168	20165	5563
119	14512	3746	169	19878	5595
120	13157	3768	170	21622	5658
121	14821	3804	171	20745	5667
122	15667	3841	172	21089	5704
123	15105	3901	173	21017	5753
124	13534	3952	174	24277	5774
125	15808	4006	175	22158	5905
126	16581	4061	176	24287	6003
127	15151	4113	177	24913	6026
128	15491	4172	178	26456	6095
129	14836	4172	179	29377	6193
130	14427	4183	180	24796	6212
131	14490	4205	181	23439	6246
132	17374	4345	182	23065	6262
133	16185	4355	183	22705	6345
134	15915	4384	184	22665	6346
135	17517	4392	185	27036	6476
136	16417	4427	186	24744	6540
137	17285	4448	187	26233	6567
138	16657	4449	188	24137	6603
139	16288	4462	189	23653	6671
140	16776	4476	190	28679	6722
141	16667	4521	191	30494	6845
142	17590	4522	192	24816	6881
143	19517	4529	193	28393	6905
144	19672	4531	194	26614	7013
145	16098	4536	195	27753	7084
146	17696	4584	196	28262	7124
147	18390	4613	197	26356	7146
148	16533	4638	198	26552	7319
149	21733	4709	199	29747	7586
150	17879	4760	200	31945	7612

Unit No.	y	x	Unit No.	y	x
201	27303	7634	253	46783	13232
202	28636	7708	254	48066	13460
203	28211	7737	255	47948	13509
204	28758	7899	256	49219	13713
205	29894	7927	257	52850	13873
206	34707	7931	258	57914	14044
207	30767	7975	259	56470	14737
208	27759	7986	260	55535	14985
209	28897	8165	261	60498	15345
210	31624	8244	262	60104	16178
211	33657	8333	263	66959	16432
212	33107	8344	264	66977	16602
213	31422	8632	265	62571	16659
214	40837	8645	266	67431	17031
215	38084	8679	267	65589	17185
216	35452	8973	268	65522	17427
217	33527	9143	269	61135	17453
218	34805	9290	270	61897	17531
219	33309	9294	271	72290	17799
220	35713	9642	272	69255	19443
221	33994	9668	273	66833	19447
222	35976	9866	274	66381	19646
223	38373	10051	275	72769	19877
224	35158	10107	276	77025	19894
225	39217	10143	277	88020	19999
226	40079	10205	278	76391	20142
227	35908	10208	279	72440	20396
228	38514	10392	280	83368	20618
229	37646	10445	281	70532	20932
230	35829	10708	282	80574	21649
231	37831	10763	283	78586	22068
232	39429	11263	284	80681	23820
233	41895	11567	285	84611	23920
234	40506	11667	286	97854	27855
235	40213	11686	287	113249	30996
236	43906	11687	288	105187	31228
237	45025	11805	289	129348	33856
238	43292	11813	290	125954	34753
239	47783	11872	291	121597	35040
240	48724	12046	292	126286	38413
241	46264	12067	292	138164	40118
242	40890	12271	294	151060	42682
243	44599	12343	295	154543	43314
244	47232	12374	296	157041	45476
245	43590	12619	297	166702	46288
246	44509	12634	298	184320	53165
247	45135	12660	299	184306	54151
248	52364	12803	300	205576	55337
249	45728	12854	301	203805	58916
250	49284	12980	302	238149	69128
251	53649	13114	303	251924	73013
252	47838	13179	304	266623	76877

Appendix B.4. Counties 70 Population

Unit No.	y	x	Unit No.	y	x
1	1924	482	51	8219	1964
2	2180	559	52	7567	1971
3	2208	748	53	8309	1985
4	2342	753	54	7709	1989
5	3087	830	55	11254	2095
6	2423	897	56	11984	2107
7	3639	965	57	6885	2119
8	3725	990	58	7850	2130
9	3867	1072	59	8124	2159
10	4319	1143	60	9550	2180
11	3790	1161	61	8054	2198
12	3582	1182	62	7626	2215
13	8824	1290	63	7649	2219
14	5644	1307	64	8919	2234
15	5073	1313	65	9384	2240
16	4576	1337	66	7965	2263
17	4994	1343	67	8627	2279
18	5674	1367	68	8442	2279
19	6887	1404	69	10041	2294
20	5895	1412	70	7954	2383
21	5323	1446	71	8253	2388
22	5428	1450	72	8852	2422
23	5938	1457	73	9375	2422
24	5179	1494	74	8245	2430
25	5837	1522	75	9561	2462
26	6511	1529	76	13518	2494
27	5481	1532	77	9276	2505
28	5645	1559	78	9456	2519
29	5843	1591	79	11310	2530
30	5710	1630	80	9754	2533
31	6551	1633	81	9305	2553
32	7365	1651	82	10493	2626
33	7887	1681	83	9838	2694
34	6396	1695	84	17909	2702
35	6607	1710	85	12978	2758
36	6815	1734	86	10465	2764
37	6804	1765	87	11496	2796
38	7046	1791	88	9630	2857
39	7533	1799	89	10146	2892
40	7814	1800	90	8485	2902
41	6261	1806	91	10765	2926
42	6893	1826	92	11875	2957
43	6526	1837	93	11970	2968
44	8136	1839	94	10728	2978
45	7157	1850	95	10330	2985
46	6460	1852	96	10141	3060
47	6329	1857	97	13507	3089
48	6863	1861	98	11285	3152
49	7273	1880	99	12069	3199
50	6548	1916	100	11493	3218

Unit No.	y	x	Unit No.	y	x
101	12650	3239	151	19365	4861
102	12482	3245	152	19019	4884
103	13861	3254	153	18054	4887
104	11306	3265	154	19886	4907
105	16873	3312	155	17277	4910
106	22314	3354	156	18826	4942
107	15227	3366	157	17208	4976
108	11660	3390	158	17975	4982
109	15483	3423	159	24122	5014
110	12662	3453	160	19036	5045
111	15787	3466	161	20038	5110
112	14979	3491	162	20875	5137
113	17421	3581	163	18013	5151
114	13604	3587	164	19571	5242
115	12492	3662	165	23348	5365
116	14788	3694	166	23426	5530
117	13287	3714	167	23078	5531
118	13425	3734	168	23729	5563
119	14393	3746	169	20497	5595
120	12598	3768	170	22243	5658
121	12458	3804	171	25852	5667
122	15602	3841	172	28181	5704
123	13563	3901	173	20496	5753
124	12934	3952	174	20441	5774
125	15178	4006	175	23643	5905
126	14635	4061	176	31756	6003
127	15758	4113	177	26070	6026
128	19290	4172	178	22678	6095
129	15617	4172	179	25451	6193
130	16762	4183	180	23211	6212
131	15835	4205	181	31191	6246
132	15844	4345	182	26332	6262
133	16287	4355	183	30550	6345
134	19040	4384	184	24357	6346
135	23705	4392	185	24565	6476
136	15556	4427	186	21788	6540
137	16968	4448	187	25754	6567
138	28496	4449	188	25537	6603
139	15711	4462	189	23276	6671
140	19127	4476	190	26193	6722
141	18893	4521	191	28656	6845
142	16967	4522	192	25648	6881
143	15724	4529	193	26985	6905
144	18787	4531	194	29365	7013
145	19772	4536	195	27431	7084
146	17682	4584	196	25999	7124
147	17922	4613	197	30117	7146
148	18710	4638	198	30390	7319
149	18251	4709	199	28936	7586
150	17766	4760	200	30071	7612

Unit No.	y	x	Unit No.	y	x
201	31289	7634	253	43867	13232
202	32537	7708	254	56335	13460
203	32546	7737	255	51121	13509
204	32014	7899	256	58531	13713
205	29215	7927	257	52657	13873
206	33340	7931	258	52704	14044
207	29697	7975	259	55908	14737
208	29405	7986	260	57885	14985
209	28558	8165	261	58119	15345
210	32160	8244	262	88232	16178
211	33517	8333	263	67461	16432
212	34550	8344	264	69011	16602
213	31795	8632	265	61405	16659
214	34140	8645	266	69424	17031
215	55725	8679	267	77409	17185
216	41579	8973	268	71402	17427
217	31839	9143	269	76049	17453
218	36317	9290	270	71580	17531
219	33562	9294	271	76467	17799
220	35772	9642	272	72027	19443
221	33068	9668	273	73230	19447
222	38086	9866	274	71384	19646
223	61273	10051	275	86071	19877
224	39115	10107	276	81982	19894
225	43108	10143	277	83814	19999
226	37858	10205	278	79978	20142
227	43461	10208	279	89592	20396
228	39161	10392	280	88838	20618
229	40378	10445	281	81869	20932
230	42376	10708	282	90418	21649
231	49583	10763	283	94582	22068
232	41244	11263	284	87141	23820
233	54738	11567	285	94969	23920
234	49163	11667	286	104500	27855
235	41956	11686	287	195718	30996
236	53086	11687	288	124612	31228
237	49173	11805	289	179698	33856
238	71992	11813	290	146533	34753
239	44581	11872	291	142797	35040
240	46503	12046	292	141382	38413
241	97280	12067	293	139979	40118
242	57198	12271	294	166024	42682
243	46014	12343	295	170517	43314
244	47755	12374	296	213511	45476
245	49084	12619	297	196385	46288
246	46843	12634	298	182929	53165
247	50438	12660	299	208483	54151
248	52930	12803	300	228230	55337
249	54223	12854	301	234744	58916
250	56434	12980	302	277761	69128
251	51886	13144	303	409644	73013
252	52855	13179	304	347732	76877

Appendix B.5. Labor Force Population

This is a clustered population of individuals extracted from the September 1976 Current Population Survey (CPS) in the United States and is used with permission of the U.S. Bureau of Labor Statistics and the U.S. Census Bureau. The clusters are compact geographic areas used as one of the stages of sampling in the CPS and are typically composed of about four nearby households. The units within clusters for this illustrative population are individual persons. Table B.3 gives population summary statistics on the auxiliary variables.

Variable names and explanations:

h = stratum

hsub = substratum (each stratum contains two substrata, denoted by *hsub*, which can be used for studying the grouped balanced half-sample variance estimation method discussed in Chapter 11)

Cluster = cluster (or segment) number

Person = person number

age = age of person

agecat = age category

1 = 19 years and under

2 = 20–24

3 = 25–34

4 = 35–64

5 = 65 years and over

race: 1 = non-Black; 2 = Black

sex: 1 = male; 2 = female

HoursPerWk: Usual number of hours worked per week

WklyWage: Usual amount of weekly wages (in 1976 U.S. dollars)

y: an artificial variable generated to follow model (8.1.1) in the text for the special case of $\sigma_i^2 = \sigma^2$ and $\rho_i = \rho$.

Table B.3. Numbers of Persons by Age/Race/Sex Group

	Non-Black		Black		
Age	Male	Female	Male	Female	Total
19 and under	13	29	2	3	47
20–24	22	22	6	4	54
25–34	65	56	7	1	129
35–64	103	110	8	15	236
65 and over	4	5	2	1	12
Total	207	222	25	24	478

h	hsub	Cluster	Person	age	agecat	race	sex	HoursPerWk	WklyWage	y
1	1	1	1	22	2	2	1	40	160	52.60
1	1	1	2	53	4	2	1	40	224	48.97
1	1	1	3	20	2	2	2	40	164	56.15
1	1	1	4	19	1	2	1	40	134	36.25
1	1	1	5	24	2	2	1	40	146	60.03
1	1	2	6	28	3	1	1	40	320	50.42
1	1	2	7	32	3	1	1	40	300	45.67
1	1	2	8	42	4	1	1	48	396	48.49
1	1	3	9	74	5	1	1	40	150	24.77
1	1	3	10	40	4	1	2	20	70	44.56
1	1	3	11	17	1	1	1	18	60	30.06
1	1	4	12	30	3	1	1	52	289	44.49
1	1	4	13	26	3	1	2	25	84	43.97
1	1	4	14	27	3	1	1	40	240	41.19
1	1	4	15	53	4	1	1	40	560	52.11
1	1	4	16	44	4	1	2	45	300	32.43
1	1	5	17	25	3	1	2	44	310	49.21
1	1	5	18	24	2	1	2	40	380	48.83
1	1	6	19	40	4	1	1	40	480	57.42
1	1	6	20	42	4	1	2	40	500	67.19
1	1	6	21	31	3	1	2	48	289	59.21
1	1	6	22	32	3	1	1	45	495	47.81
1	1	7	23	39	4	1	1	40	500	44.35
1	1	7	24	41	4	1	1	40	500	63.60
1	1	7	25	37	4	1	2	40	178	68.44
1	1	8	26	20	2	2	2	25	84	53.37
1	1	8	27	48	4	2	2	40	225	73.71
1	1	8	28	17	1	2	1	31	112	71.17
1	1	8	29	66	5	2	1	10	115	70.55
1	1	9	30	39	4	1	1	40	510	36.42
1	1	9	31	61	4	1	2	40	280	50.78
1	1	9	32	58	4	1	2	45	250	38.84
1	1	10	33	29	3	1	1	40	380	40.53
1	1	10	34	25	3	1	2	36	207	41.19
1	1	10	35	25	3	1	1	59	440	52.10
1	1	11	36	23	2	1	2	40	134	27.87
1	1	11	37	33	3	1	1	72	250	34.75
1	1	11	38	33	3	1	2	40	210	48.75
1	1	12	39	52	4	1	2	40	300	48.83
1	1	12	40	22	2	1	1	32	128	63.80
1	1	13	41	66	5	1	1	40	240	33.33
1	1	13	42	58	4	1	2	30	104	30.34
1	1	13	43	27	3	1	2	35	175	29.63
1	1	13	44	33	3	1	1	40	418	58.12
1	1	13	45	33	3	1	2	40	558	36.82
1	1	14	46	45	4	1	1	40	560	49.08
1	1	14	47	18	1	1	1	48	336	50.11
1	1	14	48	47	4	1	1	40	400	38.96
1	1	15	49	36	4	1	1	40	480	48.47
1	1	15	50	34	3	1	1	40	420	42.55

h	hsub	Cluster	Person	age	agecat	race	sex	HoursPerWk	WklyWage	y
1	1	15	51	28	3	1	2	40	275	39.56
1	1	15	52	37	4	1	1	45	325	57.94
1	1	16	53	52	4	1	2	32	224	45.73
1	1	16	54	21	2	1	1	40	280	57.72
1	1	16	55	22	2	1	1	40	290	49.32
1	1	16	56	22	2	1	2	40	160	61.96
1	1	16	57	28	3	1	2	40	190	47.27
1	1	17	58	37	4	1	1	50	669	59.63
1	1	17	59	37	4	1	2	15	60	48.70
1	1	17	60	34	3	1	1	49	375	51.18
1	1	18	61	26	3	1	1	40	312	66.67
1	1	18	62	58	4	1	1	40	256	57.52
1	1	19	63	45	4	2	1	40	400	66.11
1	1	19	64	41	4	2	2	40	180	45.09
1	1	19	65	20	2	2	2	40	160	41.30
1	1	19	66	41	4	2	1	40	420	72.00
1	1	20	67	37	4	1	1	40	400	41.21
1	1	20	68	28	3	1	2	40	254	32.12
1	1	20	69	23	2	1	1	60	500	55.67
1	1	21	70	45	4	1	1	40	300	48.72
1	1	21	71	47	4	1	1	40	664	56.68
1	1	21	72	20	2	1	1	40	200	57.31
1	1	21	73	46	4	1	2	40	135	60.31
1	1	22	74	58	4	1	2	40	40	55.20
1	1	22	75	37	4	1	2	48	175	40.20
1	1	22	76	41	4	1	2	40	204	37.84
1	1	22	77	42	4	1	1	40	440	72.08
1	1	23	78	32	3	1	1	40	545	55.51
1	1	23	79	58	4	1	1	40	350	35.95
1	1	23	80	52	4	1	2	25	93	51.03
1	1	24	81	43	4	1	1	40	150	52.98
1	1	24	82	35	4	1	2	35	72	31.82
1	1	24	83	29	3	1	1	40	224	47.01
1	1	24	84	35	4	1	2	50	342	38.71
1	1	24	85	25	3	1	1	40	520	39.12
1	1	24	86	25	3	1	2	35	122	65.80
1	1	25	87	60	4	1	1	40	400	53.77
1	1	25	88	54	4	1	2	43	101	59.64
1	1	26	89	31	3	1	2	40	419	69.37
1	1	26	90	25	3	2	1	40	360	69.31
1	1	26	91	24	2	2	1	40	250	60.40
1	1	26	92	51	4	1	1	50	450	38.66
1	1	26	93	47	4	1	2	40	210	47.26
1	1	26	94	25	3	1	2	40	246	39.07
1	1	27	95	37	4	1	2	40	150	44.65
1	1	27	96	19	1	1	2	40	180	58.80
1	1	27	97	34	3	1	1	40	500	45.55
1	1	28	98	60	4	1	1	38	257	54.36
1	1	28	99	32	3	1	1	40	375	60.99
1	1	28	100	36	4	1	1	28	300	53.61

h	hsub	Cluster	Person	age	agecat	race	sex	HoursPerWk	WklyWage	y
1	1	28	101	30	3	1	2	50	400	63.45
1	2	29	102	34	3	1	1	37	222	42.15
1	2	29	103	43	4	1	1	40	900	51.74
1	2	29	104	34	3	1	1	40	824	54.96
1	2	30	105	29	3	1	2	40	356	38.73
1	2	30	106	28	3	1	1	47	450	36.18
1	2	30	107	27	3	1	2	43	450	51.54
1	2	30	108	34	3	1	2	38	375	49.31
1	2	30	109	32	3	1	1	40	500	52.62
1	2	30	110	31	3	1	2	30	180	47.69
1	2	31	111	32	3	1	1	45	376	51.24
1	2	31	112	30	3	1	2	40	138	59.71
1	2	31	113	59	4	1	2	36	134	46.96
1	2	31	114	33	3	1	1	40	353	42.59
1	2	31	115	32	3	1	2	40	400	58.22
1	2	32	116	41	4	1	2	40	315	41.39
1	2	32	117	20	2	1	1	40	180	45.43
1	2	32	118	18	1	1	2	35	100	39.55
1	2	32	119	26	3	1	1	40	542	49.03
1	2	33	120	66	5	2	1	40	214	43.42
1	2	33	121	54	4	2	2	15	68	33.56
1	2	33	122	65	5	2	2	40	54	48.61
1	2	33	123	44	4	2	2	40	154	38.57
1	2	33	124	18	1	2	2	40	134	65.28
1	2	34	125	61	4	1	2	32	165	35.93
1	2	34	126	17	1	1	2	20	80	40.45
1	2	34	127	44	4	2	2	12	42	46.04
1	2	34	128	40	4	1	2	30	162	44.07
1	2	34	129	33	3	1	1	45	200	54.67
1	2	34	130	15	1	1	2	12	28	38.66
1	2	34	131	20	2	1	2	50	150	50.03
1	2	35	132	59	4	1	1	40	331	69.45
1	2	35	133	26	3	1	1	40	250	64.47
1	2	35	134	39	4	1	2	35	280	50.01
1	2	35	135	34	3	1	2	48	325	61.82
1	2	35	136	87	5	1	2	42	100	64.94
1	2	36	137	22	2	2	2	32	191	62.08
1	2	36	138	38	4	1	1	40	320	43.42
1	2	36	139	17	1	1	2	40	158	39.29
1	2	36	140	15	1	1	2	25	84	58.00
1	2	37	141	31	3	1	1	50	625	28.76
1	2	37	142	39	4	1	1	56	336	68.75
1	2	37	143	34	3	1	2	40	350	54.52
1	2	37	144	50	4	1	2	40	156	36.06
1	2	37	145	30	3	1	1	40	250	42.41
1	2	37	146	26	3	1	2	40	160	48.30
1	2	37	147	29	3	1	2	40	280	43.07
1	2	38	148	34	3	1	1	40	250	48.77
1	2	38	149	61	4	1	2	45	215	33.83
1	2	38	150	42	4	1	1	60	649	48.12

h	hsub	Cluster	Person	age	agecat	race	sex	HoursPerWk	WklyWage	y
1	2	38	151	37	4	1	2	40	210	46.25
1	2	39	152	50	4	1	1	40	210	68.25
1	2	39	153	47	4	1	2	40	130	57.18
1	2	40	154	40	4	1	2	45	338	55.39
1	2	40	155	44	4	1	1	40	400	50.37
1	2	40	156	37	4	1	2	20	80	40.66
1	2	40	157	17	1	1	2	10	34	50.92
1	2	41	158	61	4	1	1	55	450	51.54
1	2	41	159	59	4	1	2	10	50	75.94
1	2	41	160	46	4	1	2	40	360	53.59
1	2	41	161	26	3	1	2	40	180	50.21
1	2	41	162	75	5	1	2	15	90	57.80
1	2	42	163	48	4	1	1	40	280	41.16
1	2	42	164	49	4	1	2	38	275	53.36
1	2	42	165	27	3	1	1	40	227	39.71
1	2	43	166	31	3	1	1	40	400	41.89
1	2	43	167	19	1	1	2	40	300	32.30
1	2	43	168	45	4	1	2	60	635	47.77
1	2	43	169	25	3	1	2	10	45	51.35
1	2	44	170	36	4	1	1	40	999	46.58
1	2	44	171	34	3	1	1	50	500	46.78
1	2	44	172	34	3	1	2	40	520	60.85
1	2	44	173	17	1	1	2	20	85	46.65
1	2	45	174	41	4	1	2	12	40	49.37
1	2	45	175	24	2	1	1	40	170	52.73
1	2	45	176	24	2	1	2	60	255	73.60
1	2	45	177	44	4	1	1	35	150	57.09
1	2	45	178	32	3	1	2	35	150	58.58
1	2	46	179	17	1	1	1	15	75	48.28
1	2	46	180	25	3	1	2	40	260	58.32
1	2	46	181	27	3	1	1	40	500	67.64
1	2	46	182	27	3	1	1	48	312	47.95
1	2	47	183	50	4	1	2	41	495	58.09
1	2	47	184	48	4	1	1	25	156	60.46
1	2	48	185	35	4	1	1	40	540	66.00
1	2	48	186	34	3	1	2	40	479	54.12
1	2	48	187	29	3	1	2	20	123	34.65
1	2	49	188	16	1	1	2	5	21	51.72
1	2	49	189	20	2	1	2	8	31	77.55
1	2	49	190	44	4	1	1	40	800	62.83
1	2	49	191	18	1	1	2	40	200	47.02
1	2	49	192	52	4	1	1	40	400	50.51
1	2	49	193	51	4	1	2	20	100	50.97
1	2	50	194	29	3	1	1	30	30	36.20
1	2	50	195	29	3	1	1	60	161	39.85
1	2	50	196	33	3	1	1	60	161	34.72
1	2	50	197	24	2	1	1	60	161	59.46
1	2	50	198	27	3	1	2	60	211	36.50
1	2	50	199	30	3	1	2	60	184	49.44
1	2	51	200	24	2	1	2	40	200	56.15

h	hsub	Cluster	Person	age	agecat	race	sex	HoursPerWk	WklyWage	y
1	2	51	201	19	1	1	2	40	264	24.46
1	2	51	202	26	3	1	2	40	200	45.53
1	2	51	203	19	1	1	2	40	240	22.76
1	2	51	204	24	2	1	2	20	160	28.53
1	2	51	205	40	4	1	2	40	928	41.12
1	2	51	206	43	4	1	1	60	600	43.77
1	2	51	207	41	4	1	2	35	350	41.40
1	2	51	208	18	1	1	1	40	200	46.31
1	2	51	209	17	1	1	2	20	95	40.62
1	2	51	210	52	4	1	2	30	321	38.30
2	1	52	211	54	4	1	1	40	440	45.26
2	1	52	212	51	4	1	2	25	200	49.52
2	1	52	213	55	4	1	2	50	600	48.71
2	1	53	214	50	4	1	1	40	640	48.36
2	1	53	215	38	4	1	2	40	450	53.75
2	1	54	216	29	3	1	1	50	638	52.41
2	1	54	217	34	3	1	2	40	280	52.28
2	1	55	218	27	3	1	1	40	520	48.49
2	1	55	219	42	4	1	1	58	637	39.21
2	1	55	220	41	4	1	2	18	113	44.01
2	1	55	221	64	4	1	1	16	100	54.64
2	1	56	222	23	2	1	1	24	108	56.43
2	1	56	223	18	1	1	2	18	81	61.28
2	1	56	224	18	1	1	2	17	77	61.35
2	1	56	225	57	4	1	2	5	21	58.34
2	1	57	226	45	4	1	1	40	650	48.25
2	1	57	227	41	4	1	2	22	132	39.42
2	1	58	228	46	4	1	1	34	600	41.30
2	1	58	229	45	4	1	2	60	380	54.90
2	1	58	230	17	1	1	2	20	85	53.30
2	1	58	231	23	2	1	1	40	170	58.90
2	1	58	232	56	4	1	1	40	640	55.43
2	1	59	233	35	4	1	1	40	610	52.61
2	1	59	234	59	4	1	1	40	500	57.53
2	1	59	235	27	3	1	1	40	380	51.05
2	1	60	236	35	4	1	1	50	561	72.53
2	1	60	237	27	3	1	2	60	360	44.23
2	1	60	238	40	4	1	1	40	730	44.64
2	1	61	239	65	5	1	1	32	124	40.66
2	1	61	240	43	4	1	2	40	150	36.72
2	1	61	241	24	2	1	2	40	330	37.27
2	1	61	242	50	4	1	2	35	175	57.98
2	1	61	243	51	4	1	1	38	100	52.20
2	1	62	244	46	4	1	1	40	560	40.89
2	1	62	245	50	4	1	1	40	200	48.87
2	1	63	246	24	2	1	1	44	750	54.46
2	1	63	247	28	3	1	2	42	342	42.22
2	1	64	248	43	4	1	2	50	500	43.30
2	1	64	249	38	4	1	1	50	600	51.98
2	1	64	250	19	1	1	2	35	121	62.82

h	hsub	Cluster	Person	age	agecat	race	sex	HoursPerWk	WklyWage	y
2	1	65	251	36	4	1	2	40	275	47.02
2	1	65	252	28	3	1	1	40	210	44.04
2	1	65	253	34	3	1	1	44	450	70.07
2	1	65	254	35	4	1	2	32	200	33.42
2	1	65	255	17	1	1	2	16	32	38.90
2	1	65	256	16	1	1	2	10	20	50.72
2	1	65	257	23	2	1	1	30	210	27.30
2	1	65	258	22	2	1	2	40	352	49.89
2	1	66	259	47	4	1	1	40	240	65.01
2	1	66	260	47	4	1	2	20	350	70.08
2	1	66	261	38	4	1	1	40	560	49.30
2	1	66	262	39	4	1	2	30	140	32.80
2	1	67	263	51	4	1	1	30	285	43.87
2	1	67	264	43	4	1	2	99	310	58.66
2	1	67	265	53	4	1	1	40	300	63.03
2	1	67	266	24	2	1	2	30	131	39.43
2	1	67	267	31	3	1	2	35	131	51.21
2	1	67	268	35	4	1	1	37	130	56.19
2	1	68	269	77	5	1	2	30	120	46.54
2	1	68	270	64	4	1	2	30	135	59.44
2	1	69	271	40	4	1	1	40	210	51.96
2	1	69	272	39	4	1	2	40	310	57.32
2	1	69	273	50	4	1	2	40	300	69.38
2	1	69	274	51	4	1	1	40	575	52.68
2	1	69	275	34	3	1	2	40	375	47.34
2	1	69	276	17	1	1	2	15	23	60.04
2	1	70	277	30	3	2	1	40	224	70.53
2	1	70	278	37	4	2	2	15	56	52.65
2	1	70	279	29	3	1	1	40	204	46.27
2	1	70	280	48	4	1	2	40	238	63.60
2	1	70	281	26	3	1	1	40	280	44.75
2	1	70	282	41	4	1	1	40	577	72.11
2	1	70	283	39	4	1	2	30	189	57.96
2	1	71	284	34	3	1	1	40	600	50.53
2	1	71	285	36	4	1	2	40	385	50.49
2	1	72	286	19	1	1	2	40	180	65.61
2	1	72	287	53	4	1	1	40	999	49.05
2	1	72	288	48	4	1	2	40	200	57.93
2	1	73	289	29	3	1	1	40	320	41.25
2	1	73	290	23	2	1	2	40	320	45.00
2	1	73	291	35	4	1	2	40	240	66.13
2	1	73	292	41	4	1	2	48	324	48.80
2	1	73	293	41	4	1	2	40	225	51.59
2	1	74	294	69	5	1	1	40	999	57.41
2	1	74	295	40	4	1	1	45	500	45.81
2	1	74	296	41	4	1	1	40	480	27.47
2	1	74	297	48	4	1	2	50	800	51.31
2	1	75	298	37	4	1	1	40	900	47.40
2	1	75	299	33	3	1	1	40	999	39.71
2	1	75	300	60	4	1	1	36	438	46.94

h	hsub	Cluster	Person	age	agecat	race	sex	HoursPerWk	WklyWage	y
2	1	75	301	46	4	1	2	35	448	67.29
2	1	75	302	18	1	1	2	40	240	55.22
2	1	75	303	50	4	1	2	50	575	67.91
2	1	76	304	59	4	2	2	40	327	50.46
2	1	76	305	35	4	1	1	40	509	56.77
2	1	76	306	30	3	1	2	40	294	27.56
2	1	76	307	38	4	1	2	40	410	66.83
2	1	77	308	47	4	2	1	40	150	38.23
2	1	77	309	49	4	1	2	40	134	51.87
2	1	77	310	21	2	1	2	40	140	46.32
2	1	77	311	20	2	1	1	45	360	45.45
2	1	77	312	25	3	1	1	40	334	43.09
2	1	77	313	25	3	1	2	35	70	34.67
2	1	77	314	34	3	1	1	40	250	45.63
2	1	77	315	39	4	1	1	50	425	46.54
2	1	77	316	37	4	1	2	40	325	51.17
2	1	77	317	46	4	1	1	58	195	18.21
2	1	77	318	43	4	1	2	58	195	31.65
2	1	77	319	22	2	1	1	40	180	48.22
2	1	77	320	19	1	1	1	40	224	62.05
2	2	78	321	60	4	2	1	40	444	52.61
2	2	78	322	43	4	2	2	40	413	62.58
2	2	78	323	25	3	2	1	38	154	55.57
2	2	79	324	18	1	2	2	40	152	69.17
2	2	79	325	49	4	2	2	40	152	41.59
2	2	79	326	32	3	2	1	40	154	73.63
2	2	80	327	22	2	1	2	40	226	76.20
2	2	80	328	29	3	1	1	43	383	47.95
2	2	80	329	30	3	1	2	20	145	45.94
2	2	80	330	46	4	1	1	40	440	33.71
2	2	80	331	46	4	1	2	40	600	49.19
2	2	80	332	39	4	1	1	60	700	51.25
2	2	80	333	33	3	1	2	47	250	43.24
2	2	80	334	38	4	1	2	40	280	57.16
2	2	80	335	17	1	1	1	40	200	35.50
2	2	81	336	26	3	1	1	40	450	41.78
2	2	81	337	25	3	1	2	40	450	65.63
2	2	82	338	37	4	1	1	40	328	59.60
2	2	82	339	35	4	1	2	40	304	40.94
2	2	83	340	26	3	1	1	55	413	55.68
2	2	83	341	22	2	1	2	12	48	31.53
2	2	84	342	40	4	1	2	25	88	27.92
2	2	84	343	40	4	1	1	50	275	49.24
2	2	84	344	17	1	1	1	40	140	37.92
2	2	84	345	20	2	1	2	40	250	62.83
2	2	85	346	51	4	1	2	40	400	44.49
2	2	85	347	47	4	1	2	40	420	24.75
2	2	85	348	17	1	1	2	18	66	51.95
2	2	85	349	42	4	1	2	40	236	34.88
2	2	86	350	43	4	1	2	40	200	25.01

h	hsub	Cluster	Person	age	agecat	race	sex	HoursPerWk	WklyWage	y
2	2	86	351	40	4	1	1	40	340	55.61
2	2	86	352	24	2	1	1	40	230	53.95
2	2	86	353	33	3	1	1	50	300	51.88
2	2	86	354	28	3	1	1	40	400	55.43
2	2	86	355	28	3	1	2	40	230	24.70
2	2	87	356	42	4	1	1	40	799	30.92
2	2	87	357	36	4	1	2	42	210	52.85
2	2	87	358	17	1	1	2	20	100	41.82
2	2	87	359	60	4	1	1	48	200	44.41
2	2	87	360	54	4	1	2	24	100	40.16
2	2	88	361	42	4	1	2	50	221	51.90
2	2	88	362	38	4	1	1	50	577	45.65
2	2	89	363	42	4	2	2	34	139	54.40
2	2	89	364	17	1	2	2	20	67	51.96
2	2	89	365	21	2	2	1	40	134	63.90
2	2	89	366	20	2	2	1	40	253	65.29
2	2	90	367	55	4	1	1	40	200	51.80
2	2	90	368	60	4	1	2	40	220	61.68
2	2	90	369	55	4	1	2	40	480	65.86
2	2	90	370	73	5	1	2	40	134	59.51
2	2	90	371	21	2	1	2	40	120	55.30
2	2	91	372	35	4	1	2	40	260	60.70
2	2	91	373	17	1	1	2	15	50	51.93
2	2	91	374	32	3	1	1	42	500	48.47
2	2	91	375	42	4	1	2	40	212	58.43
2	2	91	376	56	4	1	2	40	325	53.02
2	2	91	377	24	2	1	1	50	158	65.18
2	2	92	378	19	1	1	2	30	101	50.81
2	2	92	379	60	4	1	2	20	101	41.31
2	2	92	380	40	4	1	1	85	850	45.78
2	2	92	381	18	1	1	1	15	60	45.87
2	2	93	382	58	4	2	2	15	50	70.13
2	2	93	383	47	4	1	1	40	392	49.44
2	2	93	384	28	3	1	2	40	134	44.57
2	2	94	385	58	4	1	1	40	40	39.07
2	2	94	386	37	4	1	2	45	257	43.01
2	2	94	387	38	4	1	1	40	310	43.16
2	2	94	388	61	4	1	1	45	287	48.73
2	2	94	389	19	1	1	2	15	48	66.96
2	2	94	390	34	3	1	2	24	176	34.35
2	2	94	391	24	2	1	1	40	200	61.12
2	2	95	392	42	4	1	1	45	800	52.63
2	2	95	393	44	4	1	2	45	646	46.90
2	2	95	394	33	3	1	1	40	580	44.23
2	2	95	395	51	4	1	1	50	350	28.71
2	2	95	396	53	4	1	2	38	184	53.17
2	2	95	397	26	3	1	1	15	53	67.06
2	2	95	398	23	2	1	2	15	53	56.63
2	2	95	399	32	3	1	1	45	250	44.51
2	2	96	400	46	4	2	2	40	200	33.67

h	hsub	Cluster	Person	age	agecat	race	sex	HoursPerWk	WklyWage	y
2	2	96	401	68	5	1	2	15	50	49.75
2	2	97	402	47	4	1	1	40	400	44.12
2	2	97	403	44	4	1	2	20	80	64.96
2	2	97	404	20	2	1	1	40	320	57.14
2	2	97	405	34	3	1	1	40	337	64.14
2	2	97	406	28	3	1	2	15	120	43.92
2	2	97	407	62	4	1	1	60	650	57.43
2	2	98	408	24	2	1	2	40	146	38.65
2	2	98	409	27	3	2	1	50	300	31.47
2	2	98	410	50	4	1	1	50	999	41.04
2	2	98	411	58	4	1	1	40	999	51.77
2	2	99	412	46	4	1	1	43	226	69.62
2	2	99	413	25	3	1	2	20	70	38.39
2	2	100	414	24	2	1	2	40	346	39.77
2	2	100	415	46	4	1	2	40	420	62.85
2	2	101	416	37	4	1	1	40	500	55.69
2	2	101	417	34	3	1	2	40	388	42.35
2	2	102	418	54	4	2	1	50	400	53.15
2	2	102	419	30	3	2	1	40	220	45.17
2	2	102	420	26	3	2	2	40	134	27.42
2	2	103	421	59	4	2	2	32	107	25.15
2	2	103	422	50	4	2	2	42	491	49.74
3	1	104	423	49	4	1	1	72	486	46.50
3	1	104	424	24	2	1	1	35	168	43.96
3	1	104	425	45	4	1	1	40	356	39.83
3	1	104	426	48	4	1	2	40	190	49.88
3	1	104	427	19	1	1	1	14	68	36.92
3	1	105	428	40	4	1	1	40	423	46.79
3	1	105	429	40	4	1	2	40	308	36.22
3	1	105	430	47	4	2	2	40	268	37.35
3	1	105	431	58	4	2	1	40	258	38.77
3	1	105	432	24	2	2	1	40	276	62.87
3	1	106	433	58	4	2	1	40	228	57.69
3	1	106	434	61	4	1	2	78	250	60.33
3	1	106	435	61	4	1	2	45	308	54.94
3	1	106	436	25	3	2	1	40	268	55.44
3	1	107	437	29	3	1	2	15	59	32.02
3	1	107	438	50	4	1	1	40	240	44.37
3	1	107	439	50	4	1	2	40	212	46.27
3	1	107	440	24	2	1	1	30	131	43.18
3	1	107	441	20	2	1	2	25	86	49.79
3	1	108	442	60	4	1	1	50	225	35.45
3	1	108	443	64	4	1	2	50	225	54.79
3	1	108	444	17	1	1	1	26	87	74.05
3	1	108	445	47	4	1	1	40	200	70.33
3	1	108	446	48	4	1	2	24	145	54.28
3	1	108	447	33	3	1	1	48	340	57.96
3	1	108	448	58	4	1	2	60	246	49.44
3	1	109	449	34	3	1	2	36	535	38.58
3	1	109	450	33	3	1	1	40	600	69.20

h	hsub	Cluster	Person	age	agecat	race	sex	HoursPerWk	WklyWage	y
3	1	109	451	32	3	1	2	12	72	44.03
3	1	109	452	29	3	1	1	40	400	43.91
3	1	109	453	29	3	1	2	40	360	34.07
3	1	109	454	32	3	1	1	40	596	55.52
3	1	110	455	59	4	1	2	38	279	54.49
3	1	110	456	32	3	1	2	30	150	52.00
3	1	110	457	35	4	1	1	40	190	55.88
3	1	110	458	35	4	1	2	40	200	43.44
3	1	110	459	16	1	1	2	40	100	45.36
3	1	111	460	39	4	1	1	40	351	61.53
3	1	111	461	35	4	1	2	40	351	63.62
3	1	111	462	38	4	1	1	45	608	27.97
3	1	111	463	35	4	1	2	50	250	55.97
3	1	111	464	17	1	1	1	30	101	29.86
3	1	111	465	42	4	1	2	40	200	38.44
3	1	111	466	19	1	1	1	40	200	53.72
3	1	112	467	30	3	1	1	50	250	31.49
3	1	112	468	31	3	1	2	40	184	46.99
3	1	113	469	24	2	1	1	15	54	57.68
3	1	113	470	23	2	1	1	32	144	64.65
3	1	113	471	20	2	1	2	25	113	51.43
3	1	114	472	36	4	1	2	40	240	58.82
3	1	114	473	34	3	1	1	40	360	50.06
3	1	115	474	44	4	1	1	30	242	52.19
3	1	115	475	19	1	1	1	10	35	47.61
3	1	115	476	41	4	1	1	40	500	50.43
3	1	115	477	59	4	1	2	5	165	55.57
3	1	115	478	60	4	1	1	48	375	56.40

Appendix B.6. Third Grade Population

The Third Grade population consists of 2427 students who participated in the Third International Mathematics and Science Study (Caslyn, Gonzales, Frase, 1999). The methods used in conducting the original study are given in TIMSS International Study Center (1996). The population used in this book consists of only students from the United States. Clusters are schools while units within clusters are the students. Because the population is large, it is not listed here but is available at the John Wiley Worldwide Web site: www.wiley.com/products/subjects/mathematics/features/software_supplem_math.html.

Variable names and explanations:

region = geographic region of the U.S.
- 1 = Northeast
- 2 = South
- 3 = Central
- 4 = West

sch.id = school identifier (1–135)

stud.id = student identifier (1–2427)
sex
1 = female
2 = male
language = language of test spoken at home
1 = always
2 = sometimes
3 = never
math = mathematics test score
ethnicity
1 = White, non-Hispanic
2 = Black
3 = Hispanic
4 = Asian
5 = Native American
6 = Other
science = science test score
community = type of location of school
2 = village or rural area
3 = outskirts of a town or city
4 = close to center of a town or city
enrollment = number of students in entire school

APPENDIX C

S-PLUS FUNCTIONS

This appendix lists several functions, written in S-PLUS™, that perform various sampling and estimation tasks. Becker, Chambers, and Wilks (1988) and Venables and Ripley (1997) are basic references for S-PLUS and the S language itself. We hope that these functions will prove useful to teachers, students, or researchers, but make no claims as to their efficiency or generality. Readers can use them as a starting point for developing their own special purpose sampling code. An alphabetical list of the functions and a brief description follow.

The electronic bulletin board sponsored by the Survey Research Methods Section (SRMS) of the American Statistical Association is another resource for information on sampling software and for more general survey-related issues. To subscribe to the list, send an email to listserv@umdd.umd.edu.

Function	Description
`ahr`	Compute the coefficient vector, diagonal of the hat matrix, and vector of residuals used in robust variance estimators (see Chapter 5). The function assumes that the matrix **X** has full column rank.
`clus.parm.est.anova`	Compute analysis of variance estimates of cluster model parameters (see Chapter 8). The population is assumed to be the Labor Force population in Appendix B but can be adapted for others. The function assumes that a sample of clusters has been selected and takes as input the identification numbers of the sample clusters. A simple random sample of units is selected within each cluster.
`clus.sam`	Select a two-stage cluster sample after randomizing order of the clusters. Clusters can be selected with probabilities proportional to the number of units per cluster or by simple random sampling without

Function	Description
	replacement. Subsamples of units within clusters are selected by simple random sampling without replacement.
`ginv`	Generalized inverse of a matrix used for finding parameter solutions in models that are not full rank.
`pps.fcn`	Select a sample of fixed size with probability proportional to a measure of size (see Chapter 3). Population should be sorted in desired order before calling this function.
`pps.random.fcn`	P1: random method from Chapter 3. Select a sample of fixed size with probability proportional to a measure of size after sorting the population in a random order.
`pps.strat.fcn`	Select a sample that approximates probability proportional to size by forming n strata with equal total measures of size and selecting one unit from each stratum at random.
`restrict.srs`	Select a balanced sample using simple random sampling without replacement. Balance is obtained on the first two moments of a single auxiliary x. The degree of balance is controlled by the parameters `E1` and `E2`.
`restrict.pps`	Select a weighted balanced sample using probability proportional to size sampling. Balance is obtained on the first two moments of a single auxiliary x. The degree of balance is controlled by the parameters `E1` and `E2`.
`strat.opt.alloc.x`	Explore different allocations to strata in an example from Chapter 6.
`T.blu`	Compute the *BLU* predictor under a working model $M(\mathbf{X}:\mathbf{W}^{-1})$
`vD`	Compute the robust variance estimator v_D (see Chapter 5)
`vH`	Compute the robust variance estimator v_H (see Chapter 5)
`vJ`	Compute the robust variance estimator v_J (see Chapter 5)
`vJ.star`	Compute the robust variance estimator v_J^* (see Chapter 5)
`vR`	Compute the robust variance estimator v_R (see Chapter 5)
`weed.high`	Identify certainty units in a *pps* sample.

ahr—computes the coefficient vector, diagonal of the hat matrix, and vector of residuals used in robust variance estimators (see Chapter 5). The function assumes that the matrix **X** has full column rank.

```
function(X, W, sam, y)
{
#     ahr
#     Compute a, h, r, vectors for use in robust variance estimates
#     X = population X matrix
#     W = population vector of regression weights
#           (typically 1/v for some vector
#           of working model variances)
#     sam = vector of sample unit numbers
#     y = population y vector
#           t(Xs) * Ws * Xs must be non-singular
#
      X <- as.matrix(X)
      Xr <- as.matrix(X[-sam, ])
      one.r <- rep(1, nrow(Xr))
      Xs <- as.matrix(X[sam, ])
      Ws <- W[sam]
      ys <- y[sam]
      A <- (t(Xs) * Ws) %*% Xs
      Ainv <- solve(A)
      a <- one.r %*% Xr %*% Ainv %*% (t(Xs) * Ws)
      H <- Xs %*% Ainv %*% (t(Xs) * Ws)
      h <- diag(H)
      r <- (diag(length(sam)) - H) %*% ys
      list(a = as.vector(a), h = as.vector(h), r = as.vector(r))
}
```

clus.parm.est.anova—compute analysis of variance estimates of cluster model parameters (see Chapter 8). The population is assumed to be the Labor Force population in Appendix B but can be adapted for others. The function assumes that a sample of clusters has been selected and takes as input the identification numbers of the sample clusters. A simple random sample of units is selected within each cluster.

```
function(clus.sam, mi.sam, full.enum = F, anal.var = ''y'')
{
#     clus.parm.est.anova
#     Compute estimates of model parameters from a cluster sample
#         using ANOVA formulas.
#     clus.sam = vector of nos. of sample clusters
#     cluspop = clustered population, e.g. Labor Force
#     mi.sam = vector of subsample sizes
#         (adjusted below for small clusters)
#     full.enum = T if clusters are fully enumerated, F if not
#     anal.var = name of analysis variable in cluspop
```

```
#
#____________________________________________________
#    Extract cluster sample data
#____________________________________________________
     n <- length(clus.sam)
     sam.pop.rows <- match(cluspop[, ''SegNo''], clus.sam,
       nomatch = 0)
     sam.pop.rows[sam.pop.rows > 0] <- 1
     clus.sam.data <- cluspop[sam.pop.rows == 1, ]
     Mi.vec <- tapply(rep(1, length(cluspop[, 1])),
         cluspop[, ''SegNo''], sum)
     Mi.sam <- Mi.vec[clus.sam]
     if(!full.enum) {
         mi.sam <- pmin(mi.sam, Mi.sam)
     }
     else {
         mi.sam <- Mi.sam
     }
     subsam.vec <- NULL
     for(i in 1:n) {
         subsam <- sort(sample(1:Mi.sam[i], mi.sam[i]))
         match.vec <- match(1:Mi.sam[i], subsam, nomatch = 0)
         match.vec[match.vec > 0] <- 1
         subsam.vec <- c(subsam.vec, match.vec)
     }
     subsam.data <- clus.sam.data[subsam.vec == 1, ]
     ybars <- tapply(subsam.data[, anal.var],
         list(subsam.data[, ''SegNo'']), mean)
     ybar.bar <- mean(subsam.data[, anal.var]) #
#
#____________________________________________________
#    ANOVA estimates of model parameters
______________________________________________________
      B <- sum(mi.sam * (ybars - ybar.bar)^2) / (n - 2)
      W <- sum( (subsam.data[, anal.var] -
             rep(ybars, mi.sam))^2/rep(mi.sam-1, mi.sam))/n
      D <- (sum(mi.sam) - (2 * sum(mi.sam^2))/sum(mi.sam))/(n-2)
      sigma2 <- (B + (D - 1) * W)/D
      rho <- (B - W)/(B + (D - 1) * W)
      list(clus.sam = clus.sam, sigma2 = sigma2, rho = rho)
}
```

`clus.sam`—Select a two-stage cluster sample after randomizing order of the clusters. Clusters can be selected with probabilities proportional to the number of units per cluster or by simple random sampling without replacement. Subsamples of units within clusters are selected by simple random sampling without replacement.

```
function(pop, clus.id, n, mi.sam, sel.meth, seed)
{
```

```
#    clus.sam
#    Select a two-stage cluster sample after randomizing order
#        of the clusters
#    pop = population matrix
#    clus.id = name or number of column for cluster identification
#    n = sample size of clusters
#    mi.sam = n-vector of sample sizes within each cluster
#        mi.sam is adjusted, if necessary, to be less than or equal
#        to the total size of each sample cluster.
#    sel.meth = ''pps'' for pps cluster sample
#             = ''srs'' for simple random sample of clusters
#    seed = seed for random no. generator, use to reproduce same
#       sample
#
     set.seed(seed)
     Mi.vec <- table(pop[ , clus.id] )
     M <- sum(Mi.vec)
     N <- length(Mi.vec)      #
#____________________________________________________________
#    Select sample of clusters
#____________________________________________________________
     if(sel.meth == ''pps'')  {
         cl.sam <- pps.random.fcn(Mi.vec, n)
     }
     if(sel.meth == ''srs'')  {
         cl.sam <- sort(sample(1:N, n, replace = F))
     }
     sam <- match(pop[ , clus.id], cl.sam, nomatch = 0)
     sam[sam > 0] <- 1
     sam.rows <- (1:M) [sam == 1]
     cl.sam.data <- pop[sam ==1,  ]
     Mi.sam <- Mi.vec[cl.sam]
     mi.sam <- pmin(Mi.sam, mi.sam)
#____________________________________________________________
#    Select subsamples from sample clusters
#____________________________________________________________
     subsam.vec <- NULL
     for(i in 1:n)  {
         subsam <- sort(sample(1:Mi.sam[i], mi.sam[i] ))
         match.vec <- match(1:Mi.sam[i], subsam, nomatch = 0)
         match.vec[match.vec > 0] <- 1
         subsam.vec <- c(subsam.vec, match.vec)
     }
     subsam.data <- cl.sam.data[subsam.vec == 1,  ]
     sam.rows <- sam.rows[subsam.vec == 1]
     list(''Pop rows in sample'' = sam.rows,
         ''Data for sample units'' = subsam.data)
}
```

`ginv`—generalized inverse of a matrix used for finding parameter solutions in models that are not full tank.

```
function(X, tol = sqrt(.Machine$double.eps))
{
#     ginv
#     Generalized inverse of a matrix X such that XGX=X
      svdX <- svd(X)
      NotZero <- svdX$d > tol * svdX$d[1]
      svdX$v[, NotZero] %*% ((1/svdX$d[NotZero]) * t(svdX$u[,
            NotZero]))
}
```

`pps.fcn`—select a sample of fixed size with probability proportional to a measure of size (see Chapter 3). The population should be sorted in the desired order before calling this function. If certainty units are in the population, the function `weed.high` should be called first.

```
function(x, n)
{
#     pps.fcn
#     Select pps sample of size n using Hartley-Rao algorithm.
#         Returns indices in population of sample units.
#         If any units are certainties, then NULL vector is
#           returned.
#     x = population vector of sizes
#     n = sample size
#
      N <- length(x)
      if(n > N) {
          stop(''Sample size > pop size.'')
      }
      cumsums <- cumsum(x)
      Skip <- cumsums[N]/n
      if(max(x) > Skip) {
          indices <- NULL
      }
      else {
          R <- runif(1, 0, Skip)
          u <- R + Skip * (0:(n - 1))
          indices <- N + 1 - outer(u, cumsums, ''<='') %*%
              matrix(rep(1, N), ncol = 1)
      }
      indices
}
```

`pps.random.fcn`—select a sample of fixed size with probability proportional to a measure of size after sorting the population in a random order (see Chapter 3). If certainty units are in the population, the function `weed.high` should be called first.

```
function(x, n)
{
#     pps.random.fcn
#     Select pps sample of size n using Hartley-Rao algorithm after
#         randomizing population order. Returns indices in
#         population of sample units.
#     x = population vector of sizes
#     n = sample size
#
      N <- length(x)
      if(n > N) {
          stop(''Sample size > pop size.'')
      }
      X <- cbind(x, 1:N)
      X.tmp <- X[sort.list(sample(1:N, N)),  ]
      smp.tmp <- pps.fcn(X.tmp[, 1], n)
      ind <- X.tmp[smp.tmp, 2]
      sort(ind)
}
```

pps.strat.fcn—select a sample that approximates probability proportional to size by stratifying the population into *n* strata and selecting one unit per stratum at random. If certainty units are in the population, the function *weed.high* should be called first.

```
function(x, n)
{
#     pps. strat.fcn
#     Select pps sample of size n by forming strata with equal total
#         measures of size and selecting one unit from each stratum
#         at random.
#
      N <- length(x)
      if(n > N) {
          stop(''Sample size > pop size.'')
      }
      cumsums <- cumsum(x)
      Skip <- cumsums[N]/n
      R <- runif(n, 0, Skip)
      u <- R + Skip * (0:(n - 1))
      indices <- N + 1 - outer(u, cumsums, ''<='') %*%
                  matrix(rep(1, N), ncol = 1)
#
#     Guarantee unique sample units in case of duplicate measures
#         of size
      while(length(unique(indices))  != length(indices))  {
          a <- indices[1:(n - 1)]
          b <- indices[2:n]
          a[a == b] <- a[a == b] - 1
```

```
        indices <- c(a, indices[n])
    }
    indices
}
```

`restrict.srs`—select a balanced sample using simple random sampling without replacement. Balance is obtained on the first two moments of a single auxiliary x. The degree of balance is controlled by the parameters `E1` and `E2`.

```
function(x, n, E1, E2)
{
#    restrict.srs
#    Select an srswor sample balanced on two moments of a
#        scalar variable x
#    x = population x vector
#    n = sample size
#    E1 = restriction parameter for 1st moment of x
#    E2 = restriction parameter for 2nd moment of x
#            E1 = E2 = 0.125 will reject about 90% of samples
#
    N <- length(x)
    if(n > N) {
        stop(''Sample size > pop size.'')
    }
    xbar.pop <- mean(x)
    x2bar.pop <- mean(x^2)
    xstd.pop <- sqrt(var(x))
    x2std.pop <- sqrt(var(x^2))  #
#
    done <- F
    NoSamsTried <- 0
    while(!done) {
        NoSamsTried <- NoSamsTried + 1
        sam <- sample(1:N, n, replace = F)
        x.sam <- x[sam]
        t1 <- (sqrt(n) * (mean(x.sam) - xbar.pop))/xstd.pop
        t2 <- (sqrt(n) * (mean(x.sam^2) - x2bar.pop))/x2std.pop
        if(abs(t1) <= E1 & abs(t2) <= E2) {
            done <- T
        }
    }
    list(NoSamsTried = NoSamsTried, sam = sort(sam))
}
```

`restrict.pps`—Select a weighted balanced sample using probability proportional to size sampling. Balance is obtained on the first two moments of a single auxiliary x. The degree of balance is controlled by the parameters `E1` and `E2`. If certainty units are in the population, the function `weed.high` should be called first.

```
function(x, n, E1, E2)
{
#    restrict.pps
#
#    Select a restricted pps sample. Units are sorted in a random
#    order, a random start pps sample is selected and checked for
#    nearness to pi-balance on x (1st 2 moments).
#    x = population x vector
#    n = sample size
#    E1 = restriction parameter for 1st moment of x
#    E2 = restriction parameter for 2nd moment of x
#            E1 = E2 = 0.125 will reject about 90% of samples
#
     N <- length(x)
     if(n > N) {
         stop(''Sample size > pop size.'')
     }
     pop.xstats <- c(mean(x), mean(x^2), mean(x^3))
     pi.bal <- c(pop.xstats[2]/pop.xstats[1],
         pop.xstats[3]/pop.xstats[1])
     x.total <- sum(x)
     pk <- x/sum(x)      # 1-draw probs
#
#    Compute standard deviations in with-replacement pps sampling
     SD.xbars <- sqrt(sum(pk * (x - pi.bal[1])^2))
     SD.x2bars <- sqrt(sum(pk * (x^2 - pi.bal[2])^2))      #
#
     NoSamsTried <- 0
     done <- F
     while(!done) {
         NoSamsTried <- NoSamsTried + 1
         sam <- pps.random.fcn(x, n)
         t1 <- (sqrt(n) * (mean(x[sam]) - pi.bal[1]))/SD.xbars
         t2 <- (sqrt(n) * (mean(x[sam]^2) - pi.bal[2]))/SD.x2bars
         if(abs(t1) <= E1 & abs(t2) <= E2) {
             done <- T
         }
     }
     list(NoSamsTried = NoSamsTried, sam = sam)
}
```

strat.opt.alloc.x—used for example in Chapter 6 to illustrate the effects of different allocations to strata under a model with $E_M(Y_{hi}) = \beta_0 + \beta_1 x_{hi} + \beta_2 x_{hi}^2$, $\text{var}_M(Y_{hi}) = \sigma^2 x_{hi}^2$ and $\sigma^2 = 1$.

```
function(cost, Nh = c(680, 552, 429, 363, 251), xbarh = c(8.7, 15.5,
76.1, 190.8, 621.6), ch = c(10, 20,
```

```
    30, 50, 70))
{
#   strat.opt.alloc.x
#       Used for example in stratification chapter to illustrate
#       allocation to strata under model
#       E(Y) = b0 + b1*x + b2*x^2,
#       var(Y) =x^2 with strat wtd-balanced sample.
#
    set.seed(313)
    x2barh <- round(runif(length(Nh), 1, 1.2) * xbarh^2, 1)  #
    cat(''Nh = '', Nh, ''\n'')
    strat.var <- function(Nh, nh, xbarh, x2barh)
    {
        sum((Nh * xbarh)^2/nh - Nh * x2barh)
    }
    n.cost <- round((cost * sum((Nh * xbarh)/sqrt(ch)))/sum(Nh *
              xbarh * sqrt(ch)), 0)
    nh.opt.cost <- round(((n.cost * Nh * xbarh)/sqrt(ch))/
             sum((Nh * xbarh)/sqrt(ch)), 0)
    v.opt.cost <- strat.var(Nh, nh.opt.cost, xbarh, x2barh)
    cost.opt <- sum(ch * nh.opt.cost)  #
#
#__Allocate n.cost optimally but ignore stratum costs__
    nh.opt.nocost <- round(((n.cost * Nh * xbarh))/
             sum((Nh * xbarh)), 0)
    v.opt.nocost <- strat.var(Nh, nh.opt.nocost, xbarh, x2barh)
    cost.opt.nocost <- sum(ch * nh.opt.nocost)    #
#
#__Allocate n.cost proportionally for comparison__
    nh.prop <- round((n.cost * Nh)/sum(Nh), 0)
    v.prop <- strat.var(Nh, nh.prop, xbarh, x2barh)
    cost.prop <- sum(ch * nh.prop)    #
#
    nh.prop.adj <- round((nh.prop * cost)/cost.prop, 0)
    v.prop.adj <- strat.var(Nh, nh.prop.adj, xbarh, x2barh)
    cost.prop.adj <- sum(ch * nh.prop.adj)    #
#
#__Equal allocation__
    nh.equal <- round(n.cost/length(Nh), 0)
    v.equal <- strat.var(Nh, nh.equal, xbarh, x2barh)
    cost.equal <- sum(ch * nh.equal)    #
#
    nh.equal.adj <- round((nh.equal * cost)/cost.equal, 0)
    v.equal.adj <- strat.var(Nh, nh.equal.adj, xbarh, x2barh)
    cost.equal.adj <- sum(ch * nh.equal.adj)  #
#
    nh <- cbind(nh.opt.cost, nh.opt.nocost, nh.prop, nh.prop.adj,
              nh.equal, nh.equal.adj)
    nh <- rbind(nh, apply(nh, 2, sum))
```

```
    dimnames(nh) <- list(NULL, c(''opt.cost'', ''opt.nocost'',
                ''prop'', ''prop.adj'', ''equal'', ''equal.adj'')
        )
    rtv <- sqrt(c(v.opt.cost, v.opt.nocost, v.prop, v.prop.adj,
                v.equal, v.equal.adj))
    tot.cost <- c(cost.opt, cost.opt.nocost, cost.prop,
                cost.prop.adj, cost.equal, cost.equal.adj)
    parms <- cbind(Nh, xbarh, ch, round((Nh * xbarh)/sqrt(ch)/sum
                ((Nh * xbarh)/sqrt(ch)), 2),
                round((Nh * xbarh)/sum(Nh * xbarh), 2))
    dimnames(parms) <- list(NULL, c(''Nh'', ''xbarh'', ''ch'',
                ''Nh*xbarh/sqrt(ch)'', ''Nh*xbarh''))
    list(parms, nh = nh, ''total costs'' = tot.cost,
                rtv = round(rtv, 2))
}
```

T.blu—compute the *BLU* predictor under a working model $M(\mathbf{X}:\mathbf{W}^{-1})$. The entire population matrices, **X** and **W**, and population vector **y** are inputs.

```
function(X, W, sam, y)
{
#   T.blu
#   Compute the BLU predictor under a working M(X: V = W^(-1))
#   Assumes population variance matrix is diagonal
#   X = matrix of auxiliaries for entire population
#   W = diagonal of inverse of variance matrix for entire
#   population,
#   sam = vector of sample unit numbers
#   y = y-vector for entire population
#
    X <- as.matrix(X)
    Xs <- as.matrix(X[sam,  ])
    Xr <- as.matrix(X[ - sam,  ])
    Ws <- W[sam]
    A <- (t(Xs) * Ws) %*% Xs
    Ainv <- ginv(A)
    beta <- Ainv %*% (t(Xs) * Ws) %*% y[sam]
    T.BLU <- sum(y[sam]) + sum(Xr %*% beta)
    T.BLU
}
```

vD—compute the robust variance estimator v_D from Chapter 5. This function uses the output of the ahr function.

```
function(a, h, r, W, sam, tol = sqrt(.Machine$double.eps))
{
#   vD
#   Robust estimate of variance
#   a, h, r = outputs from ahr function
```

```
#    W = diagonal weight matrix for entire population,
#         should agree with input to ahr function
#    sam = vector of sample unit numbers

     Zero <- (1 - h) < tol
     h[Zero) <- 0
     psi <- r^2/(1 - h)
     A <- sum(a^2 * psi)
     B <- sum(1/W[ - sam])/sum(1/W[sam]) * sum(psi)
     A + B
}
```

vH—compute the robust variance estimator v_H from Chapter 5. This function uses the output of the ahr function.

```
function(a, h, r, W, sam, tol = sqrt(.Machine$double.eps))
{
#    vH
#    Robust estimate of variance
#    a, h, r = outputs from ahr function
#    W = diagonal weight matrix for entire population,
#         should agree with input to ahr function
#    sam = vector of sample unit numbers

     Zero <- (1 - h) < tol
     h[Zero] <- 0
     psi <- (r^2 * sum(a^2 * (1/W[sam])))/
          sum(a^2 * (1/W[sam]) * (1 - h))
     A <- sum(a^2 * psi)
     B <- sum(1/W[ - sam])/sum(1/W[sam]) * sum(psi)
     A + B
}
```

vJ—compute the robust variance estimator v_J from Chapter 5. This function uses the output of the ahr function.

```
function(a, h, r, W, sam, tol = sqrt(.Machine$double.eps))
{
#    vJ
#    Jackknife estimate of variance
#    a, h, r = outputs from ahr function
#    W = diagonal weight matrix for entire population,
#         should agree with input to ahr function
#    sam = vector of sample unit numbers
     n <- length(r)
     Zero <- (1 - h) < tol
     h[Zero] <- 0
     psi <- r^2/(1 - h)^2
     A <- sum(a^2 * psi)
     B <- sum(1/W[ - sam])/sum(1/W[sam]) * sum(psi)
```

```
    C <- (sum((a * r)/(1 - h)))^2/length(sam)
    (n - 1)/n * (A - C) + B
}
```

vJ.star—compute the robust variance estimator v_J^* from Chapter 5. This function uses the output of the ahr function.

```
function(a, h, r, W, sam, tol = sqrt(.Machine$double.eps))
{
#   vJ.star
#   Robust estimate of variance
#   a, h, r = outputs from ahr function
#   W = diagonal weight matrix for entire population,
#        should agree with input to ahr function
#   sam = vector of sample unit numbers

    Zero <- (1 - h) < tol
    h[Zero] <- 0
    psi <- r^2/(1 - h)^2
    A <- sum(a^2 * psi)
    B <- sum(1/W[ - sam])/sum(1/W[sam]) * sum(psi)
    A + B
}
```

vR—compute the robust variance estimator v_R from Chapter 5. This function uses the output of the ahr function.

```
function(a, r, W, sam)
{
#   vR
#   Robust sandwich estimate of variance
#   a, r = outputs from ahr function
#   W = diagonal weight matrix for entire population,
#        should agree with input to ahr function
#        sam = vector of sample unit numbers
#
    psi <- r^2
    A <- sum(a^2 * psi)
    B <- sum(1/W[ - sam])/sum(1/W[sam]) * sum(psi)
    A + B
}
```

weed.high—identify certainties in a *pps* sample. Return the indices of the certainty units in the population and their measures of size.

```
function(x, n)
{
#   weed.high
#   Identify certainty units in a pps sample
```

```
#     Returns indices in population of certainty units.
#     x = population vector of sizes (assumed to be positive)
#     n = sample size
#
      N <- length(x)
      if(n > N)  {
            stop(''Sample size > pop size.'')
      }
      if(n == N)  {
            stop(''Sample size = pop size.  All units are
            certainties.'')
      }
      X <- cbind(x, 1:N)
      X <- X[sort.list( - x),  ]
      skip <- sum(X[, 1])/n
      if(max(x) < skip)  {
            indices <- NULL
            cat(''No x's large enough to be certainties.\n'')
      }
      else  {
            done <- F
            certs <- NULL
            X0 <- X
            X.certs <- NULL
            while(!done)  {
                i <- max((1:N) [X0[, 1] >= skip])
                X.certs <- rbind(X.certs, X0[1:i,  ])
                if((i + 1) <= N)  {
                    X0 <- X0[(i + 1):N,  ]
                    N <- length(X0[, 1])
                    n0 <- n - nrow(X.certs)
                    skip <- sum(X0[, 1])/n0
                    if(max[X0[, 1]) < skip)  {
                        done <- T
                    }
                }
            }
            dimnames(X.certs) <- list(NULL, c(''x'', ''pop index''))
            list(''Matrix of certainties'' = X.certs)
      }
}
```

Bibliography

Agresti, A. (1990), *Categorical Data Analysis*, New York: Wiley.

Arthanari, T. S., and Dodge, Y. (1981), *Mathematical Programming in Statistics*, New York: Wiley.

Bardsley, P., and Chambers, R. L. (1984), Multipurpose Estimation from Unbalanced Samples, *Applied Statistics*, **33**, 290–299.

Basu, D. (1969), Role of the Sufficiency and Likelihood Principles in Sample Survey Theory, *Sankhyā A*, **31**, 441–454.

Basu, D. (1971), An Essay on the Logical Foundations of Survey Sampling in Godambe, V. P., and Sprott, D. A. (eds.), *Foundations of Statistical Inference*, Toronto: Holt, Rinehart, and Winston, pp. 203–242.

Basu, D. (1980), Randomization Analysis of Experimental Data: The Fisher Randomization Test, *Journal of the American Statistical Association*, **75**, 575–595.

Bates, D. M., and Pinheiro, J. C. (1998), Computational Methods for Multilevel Modeling, University of Wisconsin technical report, http://franz.stat.wisc.edu/pub/NLME/CompMulti.pdf

Bates, D. M., and Watts, D. G. (1988), *Nonlinear Regression Analysis and Its Applications*, New York: Wiley.

Becker, R. A., Chambers, J. M., and Wilks, A. R. (1988), *The New S Language*, Pacific Grove, CA: Wadsworth & Brooks/Cole.

Belsley, D. A., Kuh, E., and Welsch, R. E. (1980), *Regression Diagnostics*, New York: Wiley.

Bethel, J. (1985), An Optimum Allocation Algorithm for Multivariate Surveys, *Proceedings of the Section on Survey Methods Research*, American Statistical Association, pp. 209–212.

Bethel, J. (1989), Sample Allocation in Multivariate Surveys, *Survey Methodology*, **15**, 47–57.

Bickel, P., and Doksum, K. (1977), *Mathematical Statistics: Basic Ideas and Selected Topics*, San Francisco: Holden-Day.

Biemer, P. and Stokes, L. (1985), Optimal Design of Interviewer Variance Experiments in Complex Surveys, *Journal of the American Statistical Association*, **80**, 158–166.

Biemer, P. and Stokes, L. (1989), The Optimal Design of Quality Control Samples to Detect Interviewer Cheating, *Journal of Official Statistics*, **5**, 23–39.

Biemer, P. and Stokes, L. (1991), Approaches to the Modeling of Measurement Errors, in Biemer, P., Groves, R., Lyberg, L., Mathiowetz, N., and Sudman, S. (eds.), *Measurement Errors in Surveys*, New York: Wiley, pp. 487–516.

Binder, D. (1983), On the Variances of Asymptotically Normal Estimators from Complex Surveys, *International Statistical Review*, **51**, 279–292.

Bjornstadt, J. (1990), Predictive Likelihood: A Review, (with discussion), *Statistical Science*, **5**, 242–254.

Bolfarine, H. and Zacks, S. (1992), *Prediction Theory for Finite Populations*, New York: Springer-Verlag.

Breidt, F. J. and Opsomer, J. D. (1999), Local Polynomial Regression Estimators in Survey Sampling, Technical Report, Iowa State University.

Brewer, K. R. W. (1963a), A Model of Systematic Sampling with Unequal Probabilities, *Australian Journal of Statistics*, **5**, 5–13.

Brewer, K. R. W. (1963b), Ratio Estimation and Finite Populations: Some Results Deducible from the Assumption of an Underlying Stochastic Process, *Australian Journal of Statistics*, **5**, 93–105.

Brewer, K. R. W. (1994), Survey Sampling Inference: Some Past Perspectives and Present Prospects, *Pakistan Journal of Statistics*, **10**, 15–30.

Brewer, K. R. W. (1995), Combining Design-based and Model-based Inference, in Cox, B., Binder, D. Chinappa, B., Christianson, A., Colledge, M., and Kott, P., (eds.), *Business Survey Methods*, New York: Wiley, pp. 589–606.

Brewer, K. R. W. (1999), Design-based or Prediction Inference? Stratified Random vs. Stratified Balanced Sampling, *International Statistical Review*, **67**, 35–47.

Brewer, K. R. W., and Hanif, M. (1983), *Sampling with Unequal Probabilities*, New York: Springer-Verlag.

Brewer, K. R. W., and Särndal, C. E. (1983), Six Approaches to Enumerative Survey Sampling, in Madow, W. G. and Olkin, I. (eds.), *Incomplete Data in Sample Surveys*, vol. 3, Academic Press, pp. 363–368.

Carr, D. (1994), Topics in Scientific Visualization: Using Gray in Plots, *Statistical Computing and Graphics Newsletter*, **5**, 11–14.

Carroll, R. J., and Ruppert, D. (1988), *Transformation and Weighting in Regression*, London: Chapman and Hall.

Casady, R. J., Dorfman, A., and Wang, S. (1998), Confidence Intervals for Domain Parameters When the Domain Sample Size Is Random, *Survey Methodology*, **24**, 57–67.

Casady, R. J., and Valliant, R. (1993), Conditional Properties of Post-stratified Estimators under Normal Theory, *Survey Methodology*, **19**, 183–192.

Caslyn, C., Gonzales, P., Frase, M. (1999), *Highlights from TIMSS*, National Center for Education Statistics, Washington, DC.

Chambers, R. L. (1986), Outlier Robust Finite Population Estimation, *Journal of the American Statistical Association*, **81**, 1063–1069.

Chambers, R. L. (1996a), Robust Case Weighting for Multipurpose Establishment Surveys, *Journal of Official Statistics*, **12**, 3–32.

Chambers, R. L. (1996b), *Survey Methods Part I*, unpublished course notes, University of Southhampton.

Chambers, R. L., Dorfman, A. H., and Hall, P. (1992), Properties of the Finite Population Distribution Function, *Biometrika*, **79**, 577–582.

Chambers, R. L., Dorfman, A. H., and Wehrly, T. E. (1993), Bias Robust Estimation in Finite Populations Using Nonparametric Calibration, *Journal of the American Statistical Association*, **88**, 268–277.

Chambers, R. L., and Dunstan, R. (1986), Estimating Distribution Functions from Survey Data, *Biometrika*, **73**, 597–604.

Chambers, R. L., and Kokic, P. N. (1993), Outlier Robust Sample Survey Inference, *Bulletin of the International Statistical Institute*, Invited Papers, 69–86.

Chen, G., and Chen, J. (1996), A Transformation Method for Finite Population Sampling Calibrated with Empirical Likelihood, *Survey Methodology*, **22**, 139–146.

Chew, V. (1970), Covariance Matrix Estimation in Linear Models, *Journal of the American Statistical Association*, **65**, 173–181.

Chromy, J. (1987), Design Optimization with Multiple Objectives, *Proceedings of the Section on Survey Methods Research*, American Statistical Association, pp. 194–199.

Cleveland, W. S. (1993), *Visualizing Data*, Summit, NJ: Hobart Press.

Cochran, W. G. (1946), Relative Accuracy of Systematic and Random Samples for a Certain Class of Populations, *Annals of Mathematical Statistics*, **17**, 164–177.

Cochran, W. G. (1977), *Sampling Techniques*, 3rd ed., New York: Wiley.

Cornfield, J. (1971), The University Group Diabetes Program: A Further Statistical Analysis of the Mortality Findings, *Journal of the American Medical Association*, **217**, 1676–1687.

Cox, D. R., and Hinkley, D. V. (1974), *Theoretical Statistics*, London: Chapman and Hall.

Cumberland, W. G., and Royall, R. M. (1981), Prediction Models and Unequal Probability Sampling, *Journal of the Royal Statistical Society B*, **43**, 353–367.

Dalenius, T., and Hodges, J. L., Jr. (1959), Minimum Variance Stratification, *Journal of the American Statistical Association*, **54**, 88–101.

Datta, G. S., and Ghosh, M. (1991), Bayesian Prediction in Linear Models: Applications to Small Area Estimation, *Annals of Statistics*, **19**, 1748–1770.

Davidson, R., and MacKinnon, J. G. (1993), *Estimation and Inference in Econometrics*, New York: Oxford University Press.

Deville, J. C., and Särndal, C.-E. (1992), Calibration Estimators in Survey Sampling, *Journal of the American Statistical Association*, **87**, 376–382.

Dippo, C., and Jacobs, C. (1983), Area Sample Redesign for the Consumer Price Index, *Proceedings of the Section on Survey Research Methods*, American Statistical Association, pp. 118–123.

Dippo, C. S., Fay, R. E., and Morganstein, D. R. (1984), Computing Variances from Complex Samples with Replicate Weights, *Proceedings of the Section on Survey Research Methods*, American Statistical Association, pp. 489–494.

Dorfman, A. H. (1991), Sound Confidence Intervals in the Heteroscedastic Linear Model through Releveraging, *Journal of the Royal Statistical Society* B, **53**, 441–452.

Dorfman, A. H. (1993), A Comparison of Design-Based and Model-Based Estimators of the Finite Population Distribution Function, *Australian Journal of Statistics*, **35**, 29–41.

Dorfman, A. H. (1994), Open Questions in the Application of Smoothing Methods to Finite Population Inference, *Computationally Intensive Statistical Methods, Proceedings of the 26th Symposium on the Interface,* pp. 201–205.

Dorfman, A. H., and Hall, P. (1993), Estimators of the Finite Population Distribution Function Using Nonparametric Regression, *Annals of Statistics*, **21**, 1452–1475.

Dorfman, A. H., and Valliant, R. (1993), Quantile Variance Estimators in Complex Surveys, *Proceedings of the Section on Survey Methods Research*, American Statistical Association, pp. 866–871.

Efron, B. (1981), *The Jackknife, the Bootstrap and Other Resampling Plans*, Philadelphia: Society for Industrial and Applied Mathematics.

Eicker, F. (1963), Asymptotic Normality and Consistency of the Least Squares Estimators for Families of Linear Regressions, *Annals of Mathematical Statistics*, **34**, 447–456.

Eicker, F. (1967), Limit Theorems for Regressions with Unequal and Dependent Errors, *Proceedings of the 5th Berkeley Symposium on Mathematical Statistics and Probability,* Vol. 1, Berkeley: University of California Press, pp. 59–82.

Ericksen, E., and Kadane, J. (1985), Estimating the Population in a Census Year: 1980 and Beyond, *Journal of the American Statistical Association*, **80**, 98–131.

Ericksen, E., Kadane, J., and Tukey, J. (1989), Adjusting the 1980 Census of Population and Housing, *Journal of the American Statistical Association*, **84**, 927–944.

Erickson, W. A. (1969), Subjective Bayesian Models in Sampling Finite Populations, *Journal of the Royal Statistical Society B*, **31**, 195–233.

Ernst, L. R. (1980), Comparisons of Estimators of the Mean Which Adjust for Large Values, *Sankhya*, **C42**, 1–16.

Estevao, V., Hidiroglou, M., and Särndal, C. E. (1995), Methodological Principles for a Generalized Estimation System at Statistics Canada, *Journal of Official Statistics*, **11**, 181–204.

Eubank, R. L. (1988), *Spline Smoothing and Nonparametric Regression*, New York: Marcel Dekker.

Fan, J., and Gibjels, I. (1992), Variable Bandwidth and Local Linear Regression Smoothers, *Annals of Statistics*, **20**, 2008–2036.

Fiacco, A. V., and McCormick, G. (1968), *Nonlinear Sequential Unconstrained Minimization Techniques*, New York: Wiley.

Francisco, C., and Fuller, W. (1991), Quantile Estimation with a Complex Survey Design, *Annals of Statistics*, **19**, 454–469.

Freedman, D., and Navidi, W. C. (1986), Regression Models for Adjusting the 1980 Census, *Statistical Science*, **1**, 3–39.

Fuller, W. (1987), *Measurement Error Models*, New York: Wiley.

Fuller, W., Loughin, M., and Baker, H. (1994), Regression Weighting in the Presence of Nonresponse with Application to the 1987–1988 Nationwide Food Consumption Survey, *Survey Methodology*, **20**, 75–85.

Gasser, T., and Müller, H. G. (1979), Kernel Estimation of Regression Functions, in *Smoothing Techniques for Curve Estimation* (Lecture Notes in Mathematics 757), T. Gasser and M. Rosenblatt (eds.), Berlin: Springer, pp. 23–68.

Gasser, T., and Müller, H. G. (1984), Estimating Regression Functions and Their Derivatives by the Kernel Method, *Scandanavian Journal of Statistics*, **11**, 171–185.

Gelfand, A. E., Hills, S. E., Racine-Poon, A., and Smith, A. F. M. (1990), Illustrations of Bayesian Inference in Normal Data Models Using Gibbs Sampling, *Journal of the American Statistical Associations*, **85**, 972–985.

Gelfand, A. E., and Smith, A. F. M. (1990), Sampling Based Approaches to Calculating Marginal Densities, *Journal of the American Statistical Association*, **85**, 398–409.

Gentle, J. (1998), *Numerical Linear Algebra for Applications in Statistics*, New York: Springer-Verlag.

Ghosh, M., and Meeden, G. (1997), *Bayesian Methods for Finite Population Sampling*, London: Chapman & Hall.

Ghosh, M., and Rao, J. N. K. (1994), Small Area Estimation: An Appraisal, *Statistical Science*, **9**, 55–93.

Gnedenko, B. V., and Kolgomorov, A. N. (1968), *Limit Distribution for Sums of Independent Random Variables*, (translated from the Russian by K. L. Chung), Cambridge, MA: Addison-Wesley.

Godambe, V. P. (1955), A Unified Theory of Sampling from Finite Populations, *Journal of the Royal Statistical Society* B, **17**, 269–278.

Godambe, V. P. (1966), A New Approach to Sampling from Finite Populations I: Sufficiency and Linear Estimation, *Journal of the Royal Statistical Society B*, **28**, 310–319.

Godambe, V. P. (1989), Estimation of Cumulative Distribution Function of a Survey Population, Technical Report, University of Waterloo.

Godambe, V. P., and Joshi, V. M. (1965), Admissibility and Bayes Estimation in Sampling Finite Populations–I, *Annals of Mathematical Statistics*, **36**, 1707–1723.

Godfrey, J., Roshwalb, A., and Wright, R. (1984), Model-based Stratification in Inventory Cost Estimation, *Journal of Business and Economic Statistics*, **2**, 1–9.

Goldstein, H. (1995), *Multilevel Statistical Models*, London: Halstead Press.

Gonzalez, J. F., Placek, P., and Scott, C. (1996), Synthetic Estimation in Followback Surveys at the National Center for Health Statistics, Chapter 2 in Schaible, W. (ed.), *Indirect Estimators in U.S. Federal Programs*, New York: Springer-Verlag.

Goodman, R., and Kish, L. (1950), Controlled Selection—A Technique in Probability Sampling, *Journal of the American Statistical Association*, **45**, 350–372.

Groves, R. (1989), *Survey Errors and Survey Costs*, New York: Wiley.

Gurney, M., and Jewett, R. S. (1975), Constructing Orthogonal Replications for Variance Estimation, *Journal of the American Statistical Association*, **71**, 819–821.

Gwet, J.-P., and Rivest, L.-P. (1992), Outlier Resistant Alternatives to the Ratio Estimator, *Journal of the American Statistical Association*, **87**, 1174–1182.

Hájek, J. (1960), Limiting Distributions in Simple Random Sampling from a Finite Population, *Publication of the Mathematical Institute of the Hungarian Academy of Sciences*, **5**, 361–374.

Hájek, J. (1971), Discussion of "An Essay on the Logical Foundations of Survey Sampling" by D. Basu, in Godambe, V. P., and Sprott, D. A. (eds), *Foundations of Statistical Inference*, Toronto: Holt, Rinehart, and Winston, p. 236.

Hall, D., and Severini, T. (1998), Extended Generalized Estimating Equations for Clustered Data, *Journal of the American Statistical Association*, **93**, 1365–1375.

Hampel, F. R., Ronchetti, E. M., Rousseeuw, P. J., and Stahel, W. A. (1986), *Robust Statistics: The Approach Based on Influence Functions*, New York: Wiley.

Hansen, M. H., Hurwitz, W. H., and Madow, W. G. (1953), *Sample Survey Methods and Theory*, Volume I, New York: Wiley.

Hansen, M. H., Madow, W. G., and Tepping, B. J. (1983), An Evaluation of Model-Dependent and Probability Sampling Inferences in Sample Surveys, *Journal of the American Statistical Association*, **78**, 776–793.

Hanson, R. H. (1978), *The Current Population Survey: Design and Methodology*, Technology Paper 40, Washington, DC: U.S. Bureau of the Census.

Hardle, W. (1990), *Applied Nonparametric Regression Analysis*, Cambridge: Cambridge University Press.

Hartley, H. O., and Rao, J. N. K. (1962), Sampling with Unequal Probabilities and without Replacement, *Annals of Mathematical Statistics*, **33**, 350–374.

Harville, D. (1977), Maximum Likelihood Approaches to Variance Component Estimation and to Related Problems, *Journal of the American Statistical Association*, **72**, 320–340.

Herson, J. (1976), An Investigation of Relative Efficiency of Least-Squares Prediction to Conventional Probability Sampling Plans, *Journal of the American Statistical Association*, **71**, 700–703.

Hidiroglou, M. A., and Srinath, K. P. (1981), Some Estimators of the Population Total from Simple Random Samples Containing Large Units, *Journal of the American Statistical Association*, **76**, 690–695.

Hoaglin, D. A., and Welsch, R. E. (1978), The Hat Matrix in Regression and ANOVA, *American Statistician*, **32**, 17–22, *Corrigenda*, **32**, 146.

Hogg, R. V., and Tanis, E. A. (1983), *Probability and Statistical Inference*, New York: Macmillan.

Horn, S. D., Horn, R. A., and Duncan, D. B. (1975), Estimating Heteroscedastic Variances in Linear Models, *Journal of the American Statistical Association*, **70**, 380–385.

Horvitz, D. G., and Thompson, D. J. (1952), A Generalization of Sampling Without Replacement from a Finite Universe, *Journal of the American Statistical Association*, **47**, 663–685.

Huber, P. (1964), Robust Estimation of a Location Parameter, *Annals of Mathematical Statistics*, **35**, 73–101.

Huber, P. (1967), The Behavior of Maximum Likelihood Estimates under Nonstandard Conditions, in *Proceedings of the Fifth Berkeley Symposium in Mathematical Statistics and Probability*, Berkeley: University of California Press, pp. 221–233.

Huber, P. J. (1981), *Robust Statistics*, New York: Wiley.

Isaki, C., and Fuller, W. (1982), Survey Design under the Regression Superpopulation Model, *Journal of the American Statistical Association*, **77**, 89–96.

Jayasuriya, B., and Valliant, R. (1996), An Application of Restricted Regression Estimation to Post-Stratification in a Household Survey, *Survey Methodology*, **22**, 127–137.

Jennrich, R. I. (1969), Asymptotic Properties of Nonlinear Least Squares Estimators, *Annals of Mathematical Statistics*, **40**, 633–649.

Judkins, D. (1990), Fay's Method of Variance Estimation, *Journal of Official Statistics*, **6**, 223–239.

Kalbfleisch, J. D., and Sprott, D. A. (1969), Application of Likelihood and Fiducial Probability to Sampling Finite Populations, in N. L. Johnson and H. Smith, Jr. (eds.), *New Developments in Survey Sampling*, New York: Wiley, pp. 358–389.

Kalton, G. (1977), Practical Methods for Estimating Survey Sampling Errors, *Bulletin of the International Statistical Institute*, **47**, 495–514.

Kalton, G., and Blunden, R. M. (1973), Sampling Errors in the British General Household Survey, *Bulletin of the International Statistical Institute*, **45**, 83–97.

Karmel, T. S., and Jain, M. (1987), Comparison of Purposive and Random Sampling Schemes for Estimating Capital Expenditures, *Journal of the American Statistical Association*, **82**, 52–57.

Kelly, E. J., and Cumberland, W. G. (1990), Prediction Theory Approach to Multistage Sampling when Cluster Sizes Are Unknown, *Journal of Official Statistics*, **6**, 437–449.

Kempthorne, O. (1955), The Randomization Theory of Experimental Inference, *Journal of the American Statistical Association*, **50**, 946–967.

Kennedy, W. J., and Gentle, J. E. (1980), *Statistical Computing*, New York: Marcel Dekker.

Kirkendall, N. J. (1992), When Is Model-based Sampling Appropriate for EIA Surveys? *Proceedings of the Section on Survey Methods Research*, American Statistical Association, pp. 637–642.

Kish, L. (1965), *Survey Sampling*, New York: Wiley.

Kokan, A. R. (1963), Optimum Allocation in Multivariate Surveys, *Journal of the Royal Statistical Society A*, **126**, 557–565.

Kokan, A. R., and Khan, S. (1967), Optimum Allocation in Multivariate Surveys: An Analytical Solution, *Journal of the Royal Statistical Society B*, **29**, 115–125.

Korn, E. L., and Graubard, B. I. (1999), *Analysis of Health Surveys*, New York: Wiley.

Korn, E. L., Midthune, D., and Graubard, B. I. (1997), Estimating Interpolated Percentiles from Grouped Data with Large Samples, *Journal of Official Statistics*, **13**, 385–399.

Kott, P. S. (1984), A Fresh Look at Bias-robust Estimation in a Finite Population, *Proceedings of the Section on Survey Methods Research*, American Statistical Association, pp. 176–178.

Kott, P. S. (1986), Some Asymptotic Results for the Systematic and Stratified Sampling of a Finite Population, *Biometrika*, **73**, 485–491.

Kott, P. S. (1990), Estimating the Conditional Variance of a Design Consistent Regression Estimator, *Journal of Statistical Planning and Inference*, **24**, 287–296.

Krewski, D. (1978), On the Stability of Some Replication Variance Estimators in the Linear Case, *Journal of Statistical Planning and Inference*, **2**, 45–51.

Krewski, D., and Rao, J. N. K. (1981), Inference from Stratified Samples: Properties of the Linearization, Jackknife, and Balanced Repeated Replication Methods, *Annals of Statistics*, **9**, 1010–1019.

Kuk, A. Y. C. (1993), A Kernel Method for Estimating Finite Population Distribution Functions Using Auxiliary Information, *Biometrika*, **80**, 385–392.

Kuo, L. (1988), Classical and Prediction Approaches to Estimating Distribution Functions from Survey Data, *Proceedings of the Section on Survey Research Methods*, American Statistical Association, pp. 280–285.

Lahiri, D. D. (1968), On the Unique Sample, the Surveyed One, paper presented at the Symposium on the Foundations of Survey Sampling, Chapel Hill, NC.

Lasdon, L., and Waren, A. (1978), Generalized Reduced Gradient Software for Linearly and Nonlinearly Constrained Problems, in H. Greenberg (ed.), *Design and Implementation of Optimization Software*, Alphen aan den Rijn: Sijthoff and Noordhoff.

Leaver, S. G., Weber, W. L., Cohen, M. P., and Archer, K. P. (1987), Item-Outlet Sample Redesign for the 1987 U.S. Consumer Price Index Revision, *Proceedings of the 46th Session*, Vol. LII, Book 3, International Statistical Institute, pp. 173–185.

Leaver, S. G., Johnson, W. H., Baskin, R., Scarlett, S., and Morse, R. (1996), Commodities and Services Sample Redesign for the 1998 Consumer Price Index Revision, *Proceedings of the Section on Survey Research Methods*, American Statistical Association, pp. 239–244.

Lee, H. (1991), Model-based Estimators That Are Robust to Outliers, *1991 Annual Research Conference*, Washington, DC: Bureau of the Census, pp. 178–202.

Lee, H. (1995), Outliers in Business Surveys, in Cox, B., Binder, D. Chinappa, B., Christianson, A., Colledge, M., and Kott, P. (eds.), *Business Survey Methods*, New York: Wiley, pp. 503–526.

Liang, K.-Y., and Zeger, S. (1986), Longitudinal Data Analysis Using Generalized Linear Models, *Biometrika*, **73**, 13–22.

Liang, K.-Y. Zeger, S., and Qaqish, B. (1992), Multivariate Regression Analyses for Categorical Data, *Journal of the Royal Statistical Society B*, **45**, 3–40.

Little, R. J. A. (1982), Models for Nonresponse in Sample Surveys, *Journal of the American Statistical Association*, **77**, 237–250.

Little, R. J. A. (1983), Discussion of "An Evaluation of Model-Dependent and Probability Sampling Inferences in Sample Surveys," by M. H. Hansen, W. G. Madow, and B. J. Tepping, *Journal of the American Statistical Association*, **78**, 776–807.

Longford, N. T. (1993), *Random Coefficient Models*, Oxford: Clarendon Press.

MacKinnon, J. G., and White, H. (1985), Some Heteroskedasticity Consistent Covariance Matrix Estimators with Improved Finite Sample Properties, *Journal of Econometrics*, **29**, 305–325.

Madow, W. G. (1948), On the Limiting Distributions of Estimates Based on Samples from Finite Universes, *Annals of Mathematical Statistics*, **19**, 535–545.

Mahalanobis, P. C. (1952), Some Aspects of the Design of Sample Surveys, *Sankhya*, **12**, 1–7.

Mardia, K. V., Kent, J. T., and Bibby, J. M. (1979), *Multivariate Analysis*, New York: Academic Press.

Mason, T. J., and McKay, F. W. (1974), *U.S. Cancer Mortality by County: 1950–1969*, DHEW publication no. (NIH) 74-615, Washington, DC: U.S. Government Printing Office.

McCarthy, P. J. (1966), Replication: An Approach to the Analysis of Data from Complex Surveys, *Vital and Health Statistics*, Series 2, no. 14, Washington, DC: National Center for Health Statistics, Public Health Service.

McCarthy, P. J. (1969), Pseudo-Replication: Half-Samples, *Review of the International Statistical Institute*, **37**, 239–264.

Moré, J. J., and Wright, S. J. (1993), *Optimization Software Guide*, Philadelphia: SIAM.

Mukhopadhyay, P. (1977), Robust Estimators of Finite Population Total under Certain Linear Regression Models, *Sankhya C*, **39**, 71–87.

Nadarya, E. A. (1964), On Estimating Regression, *Theory of Probability and Its Applications*, **9**, 141–142.

Narula, S., and Weistroffer, H. (1989), Algorithms for Multiple Objective Nonlinear Programming Problems, in A. Lockett and G. Islei, (eds.), *Improving Decision Making in Organizations*, Berlin: Springer-Verlag, pp. 434–443.

Nelder, J. A., and Wedderburn, R. W. M. (1972), Generalized Linear Models, *Journal of the Royal Statistical Society A*, **135**, 370–384.

Neyman, J. (1934), On the Two Different Aspects of the Representative Method: The Method of Stratified Sampling and the Method of Purposive Selection, *Journal of the Royal Statistical Society*, **109**, 558–606.

Pathak, P. K. (1966), An Estimator in πps Sampling of Multiple Characteristics, *Sankhya A*, **28**, 35–40.

Pfeffermann, D. (1993), The Role of Sampling Weights when Modeling Survey Data, *International Statistical Review*, **61**, 317–337.

Pfeffermann, D., and Sverchkov, M. (1999). Parametric and Semi-Parametric Estimation of Models Fitted to Survey Data, *Sankhya* B, 61, parts 1, 1–21.

Plackett, R. L., and Burman, J. P. (1946), The Design of Optimum Multifactorial Experiments, *Biometrika*, **33**, 305–325.

Platek, R., Rao, J. N. K., Särndal, C.-E., and M. P. Singh (1987), *Small Area Statistics: An International Symposium*, New York: Wiley.

Prentice, R., and Zhao, L. (1991), Estimating Equations for Parameters in Means and Covariances of Multivariate Discrete and Continuous Responses, *Biometrics*, **47**, 825–839.

Rao, C. R. (1973), *Linear Statistical Inference*, 2nd ed., New York: Wiley.

Rao, J. N. K. (1985), Conditional Inference in Survey Sampling, *Survey Methodology*, **11**, 15–31.

Rao, J. N. K. (1994), Estimating Totals and Distribution Functions Using Auxiliary Information at the Estimation Stage, *Journal of Official Statistics*, **10**, 153–165.

Rao, J. N. K., Kovar, J. G., and Mantel, H. J. (1990), On Estimating Distribution Functions and Quantiles from Survey Data Using Auxiliary Information, *Biometrika*, **77**, 365–375.

Rao, J. N. K., and Shao, J. (1993), On Balanced Half-sample Variance Estimation in Stratified Sampling, *Journal of the American Statistical Association*, **91**, 343–348.

Rao, J. N. K., and Shao, J. (1999), Modified Balanced Repeated for Complex Survey Data, *Biometrika*, **86**, 403–415.

Rao, J. N. K., and Wu, C. F. J. (1988), Resampling Inference with Complex Survey Data, *Journal of the American Statistical Association*, **83**, 231–241.

Rivest, L. P. (1994), Statistical Properties of Winsorized Mean for Skewed Distributions, *Biometrika*, **81**, 373–384.

Robinson, J. (1987), Conditioning Ratio Estimates under Simple Random Sampling, *Journal of the American Statistical Association*, **83**, 826–831.

Ross, A., and Chambers, R. L. (2000), *Prediction Methods for Survey Sampling*, New York: Springer-Verlag, in press.

Rousseeuw, P. (1984), Least Median of Squares Regression, *Journal of the American Statistical Association*, **79**, 1063–1069.

Royall, R. M. (1976a), Likelihood Functions in Finite Population Sampling Theory, *Biometrika*, **63**, 605–614.

Royall, R. M. (1976b), The Linear Least Squares Prediction Approach to Two-Stage Sampling, *Journal of the American Statistical Association*, **71**, 657–664.

Royall, R. M. (1976c), Current Advances in Sampling Theory: Implications for Human Observational Studies, *American Journal of Epidemiology*, **104**, 463–473.

Royall, R. M. (1979), Prediction Models in Small Area Estimation, in Steinberg, J. (ed.), *Synthetic Estimates for Small Areas*, National Institute on Drug Abuse, Research monograph 24, Washington, DC: U.S. Government Printing Office.

Royall, R. M. (1983), Finite Populations, Sampling From, in *Encyclopedia of Statistical Sciences*, Vol. 3, pp. 96–101.

Royall, R. M. (1986), The Prediction Approach to Robust Variance Estimation in Two-stage Cluster Sampling, *Journal of the American Statistical Association*, **81**, 119–123.

Royall, R. M. (1986), Model Robust Confidence Intervals Using Maximum Likelihood Estimators, *International Statistical Review*, **54**, 221–226.

Royall, R. M. (1988), The Prediction Approach to Sampling Theory, in P. R. Krishaniah and C. R. Rao (eds.), *Handbook of Statistics*, Vol. 6, Amsterdam: Elsevier Science Publishers, pp. 399–413.

Royall, R. M. (1992), Robustness and Optimal Design Under Prediction Models for Finite Populations, *Survey Methodology*, **18**, 179–185.

Royall, R. M. (1994), Discussion of "Sample Surveys 1975–1990; An Age of Reconciliation?" by T. M. F. Smith, *International Statistical Review*, **62**, 19–21.

Royall, R. M., and Cumberland, W. G. (1978), Variance Estimation in Finite Population Sampling, *Journal of the American Statistical Association*, **73**, 351–358.

Royall, R. M., and Cumberland, W. G. (1981a), An Empirical Study of the Ratio Estimator and Estimators of Its Variance, *Journal of the American Statistical Association*, **76**, 66–77.

Royall, R. M., and Cumberland, W. G. (1981b), The Finite Population Linear Regression Estimator and Estimators of Its Variance—An Empirical Study, *Journal of the American Statistical Association*, **76**, 924–930.

Royall, R. M., and Cumberland, W. G. (1985), Conditional Coverage Properties of Finite Population Confidence Intervals, *Journal of the American Statistical Association*, **80**, 355–359.

Royall, R. M., and Cumberland, W. G. (1988), Does Simple Random Sampling Provide Adequate Balance?, *Journal of the Royal Statistical Society B*, **50**, 118–124.

Royall, R. M., and Eberhardt, K. R. (1975), Variance Estimates for the Ratio Estimator, *Sankhyā, Ser. C*, **37**, 43–52.

Royall, R. M., and Herson, J. (1973a), Robust Estimation in Finite Populations I, *Journal of the American Statistical Association*, **68**, 880–889.

Royall, R. M., and Herson, J. (1973b), Robust Estimation in Finite Populations II: Stratification on a Size Variable, *Journal of the American Statistical Association*, **68**, 890–893.

Royall, R. M., and Pfeffermann (1982), Balanced Samples and Robust Bayerian Inference in Finite Population Sampling, *Biometrika*, **69**, 401–409.

Rubin, D. B. (1976), Inference and Missing Data, *Biometrika*, 581–592.

Ruppert, D., and Wand, M. P. (1994), Multivariate Locally Weighted Least Squares Regression, *Annals of Statistics*, **22**, 1346–1370.

Rust, K. (1985), Variance Estimation for Complex Estimators in Sample Surveys, *Journal of Official Statistics*, **1**, 381–397.

Rustagi, R. K. (1978), Some Theory of the Prediction Approach to Two-stage and Stratified Two-stage Sampling, unpublished Ph.D. dissertation, Ohio State University.

Särndal, C.-E. (1996), Efficient Estimators with Simple Variance in Unequal Probability Sampling, *Journal of the American Statistical Association*, **91**, 1289–1300.

Särndal, C.-E., Swensson, B., and Wretman, J. (1992), *Model Assisted Survey Sampling*, New York: Springer-Verlag.

Särndal, C.-E., and Wright, R. (1984), Cosmetic Form of Estimators in Survey Sampling, *Scandinavian Journal of Statistics*, **11**, 146–156.

SAS/STAT User's Guide (1990), Volume 2, Version 6.0, Cary NC: SAS Institute.

Savage, L. J. (1962), *The Foundations of Statistical Inference*, London: Methuen and Company.

Schaible, W. (ed.) (1996), *Indirect Estimators in U.S. Federal Programs*, New York: Springer-Verlag.

Schittkowski, K. (1985), NLPQL: A FORTRAN Subroutine Solving Constrained Nonlinear Programming Problems, *Annals of Operations Research*, **5**, 485–500.

Schott, J. R. (1997), *Matrix Analysis for Statistics*, New York: Wiley.

Scott, A. J., Brewer, K. R. W., and Ho, E. W. H. (1978), Finite Population Sampling and Robust Estimation, *Journal of the American Statistical Association*, **73**, 359–361.

Scott, A. J., and Smith, T. M. F. (1969), Estimation in Multistage Surveys, *Journal of the American Statistical Association*, **64**, 830–840.

Searle, S. R. (1971), *Linear Models*, New York: Wiley.

Searle, S. R. (1987), *Linear Models for Unbalanced Data*, New York: Wiley.

Searle, S. R., Casella, G., and McCullogh, C. E. (1992), *Variance Components*, New York: Wiley.

Searls, D. T. (1966), An Estimator Which Reduces Large True Observations, *Journal of the American Statistical Association*, **61**, 1200–1204.

Seber, G. A. F., and Wild, C. J. (1989), *Nonlinear Regression*, New York: Wiley.

Sedransk, N., and Sedransk, J. (1979), Distinguishing Among Distributions Using Data from Complex Sample Designs, *Journal of the American Statistical Association*, **74**, 754–760.

Sen, A. R. (1953), On the Estimate of the Variance in Sampling with Varying Probabilities, *Journal of the Indian Society of Agricultural Statistics*, **5**, 119–127.

Serfling, R. J. (1981), *Approximation Theorems of Mathematical Statistics*, New York: Wiley.

Shah, B., Barnwell, B., and Bieler, G. (1997), *SUDAAN User's Manual Release 7.5*, Research Triangle Park, NC: Research Triangle Institute.

Shao, J. (1994), Resampling Methods in Sample Surveys, with discussion, *Statistics*, **27**, 203–254.

Shao, J., and Tu, D. (1995), *The Jackknife and Bootstrap*, New York: Springer-Verlag.

Shao, J., and Wu, C. F. J. (1989), A General Theory for Jackknife Variance Estimation, *Annals of Statistics*, **17**, 1176–1197.

Shao, J., and Wu, C. F. J. (1992), Asymptotic Properties of the Balanced Repeated Replication Method for Sample Quantiles, *Annals of Statistics*, **20**, 1571–1593.

Silva, P. L. D. N., and Skinner, C. J. (1995), Estimating Distribution Functions with Auxiliary Information Using Poststratification, *Journal of Official Statistics*, **11**, 277–294.

Skinner, C. J., Holt, D., and Smith, T. M. F. (eds.) (1989), *Analysis of Complex Surveys*, New York: Wiley.

Smith, T. M. F. (1976), The Foundations of Survey Sampling: A Review, *Journal of the Royal Statistical Society A*, **139**, 183–204.

Smith, T. M. F. (1983), On the Validity of Inferences from Non-random Samples, *Journal of the Royal Statistical Society A*, **146**, 394–403.

Smith, T. M. F. (1984), Present Position and Potential Developments: Some Personal Views, Sample Surveys, *Journal of the Royal Statistical Society*, A, **147**, 208–221.

Smith, T. M. F. (1994), Sample Surveys 1975–1990; An Age of Reconciliation? and discussion, *International Statistical Review*, **62**, 5–34.

STATA Corporation (1997), *STATA User's Guide*, College Station, TX: STATA Press.

Statistical Sciences, Inc. (1995), *S-Plus User's Manual*, Seattle: Mathsoft, Inc.

Steinberg, J. (1979), *Synthetic Estimates for Small Areas*, National Institute on Drug Abuse, Research Monograph 24, Washington, DC: U.S. Government Printing Office.

Steuer, R. (1986), *Multiple Criteria Optimization: Theory, Computation, and Application*, New York: Wiley.

Stuart, A. (1962), *Basic Ideas of Scientific Sampling*, New York: Hafner.

Stukel, D., Hidiroglou, M., and Särndal, C.-E. (1996), Variance Estimation for Calibration Estimators: A Comparison of Jackknifing versus Taylor Series Linearization, *Survey Methodology*, **22**, 117–125.

Sugden, R. A., and Smith, T. M. F. (1984), Ignorable and Informative Designs in Survey Sampling Inference, *Biometrika*, **71**, 495–506.

Sukhatme, P. V., and Sukhatme, B. V. (1970), *Sampling Theory of Surveys with Applications*, Ames: Iowa State University Press.

Tallis, G. W. (1978), Note on Robust Estimation in Finite Populations, *Sankhya C*, **40**, 136–138.

Tam, S. M. (1995), Optimal and Robust Strategies for Cluster Sampling, *Journal of the American Statistical Association*, **90**, 379–382.

Thompson, M. E. (1997), *Theory of Sample Surveys*, London: Chapman & Hall.

Thompson, S. K. (1992), *Sampling*, New York: Wiley.

TIMSS International Study Center (1996), *Third International Mathematics and Science Study: Technical Report, Volume 1 Design and Development*, Chestnut Hill, MA: Boston College.

Tukey, J. W. (1960), A Survey of Sampling from Contaminated Distibutions, in I. Olkin, S. Ghurye, W. Hoeffding, W. Madow, and H. Mann, (eds.), *Contributions to Probability and Statistics*, Stanford, CA: Stanford University Press, pp. 448–485.

U.S. Department of Labor (1995), *Report on the American Workforce*, Washington, DC: Government Printing Office.

Valliant, R. (1985), Nonlinear Prediction Theory and the Estimation of Proportions in a Finite Population, *Journal of the American Statistical Association*, **80**, 631–641.

Valliant, R. (1986), Mean Squared Error Estimation in Finite Populations under Nonlinear Models, *Communications in Statistics A*, **15**, 1975–1993.

Valliant, R. (1987a), Conditional Properties of Some Estimators in Stratified Sampling, *Journal of the American Statistical Association*, **82**, 509–519, and Correction, **85**, 272.

Valliant, R. (1987b), Some Prediction Properties of Balanced Half-sample Variance Estimators in Single-stage Sampling, *Journal of the Royal Statistical Society B*, **49**, 68–81.

Valliant, R. (1987c), Generalized Variance Functions in Stratified Two-stage Sampling, *Journal of the American Statistical Association*, **82**, 499–508.

Valliant, R. (1990), Comparisons of Variance Estimators in Stratified Random and Systematic Sampling, *Journal of Official Statistics*, **6**, 115–131.

Valliant, R. (1993), Poststratification and Conditional Variance Estimation, *Journal of the American Statistical Association*, **88**, 89–96.

Valliant, R. (1996), Limitations of Balanced Half-Sampling, *Journal of Official Statistics*, **12**, 225–240.

Valliant, R., and Gentle, J. (1997), An Application of Mathematical Programming to a Sample Allocation Problem, *Computational Statistics and Data Analysis*, **25**, 337–360.

Venables, W. N., and Ripley, B. D. (1997), *Modern Applied Statistics with S-Plus*, 2nd ed., New York: Springer-Verlag.

Wang, S., and Dorfman, A. H. (1996), A New Estimator for the Finite Population Distribution Function, *Biometrika*, **83**, 639–652.

Watson, G. S. (1964), Smooth Regression Analysis, *Sankhya*, Series A, **26**, 359–372.

Weistroffer, H., and Narula, S. (1991), The Current State of Nonlinear Multiple Criteria Decision Making, in G. Fandel and H. Gehring (eds.), *Operations Research*, Berlin: Springer-Verlag, pp. 109–119.

White, H. (1982), Maximum Likelihood Estimation of Misspecified Models, *Econometrica*, **50**, 1–25.

Williams, J., Brown, E., and Zion, G. (1993), The Challenge of Redesigning the Consumer Price Index Area Sample, *Proceedings of the Section on Survey Research Methods*, American Statistical Association, pp. 200–205.

Winsten, C. G. (1965), Discussion of "The Planning of Observational Studies of Human Populations," by W. G. Cochran, *Journal of the Royal Statistical Society A*, **128**, 234–265.

Wolter, K. (1985), *Introduction to Variance Estimation*, New York: Springer-Verlag.

Woodruff, R. S. (1952), Confidence Intervals for Medians and Other Position Measures, *Journal of the American Statistical Association*, **47**, 635–646.

Wright, R. (1983), Finite Population Sampling with Multivariate Auxiliary Information, *Journal of the American Statistical Association*, **78**, 879–884.

Wu, C. J. F. (1991), Balanced Repeated Replications Based on Mixed Orthogonal Arrays, *Biometrika*, **78**, 181–188.

Yates, F., and Grundy, P. M. (1953), Selection Without Replacement from Within Strata with Probability Proportional to Size, *Journal of the Royal Statistical Society B*, **15**, 235–261.

Yung, W., and Rao, J. N. K. (1996), Jackknife Linearization Variance Estimators under Stratified Multi-stage Sampling, *Survey Methodology*, **22**, 23–32.

Zeger, S., and Liang, K.-Y. (1986), Longitudinal Data Analysis for Discrete and Continuous Outcomes, *Biometrics*, **42**, 121–130.

Answers to Select Exercises

CHAPTER 2

Exercise 2.1

Answers: ratio model, $b = 2.578755$; regression model, $a = 93.026748$, $b = 2.200135$.

Exercise 2.2

Answers: $T = 249{,}004$; $\hat{T}_R = 278{,}392$; $\hat{T}_{LR} = 274{,}077$.

Exercise 2.3

Answers: ratio model, $b = 3.280884$; regression model, $a = 51.259080$, $b = 3.114242$.

Exercise 2.4

Answers: $\hat{T}_0 = 396{,}616$; $\hat{T}_R = 354{,}191$; $\hat{T}_{LR} = 356{,}346$.

Exercise 2.5

Answers: $(n^{-1} - N^{-1})\sigma^2/\mu^2 = k^2$, where k is the desired *cv*. Solve for $n \cong 40$.

Exercise 2.11

Answers: (a) $\hat{\bar{Y}} = \bar{Y}_s$, $\operatorname{var}(\bar{Y}_s - \bar{Y}) = \dfrac{\sigma^2}{n}\left(1 - \dfrac{n}{N}\right)(1 - \rho)$;

(b) $\hat{\mu} = \bar{Y}_s$, $\operatorname{var}_M(\hat{\mu}) = \dfrac{\sigma^2}{n}[1 + (n-1)\rho]$.

Exercise 2.12

Answer: $\hat{\beta} = \Sigma_s (x_i Y_i/v_i)/\Sigma_s (x_i^2/v_i)$; $\hat{T} = n\bar{Y}_s + \hat{\beta}(N - n)\bar{x}_r$ in general

$v(x) = 1$: $\hat{\beta} = 2.368178$, $\hat{T} = 256{,}176.4$

$v(x) = x^2$: $\hat{\beta} = 2.811252$; $\hat{T} = 302{,}920.3$.

Exercise 2.13

Answers: $\hat{\beta}_0 = \bar{Y}_w - \hat{\beta}_1 \bar{x}_w$, $\hat{\beta}_1 = \Sigma_s (x_i - \bar{x}_w)(Y_i - \bar{Y}_w) v_i^{-1} / \Sigma_s (x_i - \bar{x}_w)^2 v_i^{-1}$;

$\hat{T} = n\bar{Y}_s + (N - n)[\bar{Y}_w + \hat{\beta}_1(\bar{x}_r - \bar{x}_w)]$ in general

Exercise 2.1 sample, $v(x) = 1$: $\hat{\beta} = 2.200135$, $\hat{T} = 274{,}077$ (same as Exercise 2.2)

$v(x) = x^2$: $\hat{\beta} = 2.697431$; $\hat{T} = 293{,}684.6$

Exercise 2.3 sample, $v(x) = 1$: $\hat{\beta} = 3.114242$; $\hat{T} = 356{,}346$ (same as Exercise 2.4).
$v(x) = x^2$: $\hat{\beta} = 3.538443$; $\hat{T} = 376{,}470.4$

CHAPTER 4

Exercise 4.11b

Answer: Under unweighted balance: 78 and 134, under root(x^2) balance: 152 and 84; use unweighted balance and $n = 134$.

CHAPTER 5

Exercise 5.16

Answers: (a) $\hat{T}(x^{1/2}, x{:}x) = 12{,}927.94$; $\sqrt{v_R} = 857.25$, $\sqrt{v_D} = 908.60$, $\sqrt{v_H} = 917.88$, $\sqrt{v_{J*}} = 966.51$, $\sqrt{v_J} = 966.27$.
(b) leverages: 0.219, 0181, 0.202, 0.137, 0.104, 0.053, 0.041, 0033, 0.021, 0.045, 0.146, 0.106, 0.140, 0.102, 0.088, 0.068, 0.080, 0.065, 0.156, 0.012.

Exercise 5.17

Answers: (a) $\hat{T}(x, x^2{:}x^2) = 344{,}149.80$; $\sqrt{v_R} = 32{,}578.48$, $\sqrt{v_D} = 45{,}406.20$, $\sqrt{v_H} = 48{,}713.08$, $\sqrt{v_{J*}} = 66{,}197.29$, $\sqrt{v_J} = 65{,}856.03$.
(b) Leverages: 0.108, 0.070, 0.022, 0.019, 0.011, 0.003, 0.003, 0.004, 0.001, 0.001, 0.592, 0.195, 0.051, 0.031, 0.030, 0.052, 0.200 0.180, 0.225, 0.204. Sample unit 11 with leverage of 0.592 is high leverage and has a large effect on some standard error estimates, especially $\sqrt{v_{J*}}$ and $\sqrt{v_J}$.

CHAPTER 6

Exercise 6.4

Answers:

	Stratum Samples n_h			
Stratum	(1) Optimal with Cost Constraint	(2) Neyman Allocation	(3) Proportional Allocation	(4) Equal Allocation
1	28	26	17	11
2	6	9	12	11
3	15	12	11	11
4	4	5	7	11
5	24	13	6	11
Total sample	77	65	53	55
Total cost	2105	2135	2095	2090
$\sqrt{\text{var}_M(\hat{T} - T)}/10^3$	835.8	888.0	1125.2	1188.7

Exercise 6.5

Answers: Set $(1 - n_h/N_h)\sigma_h/(n_h \bar{Y}_h^2) = (0.10)^2$ and solve for the required sample size in strata 1 and 2. Compute the cost for strata 1 and 2 for these sample sizes. The budget for strata 3–5 will be \$5000 minus the amount spent on strata 1 and 2. Compute the optimum total sample for strata 3–5 using Corollary 6.1.1 and allocate it using Theorem 6.1.1. The resulting allocation is

h	1	2	3	4	5
n_h	48	27	28	7	47

for a total sample of $n = 157$ and a cost of \$4985.

Exercise 6.10

Answer: $n^*/n = (\Sigma_h N_h \sqrt{\bar{x}_h}/N\sqrt{\bar{x}})^2$.

Exercise 6.11

Answer: $n = (\Sigma_h N_h^2 \sigma_h^2/w_h)/(V_0 + \Sigma_h N_h \sigma_h^2)$ with $w_h = N_h\sigma_h/\Sigma_h N_h \sigma_h$ for Neyman allocation, $w_h = N_h/N$ for proportional, and $w_h = 1/H$ for equal.

Exercise 6.12

Answer:

Stratum	Stratum Samples n_h			
	(1) Optimal with Cost Constraint	(2) Neyman Allocation	(3) Proportional Allocation	(4) Equal Allocation
1	2	5	15	13
2	3	7	12	13
3	27	27	9	13
4	70	57	8	13
5	222	128	5	13
Total sample	324	224	49	65
Total cost	5280	5280	5280	5330
$\sqrt{\text{var}_M(\hat{T} - T)}/10^3$	11.9	14.6	74.0	47.0

Exercise 6.13

Answer: $\dfrac{n_h}{n} = \dfrac{\sigma_h \bar{x}_h}{\sigma_1 \bar{x}_1 + \sigma_2 \bar{x}_2}$.

Exercise 6.14

Answers: (a) $n_h = (C^* - C_0)\dfrac{\sigma_h / c_h^{1/2}}{\Sigma_h \sigma_h c_h^{1/2}}$, $\quad n = (C^* - C_0)\dfrac{\Sigma_h \sigma_h / c_h^{1/2}}{\Sigma_h \sigma_h c_h^{1/2}}$;

$$V_{\min} = 2\left(\sum_h \sigma_h c_h^{1/2}\right)^2 \Big/ H(C^* - C_0); \quad \text{(b)}\ n_h = \sigma_h^2 (C^* - C_0) \Big/ \left(\sum_h \sigma_h^2 c_h\right),$$

$$n = \sum_h \sigma_h^2 (C^* - C_0) \Big/ \left(\sum_h \sigma_h^2 c_h\right),\ V = 2\left(\sum_h \sigma_h^2 c_h\right) \Big/ (C^* - C_0).$$

Exercise 6.15

Answer:

Stratum	N_h	σ_h	Stratum Samples n_h (1) Optimal Allocation	(2) $\bar{V}_{\min}$ Allocation	(3) V Allocation
1. Manufacturing	680	432	56	35	29
2. Retail trade	552	229	15	12	8
3. Wholesale trade	429	344	31	31	19
4. Services	363	196	11	12	6
5. Government	251	686	28	48	74
Total sample			141	138	136
$\sqrt{\operatorname{var}_M(\hat{T}_{OS} - T)}/10^3$			67.0	72.1	83.8
$\bar{V}/10^3$			12.4	10.6	12.7

Exercise 6.17

Answers: (a) $\operatorname{var}_M(\hat{T} - T) = \dfrac{\sigma^2}{n_0}\sum_h (N_h \bar{x}_h^{(\gamma/2)})^2 - \sigma^2 N\bar{x}^{(\gamma)}$; (b) $\Sigma_h (N_h \bar{x}_h^{(\gamma/2)})^2$; (c) Add and subtract $\sigma^2 H\bar{Z}^2/n_0$ from the error variance to obtain $\operatorname{var}_M(\hat{T} - T) = \dfrac{\sigma^2}{n_0}S^2 + \dfrac{\sigma^2}{n_0}\dfrac{(N\bar{x}^{(\gamma/2)})^2}{H} - \sigma^2 N\bar{x}^{(\gamma)}$. Minimize by eliminating the term involving S.

Exercise 6.27

Answer: $\operatorname{bias}_M(v_{LCO})/N^2 \approx 2(\bar{x}_0 - \bar{x})\sum_h \dfrac{W_h^3}{n_h^2} s_{xvhs} \Big/ \sum_h \dfrac{W_h^2}{n_h} s_{xhs}^2 = O(n^{-3/2})$. The variance of $\hat{T}_{LC}$ is $O(n^{-1})$, implying that in large samples the model bias of v_{LCO} should be negligible.

CHAPTER 7

Exercise 7.3

Answers: (a)

Age Category	Male	Female
19 and under	1	2
20–24	1	1
25–34	5	4
35+	8	8

(b) solution using ginv, $\boldsymbol{\beta}^o = (138.41, 117.79, 20.63, -88.43, 19.38, 45.20, 162.26)$, $\hat{T} = 141{,}523.3$; (c) $\boldsymbol{\beta}^o = (128.11, 106.17, 21.94, -54.28, 23.22, 40.69, 118.47, -46.00, 62.50, 23.43, 66.25, -8.27, -39.27, 17.26, 52.23)$, $\hat{T} = 141{,}636$; (d) weights for part (b) $\dot{\mathbf{g}}_s = (15.10, 15.93, 27.43, 15.07, 14.81, 15.93, 15.07, 15.93, 15.07, 13.95, 15.93, 14.81, 13.95, 15.93, 15.07, 15.07, 15.93, 15.95, 15.93, 15.07, 15.95, 13.95, 15.07, 15.93, 15.07, 26.57, 14.81, 13.95, 13.95, 14.81)$; weights for part (c) $\dot{\mathbf{g}}_s = (15.00, 16.38, 26.00, 14.62, 14.25, 16.38, 14.62, 16.38, 14.62, 14.40, 16.38, 14.25, 14.40, 16.38, 14.62, 14.62, 16.38, 16.00, 16.38, 14.62, 16.00, 14.40, 14.62, 16.38, 14.62, 28.00, 14.25, 14.40, 14.40, 14.25)$.

Exercise 7.4

Answers: (a)

Age Category	Male	Female
19 and under	0	1
20–24	4	2
25–34	3	3
35+	10	7

(b) solution using *ginv*, $\boldsymbol{\beta}^o = (172.07, 147.49, 24.57, -29.88, 58.07, 143.88)$, $\hat{T} = 162{,}665.6$; (c) $\boldsymbol{\beta}^o = (157.88, 114.53, 43.35, -1.88, 52.89, 106.88, -15.53, 23.36, 106.71, 13.65, 29.54, 0.17)$, $\hat{T} = 161{,}839.4$; (d) weights for part (b) $\dot{\mathbf{g}}_s = (19.04, 12.56, 17.48, 12.32, 12.56, 12.56, 12.56, 19.04, 23.96, 17.48, 17.24, 17.48, 23.96, 17.48, 12.32, 23.96, 12.56, 19.04, 12.56, 12.32, 17.24, 12.56, 12.56, 17.48, 12.32, 12.56, 12.56, 17.48, 17.48, 17.24)$; weights for part (c) $\dot{\mathbf{g}}_s = (24.00, 11.70, 18.71, 10.75, 11.70, 11.70, 11.70, 24.00, 19.00, 18.71, 19.33, 18.71, 19.00, 18.71, 10.75, 19.00, 11.70, 24.00, 11.70, 10.75, 19.33, 11.70, 11.70, 18.71, 10.75, 11.70, 11.70, 18.71, 18.71, 19.33)$.

Exercise 7.5

Answers: (a) solution using *ginv* (hours worked per week is listed last), $\boldsymbol{\beta}^o = (128.05, 111.67, 16.38, -88.08, 16.12, 40.22, 159.79, 0.47)$, $\hat{T} = 141{,}291.8$;

(b) $\boldsymbol{\beta}^o = (111.24, 96.04, 15.20, -55.71, 18.41, 33.97, 114.57, -50.10, 61.79, 18.92, 65.43, -5.60, -43.38, 15.05, 49.14, 0.81)$, $\hat{T} = 141{,}228$; (c) weights for part (a) $\dot{\mathbf{g}}_s = (13.51, 14.90, 26.87, 13.76, 15.28, 12.12, 18.50, 14.90, 15.16, 15.53, 15.46, 15.28, 9.95, 19.09, 16.83, 15.16, 20.48, 20.23, 12.67, 15.16, 13.26, 11.35, 15.16, 14.90, 13.76, 27.13, 15.28, 15.53, 15.53, 15.28)$; weights for part (b) $\dot{\mathbf{g}}_s = (15.00, 15.66, 26.00, 12.81, 14.25, 12.64, 17.95, 15.66, 14.32, 16.51, 16.26, 14.25, 10.48, 20.19, 16.13, 14.32, 21.70, 19.77, 13.24, 14.32, 12.23, 11.98, 14.32, 15.66, 12.81, 28.00, 14.25, 16.51, 16.51, 14.25)$.

Exercise 7.11

Answer: Model a, $\hat{T} = 141{,}262.5$; model b, $\hat{T} = 141{,}993.7$; model c, $\hat{T} = 136{,}566$.

CHAPTER 8

Exercise 8.5

Answer: $E_M(\hat{T}_U - T) = N\mu(\bar{M}_s - \bar{M})/\bar{M}$.

Exercise 8.7

Answers: ANOVA estimates, $\sigma^2 = 160.9759$, $\rho = 0.285$
MLEs: $\mu = 51.69669$, $\sigma^2 = 145.9743$, and $\rho = 0.098$.

Exercise 8.10

Answer: $\hat{T}_0 = 24{,}696.35$; $\hat{T}_R = 24{,}571.43$; $\hat{T}_p = 24{,}696.35$; $\hat{T}_H = 24{,}688.51$; $\hat{T}_1 = 24{,}688.51$. The population total is $T = 23{,}754.22$.

Exercise 8.22

Answers:

(a) $\bar{m}_{\text{opt}}$: 14, 4, 2

(b) fixed cost

n_{opt}: 111, 250, 333

cv(%): 4.920, 6.883, 8.123

(c) fixed cv

n_{opt}: 27, 119, 222

cost: \$24,300, \$47,600, \$66,600.

Exercise 8.26

Answers: $E_M(\hat{T}_p - T) = M[\beta_0(\bar{M}_s^{(-1)} - 1/\bar{M}) + \beta_2(\bar{M}_s - \bar{M}^{(2)}/\bar{M})]$,
$E_M(\hat{T}_1 - T) = (M - M_s)[\beta_0(\bar{M}_s^{(-1)} - 1/\bar{M}_s) + \beta_2(\bar{M}_s - \bar{M}_r^{(2)}/\bar{M}_r)]$, and

$E_M(\hat{T}_1 - T) = (1 - n/N)E_M(\hat{T}_p - T)$. When $\bar{M}_s = \bar{M}$ and $\bar{M}_s^{(2)} = \bar{M}^{(2)}$, the biases of $\hat{T}_H$ and $\hat{T}_0$ are equal.

Exercise 8.27

Answer: $[E_M(\hat{T}_H - T)]^2 = (N-n)^2[\beta_0\{\bar{M}_s^{(-1)} - 1\} - \beta_2(\bar{M}^{(2)} - \bar{M}^2)]^2$

$\text{var}_M(\hat{T}_R - T) - \text{var}_M(\hat{T}_H - T) = \sigma^2 V_s^2[\bar{m}\rho(N-n)^2 + (1-\rho)(N^2 - n^2)]\bar{M}^2/m.$

Exercise 8.28

Answer: Use the overall selection probability $\pi_{ij} = (nM_i/M)(m_i/M_i) = nm_i/M$.

Exercise 8.29

Answer: Use the overall selection probability $\pi_{ij} = (n/N)(m/M)$ to compute $E_\pi(m_c) = f_1 f_2 M_c$, $E_\pi(\Sigma_s M_i m_{ic}/m_i) = f_1 M_c$.

Exercise 8.32

Answer: (a) math = 486.87, science = 514.89. (b) math = 485.00, science = 508.463.

CHAPTER 9

Exercise 9.8

Answer: Minimize the sum of squares $\Sigma_s w_i(\dot{v}_i - M_i\theta_1 - M_i^2\theta_2)^2$ for general w_i.

Exercise 9.10

Answer:

	$\hat{T}_R$	$\hat{T}_p$
$\hat{T}$	22,147.044	22,655.527
v_0	1,437,925.288	1,316,521.487
$N\hat{B}_1/f$	1,512,826.041	1,386,663.436
$-2N\hat{B}_2$	−152,342.721	−150,098.175
$N\hat{B}_3$	77,441.967	79,956.226
v_1	1,734,760.535	1,559,996.366
v_2	1,297,934.220	1,247,997.093
$\tilde{v}_0$	1,406,796.574	1,290,831.341

CHAPTER 10

Exercise 10.5

Answer: The value of Q that minimizes $\Sigma_\alpha[\hat{\theta}^{(\alpha)} - Q]^2$ is the mean of the $\hat{\theta}^{(\alpha)}$.

Exercise 10.6

Answer: No conditions HS-1 and HS-2 are not satisfied. g_{hi} depends on the units in the sample, in general.

Exercise 10.7

Answer: $\hat{T}_0 = 3{,}200$
Half-sample ests: 2,700 2,950 3,150 3,200 3,850 3,000 3,900 2,850
$v_{BHS} = v_0 = 173{,}750$

Exercise 10.8

Answer: $\hat{p} = 0.5156$
$\hat{p}^{(\alpha)} = 0.7592$ 0.5932 0.4603 0.3906 0.4675 0.6667 0.3333 0.5614
$v_{BHS}(\hat{p}) = 0.01806$

CHAPTER 11

Exercise 11.1

Answer: Bias-corrected estimator is $\hat{T} = \Sigma_s Y_i + (N - n)(\tilde{\mu} + \bar{r}_s) = \hat{T}_0$, which is not resistant.

Exercise 11.3

Answers: $\tilde{\mu}_1$ bisquare = 10.62; $\tilde{\mu}_1$ huber = 11.34; $\bar{x}_s = 13.19$
$\tilde{T}$ bisquare = 5373.85; $\tilde{T}$ huber = 5716.60; $\hat{T}_0 = 6594.6$

Exercise 11.4

Answers: (a) $\hat{T}_R = 11{,}504{,}063$

(b)	$\tilde{T}_1$	$\tilde{T}_2$
bisquare	10,729,287	10,619,415
Huber	11,117,662	11,135,362

Population total = 11,243,111
(c) $\hat{\beta}$ $M(0,1{:}x) = 4.2371$; $\hat{\beta}$ bisquare = 3.8834; $\hat{\beta}$ Huber = 4.0897

Exercise 11.5

Answers: (a) $\hat{T}_R = 16{,}006{,}287$

(b)	$\tilde{T}_1$	$\tilde{T}_2$
bisquare	11,329,449	11,150,142
Huber	12,345,463	12,722,558

Exercise 11.11

Answer: $\text{var}_M[\hat{F}_N(t) - F_N(t)] = \sum_h \frac{N_h^2(1 - f_h)}{n_h} p_h(t)[1 - p_h(t)]$ and

$$\text{var}_M[\hat{F}_N(t) - F_N(t)] = \sum_h \frac{N_h^2(1 - f_h)}{n_h - 1} p_{hs}(t)[1 - p_{hs}(t)] \text{ where } f_h = n_h/N_h.$$

Exercise 11.14

Answers: quantile estimates, 25th, 50th, and 75th: 10.552, 24.413, 55.230; population quartiles are 11, 22, 48.

Exercise 11.18

Answer:

$$\theta^*_{cd} = \frac{\text{cov}_M(\bar{Y}_{s_{cl}} - \dot{\mu}_c, \bar{Y}_{r_{cl}} - \dot{\mu}_c)}{\text{var}_M(\bar{Y}_{s_{cl}} - \dot{\mu}_c)}.$$

Exercise 11.20

Answers: (a) $\alpha_{opt} = \text{cov}(X - Y, Z\text{-}Y)/\text{var}(X - Y)$

(b) $2\alpha_{opt} - 1 < \alpha < 2\alpha_{opt}$

Exercise 11.22

Answers: If the variance is constant for counties, it will be proportional to C_d for state d.

Author Index

Italicized number/letters indicate page/first author in Bibliography. See, e.g., Archer below.

Subject Index

WILEY SERIES IN PROBABILITY AND STATISTICS
ESTABLISHED BY WALTER A. SHEWHART AND SAMUEL S. WILKS

Editors
Noel A. C. Cressie, Nicholas I. Fisher, Iain M. Johnstone, J. B. Kadane, David W. Scott, Bernard W. Silverman, Adrian F. M. Smith, Jozef L. Teugels; Vic Barnett, Emeritus, Ralph A. Bradley, Emeritus, J. Stuart Hunter, Emeritus, David G. Kendall, Emeritus

Probability and Statistics Section

*ANDERSON · The Statistical Analysis of Time Series
ARNOLD, BALAKRISHNAN, and NAGARAJA · A First Course in Order Statistics
ARNOLD, BALAKRISHNAN, and NAGARAJA · Records
BACCELLI, COHEN, OLSDER, and QUADRAT · Synchronization and Linearity: An Algebra for Discrete Event Systems
BARNETT · Comparative Statistical Inference, *Third Edition*
BASILEVSKY · Statistical Factor Analysis and Related Methods: Theory and Applications
BERNARDO and SMITH · Bayesian Statistical Concepts and Theory
BILLINGSLEY · Convergence of Probability Measures, *Second Edition*
BOROVKOV · Asymptotic Methods in Queuing Theory
BOROVKOV · Ergodicity and Stability of Stochastic Processes
BRANDT, FRANKEN, and LISEK · Stationary Stochastic Models
CAINES · Linear Stochastic Systems
CAIROLI and DALANG · Sequential Stochastic Optimization
CONSTANTINE · Combinatorial Theory and Statistical Design
COOK · Regression Graphics
COVER and THOMAS · Elements of Information Theory
CSÖRGŐ and HORVÁTH · Weighted Approximations in Probability Statistics
CSÖRGŐ and HORVÁTH · Limit Theorems in Change Point Analysis
*DANIEL · Fitting Equations to Data: Computer Analysis of Multifactor Data, *Second Edition*
DETTE and STUDDEN · The Theory of Canonical Moments with Applications in Statistics, Probability, and Analysis
DEY and MUKERJEE · Fractional Factorial Plans
*DOOB · Stochastic Processes
DRYDEN and MARDIA · Statistical Shape Analysis
DUPUIS and ELLIS · A Weak Convergence Approach to the Theory of Large Deviations
ETHIER and KURTZ · Markov Processes: Characterization and Convergence
FELLER · An Introduction to Probability Theory and Its Applications, Volume I, *Third Edition,* Revised; Volume II, *Second Edition*
FULLER · Introduction to Statistical Time Series, *Second Edition*
FULLER · Measurement Error Models
GHOSH, MUKHOPADHYAY, and SEN · Sequential Estimation
GIFI · Nonlinear Multivariate Analysis
GUTTORP · Statistical Inference for Branching Processes
HALL · Introduction to the Theory of Coverage Processes
HAMPEL · Robust Statistics: The Approach Based on Influence Functions
HANNAN and DEISTLER · The Statistical Theory of Linear Systems
HUBER · Robust Statistics

*Now available in a lower priced paperback edition in the Wiley Classics Library.

Probability and Statistics (Continued)

HUSKOVA, BERAN, and DUPAC · Collected Works of Jaroslav Hajek—with Commentary
IMAN and CONOVER · A Modern Approach to Statistics
JUREK and MASON · Operator-Limit Distributions in Probability Theory
KASS and VOS · Geometrical Foundations of Asymptotic Inference
KAUFMAN and ROUSSEEUW · Finding Groups in Data: An Introduction to Cluster Analysis
KELLY · Probability, Statistics, and Optimization
KENDALL, BARDEN, CARNE, and LE · Shape and Shape Theory
LINDVALL · Lectures on the Coupling Method
McFADDEN · Management of Data in Clinical Trials
MANTON, WOODBURY, and TOLLEY · Statistical Applications Using Fuzzy Sets
MORGENTHALER and TUKEY · Configural Polysampling: A Route to Practical Robustness
MUIRHEAD · Aspects of Multivariate Statistical Theory
OLIVER and SMITH · Influence Diagrams, Belief Nets and Decision Analysis
*PARZEN · Modern Probability Theory and Its Applications
PRESS · Bayesian Statistics: Principles, Models, and Applications
PUKELSHEIM · Optimal Experimental Design
RAO · Asymptotic Theory of Statistical Inference
RAO · Linear Statistical Inference and Its Applications, *Second Edition*
RAO and SHANBHAG · Choquet-Deny Type Functional Equations with Applications to Stochastic Models
ROBERTSON, WRIGHT, and DYKSTRA · Order Restricted Statistical Inference
ROGERS and WILLIAMS · Diffusions, Markov Processes, and Martingales, Volume I: Foundations, *Second Edition;* Volume II: Îto Calculus
RUBINSTEIN and SHAPIRO · Discrete Event Systems: Sensitivity Analysis and Stochastic Optimization by the Score Function Method
RUZSA and SZEKELY · Algebraic Probability Theory
SCHEFFE · The Analysis of Variance
SEBER · Linear Regression Analysis
SEBER · Multivariate Observations
SEBER and WILD · Nonlinear Regression
SERFLING · Approximation Theorems of Mathematical Statistics
SHORACK and WELLNER · Empirical Processes with Applications to Statistics
SMALL and McLEISH · Hilbert Space Methods in Probability and Statistical Inference
STAPLETON · Linear Statistical Models
STAUDTE and SHEATHER · Robust Estimation and Testing
STOYANOV · Counterexamples in Probability
TANAKA · Time Series Analysis: Nonstationary and Noninvertible Distribution Theory
THOMPSON and SEBER · Adaptive Sampling
WELSH · Aspects of Statistical Inference
WHITTAKER · Graphical Models in Applied Multivariate Statistics
YANG · The Construction Theory of Denumerable Markov Processes

Applied Probability and Statistics Section

ABRAHAM and LEDOLTER · Statistical Methods for Forecasting
AGRESTI · Analysis of Ordinal Categorical Data
AGRESTI · Categorical Data Analysis

*Now available in a lower priced paperback edition in the Wiley Classics Library.

Applied Probability and Statistics (Continued)

ANDERSON, AUQUIER, HAUCK, OAKES, VANDAELE, and WEISBERG · Statistical Methods for Comparative Studies
ARMITAGE and DAVID (editors) · Advances in Biometry
*ARTHANARI and DODGE · Mathematical Programming in Statistics
ASMUSSEN · Applied Probability and Queues
*BAILEY · The Elements of Stochastic Processes with Applications to the Natural Sciences
BARNETT and LEWIS · Outliers in Statistical Data, *Third Edition*
BARTHOLOMEW, FORBES, and McLEAN · Statistical Techniques for Manpower Planning, *Second Edition*
BASU and RIGDON · Statistical Methods for the Reliability of Repairable Systems
BATES and WATTS · Nonlinear Regression Analysis and Its Applications
BECHHOFER, SANTNER, and GOLDSMAN · Design and Analysis of Experiments for Statistical Selection, Screening, and Multiple Comparisons
BELSLEY · Conditioning Diagnostics: Collinearity and Weak Data in Regression
BELSLEY, KUH, and WELSCH · Regression Diagnostics: Identifying Influential Data and Sources of Collinearity
BHAT · Elements of Applied Stochastic Processes, *Second Edition*
BHATTACHARYA and WAYMIRE · Stochastic Processes with Applications
BIRKES and DODGE · Alternative Methods of Regression
BLISCHKE AND MURTHY · Reliability: Modeling, Prediction, and Optimization
BLOOMFIELD · Fourier Analysis of Time Series: An Introduction, *Second Edition*
BOLLEN · Structural Equations with Latent Variables
BOULEAU · Numerical Methods for Stochastic Processes
BOX · Bayesian Inference in Statistical Analysis
BOX and DRAPER · Empirical Model-Building and Response Surfaces
*BOX and DRAPER · Evolutionary Operation: A Statistical Method for Process Improvement
BUCKLEW · Large Deviation Techniques in Decision, Simulation, and Estimation
BUNKE and BUNKE · Nonlinear Regression, Functional Relations and Robust Methods: Statistical Methods of Model Building
CHATTERJEE and HADI · Sensitivity Analysis in Linear Regression
CHERNICK · Bootstrap Methods: A Practitioner's Guide
CHILÈS and DELFINER · Geostatistics: Modeling Spatial Uncertainty
CHOW and LIU · Design and Analysis of Clinical Trials: Concepts and Methodologies
CLARKE and DISNEY · Probability and Random Processes: A First Course with Applications, *Second Edition*
*COCHRAN and COX · Experimental Designs, *Second Edition*
CONOVER · Practical Nonparametric Statistics, *Second Edition*
CORNELL · Experiments with Mixtures, Designs, Models, and the Analysis of Mixture Data, *Second Edition*
*COX · Planning of Experiments
CRESSIE · Statistics for Spatial Data, *Revised Edition*
DANIEL · Applications of Statistics to Industrial Experimentation
DANIEL · Biostatistics: A Foundation for Analysis in the Health Sciences, *Sixth Edition*
DAVID · Order Statistics, *Second Edition*
*DEGROOT, FIENBERG, and KADANE · Statistics and the Law
DODGE · Alternative Methods of Regression
DOWDY and WEARDEN · Statistics for Research, *Second Edition*
DUNN and CLARK · Applied Statistics: Analysis of Variance and Regression, *Second Edition*
*ELANDT-JOHNSON and JOHNSON · Survival Models and Data Analysis
*FLEISS · The Design and Analysis of Clinical Experiments

*Now available in a lower priced paperback edition in the Wiley Classics Library.

Applied Probability and Statistics (Continued)

FLEISS · Statistical Methods for Rates and Proportions, *Second Edition*
FLEMING and HARRINGTON · Counting Processes and Survival Analysis
GALLANT · Nonlinear Statistical Models
GLASSERMAN and YAO · Monotone Structure in Discrete-Event Systems
GNANADESIKAN · Methods for Statistical Data Analysis of Multivariate Observations, *Second Edition*
GOLDSTEIN and LEWIS · Assessment: Problems, Development, and Statistical Issues
GREENWOOD and NIKULIN · A Guide to Chi-Squared Testing
*HAHN · Statistical Models in Engineering
HAHN and MEEKER · Statistical Intervals: A Guide for Practitioners
HAND · Construction and Assessment of Classification Rules
HAND · Discrimination and Classification
HEIBERGER · Computation for the Analysis of Designed Experiments
HEDAYAT and SINHA · Design and Inference in Finite Population Sampling
HINKELMAN and KEMPTHORNE: · Design and Analysis of Experiments, Volume 1: Introduction to Experimental Design
HOAGLIN, MOSTELLER, and TUKEY · Exploratory Approach to Analysis of Variance
HOAGLIN, MOSTELLER, and TUKEY · Exploring Data Tables, Trends and Shapes
HOAGLIN, MOSTELLER, and TUKEY · Understanding Robust and Exploratory Data Analysis
HOCHBERG and TAMHANE · Multiple Comparison Procedures
HOCKING · Methods and Applications of Linear Models: Regression and the Analysis of Variables
HOGG and KLUGMAN · Loss Distributions
HOSMER and LEMESHOW · Applied Logistic Regression, *Second Edition*
HØYLAND and RAUSAND · System Reliability Theory: Models and Statistical Methods
HUBERTY · Applied Discriminant Analysis
HUNT and KENNEDY · Financial Derivatives in Theory and Practice
JACKSON · A User's Guide to Principle Components
JOHN · Statistical Methods in Engineering and Quality Assurance
JOHNSON · Multivariate Statistical Simulation
JOHNSON and KOTZ · Distributions in Statistics
JOHNSON, KOTZ, and BALAKRISHNAN · Continuous Univariate Distributions, Volume 1, *Second Edition*
JOHNSON, KOTZ, and BALAKRISHNAN · Continuous Univariate Distributions, Volume 2, *Second Edition*
JOHNSON, KOTZ, and BALAKRISHNAN · Discrete Multivariate Distributions
JOHNSON, KOTZ, and KEMP · Univariate Discrete Distributions, *Second Edition*
JUREČKOVÁ and SEN · Robust Statistical Procedures: Aymptotics and Interrelations
KADANE · Bayesian Methods and Ethics in a Clinical Trial Design
KADANE AND SCHUM · A Probabilistic Analysis of the Sacco and Vanzetti Evidence
KALBFLEISCH and PRENTICE · The Statistical Analysis of Failure Time Data
KELLY · Reversability and Stochastic Networks
KHURI, MATHEW, and SINHA · Statistical Tests for Mixed Linear Models
KLUGMAN, PANJER, and WILLMOT · Loss Models: From Data to Decisions
KLUGMAN, PANJER, and WILLMOT · Solutions Manual to Accompany Loss Models: From Data to Decisions
KOTZ, BALAKRISHNAN, and JOHNSON · Continuous Multivariate Distributions, Volume 1, *Second Edition*
KOVALENKO, KUZNETZOV, and PEGG · Mathematical Theory of Reliability of Time-Dependent Systems with Practical Applications
LACHIN · Biostatistical Methods: The Assessment of Relative Risks

*Now available in a lower priced paperback edition in the Wiley Classics Library.

Applied Probability and Statistics (Continued)

LAD · Operational Subjective Statistical Methods: A Mathematical, Philosophical, and Historical Introduction
LANGE, RYAN, BILLARD, BRILLINGER, CONQUEST, and GREENHOUSE · Case Studies in Biometry
LAWLESS · Statistical Models and Methods for Lifetime Data
LEE · Statistical Methods for Survival Data Analysis, *Second Edition*
LEPAGE and BILLARD · Exploring the Limits of Bootstrap
LINHART and ZUCCHINI · Model Selection
LITTLE and RUBIN · Statistical Analysis with Missing Data
LLOYD · The Statistical Analysis of Categorical Data
MAGNUS and NEUDECKER · Matrix Differential Calculus with Applications in Statistics and Econometrics, *Revised Edition*
MALLER and ZHOU · Survival Analysis with Long Term Survivors
MANN, SCHAFER, and SINGPURWALLA · Methods for Statistical Analysis of Reliability and Life Data
McLACHLAN and KRISHNAN · The EM Algorithm and Extensions
McLACHLAN · Discriminant Analysis and Statistical Pattern Recognition
McNEIL · Epidemiological Research Methods
MEEKER and ESCOBAR · Statistical Methods for Reliability Data
*MILLER · Survival Analysis, *Second Edition*
MONTGOMERY and PECK · Introduction to Linear Regression Analysis, *Second Edition*
MYERS and MONTGOMERY · Response Surface Methodology: Process and Product in Optimization Using Designed Experiments
NELSON · Accelerated Testing, Statistical Models, Test Plans, and Data Analyses
NELSON · Applied Life Data Analysis
OCHI · Applied Probability and Stochastic Processes in Engineering and Physical Sciences
OKABE, BOOTS, and SUGIHARA · Spatial Tesselations: Concepts and Applications of Voronoi Diagrams
PANKRATZ · Forecasting with Dynamic Regression Models
PANKRATZ · Forecasting with Univariate Box-Jenkins Models: Concepts and Cases
PIANTADOSI · Clinical Trials: A Methodologic Perspective
PORT · Theoretical Probability for Applications
PUTERMAN · Markov Decision Processes: Discrete Stochastic Dynamic Programming
RACHEV · Probability Metrics and the Stability of Stochastic Models
RÉNYI · A Diary on Information Theory
RIPLEY · Spatial Statistics
RIPLEY · Stochastic Simulation
ROLSKI, SCHMIDLI, SCHMIDT, and TEUGELS · Stochastic Processes for Insurance and Finance
ROUSSEEUW and LEROY · Robust Regression and Outlier Detection
RUBIN · Multiple Imputation for Nonresponse in Surveys
RUBINSTEIN · Simulation and the Monte Carlo Method
RUBINSTEIN and MELAMED · Modern Simulation and Modeling
RYAN · Statistical Methods for Quality Improvement, *Second Edition*
SCHIMEK · Smoothing and Regression: Approaches, Computation, and Application
SCHUSS · Theory and Applications of Stochastic Differential Equations
SCOTT · Multivariate Density Estimation: Theory, Practice, and Visualization
*SEARLE · Linear Models
SEARLE · Linear Models for Unbalanced Data
SEARLE, CASELLA, and McCULLOCH · Variance Components
SENNOTT · Stochastic Dynamic Programming and the Control of Queueing Systems
STOYAN, KENDALL, and MECKE · Stochastic Geometry and Its Applications, *Second Edition*

*Now available in a lower priced paperback edition in the Wiley Classics Library.

Applied Probability and Statistics (Continued)

STOYAN and STOYAN · Fractals, Random Shapes and Point Fields: Methods of Geometrical Statistics
THOMPSON · Empirical Model Building
THOMPSON · Sampling
THOMPSON · Simulation: A Modeler's Approach
TIJMS · Stochastic Modeling and Analysis: A Computational Approach
TIJMS · Stochastic Models: An Algorithmic Approach
TITTERINGTON, SMITH, and MAKOV · Statistical Analysis of Finite Mixture Distributions
UPTON and FINGLETON · Spatial Data Analysis by Example, Volume 1: Point Pattern and Quantitative Data
UPTON and FINGLETON · Spatial Data Analysis by Example, Volume II: Categorical and Directional Data
VAN RIJCKEVORSEL and DE LEEUW · Component and Correspondence Analysis
VIDAKOVIC · Statistical Modeling by Wavelets
WEISBERG · Applied Linear Regression, *Second Edition*
WESTFALL and YOUNG · Resampling-Based Multiple Testing: Examples and Methods for *p*-Value Adjustment
WHITTLE · Systems in Stochastic Equilibrium
WOODING · Planning Pharmaceutical Clinical Trials: Basic Statistical Principles
WOOLSON · Statistical Methods for the Analysis of Biomedical Data
*ZELLNER · An Introduction to Bayesian Inference in Econometrics

Texts and References Section

AGRESTI · An Introduction to Categorical Data Analysis
ANDERSON · An Introduction to Multivariate Statistical Analysis, *Second Edition*
ANDERSON and LOYNES · The Teaching of Practical Statistics
ARMITAGE and COLTON · Encyclopedia of Biostatistics: Volumes 1 to 6 with Index
BARTOSZYNSKI and NIEWIADOMSKA-BUGAJ · Probability and Statistical Inference
BENDAT and PIERSOL · Random Data: Analysis and Measurement Procedures, *Third Edition*
BERRY, CHALONER, and GEWEKE · Bayesian Analysis in Statistics and Econometrics: Essays in Honor of Arnold Zellner
BHATTACHARYA and JOHNSON · Statistical Concepts and Methods
BILLINGSLEY · Probability and Measure, *Second Edition*
BOX · R. A. Fisher, the Life of a Scientist
BOX, HUNTER, and HUNTER · Statistics for Experimenters: An Introduction to Design, Data Analysis, and Model Building
BOX and LUCEÑO · Statistical Control by Monitoring and Feedback Adjustment
BROWN and HOLLANDER · Statistics: A Biomedical Introduction
CHATTERJEE and PRICE · Regression Analysis by Example, *Third Edition*
COOK and WEISBERG · Applied Regression Including Computing and Graphics
COOK and WEISBERG · An Introduction to Regression Graphics
COX · A Handbook of Introductory Statistical Methods
DILLON and GOLDSTEIN · Multivariate Analysis: Methods and Applications
*DODGE and ROMIG · Sampling Inspection Tables, *Second Edition*
DRAPER and SMITH · Applied Regression Analysis, *Third Edition*
DUDEWICZ and MISHRA · Modern Mathematical Statistics
DUNN · Basic Statistics: A Primer for the Biomedical Sciences, *Second Edition*
EVANS, HASTINGS, and PEACOCK · Statistical Distributions, *Third Edition*
FISHER and VAN BELLE · Biostatistics: A Methodology for the Health Sciences

*Now available in a lower priced paperback edition in the Wiley Classics Library.

Texts and References (Continued)

FREEMAN and SMITH · Aspects of Uncertainty: A Tribute to D. V. Lindley
GROSS and HARRIS · Fundamentals of Queueing Theory, *Third Edition*
HALD · A History of Probability and Statistics and their Applications Before 1750
HALD · A History of Mathematical Statistics from 1750 to 1930
HELLER · MACSYMA for Statisticians
HOEL · Introduction to Mathematical Statistics, *Fifth Edition*
HOLLANDER and WOLFE · Nonparametric Statistical Methods, *Second Edition*
HOSMER and LEMESHOW · Applied Logistic Regression, *Second Edition*
HOSMER and LEMESHOW · Applied Survival Analysis: Regression Modeling of Time to Event Data
JOHNSON and BALAKRISHNAN · Advances in the Theory and Practice of Statistics: A Volume in Honor of Samuel Kotz
JOHNSON and KOTZ (editors) · Leading Personalities in Statistical Sciences: From the Seventeenth Century to the Present
JUDGE, GRIFFITHS, HILL, LÜTKEPOHL, and LEE · The Theory and Practice of Econometrics, *Second Edition*
KHURI · Advanced Calculus with Applications in Statistics
KOTZ and JOHNSON (editors) · Encyclopedia of Statistical Sciences: Volumes 1 to 9 with Index
KOTZ and JOHNSON (editors) · Encyclopedia of Statistical Sciences: Supplement Volume
KOTZ, REED, and BANKS (editors) · Encyclopedia of Statistical Sciences: Update Volume 1
KOTZ, REED, and BANKS (editors) · Encyclopedia of Statistical Sciences: Update Volume 2
LAMPERTI · Probability: A Survey of the Mathematical Theory, *Second Edition*
LARSON · Introduction to Probability Theory and Statistical Inference, *Third Edition*
LE · Applied Categorical Data Analysis
LE · Applied Survival Analysis
MALLOWS · Design, Data, and Analysis by Some Friends of Cuthbert Daniel
MARDIA · The Art of Statistical Science: A Tribute to G. S. Watson
MASON, GUNST, and HESS · Statistical Design and Analysis of Experiments with Applications to Engineering and Science
MURRAY · X-STAT 2.0 Statistical Experimentation, Design Data Analysis, and Nonlinear Optimization
PURI, VILAPLANA, and WERTZ · New Perspectives in Theoretical and Applied Statistics
RENCHER · Linear Models in Statistics
RENCHER · Methods of Multivariate Analysis
RENCHER · Multivariate Statistical Inference with Applications
ROSS · Introduction to Probability and Statistics for Engineers and Scientists
ROHATGI · An Introduction to Probability Theory and Mathematical Statistics
RYAN · Modern Regression Methods
SCHOTT · Matrix Analysis for Statistics
SEARLE · Matrix Algebra Useful for Statistics
STYAN · The Collected Papers of T. W. Anderson: 1943–1985
TIERNEY · LISP-STAT: An Object-Oriented Environment for Statistical Computing and Dynamic Graphics
WONNACOTT and WONNACOTT · Econometrics, *Second Edition*
WU and HAMADA · Experiments: Planning, Analysis, and Parameter Design Optimization

*Now available in a lower priced paperback edition in the Wiley Classics Library.

WILEY SERIES IN PROBABILITY AND STATISTICS

ESTABLISHED BY WALTER A. SHEWHART AND SAMUEL S. WILKS

Editors

Robert M. Groves, Graham Kalton, J. N. K. Rao, Norbert Schwarz, Christopher Skinner

Survey Methodology Section

BIEMER, GROVES, LYBERG, MATHIOWETZ, and SUDMAN · Measurement Errors in Surveys
COCHRAN · Sampling Techniques, *Third Edition*
COUPER, BAKER, BETHLEHEM, CLARK, MARTIN, NICHOLLS, and O'REILLY (editors) · Computer Assisted Survey Information Collection
COX, BINDER, CHINNAPPA, CHRISTIANSON, COLLEDGE, and KOTT (editors) · Business Survey Methods
*DEMING · Sample Design in Business Research
DILLMAN · Mail and Telephone Surveys: The Total Design Method, *Second Edition*
GROVES and COUPER · Nonresponse in Household Interview Surveys
GROVES · Survey Errors and Survey Costs
GROVES, BIEMER, LYBERG, MASSEY, NICHOLLS, and WAKSBERG · Telephone Survey Methodology
*HANSEN, HURWITZ, and MADOW · Sample Survey Methods and Theory, Volume 1: Methods and Applications
*HANSEN, HURWITZ, and MADOW · Sample Survey Methods and Theory, Volume II: Theory
KISH · Statistical Design for Research
*KISH · Survey Sampling
KORN and GRAUBARD · Analysis of Health Surveys
LESSLER and KALSBEEK · Nonsampling Error in Surveys
LEVY and LEMESHOW · Sampling of Populations: Methods and Applications, *Third Edition*
LYBERG, BIEMER, COLLINS, de LEEUW, DIPPO, SCHWARZ, TREWIN (editors) · Survey Measurement and Process Quality
SIRKEN, HERRMANN, SCHECHTER, SCHWARZ, TANUR, and TOURANGEAU (editors) · Cognition and Survey Research
VALLIANT, DORFMAN, and ROYALL · A Finite Population Sampling and Inference

*Now available in a lower priced paperback edition in the Wiley Classics Library.